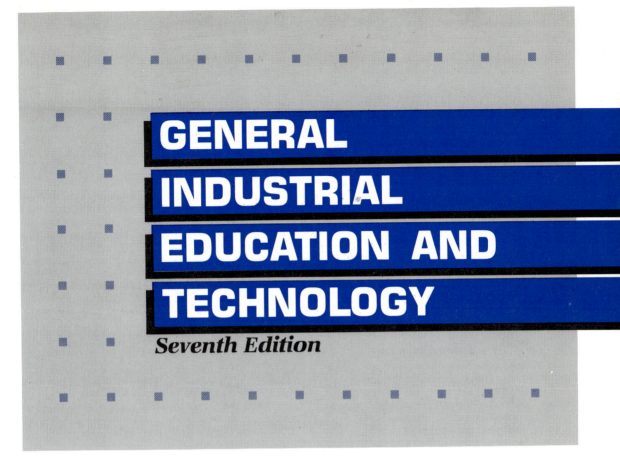

GENERAL
INDUSTRIAL
EDUCATION AND
TECHNOLOGY

Seventh Edition

Chris H. Groneman
John L. Feirer

McGraw-Hill Book Company
New York   Atlanta   Dallas   St. Louis   San Francisco
Auckland   Bogotá   Guatemala   Hamburg   Johannesburg   Lisbon
London   Madrid   Mexico   Montreal   New Delhi   Panama   Paris
San Juan   São Paulo   Singapore   Sydney   Tokyo   Toronto

McGraw-Hill Publications in Industrial Education
Chris H. Groneman, Consulting Editor

Books in Series
GENERAL INDUSTRIAL EDUCATION AND
   TECHNOLOGY • Groneman and Feirer
GENERAL METALS • Feirer
GENERAL POWER MECHANICS • Crouse,
   Worthington, Margules, and Anglin
GENERAL WOODWORKING • Groneman
GETTING STARTED IN DRAWING AND PLANNING
   • Groneman and Feirer
GETTING STARTED IN ELECTRICITY AND
   ELECTRONICS • Groneman and Feirer
GETTING STARTED IN METALWORKING •
   Groneman and Feirer
GETTING STARTED IN WOODWORKING •
   Groneman and Feirer
TECHNICAL ELECTRICITY AND ELECTRONICS •
   Buban and Schmitt
TECHNICAL WOODWORKING • Groneman and
   Glazener
UNDERSTANDING ELECTRICITY AND
   ELECTRONICS • Buban and Schmitt

Sponsoring Editor: Don Hepler
Editing Supervisor: Sharon E. Kaufman
Design and Art Supervisor: Karen Tureck
Production Supervisor: Laurence Charnow

Interior Design: Keithley and Associates Inc.
Cover Design: Renee Kilbride Edelman
Cover Photographer: Ken Karp

Acknowledgments

Sincere thanks are extended to the many associations, governmental agencies, corporations, and companies that have provided information, illustrations, photographs, and other pertinent data relevant to this seventh edition. Acknowledgment is generally noted throughout this text where such materials are used. For purposes of unity, continuity, simplicity, and clarity, individual line credits have not been cited in the units pertaining to processes.

Special recognition is given to the following individuals for their assistance: Dr. Ervin Dennis, University of Northern Iowa, for the section pertaining to graphic arts; Dr. Robert Magowan, Memphis State University, for materials in the several units relating to polyester plastics; Mr. Arthur McNichol and students, San Antonio, Texas Independent School District, for the section relating to photography; Dr. Darrel Smith, California State College, Pennsylvania, for illustrations; and Dr. J. D. Helsel, California State College, Pennsylvania, for working drawings in the appendix on suggested project activities.

Credit should also be given to Stanley Tools, Porter-Cable (formerly Rockwell), and Delta (formerly Rockwell) International, Inc., for providing many photographs and illustrations of tools, machines, and processes.

The authors also acknowledge the assistance of Mr. Edward Daves, Industrial Arts Instructor, and students of the Fresno, California, United School District. Appreciation is expressed to Mrs. Jane Feirer and Mrs. Virginia Groneman for their assistance in manuscript preparation.

Library of Congress Cataloging in Publication Data

Groneman, Chris Harold, date
   General industrial education and technology.

   (McGraw-Hill publications in industrial education)
   Includes index.
   SUMMARY: A junior high industrial arts textbook introducing the basics of drafting, graphic art, woodworking, electronics, metalworking, photography, plastics, electricity, power mechanics, and transportation. Also provides information on careers in each of these areas.
      1. Industrial arts—Juvenile literature.
      2. Manual training—Juvenile literature.
   [1. Industrial arts] I. Feirer, John Louis. II. Title.
   TT165.G74   1986   600   85-4857

General Industrial Education and Technology, Seventh Edition

ISBN 0-07-025023-5

Contents

Preface

The seventh edition of *General Industrial Education and Technology* offers a comprehensive program that introduces students to technology at all levels of learning. The courses in which this text can be used include "Introduction to Technology," "Industrial Technology," "General Industrial Arts," "General Industrial Education," and "Comprehensive General Shop." It can serve in a series of laboratory experiences following the cluster approach to include visual communications, production, power, energy, and transportation technologies.

The contents of this completely revised edition has been restructured to include material that reflects current changes in the field of industry and technology. Metric units are given first following the customary, or inch, dimensions in most of this text. In Sections 5 and 9 metric units are used, following standard industry practice. The United States and Canada are committed to change from the customary (English or Imperial) system to SI (Système International), or the metric, system as soon as possible. The need for this changeover has been brought about by the desire to standarize the measuring system throughout the world so that trade with other countries can be carried on with less confusion.

A new unit on high technology (including computers and robotics) and other sophisticated technological concepts has been added. Each unit has been updated with new information and illustrations. Current developments in industry and technology have increased the need for a broader understanding and appreciation of the complexity of industry and its technological impact on everyday living. This text is designed to motivate students' interests. Students acquire a basic understanding of the many career possibilities open to them. They consider and explore each technological area through hands-on activities.

The text pinpoints the vital topic of recycling industrial materials. Each student should understand the question of ecology and develop an awareness of ecological and environmental problems.

Each instructional section provides comprehensive activities for 9, 18, 36, or 40 weeks of technological exploration. The problems and discussion topics at the end of each section give students an opportunity to review and reinforce the learning experience.

The use of a second color accents the illustrations. It also emphasizes important points and safety precautions. Many photographs and line drawings show the techniques and materials of industry in a realistic manner.

General Industrial Education and Technology, Seventh Edition, provides introductory industrial and technology students with an interesting, complete, challenging, and modern textbook that will enhance their understanding and appreciation of the role of American technological industry in modern society.

The authors and publisher have welcomed the many constructive comments and criticisms offered by teachers who have used the first six editions of this text. They are interested in receiving observations regarding this latest edition.

Chris H. Groneman
John L. Feirer

About the Authors

Chris H. Groneman received his B.S. and M.S. degrees from Kansas State College, Pittsburg, Kansas, and his doctorate from the Pennsylvania State University. His teaching experiences include several years in junior and senior high schools and junior colleges. His university teaching and administrative years were spent at East Texas State University, Commerce; California State University, Fresno; the University of Hawaii, Honolulu; and Texas A & M University, College Station, where he was head of the Industrial Education Department for many years.

He is a member of, and has held office in, numerous professional organizations. He was honored by his alma mater with the Meritorious Achievement Citation, the highest recognition that Kansas State College can confer on one of its graduates.

Dr. Groneman is Consulting Editor of the McGraw-Hill Publications in Industrial Education series, of which *General Industrial Education and Technology* is a part. He is also the author of *General Woodworking* and the coauthor of *Technical Woodworking*, two other titles in the series.

John L. Feirer completed his B.S. degree at the University of Wisconsin, Stout; his M.S. degree at the University of Minnesota; and his doctoral degree at the University of Oklahoma. He has teaching experience in junior and senior high schools in Wisconsin, Minnesota, and Michigan. Dr. Feirer was professor and head of the Industrial Education and Technology department at Western Michigan University, Kalamazoo. He is now a *Distinguished Faculty Scholar* on campus.

He was executive editor of *Industrial Education* magazine. Dr. Feirer is active in industrial arts organizations and has written widely for the field. He has received the honorary citation from Epsilon Pi Tau and has been presented the Apollo Achievement Award for his work with the National Aeronautics and Space Administration. Dr. Feirer is the author of the McGraw-Hill text *Machine Tool Metalworking*. He is also an author or coauthor of several books in the McGraw-Hill Publications in Industrial Education series, including *General Metals*.

DEVELOPMENT OF AMERICAN INDUSTRIAL TECHNOLOGY

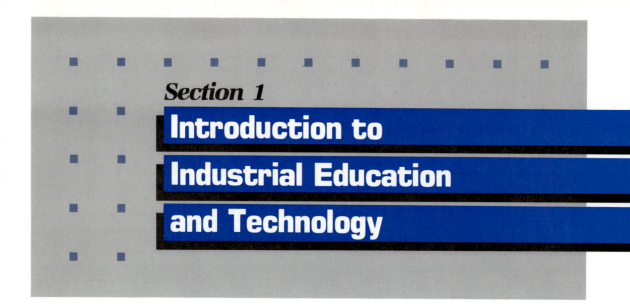

Introduction to Industrial Education and Technology

Unit 1

General Industrial Education and Technology

General industrial education and technology is an area in which you will learn about the technology of industry. You will explore and study how industry operates and how the various segments (parts) relate to your life. Our discussion of career opportunities will give you a glimpse into future possibilities so that you can make an informed choice of a career in industry. You may also discover an interesting hobby which will provide an avocational interest later in your life.

Industrial materials and processes offer challenging project activities in drafting, graphic arts, and photography through the *visual communications technologies*; woodwork, metalwork, and plastics through the many *production technologies*; and electrical systems, power, energy, and transportation through the *energy technologies* (Fig. 1-1).

Valuable information about manufacturing and construction, along with other important topics related to the growth of American industry, its organization, ecol-

Fig. 1-1. General industrial education and technology is a laboratory in which you investigate many industrial activities.

ogy and environmental studies, and industrial computerization, will give you an overall understanding of the importance of industrial technology in the United States. Millions of people daily produce industrial products from materials with which you will experiment.

What You Will Learn in General Industrial Education and Technology

In your study of general industrial education and technology, you will:

1. *Develop an interest in industry and technology.* Industrial workers use many different materials. Woods, plastics, metals, and electrical parts are just a few of them. You should understand the many uses for these materials. Knowing their uses will help you understand industrial life, its problems, and its opportunities. You will also learn to understand, use, and control technology.

2. *Develop an awareness of career opportunities in industry.* Also, select, prepare for, and enter training programs in existing and emerging occupations (Fig. 1-2). Career opportunities and facts about jobs are found in each section of this book. You will learn about *professions, careers, occupations,* and *jobs.* Knowing what these four terms mean will help you choose the best courses to prepare for a future job in industry.

3. *Develop safe working habits.* A good worker learns and follows safety rules. All through this book you will be reminded about how to care for your tools, your equipment, and yourself. Make sure you follow safety rules at all times. Knowing these rules will help you form good safety habits.

4. *Develop an appreciation of good design and skillful work and learn how to buy industrial products wisely.* This course will help you learn some easy rules about good

Fig. 1-2. **You will learn about many different career opportunities, including those in graphic arts. This student is learning to operate an offset press.**

design. These rules will be used when you plan your project activities. You will learn the standards of skillful work by building an activity project carefully. Knowing good design and learning about materials used in industry will also make you a wiser buyer of industrial products.

5. *Develop orderly ways to work.* Architects, engineers, physicians, dentists, lawyers, teachers, and nearly all other professional people must plan their work ahead of time. The problems and activities you will find in this book give you an opportunity to plan the way you will work. Planning ahead will save you time, effort, and materials (Fig. 1-3).

6. *Develop hand- and machine-tool skills and concepts.* These may later help you earn a living. They will also help you understand how people work in industry.

■ 3

Fig. 1-3. These students are learning drafting skills, an essential part of effective industrial planning. (McDonnell Douglas Corp.)

Fig. 1-4. Hobbies are often started in general industrial laboratories. These students are silk-screening T-shirts.

This know-how can help you fix something around your home, and it may lead you to discover and enjoy a new hobby (Fig. 1-4).

7. *Develop an awareness of the impact of technology on your personal and working lives.*

8. *Develop problem-solving and creative abilities involving materials, processes, and products of industry.*

Unit 2

Product Design

People who create new ideas for products are called *designers*. Everything that is made or manufactured, from cars to furniture, must first be designed.

Design—Formal and Informal

When you design a project activity, you plan it first. Then you may make a set of working drawings to guide you as you build the product. When you work this way, you are working with a *formal design*. (See Fig. 2-1.)

Sometimes you will work without a working drawing. When you do this, you are doing *informal designing*. (See Fig. 2-2.) In either case, you do both the planning and the building. In other words, you do

Fig. 2-1. This designer is preparing a picture presentation. (Zinc Institute, Inc.)

Fig. 2-2. This designer is working with the aid of sketches. (Zinc Institute, Inc.)

the *designing* of the project activity you are building *while you are building* it.

Design—An Individual Process

The experience of designing is not always the same because no two people will de-sign something in just the same way. Figures 2-3 and 2-4 show two clocks. Both clocks tell time but are very different because they show the design ideas of two different people.

Learning about Design

You should know about design and what kind of design will make you a good de-signer. In this section, you will learn to use your *imagination* to create products.

Fig. 2-3. A clock in a wood case design. (Howard Miller Clock Company)

Fig. 2-4. A clock in a plastic (styrene) case de-sign.

The Three Major Parts of Design

There are three parts to the design process that affect the way something is designed. They are the *creative*, the *technical*, and the *aesthetic* (beautiful) parts. All three of these can be found in each and every design.

The Creative Part of Design. The creative part of design is the idea or expression part. It is all yours—and only yours.

Figure 2-5 shows a house like most houses we see every day. Figure 2-6 shows a very different one.

When you design a product, you should show your own feelings, too. You should

Fig. 2-5. A two-story home of traditional design. (American Wood Preservers Institute)

Fig. 2-6. A contemporary home designed for passive solar energy. (Georgia-Pacific Company)

Fig. 2-7. Creativity and self-expression are involved in the construction of this sculptured form of zinc. (Zinc Institute, Inc.)

express your ideas. Figure 2-7 shows a statue made of zinc. The artist showed great imagination in its creation.

The Technical Part of Design. Changing a design idea into the product is the technical part of design. This means that a person uses all the methods needed to make a final product.

If you are designing a wooden product, for example, you must know the ways of working with wood. All products require careful planning and exact measurement. Working drawings are important parts of technical designing (Fig. 2-8).

The way a product will be used is called its *function.* Function is another part of the technical design. When you design a product, you must think about its function.

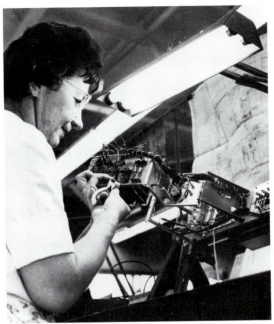

Fig. 2-8. This worker is using working drawings for inspection of a product. Inspection is an important part of the technical design of any product. (Western Electric Company)

Fig. 2-9. This office chair is designed for comfort. (Dunbar)

Fig. 2-10 a. Henry Ford's first car, the Quadricycle, did not emphasize the beauty of design. This self-propelled vehicle was built in 1896. (Ford Motor Company)

Fig. 2-10 b. This is a beautifully designed research car.

Figure 2-9 shows a chair designed for rest and comfort. This high-back chair is made so that any sitter will fit in it. Besides comfort, the chair offers good body and head support. The designer of this chair thought about function when the chair was designed.

The Aesthetic Part of Design. The last part of designing is the aesthetic (beauty or good looks) part. When a product is pleasing to the senses, it is beautiful. Things are sometimes designed to be workable first and good-looking later. See Fig. 2-10 a and b.

Remember that the aesthetic part of design is only one part. The creative and the technical parts are just as important.

The Elements of Aesthetic Design

The *elements*, or *parts*, of aesthetic design are *form, space, light* and *shadow, texture,*

Fig. 2-11. **This ship combines the six basic elements of aesthetic design.**
(Bethlehem Steel Company)

line, and *color.* Every product that is beautiful has these six elements in some combination (Fig. 2-11). If a designer carefully blends these elements in a design, the product will appeal to the senses.

Form, Space, and Light and Shadow.
Form is the mass, or shape, of a product. Form gives depth and a sense of fullness to an object. The form may be closed and solid (Fig. 2-12), closed and volume-containing (Fig. 2-13), or open and projecting (Fig. 2-14).

Space may surround a form. Or space may be enclosed by a form. Figures 2-12, 2-13, and 2-14 show these differences. Light reflects off the surface of a form. Shadows appear in areas where the light is cut *off.* Light and shadow help give a form the *appearance of depth* (how deep something seems to be).

Texture.
Texture is the surface quality of a material. Materials such as burlap have rough and dull surfaces. Others, such as glass and metal, are smooth and shiny. A texture should be pleasing to the sense of touch. Figure 2-15 shows an example of the texture found in teakwood. The rich grain pattern of the wood adds to the beauty of the bowl. Figure 2-16 shows how a designer has used textures to create contrasts. The contrasts are a part of the design of the furniture.

Fig. 2-12. **This plastic server is a closed and solid form.** (Thayer Coggin Inc.)

Fig. 2-13. These desk accessories are closed and volume-containing forms. (McDonald Products Corporation)

Fig. 2-14. This gourmet rack is an example of an open and projecting form. (Drexel, Inc.)

Fig. 2-15. Note how the texture of the wood contributes to the beauty of the tray. (Dansk Designs)

Fig. 2-16. The deeply carved design on the doors adds interest to this furniture piece. (Broyhill Furniture Industries)

Line. Line gives an object a sense of length. It can also produce a sense of movement in an object. See how the lines of the space-station model seem to say that the object has speed (Fig. 2-17).

Fig. 2-17. The lines of the space station give the feeling of speed and movement. (NASA)

Fig. 2-18. The rocking chair is made of bent-wood, which may be seen as curved lines in space. Note the feeling of back-and-forth movement in the chair seat and back.

Figure 2-18 shows a rocking chair made of laminated wood. Notice the lines of the seat and the back parts of the chair. There is a *curvilinear* (curved-line) movement in this part of the chair.

Another way to see line in aesthetic design is to study the outline and the edges of a product. Look at the contour and edge lines in Fig. 2-19.

Color. Wavelengths of light reflect from a surface and produce color. Color can be used to produce aesthetic effects. Color can be added to a product by using paint. The natural color of the material itself can also give a product a pleasing appearance. Rich tones of brown are a natural part of wood. When you work with enamels, you are adding color in the form of paint. Cool grays and blues, colors that are found naturally in steel or aluminum, can also be beautiful.

Color is very important in aesthetic design. It gives richness and variety to a product. Suppose there were no textures or patterns that added variety. A one-texture,

Fig. 2-19. The contour and edge lines of this dish are graceful and rhythmic. They give a sense of simplicity to its form. (Georg Jensen Silversmiths, Limited Design by Henning Koppel)

one-color world would be very dull. That is why designers use colors and textures to add an interesting and pleasing look to their designs.

The Basic Principles of Aesthetic Design

The six basic *principles* of aesthetic design are *unity*, *variety*, *emphasis*, *balance*, *repetition*, and *rhythm*. If these basic principles are used properly, a product will be well designed.

Unity, Repetition, and Rhythm. Look at Fig. 2-20. The designer has used the basic principles of unity, repetition, and rhythm.

Unity is the most important principle. With *unity* (a "oneness"), the designer produces a wholeness in the design of a product. In Fig. 2-20, the designer has created

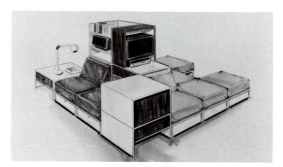

Fig. 2-20. The basic principles of unity, repetition, and rhythm are illustrated here.

unity by using rectangular lines that repeat themselves. And *planes* (levels) in space are also repeated. There are unity and *repetition* (repeating) here at the same time.

The repetition of curved lines and planes in space leads to another important basic principle: *rhythm*. Rhythm is the feeling of movement that repeats, or goes on over and over again. In Fig. 2-20, repetition creates a pleasing sense of rhythm throughout the forms.

Variety, Emphasis, and Balance. A designer must also use the principle of variety, which keeps a product from being dull or boring. Variety produces contrast and interest. See how the designer has varied or changed the textures and lines in Fig. 2-21. The smooth-textured surface of the headboard and chests makes a good contrast to the dark, rough texture of the bed cover.

When a designer wants to get people to look at one particular part of a product, the principle of *emphasis* is used. In Fig. 2-21, the emphasis, or stress, is on the headboard, cabinets, and chests. These parts are larger and more outstanding than the bed alone.

Without balance, no product would be pleasing to look at. It would not be as useful, either. *Balance* comes from making sure a design has equal weight that you can see. Figure 2-22 shows a chair that has a balanced appearance and is balanced to sit in. The line drawn through it helps you see that the chair has *formal balance*. That means that there are two equal parts.

Secondary Principles of Aesthetic Design

In addition to the six basic principles of aesthetic design, there are some *secondary* (lesser) principles that are important for designers to use. Two of these are the materials used in design and the surface decoration.

Materials in Aesthetic Design. Materials play an important part in the design of a product. Materials add to what a product does and how it looks. For example, a coffeepot may be made of heat-resistant glass. It is not the same, either in appearance or the way it can be used, as a coffeepot made of chrome-plated steel. A good designer makes the best use of materials. Figure 2-23 shows a potter's wheel.

Fig. 2-21. The basic design principles of variety, emphasis, and balance are applied in this bedroom set. (Broyhill Furniture Industries)

Fig. 2-22. A line drawn through the middle of this chair shows the formal balance of the piece.

Fig. 2-23. **The skillful hands of a potter work the clay on the potter's wheel into an interesting and creative design.** (Josiah Wedgwood & Sons, Incorporated)

Fig. 2-24. **This hall chest is decorated by the use of a silk screen process.** (John Widdicomb Co.)

Surface Decoration. A designer should decorate the surface of a product when decoration will add to its beauty. But a designer should not decorate the surface just to cover up mistakes. Poor materials or poor work cannot really be hidden by surface decoration.

Figure 2-24 shows the details of surface decoration. Notice how well the surface decoration adds to the overall form of this chest. Surface decoration used like this adds to the good looks of a product.

Unit 3

Planning your Activity Project

When you plan a project activity, first think about what it must do—how it will be used. Keep the *function* of the product in mind as you plan. You should plan with materials and equipment in mind. Be sure they are on hand and easy to get or use. Figure out the total cost. Know what skills and concepts are needed to complete your activity.

Basic Ideas

Finding a basic idea is sometimes the hardest part of planning a project. You can get some ideas by looking through magazines. You can study catalogs. Visiting stores may also give you some ideas. Your teacher may also offer ideas to you. Ideas for suggested project activities are also provided at the back of this text.

Preliminary Sketches

Planning is *organized thinking*. Sketching is one of the best ways to organize your thoughts—to put your thoughts in order. Sketching makes you organize your thoughts on paper.

Your first sketches may be very rough (Fig. 3-1). But as ideas become clearer, your sketches will get better. They will be more complete. Your teacher should look at your first sketches and give you some ideas that may help make them better. It is a lot easier to change a sketch than to change a design once you have started the real work.

When the design is in sketch form, you can change it easily enough. And you can also change sizes, proportions, and overall make-up, too.

Materials

You should decide what materials will be best for your design. The type of material should be thought about—hard or soft wood, or aluminum or wrought iron, for example.

Drawings

You must be sure that the overall dimensions are what you want. Then you must prepare a *working drawing*. This drawing should have the *exact dimensions* (measurements). The details for construction are also needed. You will use the working drawing while you are actually making the product.

You may also want to make a *pictorial sketch* (Fig. 3-2). This kind of drawing will show what your project will look like in three-dimensional form. Reduced-scale models will also show the three dimensions in the proportions that the finished, full-scale project will have.

Bill of Materials and Cost

After you have finished your working drawing, you should make a bill of materials. List the sizes you will need. List how much of each kind of material is needed. From this listing, you can figure out how much the project activity will cost. You

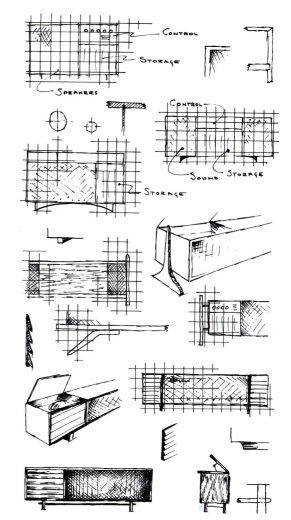

Fig. 3-1. Preliminary sketches for a project activity.

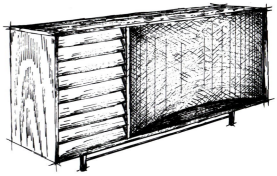

Fig. 3-2. Pictorial sketches for a project activity.

STUDENT'S PLAN SHEET

Student's name _____ Class _____

Name of project _____ Date started _____ Date completed _____

Estimated time _____ Actual time _____

Personal efficiency: estimated time ÷ actual time = _____ %

Source of drawing _____

Materials Required

No. of pieces	Description and size of piece	Kind of materials	Unit cost	Extended cost

Tools:
1.   5.   9.
2.   6.   10.
3.   7.   11.
4.   8.   12.

Order of Procedure:
1.
2.
3.
4.
5.
6.
7.
8.
9.
10.
11.
12.
13.

Approved _____

Fig. 3-3. A typical student's plan sheet.

can find out by multiplying the *unit cost* (the price of each item) by the number of items you will need.

Planning Your Procedure

You should list the operations needed to do your project activity. You should also list the order of doing them. You should plan what operations will come before others. This will save you a lot of time and work. And it will do away with mistakes as you work along. You can also make a list of tools and machinery. Listing tools and ma-

chinery will tell you what skills are needed to complete the project activity.

It is helpful to figure out how much time is needed for each step. Then you will be sure that your project will be finished at the right time.

The student's plan sheet (Fig. 3-3) shows what items should be thought about in planning a project activity. There are many kinds of plan forms. But most of them ask for the kind of information in the form shown in Fig. 3-3. Be sure to get your teacher's permission before starting your project activity.

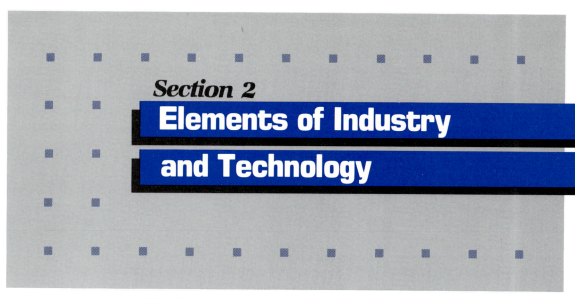

Elements of Industry
and Technology

Unit 4

Growth of American
Industry and Technology

In this unit and the next units, you will learn about industry and technology. American industry involves the study of tools, materials, processes, and products. Technology is a combination of people's experiences with and discoveries of tools, machines, and processes. These add to people's ability to *produce* and *distribute* (make and sell) what other people need. Changes come about in technology as people learn more about natural resources and how they can be used best. The use of new machines, materials, processes, and products is part of technological change.

Earliest History

People have always been *technologists*— in other words, tool-using creatures. From earliest times, people designed tools. Tools were used to help people make use of the earth's natural resources. In earliest times, tools were made of rocks or animal bones. They were used for sawing, drilling, hunting, farming, chopping, slicing, scraping, and measuring. Later on, tools were made from bronze. Still later, people learned to make tools of copper, iron, and steel.

As people learned more, they developed simple machines to help them. These machines saved people's energy. Things could be made much faster with simple machines. People learned how to become farmers, builders, mechanics, and, much later, scientists.

Our space-age technology could not exist if the inclined plane, the wedge, the screw, the wheel and axle, the lever, and the pulley had not been thought up many, many years ago. Technological growth was slow at first. But each new idea and discovery was very important (Fig. 4-1 a, b, c, and d).

As technology grew, people learned that they had to divide up all the jobs that had to be done. One person could not do everything that had to be done. One person could not make everything that was needed. People joined together the tools and strengths of many people to make goods they all needed. People discovered that they could have more if they planned for the future. The first *capitalists* were those people who had saved up much money or other wealth. Savings could be used to get tools, materials, and buildings where workers did their jobs. As more and more people worked together, using tools and machines, it became important to have managers. Managers make up a group called *management*. Management decides how to use *capital* (money or other wealth) and workers.

Technology slowly grew. It became necessary to divide skills into certain basic trades and crafts. One person *specialized* (worked best) in one trade. Another specialized in another trade. As this happened, people began to live closer together. First they lived in villages, then in towns, and then in larger cities and urban areas.

It would be hard to think of living without books, magazines, and newspapers. And yet, until about 1450, there were very few books. Books at that time were copied by hand—word by word. The great change in bookmaking came over 500 years ago. Johann Guttenberg, of Mainz, Germany thought up the idea of *movable type*.

Guttenberg thought of making letters of the alphabet on blocks of wood, with one

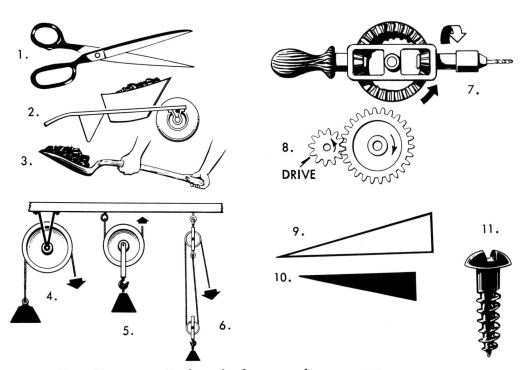

Fig. 4-1 a. All machines, no matter how simple or complex, are a combination of such simple machines as the lever, inclined plane, wedge, screw, wheel, axle, and pulley. Can you find these machines in the common tools shown?

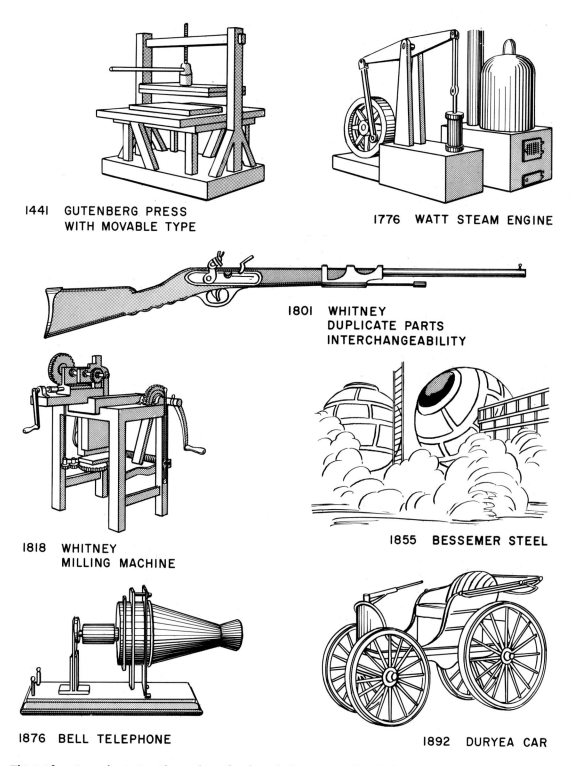

1441 GUTENBERG PRESS
WITH MOVABLE TYPE

1776 WATT STEAM ENGINE

1801 WHITNEY
DUPLICATE PARTS
INTERCHANGEABILITY

1818 WHITNEY
MILLING MACHINE

1855 BESSEMER STEEL

1876 BELL TELEPHONE

1892 DURYEA CAR

Fig. 4-1 b.   Some important inventions that have led to our modern industrial society.

Fig. 4-1 c. **Assembling the most complex machine ever built—the space shuttle.** (NASA)

Fig. 4-1 d. **The visual (video) telephone is an invention that is over 20 years old, but to date it has not been developed commercially.** (Western Electric Company)

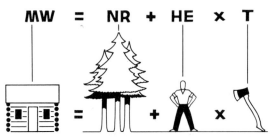

Fig. 4-2. **The basic law of production is illustrated here. People's material welfare equals the natural resources at their disposal plus their human energy (muscular and mental) multiplied by the efficiency of the tools available for their use.**

letter per block. In this way, he could set up the blocks to print words on a page. In 1450, he printed the famous Gutenberg Bible. His printing process was crude. But it was the first method of *duplicating* (making more than one copy) the printed page in large numbers. Soon printing presses began to turn out all kinds of books.

Another important early *booklet* (small book) was published in 1507. It contained letters written by Amerigo Vespucci, an Italian explorer. Vespucci's letters described a new continent. The publisher of the letters said that the new continent should be called *America*, in honor of Vespucci. Thus, America got its name partly because of the discovery of the first mass-production machine, the printing press. And it just so happened that the United States of America was much later to become the home of mass production.

Even in colonial America, technology was very simple. Many families made all their own clothing, housing, and tools. They bought only a few basic weapons and certain metal items that they could not make themselves. As families and communities grew, colonial Americans traded for food, tools, and things they could not make. Some of the things they could not make were glass, gun-powder, and metal blades.

The products people needed—their *material welfare* (MW)—were made from *natural resources* (NR) with *human energy* (HE), using simple hand *tools* (T) (Fig. 4-2).

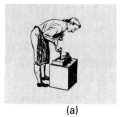 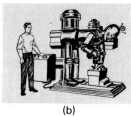

(a)   (b)

Fig. 4-3. (*a*) **People in colonial times worked with materials using their muscles and hand tools. They tired easily, and great skill was needed. The result was limited output. Even with great skill, no two pieces could be made exactly alike.** (*b*) **Workers today use power machine tools. These machines make it possible to build all other machines and to produce all the goods that people need.**

Today, the millions of products people make still require the same three things: natural resources, human energy, and tools. The great difference today is in the kinds of tools people have to work with (Fig. 4-3 a and b). Henry Ward Beecher said, "A *hand tool* is but the extension of a hand, and a machine is but a complicated tool." All machine tools are made up of the same basic parts (Fig. 4-4):

1. A *frame* to hold the parts together
2. A *source of power*, such as a motor or engine
3. A *method of transmitting the power*, such as belts, pulleys, gears, and hydraulic systems
4. An *operating mechanism* for moving the tool-holder up and down or changing the direction of power
5. *Tool-holding devices*, such as chucks and mandrels
6. *Work-holding devices*, such as vises, jigs, and fixtures.

Each machine may seem to be very different. But they all have the same basic parts. The basic parts are called by different names.

Technology did not really begin to develop rapidly until the *industrial revolu-* tion, in the late 1700s. Better sources of power were developed first. The most important was the *steam engine* that James Watt invented. This was used to run equipment in factories.

Modern technology in the United States goes back to the ideas of *Eli Whitney*, the inventor of the cotton gin. Whitney's cotton gin was an important invention. But it was not nearly as important as his ideas about *mass production*. Before Whitney's time, *gunsmiths* (gun makers) had to make and fit all muskets one at a time. No two muskets were exactly alike. The parts were not *interchangeable*. In other words, a part from one gun could not be used in another gun.

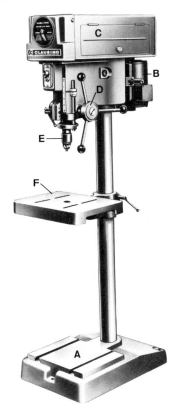

Fig. 4-4. Basic elements of all machines: (*a*) frame, (*b*) source of power, (*c*) power transmission, (*d*) operating mechanism, (*e*) tool-holding devices, and (*f*) work-holding devices. (Clausing Manufacturing Corporation)

Eli Whitney was not happy with the way guns were made—one at a time, piece by piece. In 1798, he suggested to the U.S. government that he could produce from 10 000 to 15 000 muskets in just two years. Washington leaders thought that Whitney would have to hire many gunsmiths to finish the work so quickly—about 16 guns each day. But the officials gave Whitney a contract to do the job anyway.

Whitney, however, had a whole new idea about producing the guns. He believed he could *mass-produce* the muskets in this way:

1. He would build special machines to do certain cutting and machining operations. For example, Whitney developed the first milling machine.

2. He would use special kinds of *labor* (workers) who would not have to be as skilled as master gunsmiths.

3. He would design *jigs* and *fixtures* for holding and machining the parts.

4. He would develop a system of measurement so that all parts would be *interchangeable.*

All these ideas were used by Whitney. But he did not really know how much "tooling-up" time was needed. At the end of the first year, he had produced very few muskets. The government leaders called him to Washington to ask why he was running behind time. The officials were shocked at what Whitney showed them.

Whitney took a large box with him. Inside, there were parts for 10 muskets: 10 barrels, 10 stocks, 10 triggers, and so on. He placed all the parts in 10 piles, one kind of part to a pile. Then he asked an official to put together a musket, using a part from each pile.

The plan worked. And the *interchangeability* of parts showed how many things could be produced in large numbers.

Whitney was given full approval. However, it took him two years to tool up to make the 10 000 muskets. But once this was done, he could have made 20 000, 30 000, or 40 000 in a very short time.

In those times, people found it very hard to understand such new ideas. But as time went on, Whitney's ideas changed people's lives. Everything from paper clips to the modern automobile was later made by mass production.

Whitney was a pioneer in industry. Once he got his factory plan going, other companies took up his ideas. And the ideas spread very fast. In fact, it can be said that *Whitney was responsible for the age of mass production.* Remember, mass production has two basic steps. First, a large number of *interchangeable parts* are made on machines. Then those parts are put together on some kind of *assembly line.*

Some of the ideas first used by Whitney and later used in all mass manufacturing include the following:

1. Machines can be built to make different parts of a product. Each piece of a product can be made by a certain machine. General-purpose machines, such as lathes, drill presses, and grinders, can be used. But at other times, special-purpose machines are needed.

2. There are *specialized devices* used by most manufacturers (Fig. 4-5). These include the following:

a. *Jigs.* A jig holds and locates the workpiece. It also guides, controls, and limits the cutting tool. A *drilling jig* is a good example. The term *jig* is also used for a device used to assemble (put together) a unit. Eli Whitney was the first to use jigs. It has been said that the name came from the fact that one of his workers called the device a *thingumajig.* Later the name was shortened to *jig* (Fig. 4-6).

b. *Fixtures.* A fixture is a work-holding device. It is used on a machine tool for machining duplicate pieces. A fixture holds the workpiece in a fixed position

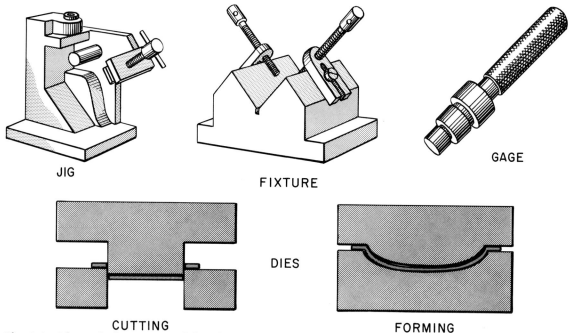

JIG

FIXTURE

GAGE

DIES

CUTTING

FORMING

Fig. 4-5. These devices are used for all types of manufacturing.

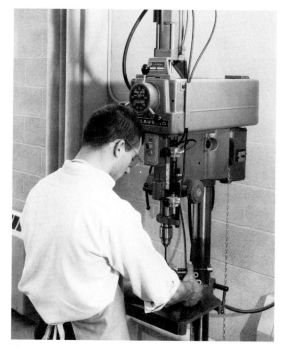

Fig. 4-6. A drilling jig in use. (Clausing Manufacturing Corporation)

in relation to the cutting tool. It is designed so that the workpiece can be put in and taken out rapidly. Fixtures are used for such operations as milling, grinding, and cutting (Fig. 4-7).

Fig. 4-7. A fixture is used to hold a metal part as the milling is done.

Fig. 4-8. This gage is used to check the diameter of a metal part.

c. *Gages.* A gage is a tool or instrument for measuring or checking the size of parts. A gage shows whether the dimensions are correct (Fig. 4-8).

d. *Dies.* A die is a tool for cutting and forming materials. The *cutting die* is much like a paper punch that cuts holes in paper. When the punched-out piece is the part that is worked on more, the operation is called *blanking*. When the blanked-out part is scrap material and the metal around it is used, the operation is called *piercing*. A *forming* die is one that is used to give shape to a material. Dies are used in large mechanical or hydraulic presses. The process of shaping a part with a die is called *drawing*, *pressing*, or *stamping* (Fig. 4-9).

3. Eli Whitney knew that he must have a system of measuring and gaging parts. In that way, they would be interchangeable. As early as 1776, James Watt, the inventor of

Fig. 4-9. This huge press has dies that form metal. (Aluminum Company of America)

the steam engine, used the thickness of a coin as the standard for cylinders. This was equal to 1/5 inch (5 millimeters). Today, parts of modern jet engines are made to an accuracy of 7/1 000 000 inch (0.000178 millimeter).

Eli Whitney thought there should be common standards of measurement. He knew that gages had to be developed if all parts were to be interchangeable (Fig. 4-10). All *quality control* in modern industry is based on this idea. In other words, a part must be made to a certain size with a

Fig. 4-10. Using an electronic measuring tool to check a manufactured part.

tolerance (margin of error). This tolerance is the amount of *leeway* (room) that is allowed above or below the basic size or dimension.

Suppose, for example, that a part is to be made 4 inches in size plus (+) or minus (−) 0.005 inch. This means that the smallest a part can be is 3.995 inches and that the largest it can be is 4.005 inches. Consider another example. A part may be made to 150 millimeters plus or minus 0.05 millimeter. Then the smallest the part may be is 149.95 millimeters. The largest it can be is 150.05 millimeters. The *limits* are the largest and the smallest dimensions that the tolerance allows. In order to check these limits, many types of gages had to be designed. Industry today uses fixed, indicating, and electronic gages to check parts for size.

The ideas of Eli Whitney were never fully used until the invention of the automobile. In 1896, R. E. Olds started a car-manufacturing plant. The plant made full use of the *assembly line*. Olds bought most of the parts from other manufacturers. Then workers assembled the parts along a pro-

duction line. Henry M. Leland, a follower of Whitney's New England school of gun makers, used the ideas of *accuracy* and *interchangeability*. These ideas helped produce better engines for cars. However, it was Henry Ford who made use of all the ideas of mass production best. He became the first master of production.

The present age of science and technology began with World War II. Importance was placed on *research* and *development*. This age has produced such things as solid-state electronic devices, nuclear energy, jet propulsion, space technology, and the automation of industrial processes (Fig. 4-11).

The term *automation* is given to all types of technological change that allow the best use of labor. However, the technical definition includes automatic controls, electronic computers, highly automatic transfer machines, and new ways of management. As you can see, the industrial revolution is still going on.

Fig. 4-11. Research and development has improved the picture tube used in television sets. The tube on the right was used in the first commercial television sets produced in 1954. At the left, a tube that uses the smaller, lighter solid state electronics is brighter and sharper. (RCA)

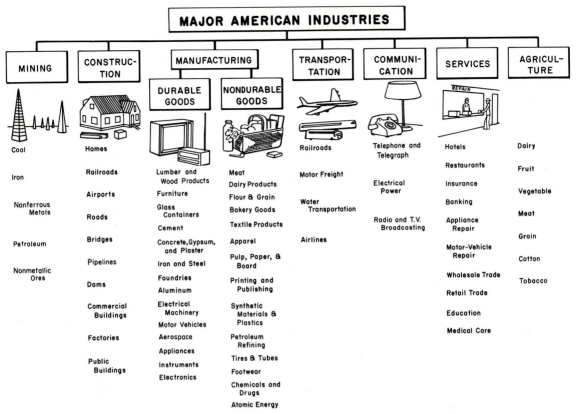

MAJOR AMERICAN INDUSTRIES

MINING	CONSTRUC-TION	MANUFACTURING		TRANSPOR-TATION	COMMUNI-CATION	SERVICES	AGRICUL-TURE
		DURABLE GOODS	NONDURABLE GOODS				

MINING	CONSTRUC-TION	DURABLE GOODS	NONDURABLE GOODS	TRANSPOR-TATION	COMMUNI-CATION	SERVICES	AGRICUL-TURE
Coal	Homes	Lumber and Wood Products	Meat	Railroads	Telephone and Telegraph	Hotels	Dairy
Iron	Railroads	Furniture	Dairy Products	Motor Freight	Electrical Power	Restaurants	Fruit
Nonferrous Metals	Airports	Glass Containers	Flour & Grain	Water Transportation	Radio and T.V. Broadcasting	Insurance	Vegetable
Petroleum	Roads	Cement	Bakery Goods	Airlines		Banking	Meat
Nonmetallic Ores	Bridges	Concrete, Gypsum, and Plaster	Textile Products			Appliance Repair	Grain
	Pipelines	Iron and Steel	Apparel			Motor-Vehicle Repair	Cotton
	Dams	Foundries	Pulp, Paper, & Board			Wholesale Trade	Tobacco
	Commercial Buildings	Aluminum	Printing and Publishing			Retail Trade	
	Factories	Electrical Machinery	Synthetic Materials & Plastics			Education	
	Public Buildings	Motor Vehicles	Petroleum Refining			Medical Care	
		Aerospace	Tires & Tubes				
		Appliances	Footwear				
		Instruments	Chemicals and Drugs				
		Electronics	Atomic Energy				

Fig. 4-12. Some major American industries.

American Industries and Occupations

Two reasons for taking this class are to learn something about the *major American industries* and to discover your talents and abilities. In that way you will be better able to select a *work career*. There are thousands of different industries in the United States. It would be impossible to study all of them. Each year, *Fortune* magazine lists the 500 largest corporations in the United States. In this list, you will find many of the major kinds of industries such as automobile manufacture, oil production, and chemical manufacturing. Many of these are written about in the sections on industrial arts that you will study.

Industries are classified as to kind (Figs. 4-12 and 4-13). Many of the more common industries and occupations are listed in the U.S. Department of Labor book *Occupational Outlook Handbook*. Every major industry needs thousands of different workers. The U.S. Department of Labor has defined over 20 000 occupations in the *Dictionary of Occupational Titles*. These are the many different occupations in all the major occupational groups. They include the following types of occupations:

1. Professional, technical, and managerial, such as doctor, lawyer, engineer, or president of a company
2. Clerical and sales
3. Service, such as hotel clerks or taxi-cab drivers
4. Farming, fishery, forestry, and related occupations
5. Processing, such as food preservation

MAJOR INDUSTRIAL/ TECHNICAL AREAS	MATERIALS AND PROCESSES	REPRESENTATIVE INDUSTRIES
Communications Technology	Drafting	Drawing and design (This is the language of all industry)
	Graphic Arts	Pulp, paper, and board Printing and publishing
	Photography	Movies Television Advertising
Production Technology	Woods	Lumber and wood products Home construction Furniture manufacturing
	Metals	Iron and steel Aluminum Motor vehicles
	Plastics	Synthetic materials and plastic products Appliances Motor vehicles Aerospace
Energy, Power, and Transportation Technology	Energy	Electronics Telephone Construction Electric power
	Power and Transportation	Motor vehicles Aerospace Motor freight Airlines

Fig. 4-13. Here you see how many of the industries relate to technology.

6. Machine trade, such as lathe operation
7. Benchwork, such as jewelry or watch repair
8. Structural work, such as carpenter or steel worker
9. Miscellaneous (others)

For example, in a large manufacturing concern, such as a motor-vehicle (automobile) manufacturer, occupations of all kinds are needed, including engineers, drafters, accountants, salespeople, machinists, and many others.

Now let us look at some of the major groups of industries.

Mining. In this industry, natural resources are taken from the earth. They are then changed into materials that can be used in manufacturing, construction, and transportation. Some common resources are copper ore, coal, petroleum, natural gas, bauxite, and iron (Fig. 4-14).

Construction. Construction includes all the major companies that use semi-

Fig. 4-14. **Petroleum workers constructing an oil-well drilling rig.** (Reprinted by permission, Marathon Oil Company.)

Fig. 4-15. **This steel worker is guiding a beam for assembly to the skeleton of a commercial building.**

processed materials, such as cement, lumber, metal, and glass, to build homes, bridges, dams, commercial buildings, factories, and other structures (Fig. 4-15).

Manufacturing. Manufacturers use the natural resources and semiprocessed materials found both above and below the ground to make the usable goods we need. These products are divided into two types: *durable goods* (those which can last a long time) and *nondurable goods* (those which are used up or that last a fairly short period of time). (See Figs. 4-16 and 4-17.) Some of the common durable goods include motor vehicles, aerospace equipment, electrical machinery, furniture, and household appliances. Industrial arts usually concentrates on these kinds of industries. Some of the nondurable industries include such things as meat, dairy products, printed and published materials, synthetic materials, and plastic products. In industrial arts, you will study some of these industries, especially printing, publishing, and plastics.

Transportation. These industries deal mostly with the movement of goods, materials, and people from one place to another by ship, train, bus, truck, or airplane. Transportation industries will grow rapidly as world markets expand (Fig. 4-18).

Fig. 4-16. A modern steel-mill rolling plant. (United States Steel)

Fig. 4-17. An automobile assembly line.

Fig. 4-18. Modern jetliners can provide transportation for several hundred people. (McDonnell Douglas Corporation)

Communication. Industries in this group deal with supplying power, including electricity and gas. They also deal with broadcasting and transmitting messages through the use of the telephone, telegraph, radio, and television (Fig. 4-19).

Service. Service industries are those in which work is done to meet the needs of people and things. Hotels and restaurants are service industries. Appliance repair and automobile, telephone, and power maintenance are also services (Fig. 4-20).

Agriculture. This industry supplies people with foods and with many basic materials used in clothing, chemicals, plastics, and other products (Fig. 4-21).

Study of Industry

You may want to make a study of one of the major American industries. This can be an individual project. You may wish to work with several other students. A good outline for such a study would be as follows:

1. The origin and background of the industry. Here you will cover such things as the time when and the place where the industry began and the people who began it and built it up. Also included are scientific developments and the place of industry in American business.

2. Organization. This should include how one industry organizes its factory, personnel, materials, and other requirements.

3. Materials. What raw materials are needed. Where they come from.

4. Processes. How the product is made. What machinery, equipment, and tools are needed.

5. Products. What kinds of products are manufactured. How they are sold.

6. Occupations. How many people are employed. What kinds of occupations there are, and what the duties are of the people in each. Included in this might be a study of labor and management.

Fig. 4-19. Telephones being checked and tested as they come off the assembly line. (Western Electric Company)

Fig. 4-20. The service technician must always be on the job. (Gulf Oil Corp.)

Fig. 4-21. Farmers produce not only food but also many materials used in manufacturing, such as cotton, peanuts, flax, and hemp. (Caterpillar Tractor Company)

7. *Trends.* How products are changing, and what the changes mean to industry and American energy. Some changes are brought about by such things as automation and computers.

8. *Working model.* The working model of an industry may be borrowed from a large company, or it can be built as a group project. Pictures or drawings from magazines and company bulletins are good sources.

Study of Occupations and Careers

Another special student report would be a careful study of an *occupation in which you think you are interested.* This is a good way to learn about your own interests. It is also a good way to learn about some of the industries that make use of a certain occupation. The report might follow this outline:

1. Name of the occupation

2. How many are employed

3. Classification of the occupation

4. Duties of those working in the occupation; any occupation includes working with people, things, or information; some occupations are more concerned with one of these than the others

5. Working conditions

6. Income, including salaries and benefits

7. Advantages and disadvantages

8. Education needed

9. How to get started in the occupation

10. Opportunities for promotion and advancement to higher jobs.

Industry, Large and Small

When we think of American industry, we often think of the giant manufacturers of automobiles, refrigerators, airplanes—the large producers that employ thousands of people. However, most American businesses are *small*. In fact, 75 percent of the businesses in the United States employ three or fewer workers (Fig. 4-22). Only 1 percent employ 1000 or more. This is true in all kinds of industries, including manufacturing, construction, communication, transportation, and retailing. Many of the smaller companies deal in service and *retailing* (selling goods). Many of the largest companies are in areas such as metal production, petroleum refining, and the manufacture of durable goods.

Fig. 4-22. A group of workers at a small business discuss their problems.

Organization of the Industry

In order for an industry to be run successfully, there must be some kind of organization. There are three major kinds of organizations: individual ownership, partnership, and corporation.

1. *Individual ownership* is the simplest. One person can set up a company with the least amount of trouble. That person alone is responsible for the money, the employment of workers, and every other part of the business.

2. In a *partnership*, two or more people join together to run a business. It is much like individual ownership except that the partners share responsibilities. Each partner is responsible for all the other partners' *debts* (money owed). The good thing about a partnership is that there is a pooling of money, skill, and experience.

3. Almost all big companies are organized in the form of a *corporation*. This is a business in which each person *invests* (puts in) a sum of money. However, each person is not responsible for the debts of the other investors or the company. All that the investors can lose if the business fails is the sum of money each has invested. To form a corporation, two or more people join together in a business agreement. They agree to work under the laws of the state where they *incorporate* (set up the corporation). Stock is sold to raise money. An organization that includes *stockholders* (the owners), a board of directors, a president and other officers is set up. The president, vice presidents, and managers are usually called the *management* group. They are responsible for the success of the business they have formed and now run.

Small businesses that deal with tools and materials are often called *job shops.* Job shops may do a number of small jobs. They may manufacture one item or a few items at a time. Large manufacturing companies are often called *mass-production companies* or *major manufacturing concerns.*

Unit 5

High Technology

In the previous units, you learned about the history and development of technology. You discovered that from earliest times, people have always been technologists. Until recently, however, technology evolved slowly. Now we live in a world that is changing rapidly. High technology includes the developments that are on the cutting edge of these changes. In the industrial world, high technology includes

Fig. 5-1 a. The space shuttle returning from another of its successful trips into space. (Rockwell International)

such evolving technologies as microprocessors, computers, lasers, robots, fiber optics, satellites, and all types of telecommunications. These high-technology developments are being applied to all the major industries described in this unit. For example, the laser is used in producing wood products, metalworking machinery, motor vehicles, electrical and commu-

nication equipment, and business machines, to name only a few examples.

High technology can be defined as the application of information, techniques, and tools to material resources in producing the goods and services that people need. The *space shuttle* is an excellent example of a machine that includes most of the high-technology developments (Fig. 5-1 a and b). By the year 2000, technological advances in computer systems, microelectronics, communications, direct energy (lasers), robotics, and composite materials will change the way we live dramatically.

While high technology is the fastest growing area of our economy, it will account for less than 10 percent (about 7.7 million careers) of the total job force by 1995. However, all these careers will require advanced education and training (Fig. 5-2 a and b). High technology will not replace the millions of highly skilled workers who use the basic technology of tools, machines, and processes described in this book. There will be a great demand for skilled workers who can run a lathe, drive a nail, weld a part in a repair shop, repair and tune an engine, and do the millions of tasks that require technical skill.

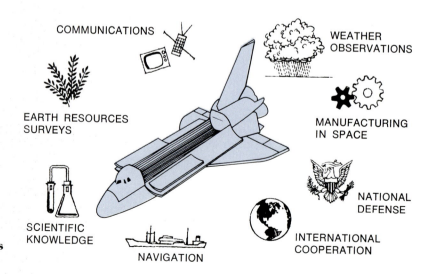

Fig. 5-1 b. Some uses of the space shuttle.

COMMUNICATIONS

WEATHER OBSERVATIONS

EARTH RESOURCES SURVEYS

MANUFACTURING IN SPACE

SCIENTIFIC KNOWLEDGE

NATIONAL DEFENSE

NAVIGATION

INTERNATIONAL COOPERATION

JOBS WITH THE LARGEST GROWTH: 1982 – 1995

OCCUPATION	INCREASE IN JOBS	
	THOUSANDS	PERCENT
Building custodians	779	27.5
Cashiers	744	47.4
Secretaries	719	29.5
General clerks, office	696	29.6
Sales clerks	685	23.5
Nurses, registered	642	48.9
Waiters, waitresses	562	33.8
Teachers, kindergarten and elementary	511	37.4
Truckdrivers	425	26.5
Nursing aides, orderlies	423	34.8
Sales reps, technical	386	29.3
Accountants, auditors	344	40.2
Auto mechanics	324	38.3
Supervisors of blue collar workers	319	26.6
Kitchen helpers	305	35.9
Guards and doorkeepers	300	47.3
Food service workers	297	36.7
Managers, store	292	30.1
Carpenters	247	28.6

THE TWENTY FASTEST GROWING OCCUPATIONS*

OCCUPATION	GROWTH	PERCENT
Computer service technicians	53,000	97
Legal assistants	43,000	94
Computer systems analysts	217,000	85
Computer programmers	205,000	77
Computer operators	160,000	76
Office machine operators	40,000	72
Physical therapy assistants	26,000	68
Electrical engineers	209,000	65
Civil engineering technicians	23,000	64
Peripheral electronic data-processing equipment operators	31,000	64
Insurance clerks, medical	53,000	62
Electrical and electronics technicians	222,000	61
Occupational therapists	15,000	60
Surveyor helpers	23,000	59
Credit clerks, banking/insurance	27,000	54
Physical therapists	25,000	54
Employment interviewers	30,000	53
Mechanical engineers	109,000	52
Mechanical engineering technicians	25,000	52
Compression and injection mold machine operators, plastics	47,000	50

*These jobs are those with a 1982 employment of 25,000 or more. Growth projects the increased number of jobs in that occupation from 1982 to 1995; percentages project increase in that occupation's work force from 1982 to 1995.
Source: U.S. Department of Labor.

Fig. 5-2 a. Study these charts carefully. Notice that from 1982 to 1995, jobs for computer service technicians will increase almost 97 percent, to a total number of about 100 000. During the same period, the number of auto mechanics will increase only half as fast, but the total number by 1995 will be almost one-third of a million, which is more than three times the number of computer service technicians. The computer service technician will be trained in a two-year technical program, while the auto mechanic must be trained either in a vocational or a technical program. Also remember that in the next few years all cars will be computer-controlled so that the auto mechanic must also understand the computer. (Source: U.S. Department of Labor)

Fig. 5-2 b. This student is preparing to be a qualified auto mechanic. (VICA)

In many manufacturing industries, high-technology developments will be applied to these processes. For example, instead of a worker using a *drill* to cut a hole, the *laser* may do the job; instead of a drafter making a drawing on a board with drafting instruments, the computer can be used to design products. While some *welding* will still be done using standard spot and arc welders, major manufacturers will use robots to do welding.

Computers[1]

Like any other tool, the computer increases your ability to do intelligent work.

[1]Some of the materials in this section, courtesy of General Electric.

It enables you to work problems more quickly, with less drudgery, and with greater accuracy.

The computer extends human brainpower just as the engine extends human muscle power. With the computer, there is a new capacity to discover, create, build, solve, and think; this will affect the way you study, work, and live your everyday life. Without computers, America's space program would not be possible. And in everyday life, computers are helping to eliminate drudgery, expand leisure, and aid progress.

One exciting thing about computers is that they are relatively new. Modern electronic computers were not invented until after World War II. Yet by 1964 there were over 18 000 computers at work in the United States, including an estimated 600 in schools.

Today, computers play a key role in finance, transportation, manufacturing, defense, science, medicine, and practically every industry and profession. They are opening challenging new careers for young men and women, for high school graduates as well as for those holding advanced college degrees.

The future clearly belongs to those who understand the computer. And, to a great extent, our future progress depends on intelligent young people who are able to put the computer to work solving our social, industrial, and scientific problems (Fig. 5-3).

Hardware. Equipment that makes up a computer system is called *hardware*. The personal computer usually consists of five basic units (Fig. 5-4).

1. An *input* device on most personal computers that looks like a typewriter keyboard; it is used to send information into the computer.

2. The *central processing unit (CPU)* plus the memory chips containing memory, control, and operating units. These units are often in the same cabinet as the key-

Fig. 5-3. **These elementary school students are learning about computers, a knowledge they must have in the years ahead.** (Gulf Oil Corporation)

board. In other computer systems, the keyboard is a separate unit.

3. A *display device* that looks like the picture tube on a television set. The computer prints out written words, graphs, and pictures on the screen. A home television set can be connected to the computer, but most computers have this unit on top of the CPU as a video monitor that is a cathode-ray tube (CRT).

4. A *disk* drive that holds the magnetic disk or cassette that stores the software

Fig. 5-4. **A personal computer including the keyboard, display device, CPU, memory unit (disk drive), and printer.** (IBM)

(program) used with the computer. The disk or cassette is the external memory that can feed information into the CPU as needed.

5. The *printer*, which is a high-speed typewriterlike device that prints the hard copy output from the computer.

The Chip. The chip is the all-important electronic building block used in computers. It is a microelectronic device less than one-eighth the size of a postage stamp, barely the size of a baby's thumbnail. The chip really is complicated electronic circuitry on a silicon base with tiny switches joined by "wires" that are etched on a thin film of metal. The chip can hold millions of electronic "parts."

About 40 years ago, the first digital computer using vacuum tubes weighed over 30 tons. A large air-conditioned room was needed to hold the equipment. Today, by contrast, a single tiny chip can perform 200 times more calculations than this first huge computer with its thousands of vacuum tubes.

In early radios, television sets, and computers, *vacuum tubes* were used to control and amplify electrical signals. These bulb-shaped tubes gave off large amounts of heat, so they burned out often (Fig. 5-5). In 1947, the *transistor* was invented. This device replaced the old vacuum tube (Fig. 5-6). Within 10 years after it was invented, the transistor consisted of a speck of silicon crystal enclosed in a can about the size of a pea. Today, transistors are smaller than a speck of dust.

A major breakthrough in technology came with the development of the silicon chip (Fig. 5-7). The *chip* with its many electronic "switches" is used in such products as computers, watches, calculators, cash registers, and items like robots and automobiles (Fig. 5-8).

Making chips constitutes what is called the semiconductor industry. Producing these chips is a very exacting process.

Fig. 5-5. The tube was large, used too much electricity, overheated, and burned out too fast.

Fig. 5-6. The transistor was smaller, lasted longer, and was not breakable. It was invented in 1947 by three scientists who won the Nobel prize for their discovery.

Fig. 5-7. Compare the size of the *chip* on the right with the *transistor* in the middle and the *tube* on the left.

Fig. 5-8. All modern calculators use the chip as the basic electronic unit.

First, engineers must design the complex circuitry that is to make up the chip. Chips are made for different purposes, and there is a wide range in the number of electronic "switches" in the final unit. This electronic circuitry is projected onto the face of a wafer (which will later become many chips). The wafers are coated with Photoresist, a light-sensitive material that makes it possible for the silicon to accept the design. The circuitry is etched onto the silicon chip. The process is much like that described in Unit 154, "Etching." The silicon wafer is then covered with a metallic coating that electrically connects the wafer's various switches. Each wafer is then cut into many chips with a *laser* beam. The chip is then mounted in a frame and sealed for protection. The frame has tiny lead wires that allow the chip to be connected to other parts of the electronic circuitry in the computer. Imagine

working on a part that is as tiny as a baby's thumbnail and using a microscope to see what is being done.

Remember, the chip with its many switches allows electricity to flow or not to flow, like the light on a reading lamp. In a computer, when the switch is "on" it represents a 1, and when the switch is "off" it represents a 0 (Fig. 5-9). You will learn more about this later when you read about computer math. These electronic switches can be turned on and off at an incredible rate of speed. A large computer, for example, can add more than 1 million four digit numbers every second.

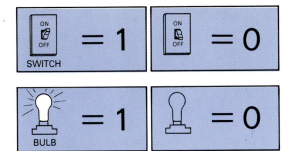

Fig. 5-9. The chip performs the same functions as an "on-off" light switch.

Computer Math (Binary Code). The computer's language is written in binary code. This is a simple way of communicating once you get the hang of it.

If you already understand binary, you know that there are really just two ideas that must be grasped.

First, whereas decimal numbers—the kind of numbers we ordinarily use for money and for metric measurement—are built on a base of 10, binary numbers are built on a base of 2. The base 10 means that when you move a digit one space to the left (and add a zero), it's worth 10 times as much. Another way of saying this is that it increases by the power of 10. With binary numbers, using a base of 2, every time you move a number to the left, it's worth 2 times as much; it increases by the power of 2.

Just as in the decimal system, which is based on the power of 10, the value increases from right to left, values in the binary system increase from right to left by the value of the base. By way of illustration:

| Decimal: | 1000 | 100 | 10 | 1 |
| Binary: | 32 | 16 | 8 | 4 | 2 | 1 |

The second thing to learn about binary numbers is that the system uses only two numbers: 1 and 0. In other words, you must forget about the numbers 2, 3, 4, 5, 6, 7, 8, and 9. They are never used in the binary system.

The question arises, How do you count to two without a 2? It's simple. As we noted, the value of the binary number increases by 2 times as you move it 1 space to the left. To get 2 in binary, then, you simply move the 1 space to the left and add zero. Thus, 10 in binary is the same as 2 in decimal.

There is an easy way to convert a binary number to decimal. Just write down the successive powers of 2, and then match up the binary number underneath. Take binary number 01100101:

0	64	32	16	8	4	2	1
0	1	1	0	0	1	0	1

$$0 + 64 + 32 \qquad + 4 \qquad + 1 = 101$$

Each digit—either 1 or 0—in a binary number is called a *bit* (short for binary digit). This is important to remember because, as we shall see, information moves through a computer a bit at a time. While the binary code may seem awkward to those who are used to the decimal system, it is simplicity itself to the computer. This is because the basic electronic unit inside the computer (the chip) can exist in only two possible states: current on or current off.

How a Computer Operates. To communicate, computers convert, or change the numbers, letters, and symbols into the binary code (two digits) of computer math. The digit 1 is like an electrical switch or light bulb that is "on," and the digit 0 is like a switch or light bulb that is "off." Most personal computers digest information in chains of eight electrical pulses called a *byte*. Just imagine these eight pulses as a bank of eight switches or electrical light bulbs. When a bulb is on in the series of eight bulbs it represents a 1, and when the bulb is off it represents a 0. These eight bulbs can be arranged in 256 different combinations. This is much more capacity than is needed to cover all the letters of the alphabet, 10 numbers, 13 punctuation marks, symbols, musical scales, and other input items that may be needed. To process the input instructions, the computer "brain" that does the switching is a central processing unit, or *CPU* (often only a single chip for most personal computers). To store excess information, memory chips are needed (Fig. 5-10).

There are two types of memory chips. The RAM (random-access memory) chip can be thought of as pages in a notebook where you write all the information that you would like to use later when performing an operation or figuring out a problem. The information in the RAM chip is there only temporarily and then is erased or destroyed just as you do with the notes in the notebook after you are finished using this information.

The ROM (read-only memory) chip is like a reference book that contains permanent information to which you will refer again and again when using the computer. The ROM information memory chip is put there by the manufacturer of the computer. The information in both RAM and ROM chips can then be used by the CPU. The CPU uses the ROM memory chip to find out how to do things, such as showing a character on the screen or sending output information. The larger the ROM memory unit is, the more functions the computer can perform. Obviously, the more capacity the RAM chip has, the more information you can store in it. For example, a popular personal computer has 128K bytes in the RAM unit and 64K bytes in the ROM unit. The 64K ROM memory chip has permanently stored 524 288 *bits* of information.

Software.   Software is the program that tells the computer what to do and how to do it. Software is written by programmers. In solving any kind of problem, a computer is helpless until it has been given a detailed set of instructions. The computer

Fig. 5-10.   **These computer terminals are connected to a central computer unit that controls the stored information in both the ROM and RAM memory units.**

can't figure out how to solve the problem. All it can do is follow directions step by step until the job is completed. This is true for any problem you try to solve with a computer, whether it's simply adding a column of numbers or sending a missle to the moon. These detailed instructions are called the computer *program.*

To write a program, you must first understand the problem to be solved. You must analyze it and break it down into its components.

One way to do this is to construct a flowchart of the problem. A flowchart is a line diagram that shows each part of the total problem as well as its logical relationship to every other part. The flowchart is similar to the diagram of a pass play that a football coach draws on the chalkboard. It shows component parts of the problem just as a play diagram indicates each blocking assignment.

To work in the computer, each part of the flowchart must be broken down into simple instructions. One part of the flowchart may require 20 or more individual steps in the computer.

The flowchart illustrated in Fig. 5-11 shows the steps in a payroll program. The boxes represent work to be done, and the diamonds are choices to be made. This

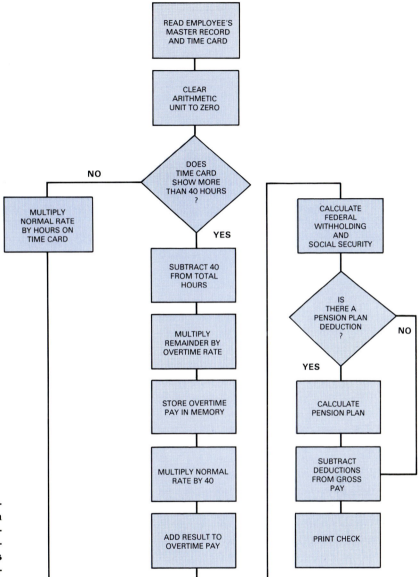

READ EMPLOYEE'S MASTER RECORD AND TIME CARD

CLEAR ARITHMETIC UNIT TO ZERO

DOES TIME CARD SHOW MORE THAN 40 HOURS ?

NO

MULTIPLY NORMAL RATE BY HOURS ON TIME CARD

YES

SUBTRACT 40 FROM TOTAL HOURS

MULTIPLY REMAINDER BY OVERTIME RATE

STORE OVERTIME PAY IN MEMORY

MULTIPLY NORMAL RATE BY 40

ADD RESULT TO OVERTIME PAY

CALCULATE FEDERAL WITHHOLDING AND SOCIAL SECURITY

IS THERE A PENSION PLAN DEDUCTION ?

NO

YES

CALCULATE PENSION PLAN

SUBTRACT DEDUCTIONS FROM GROSS PAY

PRINT CHECK

Fig. 5-11. Computer programming begins with a flowchart. This information is then converted into one of the languages that are used by the computer.

program can figure out the pay for employees who work overtime. In the third step, the program asks, "Does time card show more than 40 hours?" If the answer is "yes," the program flows forward and works a series of instructions that calculate overtime pay. If the answer is "no," the program skips these instructions and goes on to figure pay at the regular rate.

This action of skipping part of the program or of choosing which of two sets of instructions to follow is called *branching*. In a football play, for example, the halfback may be given two assignments. If the opposing linebacker rushes in, the halfback blocks that person. If the linebacker drops back, the halfback moves out a short distance to the side and becomes a pass receiver. What the halfback does follows logically from what the opposing linebacker does.

It's the same way with branching. The computer decides which of two sets of instructions to follow by asking whether the hours worked are greater than 40. It makes a simple "yes-no" decision.

In very complex problems where there are thousands of variables, there may be thousands of such choices. The computer takes them one at a time, makes a simple "yes-no" decision, and switches to the appropriate instructions.

Because it can make decisions, we say the computer has a logical capability. But it's important to realize that even in making these decisions, the computer is simply following instructions. It is the person writing the program who decides when a decision is needed and indicates which factors will determine which choice will be taken.

Sometimes programs have bugs—computer talk for errors. If you don't debug before running a program, you'll get "garbage."

Once the problem has been mapped out in a flowchart and broken down into individual simple operations, you must write the instructions telling the computer how to handle each operation. But here we run into another problem. The computer doesn't understand English. You have to write out instructions in a language, or code, that will be understood by the computer.

Here are some languages you may run into: FORTRAN (FORmula TRANslation); COBOL (COmmon Business Oriented Language); BASIC (Beginner's All-purpose Symbolic Instruction Code), which was derived from FORTRAN and COBOL; APL (A Programming Language), which is a language for mathematical concepts; and PL/I, which can handle both business and technical problems.

While there are many different computer languages, the four most popular for beginners are the following:

BASIC is used so widely, it is almost the standard microcomputer language. BASIC is popular because it is relatively easy to learn and use. A program can be written using only a few instructions. These instructions are easy to interpret since they are like English.

PASCAL is another popular computer language that was designed primarily for teaching languages. Pascal is used in schools as the language for standardized tests.

PILOT was written for use by teachers in writing tests and other programs used in teaching.

LOGO is a computer language that is useful for producing graphics on the screen.

Computers are being designed that can interpret written and spoken English language. With these computers, the operator won't need to learn a special language. These advanced computers also will be able to think and reason much like a human being.

Careers. There are many and varied careers directly and indirectly related to computers, including the following:

A. *Basic computer operations.* These are jobs that exist purely from the presence of the computer itself. Today, many of them are particularly tied to the use of larger computers for electronic data processing (EDP). Governments use them to keep track of large amounts of information in payrolls, military records, licenses, and tax rolls. Business and industry use them to handle financial records, sales, shipping, inventory, customer lists, research, and data analysis. The growing use of minicomputers and desk-top microcomputers will open up jobs in smaller businesses and industries. These machines will have to be sold, programmed, operated, and serviced just like the larger ones.

1. *Systems analysts.* Basically, these people solve a company's information problems. They analyze the problem, help choose what information is needed, decide on the kind of computer capabilities required, and then organize a system for collecting, preparing, processing, and distributing the information.

2. *Programmers.* It is the programmer's job to translate the system analyst's requirements into a step-by-step set of instructions for the computer. Programmers may also develop a set of instructions for the operators.

3. *Operators.* The operator's job is to enter the data into the computer, operate the computer, and then retrieve the information. Operators include keypunch operators, word-processing operators, data entry clerks, console operators, high-speed-printer operators, and card-type converter operators.

4. *Service technicians.* Service technicians are often known as field or customer engineers. They're responsible for installing and servicing the computer equipment. Installation can include laying cable, making electrical connections, testing the machinery, and correcting problems. Servicing includes record keeping, adjusting and oiling parts, and running special peak-capacity tests to uncover problems.

B. *Computers as components.* More and more products are being designed and built with computers as integral parts. Airline pilots change throttle settings, and a computer shows them the new fuel-consumption rate on a dial. Tailors slip wheels into sewing machines, and a computer lays down intricate stitches on the material. You push a button on your television set, and a microcomputer switches to another channel and adjusts the picture. You push some buttons on your clock ra-

Fig. 5-12. **Many large planes couldn't fly without computers.**

dio, and no matter what station you fall asleep to, a computer "remembers" to wake you on time to your favorite morning station. As manufacturers become aware of the things computers do, they are looking for ways to incorporate them in their products before a competitor does. Computers are revolutionizing products in many fields, including typing, dictation, oil exploration, appliances, instruments, electric motors and controls, medicine, automobiles, and communications (Fig. 5-12).

1. *Engineers.* The people qualified to design and build computers for making products are engineers. They are trained to find practical applications for scientific advances, develop new products, and improve systems and processes (Fig. 5-13). Certain kinds of en-

Fig. 5-13. **This engineer is doing computer-aided design. You'll learn more about this in Unit 10, "Manufacturing Technology."** (Adage, Inc.)

gineers are more important in this area, however. Industrial companies turn to research and development engineers, electronics engineers, and computer applications engineers. These people have special training that helps them determine the various ways microcomputers can be used to improve products.

2. *Technicians.* In applying computers to other products, the technician is an important member of the engineering team.

3. *Operators.* At the design stage, operators may be needed, particularly if the problem is complex enough to require input to a large computer. At the point of use, the operator may not even be aware that there is a computer involved. Tell some people who use a bank's automatic-teller machine that they have just run a computer program and they'll be amazed. It will be that way even more frequently as manufacturers simplify the application of computers to their products. You—and everyone else—will operate computers without realizing it or thinking much about it.

Robotics

Robotics is a rapidly growing field that is playing an increasing role in manufacturing and space exploration. Eventually, robots will be intelligent machines with a sense of vision, touch, and hearing. Today, robotic devices are employed in most heavy industries, including automobiles, tractors, and off-the-road vehicles. Many other industries, such as furniture manufacturing, are using more robots. Robotic devices are also used on many planetary space probes and on the space shuttle. They will eventually be used to assemble space structures. Robots are primarily machines with arms and hands that can be programmed to do a variety of tasks automatically. Robots in use in industry and space bear no resemblance to the nice hu-

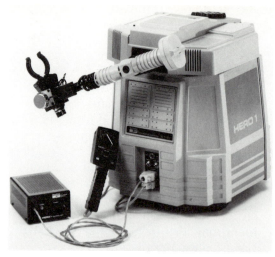

Fig. 5-14 a. This small wheeled robot was one of the first of its kind. It is used to learn how robots function. In the years ahead, this type of robot will become much more complex. (Health-Zenith)

manlike robots in the movie *Star Wars*. Robots may be an *armlike* device that moves objects from place to place, or they may be highly sophisticated *wheeled vehicles* of the kind that explore the surface of Mars or will someday be the "butler and maid" in your house (Fig. 5-14 a, b, and c).

History. The idea of developing a "mechanical person" goes far back into history. During the eighteenth century, the Japanese developed a mechanical doll that served tea to guests. Many mechanical toys, called automatons, were the forerunners of the robot. The word "robot" was first used in a play by a Czech author and playwright in 1921. The play told the story of a manufacturer who replaced human workers with artificial workers. The humanlike androids rebelled and took over the factory. The Czech word *robota* means "compulsory labor" or "serf."

In the 1930s, robots were popular characters in films and comic books. It wasn't until the beginning of the computer age in

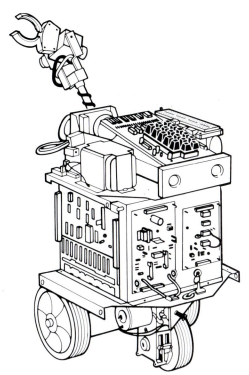

Fig. 5-14 b. The interior of the robot.

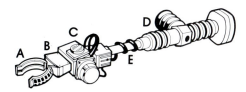

Fig. 5-14 c. Note how the arm and hand (called a gripper) can move. The arm provides five axes of motion: (a) Gripper opens and closes $3^1/_2$ inches, (b) wrist rotates 350°, (c) wrist pivots 180°, (d) arm pivots 150°, and (e) arm extends and retracts 5 inches.

the 1940s that the first real robots were developed. The first working robots were used in the radioactive sections of nuclear power plants in the late 1950s.

The first operating robots in space were used on the Surveyor spacecraft that landed on the moon in 1967. Robots were also used in the Viking spacecraft that landed on Mars. It took samples from the

Fig. 5-15. A typical heavy-duty industrial robot.

surface of Mars, placed the material into the on-board lab, and analyzed it.

Beginning in the early 1960s, robots have been being used in American automobile factories. By the mid 1990s, one automobile manufacturer alone will have over 14 000 robots in its factories.

Basic Robot. A typical robot consists of one or more manipulators (arms), end effectors (hands), controller, power supply, and an array of sensors to provide environmental feedback. Because the majority of robots today are used for industrial purposes, their classification is based on their industrial use (Fig. 5-15).

Robot Classes. The classes of robots are as follows:

Pick-and-place robot. The simplest form of robot. This robot picks up an object and places it in another location (Fig. 5-16 a, b, and c).

Servo robot. Robots of several types that employ servomechanisms for the arms and hands to alter direction in midair without tripping a mechanical switch.

Programmable robot. A servo robot that is driven by a programmable control that memorizes a sequence of movements and repeats them perpetually. This kind of robot is programmed by "walking" the arm and hand through the desired movement.

Computerized robot. A servo robot run by a computer. This kind of robot is pro-

Fig. 5-16 a. This robot is feeding a punch press. The robot moves a piece of sheet metal into the press and then removes it after it is cut. (Prab Robots)

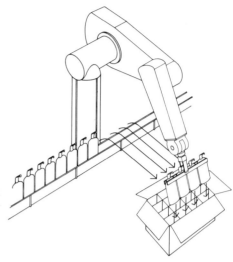

Fig. 5-16 c. This robot is packaging products as they move along the assembly line. People used to do this job.

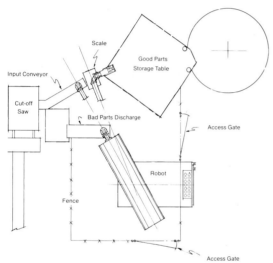

Fig. 5-16 b. This diagram shows a robot being used to feed a cut-off saw.

grammed by instructions fed into the controller electronically. "Smart" robots, as they are called, may have the ability to improve on their work instructions (Fig. 5-17).

Sensory robot. A computerized robot with one or more artificial senses to sense its environment and feed back information to the controller. The senses are usually sight or touch.

Assembly robot. A computerized robot with sensors that is designed for assembly-line jobs.

Fig. 5-17. This is a robot of a different design. It is controlled by computers. (IBM)

Primitive Robot Functions. Robots are designed primarily for manipulation purposes. The actions produced by the hand on the end of an arm are:

1. Moving
   from point to point
   following a desired trajectory (a path or curve)
   following a contoured surface
2. Changing orientation
3. Rotation

Nonservo robots are capable of point-to-point motions. For each desired motion, the manipulator moves at full speed until the limits of travel are reached. Nonservo robots are often referred to as "bang-bang" or "pick-and-place" robots. When nonservo robots reach the end of a particular motion, a mechanical stop is tripped, stopping the motion.

Power Supplies. The three main forms of power supply in use today are pneumatic, hydraulic, and electrical. With pneumatic power, pressurized gas is used to move the joint. Gas is propelled by a pump through a tube to a particular joint and then actuates movement. Air brakes on trucks are a common form of pneumatic power use. Pneumatic units are inexpensive and simple, but their movement is "squooshy." Consequently, they are usually reserved for pick-and-place robots.

Hydraulic is the most common industrial system and is capable of producing high power. The principal disadvantages of hydraulic systems are the accompanying devices, including pumps and storage tanks, and problems with system leaks.

Electric controls offer clean movements, can be precisely controlled, and are very reliable and accurate, but they do not deliver the high power controls that hydraulic can deliver.

Control. Control for robots can consist of a series of stops and limit switches that are tripped by the arm during its motion or highly complicated computer-controlled devices that use machine vision and sonic sensors. With computer-controlled robots, motions of the arm and hands are programmed. In other words, the computer "memorizes" the desired motions. Sensor devices on the arm determine the distance from the hand to the object that is to be manipulated.

Robots on the Space Shuttle. The space shuttle is a space launch system that is designed primarily for transporting large and small payloads to and from the earth. In the back of the *Orbiter* is a cargo bay that is large enough to contain one and one-half school buses. Once the *Orbiter* is in space, payloads in the bay have to be moved. Satellites, for example, have to be removed from the bay and placed into orbit. Defective satellites that are already in space must be brought back into the bay for repair or returned to earth. To handle these tasks as well as the assembly of large structures in orbit, a robot arm has been installed along the hinges of the portside (left side) cargo bay door.

The robot arm is a highly sophisticated robotic device that is similar to the human arm. Fifteen meters long, the robot arm features a shoulder, elbow, wrist, and hand, although the hand does not look at all like a human hand. The skeleton of the arm is made of lightweight graphite composite materials. Covering the skeleton are skin layers consisting of thermal blankets. The muscles driving the joints are electric actuators (motors). Built-in sensors, like nerves, sense joint positions and rates of rotation.

A robot arm on the space shuttle can pick up a satellite in space, place it in the cargo bay for repairs, and then lift the unit back out in space operating properly. The astronauts are the *mechanics* (using specially designed wrenches and other tools that will operate in a gravity-free environment). Even astronauts must have "hands

Fig. 5-18. Here, the space shuttle is using its robot arm to catch a satellite that needs repairing.

on" skills in using hand tools and machines (Fig. 5-18).

Careers in Robotics. In the years ahead, most major manufacturing concerns will use thousands of robots. Designing these robots will be the responsibility of the *engineers*. Setting up and maintaining the robots will require many *robotic technicians*. Very few robots can be purchased and directly used on a work station. Most robots must be adapted to fulfill the functions of the particular work station.

A *robot technician* must be able to select a robot design, assemble it, adapt it with standard parts, program it, and maintain it. The skills of a robotics technician are needed in industrial engineering, plant engineering, factory engineering and maintenance.

The best preparation for a career as a robot technician is a two-year program in electronics leading to full qualification as an electronic technician. The reason for this is that each robot is different and special in-plant training is needed to learn how to program and repair the specific kind of robot that is in use in each particular manufacturing facility.

Lasers

The laser, invented by C. H. Townes and Arthur Schawlow in 1958, is a great scientific discovery that has changed many processes in manufacturing, medicine, communications, and merchandising. Laser light is electromagnetic radiation; however, a beam of laser light differs greatly from light emitted by other luminous (light) sources.

The term *laser* is an abbreviation for *l*ight *a*mplification by *s*timulated *e*mission of *r*adiation. The laser produces a very narrow intensive (strong) beam of light that can be focused on a spot that is very small in diameter. The laser produces heat of over 7500°F. This concentrated energy will penetrate (go through) a variety of materials. The laser beam can be roughly compared to using a magnifying glass to direct the rays of the sun onto kindling (such as wood) to start a fire. The intense heat from the small beam of light ignites the wood.

There are five types of lasers:

1. *Solid-state rod-type lasers*, which use materials such as ruby, glass, and neodymium-YAG for laser emission (release).

2. *Semiconductor diode-type lasers*, which create laser emission by passing a current through materials such as gallium arsenide.

3. *Gas-type lasers*, in which laser emission is produced by passing a current through gases such as helium, argon, carbon dioxide, nitrogen, or xenon. The current causes the gas to ionize and radiate. The radiation oscillates (moves) with a tube provided with mirrored ends and then discharges energy from the end of the tube.

4. *Liquid-type lasers*, which consist of solutions such as coumarine or rhodamine red. In this type of laser, the materials are stimulated to emission by irradiating the liquid or dye solution with another laser beam, such as that from a ruby, neodymium-YAG, or carbon dioxide laser.

5. *Excimer laser* is an inert-gas halide, which is chemically unstable except in its excited state.

Laser beams are focused to increase the intensity of the beam by condensing it into an extremely small area. For example, when a 1/4-inch diameter laser beam is condensed by means of a convex lens to a focus of 5-mil diameter, its intensity increases more than 2000 times. (A *mil* is 0.0254 millimeter, or 1/1000 inch.) The beam is now capable of drilling a hole of 5-mil diameter through thin (2-mil) metal. Therefore, by using a convex lens, a laser beam can be concentrated on a small point and used for manufacturing processes such as cutting, welding, drilling, and to remove metal (etch) from metallic objects placed at the beam focus. In the medical and dental fields, a focused laser beam can be used for cauterizing (burning), cutting, and sterilizing human tissue or glazing human teeth in restorative (repair) dentistry.

All types of materials, including metal, wood, plastic, rubber, and cork, can be cut with the laser beam to very precise designs. In fact, the cutting is so sharp that the edges do not have to be worked in any other way, such as filing or buffing.

A laser is used for drilling many different materials, some of which cannot be drilled by any other method. For example, the laser has modernized such tedious mechanical processes as diamond drilling, since it produces a very narrow and intensive beam of light that can be focused to a spot very small in diameter. In Fig. 5-19 a you see a laser used to cut a hole in a diamond. The diamond is mounted in steel casing then pierced by repeated shots of a laser beam. The red laser beam is focused into a spot 10 mils in diameter. Because of the potential danger to a person's vision if the beam if accidentally viewed, closed circuit TV is used to monitor this type of cutting.

The laser is also widely used to etch materials such as wood and metal. Next time you visit a gift shop, inspect desk sets of hardwood that have very finely etched pictures on the surfaces. This etching was done with a laser beam. Etching can be done on all kinds of materials, including steel, plastics, wood, and aluminum.

Because the laser can be focused to a very small spot, it's use is extremely accurate as shown in Fig. 5-19 b. And since the laser beam widens very little over long distances, the laser is also used as a very accurate measuring device. Judging the distance or level of work pieces and identifying the precise height above datum points in surveying is now often done with lasers. Figure 5-20 shows a laser being used to measure the small diameter of an optic fiber strand. The scattering shown in

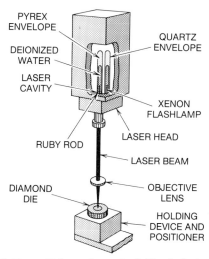

Fig. 5-19 a.   Using a laser to drill a hole in a diamond die.

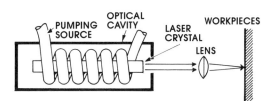

Fig. 5-19 b.   The laser can cut material very accurately. (Quasar Industries)

Fig. 5-20 is a process of extracting energy from a beam of light and re-emitting this energy at the same or at a different wavelength, in the same or different directions.

The laser is also used in communications such as the lightwave system that is replacing copper wire to carry telephone messages. (See Unit 186, "Communication Systems.") Lasers will also play an important role in the defense of our country through their use with defense satellites.

Since the laser beam can traverse (cross) great distances with little change in the radiation level, extreme intensities can be obtained thus radiation injury (especially to the eyes) can occur at great distances from the source.

As laser manufacturing tools are used more in production, installation and use of these machines becomes more important. New and unique hazards arise that are quite unlike the usual hazards of a production machine. For instance, if a machine operator misdirects or breaks a drill, only the operator may be in danger. However, if someone aims a high power laser in the wrong direction, other personnel in the area may also be injured by laser radiation. Thus, it is vital to be aware of what lasers can and cannot do, what hazards are involved, and what precautions should be taken.

Fiber Optics

People have used light to communicate for thousands of years. A light signal warned

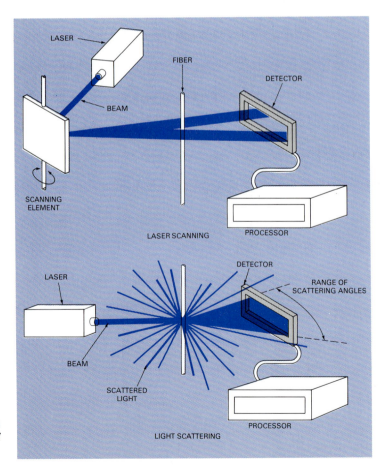

Fig. 5-20. Note how the laser can accurately measure the diameter of fiber optics material.

Paul Revere of the arrival of English forces at the start of the Revolutionary War. However, the light signal did not tell Revere the route the invaders were taking. These signals were not an accurate way to provide information. A little over a century ago it was discovered that light could be guided in different directions through a thin strip of glass. However, it was not until 1950 that the first industrial uses of optical fibers were developed. Fiber optics became an important part of cathode-ray tubes, medical microscopes, and photocopiers.

Not long ago, it was discovered that communications are possible over long distances using thin glass fibers instead of metal wires. This led to one of the most exciting developments in communication technology: *lightwave transmission*. This technology has already begun to make dramatic changes in the way information is communicated. You will learn more about this in Unit 186, "Communication Systems."

Fiber optics are also being used in jet airplanes to replace electrical wires. Fiber optics carry enormous amounts of information at high speeds on tiny pulses of light. Fiber optics look like strands of clear glass, and they're as fine as human hair. Fiber optics are ideal for planes because they require less space and weight than electrical wires (Fig. 5-21).

A fiber optics system will be able to transmit the equivalent of an encyclopedia set (24 volumes) of information (about 40 million words) in a single second.

Holograms

A hologram is a pattern on photographic film that produces a three-dimensional picture of an object. Holograms using laser light were developed in the early 1960s. If you stand in front of a hologram (whole message), you can look around the object to see what is blocked out in the head-on view of a typical picture. The image is so

Fig. 5-21. All electrical systems in military jet aircraft will eventually use fiber optics instead of metal wire for communications systems.

lifelike that it appears to be the actual object. Holograms are produced by using a laser beam to illuminate the object. The image is created by shining a similar beam of light on the hologram. The hologram changes the light to reconstruct the light patterns from the original object (Fig. 5-22 a and b).

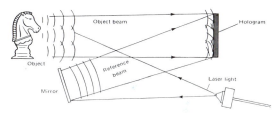

Fig. 5-22 a. How a hologram is formed. Laser light is split into a reference beam and an object beam. Reflected light from the object interferes with the reference beam, producing the interference pattern called the hologram.

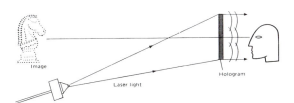

Fig. 5-22 b. How the image is reconstructed. The pattern is converted into a picture by shining laser light through the back of the hologram. Waves that emerge from front of plate combine to reconstruct a three-dimensional image of the original object.

Holographic pictures (called holograms) are so realistic that people gasp in amazement. As a person stands in front of a hologram and moves his or her head, the position of an object in the foreground relative to the background changes, just as if the person were looking at the actual scene rather than a picture.

Holograms have many uses in business and industry. In architecture and industrial design, holograms can replace scale models. The hologram of a building model can be made to appear full scale so that the client can see what will be built. Industry uses the hologram to study stress analysis and vibration analysis. Holograms can also be used to store amazing amounts of information. This will be extremely useful in computer applications.

Materials

Many of the same materials you will use to build a project activity are needed to produce high-technology products. For example, a computer is built of metal, plastics, ceramics (glass and silicon), wire, and electronic units. These materials must be cut, shaped, assembled, and finished just as you must do in building a project activity.

Industry makes use of many types of special metals, plastics, and composite materials. A *composite material* is a combination of two or more different materials that retain their own identity. Fiberglas is a familiar composite that you will use in working with plastics. This material is a mixture of glass fibers and polyester resins. See Units 99 and 100. Another common composite is plastic laminate (you may know it by the trade name Formica). Laminates are made of layers of kraft paper impregnated with resins and covered with resin-saturated pattern paper.

Newer composite materials are being used in all types of high-technology products. Composites (such as a combination of carbon and plastics) are being used as the outer skin of planes to replace aluminum and other metals. Composites and fiber optics will be used in building future military planes; this will make it impossible for radar to detect these aircraft.

Glossary

Access. The act of obtaining data stored in the computer.

Binary. Having two possible positions. The 0 corresponds to the "off" position, and the 1 to the "on" position of the electronic circuitry.

Bit. Abbreviation for binary digit, one of the two numbers—0 and 1—used to encode (convert) computer data. A bit is expressed by a high or low electrical voltage. It is the smallest unit of information a computer uses.

Byte. A group of 8 bits used to encode (convert) a single letter, number, or symbol.

Cathode-ray tube. (CRT) The video screen that displays data and graphics.

Central processing unit. (CPU) The part of the computer that controls its overall operation. The CPU carries out instructions given to the computer.

Chip. A small piece of silicon that is a total semiconductor device. It contains microminiaturized electronic circuits.

Computer. A device that stores information, manipulates data, and solves problems.

Data base. The collection of information stored in a computer.

Data-entry device. Equipment that will convert data in human-readable form into a code the computer can understand.

Data processing. Operations performed on data to achieve a particular objective.

Debug. To locate, remove, and correct mistakes in a program.

Decode. To convert data from a coded

form into a human-readable form.

Disk (or *Disc*). See Magnetic disk.

Diskette. A thin, flexible platter, similar to a 45-rpm record, coated with a magnetic substance and used to store information; a floppy disk.

Electronic Data Processing. (EDP) The processing of data mainly by electronic digital computers.

Encode. To convert data in a human-readable form into a code the computer or telephone can understand.

Erasable Programmable Read-Only Memory. (EPROM) A type of memory in which stored information can be erased by ultraviolet light beamed in a window of the chip package.

Fiber optics. A thin ribbon of glass that can be made into cables to carry light pulses over long distances. This process is called lightwave communication.

File. A block of information that makes up a group of related data on a disk that is like the files used in office work.

Floppy disk. See Diskette.

Fluid power. The transfer of power from one place to another by means of air, oil, or other liquids. When air is used it is called *pneumatics*, as in the air-powered tools used in manufacturing and servicing. When oil is used, it is called *hydraulics*, such as in the hydraulic braking system used in cars.

Gate. The controlling element of certain transistors; a logic circuit that has two or more inputs that control one output.

Hard copy. A printout on paper.

Hardware. The physical equipment of the computer system.

Input. To transfer data from an input device (such as a terminal) to the computer's memory.

Input/Output. (I/O) The passage of information into or out of the computer.

Integrated circuit. This is a semiconductor circuit that combines many electronic components in a single substrate,

usually silicon.

K. An abbreviation in the metric system for kilo (1000). A 1K memory chip, however, contains 1024 bits because it is a binary device based on the power of 2. A computer that has 16K memory has a capacity of 16 times 1024 bytes of memory.

Laser. A high-intensity beam of light that is emitted over a very narrow frequency range and can be directed with high precision.

Logic. The fundamental principles and the connection of circuit elements for computation in computers.

Magnetic disk. A flat, circular plate on which data can be recorded.

Magnetic tape. A tape on which data can be recorded.

Mainframe computer. A central computer with a huge CPU and a memory capacity that serves a number of terminals.

Main storage. The main memory of the computer. It stores data that are being processed or used currently.

Mask. A glass photographic plate that contains the circuit pattern used in the silicon-chip manufacturing process.

Memory. Computer elements that retain instructions or data used by the CPU.

Memory chip. A semiconductor device that stores information in the form of electrical charges.

Microcomputer. A type of computer that is smaller than both the minicomputer and the mainframe computer.

Microprocessor. An integrated circuit in one chip that provides functions equal to those contained in the CPU of a computer. A microprocessor interprets and carries out instructions. It usually can do arithmetic and has some memory. It is a CPU on a chip or a computer system designed around such a device.

Minicomputer. A type of computer that is larger than the microcomputer and smaller than the mainframe computer.

Multiplexer. An electronic device that combines many pulses to transmit them over long distances.

Network. A system of interconnected computer systems that uses terminals for communications.

Off-line. Not connected to the computer.

On-line. Connected to the computer.

Output. Data transferred from the computer memory to a storage or output device (such as a printer).

Peripheral. An external or remote device in a computer system. Input/output devices, such as keyboards, printers, magnetic tapes, and magnetic disks, are peripherals.

Program. A set of instructions, arranged in order, telling the computer to do a certain task or solve a problem.

Random-access memory. (RAM) A memory of which any piece of information can be stored or retrieved independently. Its contents are held only temporarily.

Read-only memory. (ROM) A memory chip in which information is permanently stored during the manufacturing process.

Remote terminal. A terminal that may be miles away from the computer and connected by telephone or other communication lines.

Robot. A reprogrammable device designed to move materials, parts, tools, or specialized devices for doing a variety of tasks.

Semiconductor. An element whose electrical conductivity is less than that of a conductor, such as copper, and greater than that of an insulator, such as glass; material chemically treated to have electronic characteristics.

Soft copy. Output that is not in printed form, such as a display on a video screen.

Transistor. A semiconductor device that acts primarily either as an amplifier or as a current switch.

Terminal. A device through which data may be entered or received; it usually is equipped with a keyboard and a display device.

Video display terminal. (VDT) See Cathode-ray tube.

Wafer. A thin disk of semiconductor material on which many chips are fabricated at one time. The chips are later separated and packaged individually.

Unit 6

The Scope of Industrial Technology

This unit gives you an idea of the ever-expanding growth of American industrial technology, (Fig. 6-1). The various *segments* (parts) of industry are discussed, with special reference to the effects of technology on career development. There are units immediately following this one and also in each section throughout this text that describe jobs, occupations, and professions and their potential influence on your selection of a career.

Progress in industry is measured by the gross national product (GNP). This is the value of goods and services produced by all of America's industries and sold to consumers, the government, and other industries. With advancing technology, a larger labor force, the availability of better-educated young adults, and enterprising management the annual gross national product will shortly be $2\frac{1}{2}$ trillion.

Fig. 6-1. Petrochemical plants make many industrial products from crude oil.

Fig. 6-3. The mobile-home manufacturing industry is providing competition for the traditional concept of on-site methods of construction, paving the way toward modular-home construction. (Skyline Mobile Homes)

The population of the United States will probably increase from the present 220 million to around 242 million persons within the next few years. This will create a tremendous consumer demand for manufactured goods, services, and construction. It will also require a vast expansion of utilities, transportation systems, and modes of travel.

Some factors that control industrial growth are measured by the desires of the

Fig. 6-2. The construction industry is undergoing a revolutionary and dramatic change in the kinds of materials used, in design, and in construction methods. (U.S. Steel Company)

consumer (buyer), increasing restrictions on private transportation owing to congested urban (city) areas, expansion of mass rapid transit, and the general attitude of the working population toward work and leisure.

Construction

The construction industry includes commercial, industrial, governmental, and private building programs. Much of this industry is undergoing a revolutionary, almost dramatic, change in the kinds of materials used, in design, and in building methods (Fig. 6-2). Construction techniques include assembly-line housing, either on the building site or at the factory, and sectional or modular units.

Wood is often supplemented or replaced by reinforced plastic wall panels or ingenious aluminum and steel wall construction. Much modern construction involves factory-cast concrete panels. A new technique is to build room units on an assembly line and install them in a building framework as complete living units.

Mobile homes are one of the fastest growing and relatively inexpensive means of housing (Fig. 6-3). Many types of these units are produced in *modules* (a packaged unit) or in sections. One of the important factors that influence the mobile-home industry is the establishment of mobile-home communities or parks. This industry lends itself to mass production, and even small plants produce as many as two completely furnished mobile homes per day. Permanent and vacation housing will increase as new technology produces more efficient mobile homes. Soon there will be 1 million mobile homes produced annually.

Building Materials

Building materials include cement, concrete block and brick, ready-mix concrete and concrete products, metal sanitary ware, plumbing fittings and fixtures, heating equipment, fabricated structural steel, wood, plastics, and other types of conventional materials. These materials will be used in greater amounts in proportion to the general expansion of the construction industry (Fig. 6-4). Occupations necessary to produce, distribute, and install them will also need more workers.

Lumber and Wood Products

Lumber and wood products industries continue to develop and expand. This is due to the efficient forestry practices begun

Fig. 6-5. The by-products of lumber production are finding new markets for consumable products. (Crown Zellerbach)

by federal and state governments. During the past 10 years there has been a great increase in the use of the by-products of lumber production (Fig. 6-5). Many of these new products, including particle board, are used more and more in the production of furniture, cabinets, store fixtures, houses, mobile homes, and industrial and commercial buildings.

Factory-built *modular units* are making an impact in home, motel, and hotel construction. The Department of Housing and Urban Development estimates that within the next few years two-thirds of all new dwellings will be factory-made (Fig. 6-6). The success in the efficient use of lumber and its by-products by the wood products industries is almost proportionate to investment in and development of research.

Fig. 6-4. Many kinds of building materials are used in the construction industry. (American Plywood Company)

Fig. 6-6. Factory-built modular units similar to this type are making an impact on hotel, motel, condominium, and home construction. (American Plywood Association)

Fig. 6-7. **Paper, paperboard, and corrugated board are direct products of forest timber.** (Weyerhaeuser Company)

Paper and Paperboard

Among the direct products of forest timber are paper and paperboard (Fig. 6-7). Technological improvements within this industry have made it profitable to increase annual exports. Packaging has become a very prominent factor in the distribution and sales of goods (Fig. 6-8).

Fig. 6-8. **The production of paper and cardboard packaging material is very important to everyday living.** (Weyerhaeuser Company)

Paper mills produce the basic material for commercial printers, newspapers, and all types of paper products. These include paper bags, sacks, and the coating and glazing of products. The markets for paper and paperboard products are expanding very rapidly because of new techniques in blending wood fibers and synthetics. Paper-mulching systems for agriculture constitute a new industry. Crops require less irrigation when they are *mulched* (roots protected).

Many companies produce *nonfiber* (pulp) building materials. American paper companies have vast investments in foreign countries where paper products are being introduced and used increasingly. This is due to an improved standard of living and a higher educational level.

One of the newer concepts in the use of wood cellulose fibers is "disposables." These are nonwoven products such as dress and work clothing, hospital gowns, sheets, pillowcases, washcloths, and towels. This is possibly the fastest-growing use of cellulose.

Printing and Publishing

Managerial planning and technological alertness are two important ingredients in the vast growth of the printing and publishing industry. There are over 1 million people employed in 40 000 printing and publishing plants in the United States (Fig. 6-9). The growth pattern of this industry is related to an increase in population to approximately 240 million within the next 10 years and to the upgrading of the educational background of this population.

Graphic arts technology has probably undergone a greater change than is realized. Automated operations in typesetting, platemaking, color proofing, and film developing are among these innovations. Computers are being used in many new ways to control various kinds of printing processes.

Fig. 6-9. There are over 40 000 printing and publishing plants in the United States. (Consolidated International, Incorporated)

The newspaper industry is a large employer among all the manufacturing industries. Practically every city and town is served by a newspaper.

Book and periodical publishing industries are expected to grow as the educational level of the population is upgraded. There are more people attending schools, technical institutes, colleges, and universities every year. Printing is not limited to newspapers, books, and periodicals. The vast increase in business forms requires more varied types of printing.

Household Consumer Durables

This category of products includes many relatively new consumer items such as microwave ovens, electric trash compactors (Fig. 6-10), continuous-cleaning ovens, garbage disposers, electric ranges that dispense hot water, dishwashers with food-warmer tops, electric towel stands, electric hibachis, men's electric hair combs, and deep-heat body massagers. These are in addition to the usual household items such as refrigerators, freezers, cooking equipment, laundry equipment, vacuum cleaners, electric housewares, furniture, mattresses, and bedsprings.

There have been innovations in low-cost molded plastics that *simulate* (imitate) expensive wood carving, new seating sus-

Fig. 6-10. Household trash compactors are an efficient means of compacting waste into smaller packages for easy disposal. (Amana Refrigeration Company)

pensions in upholstered furniture that eliminate springs, and self-adjusting casters. Automated production of rigid urethane furniture parts has made more artistic creation possible. The upsurge in mobile and modular homes has created a need for new products designed and manufactured specifically for furnishing these units. It is anticipated that durable household consumer products will virtually double in production in the next 10 years.

Primary Metals

Basic metals are processed in steel mills (Fig. 6-11), copper-wire mills, brass mills, and aluminum-producing plants (Fig. 6-12). Pollution awareness and expensive technological changes in metals production and manufacturing have increased the demand for better-trained workers. The present trend of restrictions on imports will have a decided effect in increasing the utilization of more American manufactured products made of the primary metals.

Fig. 6-11. **Sheet metal is a primary metal for thousands of products, including those of the transportation industry.** (American Iron and Steel Institute)

Fig. 6-12. **The aluminum industry is expected to expand extensively because of new technology in the transportation, building, and container industries.** (Aluminum Company of America)

The aluminum industry is expected to expand production extensively as a result of the stimulus of new developments in the automotive, building, and container industries. Copper production will experience a sharp upturn as a result of expansions and new developments in communications, electric power, transportation, and water desalinization.

Requirements for steel will increase because of consumer pressures for durable goods and new types of transportation. Ferrous castings will have a small rate of growth within the next decade; however, the high cost of pollution control may lead to the construction of larger but fewer foundries.

Metalworking Machinery

Metalworking machinery includes metal-cutting and metal-forming machine tools; special tools, dies, jigs, and fixtures; and welding equipment. Much of the growth pattern in this area will depend on governmental restrictions and import quotas.

Automated and numerically controlled systems will require more technically trained personnel (Fig. 6-13). This relates to the manufacturing of machinery and the application of these machines in industrial

Fig. 6-13. **Numerically controlled machinery is the result of a relatively new technology for increasing industrial production.** (Kearney & Trecker Corporation)

Fig. 6-14. One-inch-wide, eight-track punch tape operates the numerically controlled machine shown in Fig. 6-13. Also shown are punched numerical-control data cards and manuals. (Sundstrand Corporation)

production. Figure 6-14 shows numerical tapes, punch cards, and operating manuals for a numerically controlled system.

Special Industrial Machinery

Special industrial machines are those used in construction, mining, oil fields, scientific research (Fig. 6-15), farms, food-product manufacture, and printing. The markets in

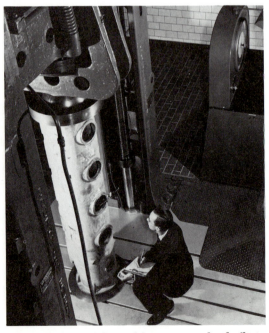

Fig. 6-15. Special machinery must be built to order for many areas of scientific research. (Aluminum Company of America)

these areas are growing. More productive farm and contruction machinery is needed. And there will be a greater use of automation and materials-handling machines for the food and textile industries.

It is anticipated that many additional types of construction machinery will have to be developed and manufactured because of advanced highway traffic control. Also influencing the need for technically advanced machines is the desire to build housing more quickly, the efforts to control pollution, the safe disposal of waste, and the public attitude toward a cleaner environment.

General Industrial Machinery

This category of equipment includes air conditioning, commercial and industrial refrigeration, industrial pumps and compressors, and materials-handling equipment. The United States is recognized as an international leader in this area of equipment production. Demand for this category of equipment is increasing as emerging countries upgrade their standard of living.

Among the kinds of materials-handling equipment are trucks and tractors (Fig. 6-16), along with conveyors and component parts for them. Other types include cranes, hoists, monorails, and elevators.

Fig. 6-16. Special materials-handling equipment, such as this tractor, is used for clearing woodlands for roadways. (International Harvester Company)

Pumps and compressors are needed for residential and business construction. These machines are used in industry, farms, construction activities, mining, oil fields, and the maritime industries.

General Mechanical Components

General *components* (parts) are valves and fittings, industrial fasteners, mechanical power-transmission equipment, antifriction bearings, and screw machine products. Usually these are components of massive equipment manufacture and represent some of the fastest-growing trends in industrial technology. Each of these categories of production is solely dependent on the increased demands for large-scale machinery and the more intricate types of control used in operating industrial plants.

The manufacture of ball and roller bearings has grown tremendously during the past 10 years. This gives an indication of expanding further as improved metals are developed to withstand heavier loads under more continuous operation.

Electronic Equipment and Components

The electronic industries have been among the most innovative and rapidly growing in the United States. Consumer electronics includes home-entertainment electronics products such as radios, television receivers, phonographs, and tape recorders. All of these products must have special electronic equipment designed for repair and testing (Fig. 6-17).

Additional equipment is necessary to meet the rising demand for telephone and telegraph services. Much money has been spent on research and development to find more original methods and devices of communication. Because of all these factors, the industrial consumption of electronic systems and equipment will continue to grow in the next decade.

Fig. 6-17. **Parts for consumer electronic appliances must be checked for defects with special electronic testing equipment.** (Westinghouse Electric Corporation)

Lighting and Wiring Equipment

The electric lighting and wiring equipment industry is an integral part of industrial, commercial, and residential structures. It includes *electric lamps* (bulbs), current-carrying and non-current-carrying wiring devices, and electrical fixtures. Adaptations of these devices are used increasingly in industrial research (Fig. 6-18).

Fig. 6-18. **An adaptation of lighting equipment is shown in this test tool, which radiates 1 million watts of power for industrial research.** (General Dynamics Corporation)

Production has become concentrated because of automation and numerical control systems. Many new lighting concepts are in the research and development stage. A small compact light source whose spectrum closely matches the color of noonday sunlight is one of the recent developments. This is a molecular arc lamp that produces fairly uniform overall colors that are close to the color spectrum of the day.

Power and Industrial Electrical Equipment

Power and industrial electrical equipment is a part of the growth of electric-power-generating facilities. Increased sales of electrical appliances and construction of new industrial plants create the need for more power. Equipment-investment programs have been formed to take care of this demand. Power boilers and nuclear reactors, distribution and speciality transformers, switchgear and switchboard apparatus, motors and generators, and industrial control systems form a part of the electrical-equipment complex (Fig. 6-19).

The Nuclear Regulatory Commission is a government agency that regulates the operation of nuclear-power generating plants. This agency approves the design and construction of new plants and maintains the operation of all existing plants in the United States.

Measurement Analysis and Control Instruments

Measurement analysis and control instruments are a necessary part of outer-space exploration, including lunar landings. These events have given great impetus to the manufacture of scientific instruments that make meteorological and geophysical measurements on the moon and on other planets. Increased attention to ecological conditions on earth has prompted the de-

Fig. 6-19. These are large electric motor stators ranging in size from 12 700 to 59 700 kW. (Westinghouse Electric Corporation)

velopment of new monitoring (checking) systems. The equipment in this sector of industry includes instruments for electrical measuring, for measuring engineering and scientific data, for measuring and controlling quantities and quality, and for automatic temperature control (Fig. 6-20). Optical instruments and lenses are part of these various instruments and controls (Fig. 6-21).

The market is expanding rapidly for instruments that measure smokestack and water wastes in order to comply with rigid pollution regulations. Much instrumentation is necessary for aircraft flight navigation, surveying, meteorological and oceanographic probing, and electronics laboratory equipment of all kinds. Clinical hospital field activity ranks as a prime growth market for instrumentation to detect diseases in the early stages.

Business Machines

The manufacturing of business machines is an essential phase of American industry. The production requires technology and expert manufacturing of sophisticated (complicated) scientific machines and cal-

Fig. 6-20. This specially built simulator chamber is used to determine the effects of vacuum or high- and low-temperature conditions on lubricants, bearings, and rubberlike materials. (McDonnell Douglas Corporation)

Fig. 6-21. An optical instrument and machine is capable of calibrating (adjusting) over 1100 measuring devices daily. (McDonnell Douglas Corporation)

Fig. 6-22. Inventory controls, business and employer records, and manufacturing programs are controlled by this data-processing center. (McDonnell Douglas Corporation)

Fig. 6-23. Computing machines provide automated accounting procedures. (McDonnell Douglas Corporation)

culators designed for special-purpose applications (Fig. 6-22).

Business-machine production includes typewriters, calculating and accounting machines, electronic computing equipment, and minicomputers (Fig. 6-23). It takes a person $1^{1}/_{2}$ minutes to multiply two five-digit numbers; a computer does it in three *nanoseconds*, which is 30 billion times as fast. Electronic computing equipment is linked closely to the rapid developments in computerized and numerical control mechanization. Computers will run factories, replace libraries, and regulate traffic, the scientists predict.

Several business-machine companies have been absorbed (taken in) by industrial conglomerates (large corporations owning many companies), thereby making competition more keen. Probably the fastest-growing segment of the business-machine market is the manufacture, distribution (selling), and use of minicomputers.

Motor Vehicles

Millions of passenger cars, buses, and trucks are made every year in the United States (Fig. 6-24). There are currently about 160 million of these vehicles operating on the streets and highways of our nation.

There are four major manufacturers (makers) of automobiles in the United States. They are called domestic-car makers. Some makers of foreign cars have

Fig. 6-24. **Millions of automobiles are produced each year.** (U.S. Department of Commerce)

compact cars. Thirty automobiles can be carried on a railroad car in a nose-down position. It is thought that several big manufacturers may begin making these railway cars. Skyways and highways are becoming more and more crowded. Railroads can carry more people and goods faster and more easily.

"People-mover" systems are also being developed. They will lessen the traffic congestion caused by cars. Some aerospace manufacturers are changing over parts of their production lines. This is being done to make vehicles for urban rapid-transit systems (Fig. 6-25).

One large aircraft company is developing vehicles with steel wheels, rubber wheels, or no wheels at all. Among these is the rail-transit system, which has the *track air-cushion vehicle* (TACV). It runs at 150 miles per hour on a thin film of air (Fig. 6-26). Another newly developed vehicle uses electromagnetic forces to support itself just a fraction of an inch over a set of guide wheels. The force also moves the vehicle.

Shipbuilding and Repair

Some countries are developing huge *maritime* (oceangoing) programs. These nations will someday handle a great amount of shipbuilding and repair. Traffic from one continent to another calls for many kinds of ships, including merchant vessels, sea barges, ore carriers, cargo ships, oil tankers, and passenger ships. Almost all these ships are ready for new kinds of designs. Nuclear power will also come into wider use as a source of energy for ships (Fig. 6-27).

Aerospace

The aerospace industry is expected to grow, because of increased airline traffic. Domestic airline-passenger traffic will double and world air cargo shipments will

built manufacturing plants in our country. There are also many foreign firms that ship passenger cars into the United States.

Research and development is an ongoing activity in the automotive industry. Work is being done on different kinds of gas-turbine, steam, electric, compressed-gas, and diesel engines. Huge sums of money are being spent to study new kinds of engines. The goals of research and development are increased safety, reduced (less) air pollution, quieter operation, fuel savings, reduced maintenance (care), and less costly production.

Railroad Cars

Railroad-car manufacturers make and deliver freight and passenger cars to railway companies. People use trains less nowadays, so production of railroad cars has decreased in the past 10 years or so. The government is hoping to add to people's interest in rail travel and to make the use of railroads grow.

A new kind of freight car is called the *Vert-a-Pack*. It is designed to carry sub-

Fig. 6-25. This is the BART (Bay Area Rapid Transit) train in operation in the San Francisco Bay region to expedite commuter urban traffic. (Rohr Industries, Incorporated)

Fig. 6-26. This aerotrain is designed and produced for express service on medium-range routes at speeds up to 150 miles per hour (240 kilometers per hour). This vehicle is supported on a thin cushion of air and is powered by a pollution-free induction motor that will straddle a vertical metal fin shaped in cross section like an inverted T. (Rohr Industries, Incorporated)

Fig. 6-27. Nuclear-powered oceangoing vessels have changed the maritime industry. (U.S. Atomic-Energy Commission)

also greatly increase in volume within the next decade (Fig. 6-28).

The U.S. government and private industry have a vast research and development program under way. This involves the design and building of prototypes of *vertical-takeoff-and-landing* (VTOL), and *short-takeoff-and-landing* (STOL) aircraft. These have tremendous potential for both civilian and military uses (Fig. 6-29).

Fig. 6-28. World air cargo shipments will increase in volume greatly during the next decade. (Western Electric Company)

Fig. 6-29. An artist's concept of the vertical-takeoff-and-landing (VTOL) aircraft. It is estimated that this vehicle will carry over 100 passengers, taking off from rooftop or ground-level heliports. (Bell Helicopter)

Outer-space exploration programs, space laboratories, satellite launchings, missile programs, and development of even more powerful fuels for rockets have opened up a new era in aerospace development. Probably the fastest-growing aspect of the aerospace industry is the increasing use of helicopters for military, commercial, and private use.

Communications

Operating *revenues* (income) from telephone and telegraphic traffic, radio and television broadcasting, and other communication services are expected to increase (Fig. 6-30). This applies to both the domestic and the international systems. The growing population will require more construction expenditures, leading to a high rate of employment in the communications industries.

Fig. 6-31. Large antennas send and receive worldwide communications with the aid of satellites. (Western Electric Company)

At present there is an expansion of data transmission and an increasing use of mobile radio. Closed-circuit television, innovations for communication in the home, telephone design, and door-answering services are expanding. Satellites and radar are used for communication throughout the world. These systems are elaborate, efficient, and far-reaching (Fig. 6-31).

There are over 9 000 commercial AM and FM broadcasting stations in the United States. Many are affiliates of the major nationwide radio and television networks. Automated programming equipment is being put into operation in radio broadcasting as rapidly as it can be developed. All phases of the communications industry give every indication of extremely rapid development of new products and increased employment.

Fig. 6-30. Communications systems such as this one make space explorations possible. (NASA)

The Impact of Technology on Industry

New materials, machines, methods, and products bring about economic and social changes. *Technological developments affect types of employment*, the *degree of excellence in skills*, *membership in unions*, and *job security*. These factors must be understood by the government, unions, company management, employees, and the general public.

Technological Changes in Productivity and Employment

The basic concept of technological change is that it permits more *efficient use of resources*, both material and human. It produces an accumulation of technological knowledge. This is the result of the work of engineers, scientists, and inventors. These *innovative* (new) concepts are put into practice only after long periods of *development*, *testing*, and *evaluation*. The innovators must convince industrial management (Fig. 7-1) of the need for a tremendous investment in new equipment and retraining of personnel. Some of the factors that must be considered before there is an investment of time and money are the attitudes of a company's management and union; federal, state, and local government codes and regulations; market prospects; competitive considerations; and financial requirements.

All of these factors and changes are measured in terms of *productivity* by employers and employees. Productivity rises as the output increases from the given input. The economic and social changes in society mean that each new generation

Fig. 7-1. **During the age of mass production, one of the ways technology increased the efficiency and volume of production was through mechanical conveyor systems.** (Westinghouse Electric Company)

must increasingly give thought to *new types* of careers. Technological changes and demands for new products have resulted in significant increases in employment. There must also be a *technological understanding* by *consumers* who use the many new products.

Development of Technology

Technological change generally leads to *higher career skill* requirements. This necessitates a higher level of education and training. The latter part of the eighteenth century was an era referred to as the *industrial revolution*. There has been a sequence of events and stages of industrial development ever since.

The first stage of the industrial revolution was called the *age of mechanization*. It began with the replacement of the handicraft worker and saw the rise of the factory system. This was brought about by power-driven machinery.

The second stage of industrial development was termed the *age of mass production*. Electric power replaced steam as the means of energy for increasing production. Mechanical conveyors carried *components* (parts) along the line of assembly, resulting in increased production.

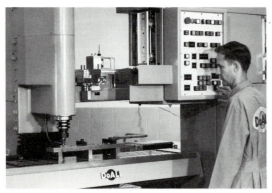

Fig. 7-2.   The age of science and technology has brought about a high degree of sophistication in automation through numerically controlled equipment. (DoAll Company)

The *age of science and technology* is the stage of industrial development in which we live. This phase began after World War II. There was a carry-over of the scientific and innovative ideas that produced the sophisticated and complicated machinery of war. Some of the unforeseen developments that *evolved* (emerged) were electronic computers, jet propulsion, nuclear energy, space technology, and automation of industrial production through numerical control. These developments created an increased need for educational and training programs for engineers, scientists, programmers, and technical planners, to mention only a few (Fig. 7-2).

Social Impact

Advances in technology create problems that require social changes. These *concepts* (ideas) include the necessity for advancement in education and training, retraining, *mobility* (movement) of the labor force, and other considerations related to social security. Social changes bring about an increased consumer literacy. There is a wider knowledge of educational requirements, materials, industrial products, and the economic impact of foreign competition.

Unit 8

Ecology and
Environmental Technology

The technology of reuse, recovery, and recycling of industry's discarded materials is emerging as an ecological solution to the world's waste-disposal problem. The United States, however, lags behind Europe in the recovery and recycling of materials. As an example, European industry

recycles over twice the amount of paper than industry does in the United States. More and more businesses are finding out that what is good on a small scale—in terms of reuse and recovery economics—is also effective for large-scale operations on a national level.

Not as well known, but widely used, is the form of recycling that involves industrial waste. The problem of hazardous (dangerous) waste is a modern, complex one, and the U.S. Office of Solid Waste estimates that excluding waste burned for fuel, only about a small percentage of the hazardous waste generated (produced) in this country is currently being recycled. Reducing or eliminating pollution at the source rather than paying to clean it up represents a means for efficient and economical recycling.

Americans produce about 750 trillion pounds of solid waste every year. Each person adds over 8 pounds of household refuse or garbage every day (Fig. 8-1). This is about 3000 pounds a year per person. The waste is made up of paper, glass, metal containers, plastics, garbage, and the like.

There are really only two places where garbage can be dumped: the land and the sea. Most of the waste is buried in sanitary landfill operations. A small percentage is burned. The rest, less than 2 percent, ends up along streets and roadways as litter or in the ocean.

Landfill operations and *incineration* (burning) are the least costly ways of getting rid of garbage. They may be, however, the most damaging to our way of life. Dumps and incinerators make air and water dirty—and they are ugly sights to see.

You may have heard the words *environment, ecology, pollution,* and *recycling.* Each is important in talking about wastes and what we should do with them.

Our *environment* is simply the world around us. It includes the air we breathe, the land we live on, and our oceans, rivers, lakes, and streams. We are beginning to learn that we must take care of the environment. If we are to be healthy, we must have clean air, pure water and soil that will allow crops to grow.

Ecology is the study of the environment. *Ecologists* are scientists who investigate how the environment supports life. They find out how plants get the water they need to grow, for example. They also study how animals of all kinds keep themselves fed. And ecologists study how people use the environment.

Pollution is another thing that ecologists study. The word *pollution* refers to anything that damages the environment or makes it dirty. The smoke from cars and factories is an example of air pollution. Oil spilled in the ocean and wastes dumped in rivers are examples of water pollution. Ecologists have found that too much pollution can damage the environment. And a badly damaged environment may not be able to supply us with our human needs.

Recycling means reusing materials that otherwise would be thrown away, buried, or burned. By reusing things, we can reduce the amount of solid waste and help keep the environment clean. Recycling can also save money and natural resources. Materials that are recovered through recycling can be used over and over again.

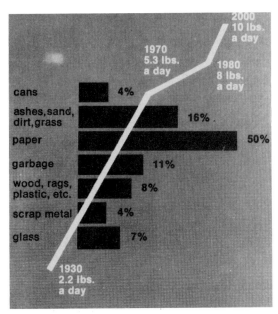

Fig. 8-1. Growth of collected household waste and a breakdown into categories. (U.S. Department of Health)

Fig. 8-2. Many civic organizations are active in the recycling movement. Here, representatives of an ecology club unload steel cans into a collection center. (Bethlehem Steel Company)

Many city governments, civic organizations (Fig. 8-2), church groups, and everyday people are working to collect trash that can be recycled. Many kinds of solid wastes called *scrap* are today being reused. For example, almost half of all the copper we use is recovered from scrap. About 30 percent of all aluminum and 20 percent of all zinc are handled this way, too. More than 50 percent of the total United States lead supply is recovered from scrap. And 25 percent of paper and paperboard comes from scrap (Fig. 8-3).

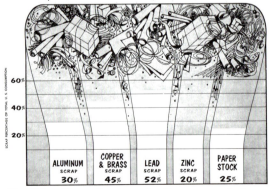

Secondary Production Accounts for a Major Portion of Raw-Material Supply

SCRAP PERCENTAGE OF TOTAL U. S. CONSUMPTION

60%
40%
20%

ALUMINUM SCRAP	COPPER & BRASS SCRAP	LEAD SCRAP	ZINC SCRAP	PAPER STOCK
30%	45%	52%	20%	25%

Fig. 8-3. Secondary production accounts for a major portion of raw-materials supply. (National Association of Secondary-Materials Industries, Incorporated)

The national stress on recycling by industries mostly concerns recycling waste paper, glass and textiles, iron and steel, and *nonferrous metals* (metals that do not contain iron, including aluminum, lead, zinc, copper, nickel alloys, and precious metals).

This national concern is shown each year during "Earth Week." The aim of Earth Week is to stress a clean and healthful environment. However, the recycling of waste is not only a civic activity. It can be a money-making activity as well.

"Recycling resources" is more than just a saying. It is the basis of a new kind of industry called the *secondary-materials industry*. This industry is involved in solid-waste management. Within the next 10 to 20 years, it could be larger than many industries now making the goods that end up as solid waste.

Recycling Iron and Steel

Recycling is not new to the steel industry. Most of the steel made in the United States comes from scrap steel. This includes *discarded* (thrown-away) machinery, scrap automobiles, railroad-car scrap, worn-out appliances, steel-mill scrap, and old steel from bridges and buildings (Fig. 8-4).

The movement for the cleanup of the environment has given a whole new meaning to recycling. A real effort is being made to recover iron and steel from litter and from burned garbage. Receiving centers are being set up all over the United States for reclaiming steel from all sources, but especially from cans. The steel industry is willing to recycle every can that is returned to its factories.

The process of recycling iron and steel scrap is as follows:

Step 1. People sort out metal cans and bring them to the *reclamation* (recycling) center.

Fig. 8-4. **Most of the steel made in the United States comes from discarded steel.** (Bethlehem Steel Corporation)

Step 2. Cans are trucked to a scrap-preparation plant near a steel plant.

Step 3. An electromagnet picks up the steel cans. This leaves nonsteel cans behind. These nonferrous cans are then taken to another scrap-processing center.

Step 4. The electromagnet drops the steel cans into a collecting bin (Fig. 8-5).

Step 5. The cans are fed to a scrap baler. About 17 000 cans make up a compressed bale.

Step 6. Bales of cans and other scrap metal are mixed together. They become

Fig. 8-6. **Bales of tin cans and other scrap metal (see Fig. 8-4) are mixed together and become the "charge" in this 250-ton basic oxygen furnace.** (Bethlehem Steel Corporation)

the *charge* (what is heated) in a 250-ton oxygen furnace (Fig. 8-6).

Step 7. The white-hot steel ingots are rolled thin. Later they are tin-coated on the ultramodern tin line. The old cans are now coils of tin-plated steel, ready for shipment to a can-manufacturing plant (Fig. 8-7).

Fig. 8-5. **The electromagnet picks up and then drops the steel cans into a collecting bin.** (Bethlehem Steel Corporation)

Fig. 8-7. **The white-hot steel ingots are rolled thin and later are tin-coated. The old cans are now gleaming coils of tin-plated steel, ready for shipment to a can manufacturing plant.** (Bethlehem Steel Corporation)

Recycling Aluminum

Fig. 8-8. **The aluminum cans drop into a bin on an electric scale. The weight is recorded in tenths of a pound and is printed on a slip that can be turned in for cash.** (Reynolds Metals Company)

Aluminum recovered from scrap generated by current operations (new scrap) accounts for about 60 percent of the total recovery from purchased scrap. Discarded aluminum products (old scrap) such as cans account for the remainder. Aluminum recovery from old scrap of about 610 000 tons in the last decade constituted about 12 percent of apparent consumption.

Aluminum is a *nonferrous metal*. It does not rust or decompose. Today, many *beverages* (drinks) come in aluminum cans. Aluminum is expensive, however, and a great deal of electricity is required to make it. That is why the recycling of aluminum has become important to the manufacturers of aluminum and to the beverage industries.

Several billion all-aluminum beverage cans are collected and recycled each year. The recycling of aluminum scrap can be a money-making program. Aluminum and beverage companies pay for each can collected (Fig. 8-8).

Figure 8-9 is a chart showing the steps in recycling nonferrous scrap metal. The

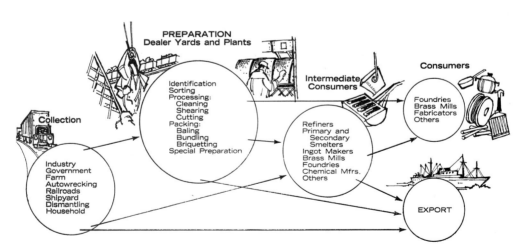

The recycling of metals conserves natural resources and eliminates mounting piles of solid waste

Fig. 8-9. **Operational sequence in recycling nonferrous scrap metal.** (National Association of Secondary-Metals Industry, Incorporated)

Fig. 8-10. As the metal hardens, impurities are skimmed off to ensure high quality. (Reynolds Metals Company)

Fig. 8-12. The storage area at the mill holds vast quantities of waste corrugated material and paper needed to keep the 200-ton-a-day recycling unit in full operation. (Crown Zellerbach)

Fig. 8-11. At the fabricating plant, the aluminum ingots go into a rolling mill where the metal is transformed into plates or sheet that is rolled into a coil. (Reynolds Metals Company)

recycling process for all aluminum items is much like that for recycling beverage cans (Figs. 8-10 and 8-11).

Recycling Paper Products

Statistics show that wastepaper utilization and recovery rates have doubled in the past 20 years. The total amount of wastepaper recovered in this period is expressed in millions of tons.

Thrown-away paper products are called *secondary fibers*. Sources are old newspapers, magazines, supermarket and department-store paper boxes, bags and bales of mixed papers, tabulating cards from offices, cartons and kraft paper from factories, tons of every variety from government agencies, and salvage from printing plants.

Wastepaper is examined, sorted, graded, processed, and baled to agree with quality-control standards of paper and paperboard manufacturers. The paper industry is expanding its use of secondary fibers so that mountains of waste can be transformed into new raw paper material (Figs. 8-12 and 8-13).

Over 12 million tons of used paper stock is collected, processed, and remanufactured each year. Most of the paperboard products, such as cartons and packaging, are made from secondary fibers.

About 60 million tons of paper and paperboard are manufactured each year. Over 80 percent of this has only a one-time use, and it quickly enters the solid-waste

Fig. 8-13. At the secondary-fiber processing plant, the wastepaper is brought into a hydrapulper and beaten with water. This is the first step toward recycling. (Crown Zellerbach)

channels. At present, less than 20 percent (about 12 million tons) of the paper and paperboard produced is made from secondary fibers (Fig. 8-14).

There has always been some reuse of wastepaper stock. However, now that the

Fig. 8-14. A paper machine manufactures a variety of kraft and other papers, utilizing recycled pulp. (Crown Zellerbach)

paper industry is faced with stricter environmental laws, there is a new interest in recycling (Fig. 8-15).

The reclamation and use of each ton of recycled paper stock saves 17 full-grown trees. Every ton of paper stock made from secondary fibers frees about 3½ acres of forest land for some other productive use for one year. The present rate of recycling paper stock can save 200 million trees per year. If the consuming industries increased the use of recycled paper stock with their raw materials to 50 percent, they would save 500 million trees (Fig. 8-16). This is equal to the combined forest area of the New England states, New Jersey, New York, Pennsylvania, and Maryland.

Plastics Recycling

The different types of plastics, including polystyrene foam, amount to less than half of 1 percent of the nation's 360 million tons of refuse that gets collected. The recycling of plastics involves finding new uses for plastic packaging and other disposable items. Packaging must be disposed of or recycled to keep it from becoming a solid-waste problem.

Since most plastics do not *decompose* (decay), this material makes a long-lasting, clean base for landfill. For example, polystyrene foam is a light and strong plastic packaging material. Because it is lightweight and can be *compacted* (squeezed down to a smaller size), it does not take up much room as landfill. And becuase it does not decompose, it does not add to air or water pollution.

At the present time, plastics is becoming a serious part of the solid-waste problem. The use of plastics is growing, and the plastics industry is spending a large amount of money on research. It is studying possible methods of recycling the various types of plastics. Plastics are used in construction, packaging, agriculture, aero-

Fig. 8-15. Paper stock is a vital recycled resource. (National Association of Secondary-Materials Industries, Incorporated)

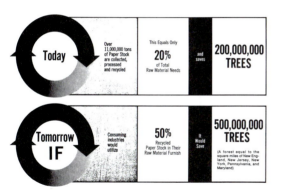

Fig. 8-16. Paper stock: solid-waste recycling and conservation in action. (National Association of Secondary-Materials Industries, Incorporated)

space, and automotive items (the average automobile has approximately 250 pounds of parts made of plastics). Plastics are also commonly found in appliances, furniture, fabrics, housewares, and toys. In schools and offices, many new and useful plastic items can be found each year. In addition, the medical profession is finding many new uses for plastics.

Several practical uses have been found for recycled plastics. For example, poly-

ethylene bottles are ground up into scrap that is used to produce high-grade agricultural drainage pipe. Another good use is to add waste plastics to other solid waste burned in incinerators. Certain plastics tend to increase the Btus (British thermal units). This creates higher temperatures, which result in a more efficient burn of the refuse. These are just the beginnings; other uses for plastic wastes will certainly be found.

Industrial Technology and the Ecological Environment

Multiple (many) environmental and ecological benefits can result from industrial creativity and technology in reprocessing hazardous waste. The Du Pont Company has invested millions of dollars in a process unit that converts (turns) iron chloride into commercial-grade ferric chloride. The ferric chloride is sold to waste-water treatment plants and water purification plants as a coagulant (something that pulls together) for suspended solids and for dewatering sewage sludge. This company not only has found an innovative way to recycle its iron chloride waste stream, it has eliminated the need for ocean disposal and has converted the waste into an essential element for water treatment sources.

Another example of this new interest is in the mining of *bauxite* (aluminum ore). Years ago, a great deal of bauxite was mined by simply stripping away the plants and soil on the surface of the earth. Huge holes were left after the bauxite was removed. The land seemed wasted. No trees or plants grew on it. Now, bauxite removal leaves a different result. The mining company reclaims the land so that it can be put to other uses. After the bauxite is removed, the huge pits are filled, and grass and trees are planted to stop soil *erosion* (runoff). The surface soil may even be treated to grow grass where cattle can graze (Fig. 8-17).

Fig. 8-17. Land for grazing cattle is one example of reclaiming strip-mined areas. (Reynolds Metals Company)

Planned forest management has been carried out for many years. It is a way of saving timber, soil, and wildlife. Figure 8-18 shows planned block cutting and the method of thinning trees to get the best growth of timber and pulp material.

Fig. 8-18. This illustrates planned block-cutting procedures in forest areas as a means of timber conservation. The cut areas are reseeded from the trees around them. (Weyerhaeuser Company)

Unit 9

Construction Technology

The construction industry involves building activities ranging from the maintenance and remodeling of a home or apartment to the construction of high-rise buildings, roads, dams, bridges, pipelines, and electrical utilities as well as many other kinds of heavy construction (Fig. 9-1 a and b). Construction projects are of three major kinds.

Heavy building construction, which sometimes is called commercial construction, includes the building of commercial, industrial, educational, institutional, transportational, and public buildings such as hospitals. These buildings make great use of steel, aluminum, concrete, brick, glass, plastics, and ceramics, with only a small amount of wood and wood products.

Fig. 9-1 a. This all-wood, solar-orientated house is typical of present-day residential construction. (Georgia-Pacific)

Fig. 9-1 b. Heavy earth-moving equipment is needed for building roads and for other heavy construction. (Caterpillar Tractor Company)

There are three ways of constructing the shell or frame of such a building. The first is a cast-in-place concrete structure. Forms are built, and then reinforcing rod is placed in the forms. Large trucks haul concrete to the site. The concrete is lifted up in buckets with a crane and placed (poured) in the forms. The shell or skeleton is built one floor at a time, and then the forms and crane are moved to complete the next level. The building may be 30, 40, 50, or more stories tall (Fig. 9-2 a and b). The second type is a framework of steel. Girders and columns are joined by riveting or welding the framework together (Fig. 9-3 a, b, and c). The third type, which is

Fig. 9-2 b. A completed structure built by the cast-in-place method.

Fig. 9-3 a. The construction site for a building of steel framework. (United States Steel)

Fig. 9-2 a. This skyscraper is being built using the cast-in-place method of construction. (Miller-Davis Company)

Fig. 9-3 b. A welder in the process of welding a girder to a column to make the framework solid. (United States Steel)

Fig. 9-3 c. Topping off the highest building in the world (the Sears Tower). Note the steel framework that can still be seen on the upper floors. Curtain walls of glass and steel already cover the lower floors. (United States Steel)

Fig. 9-4. Wood laminated beams form the shell of this unusual building. (American Plywood Association)

Fig. 9-5. Most homes are built primarily of wood and wood products. (Georgia-Pacific)

Fig. 9-6. This steel arc will be the main support for the bridge being built. (United States Steel)

used primarily for churches and smaller commercial buildings, has a frame of timbers or laminated beams (Fig. 9-4).

Residential or *light building construction* includes homes, apartments, condominiums, and small commercial structures built primarily of wood and wood products. These shelters have a framework of wood (sometimes metal framing is used) with an exterior that is usually made of wood, wood products, and sometimes brick, metal siding, or ceramic materials (Fig. 9-5).

Other heavy construction is the third type, and it includes all kinds of engineering and public works projects. Typical projects are roads, highways, bridges, dams, pipelines, electrical utilities, treatment plants, docks, piers, and military installations (Fig. 9-6). The way these projects are constructed varies widely. In road

Fig. 9-7 a. Heavy equipment such as this track loader is needed to move earth during the building of a road. (Caterpillar Tractor Company)

Fig. 9-7 b. Heavy equipment being used to pave a road with concrete. (Portland Cement Association)

construction, the land must be cleared, bridges and access roads must be built, the roadbed must be completed, and finally the finished road must be constructed of poured concrete or bituminous materials (Fig. 9-7 a, b, and c). Electrical utilities also build power plants and transmission lines. The military services may need housing, storage facilities, communication systems, runways, and many other kinds of heavy construction.

All construction projects follow these basic steps:

1. *Selecting the site.* The site must be selected and cleared. This includes bringing in earth-moving equipment and digging the opening for the foundation or basement (Fig. 9-8).

2. *Installing the foundation.* Forms must be built for the foundation and the basement walls. Reinforced cast-in-place concrete is used for most buildings, although for residential construction, a wood basement or a concrete block basement may be used (Fig. 9-9).

3. *Building the structure.* The building of the structure may include steel, concrete, wood, plastics, and many other materials, depending on the nature of the structure (Fig. 9-10).

Fig. 9-7 c. Rolling a road surface after bituminous material has been laid.

Fig. 9-8. Using heavy earth-moving equipment to excavate for a foundation. (Reynolds Aluminum)

Fig. 9-9. **Constructing a wood basement.** (American Plywood Association)

4. *Installing the utilities.* This includes electrical, heating, air-conditioning, plumbing, and communication systems (Fig. 9-11).

5. *Completing the exterior.* Upon the completion of the basic frame or shell of the building, the exterior is completed using a variety of materials (Fig. 9-12).

6. *Completing the interior.* The interior must be completed by installing the necessary walls, floors, cabinets, and any other divisions (Fig. 9-13).

7. *Finishing.* Both the interior and the exterior of the surfaces must be finished. The finish will depend on the kinds of mate-

Fig. 9-10. **A small home of truss-frame construction.**

Fig. 9-12. **Installing siding on a house.** (Simpson Company)

Fig. 9-11. **Using plastic pipe to install a plumbing system.** (ABS Institute)

Fig. 9-13. **The interior can be finished with many different materials.**

rials that are used in the structure (Fig. 9-14).

8. *Landscaping.* Landscaping includes a detailed plan for placement of trees, shrubs, and grass plots as well as a plan for their installation (Fig. 9-15).

Living Structures

For most people, the type of place they live in is very important. It may be a one-family home, small apartment, large apartment, condominium, town house, mobile home (Fig. 9-16), houseboat, or one of many other kinds. This unit deals mostly with house and apartment construction because this is the area of greatest interest to most people.

Location

It is important to select the best possible building site. The location should meet your personal needs in terms of schools, shopping, playgrounds and parks, commuting to a place of work, availability of utilities, and many other factors.

Planning a Home or Apartment

The cost of living quarters that the average individual or family can afford to own should be in the range of 2 to 2¹/₂ times the total yearly income. Of course, this amount may change with other factors, such as age, opportunity for promotion, and inflation. Planning a home starts when a designer or architect puts the plan on paper in the form of drawings called *prints*. These prints show what the structure will look like, how big it will be, how many rooms it will have, the size of each room, and the materials that will be used. Standard sets of house plans can be bought (Fig. 9-17).

A *plot* (land) plan is also needed to show the boundaries of the property and the exact place on the land where the structure will be built. Elevation plans show what

Fig. 9-14. The exterior of this home has a natural stain finish. (California Redwood Association)

Fig. 9-15. A well-landscaped apartment complex. (Shakertown Corporation)

Fig. 9-16. Mobile homes take on a permanent look when set on foundations or over garages. (American Plywood Association)

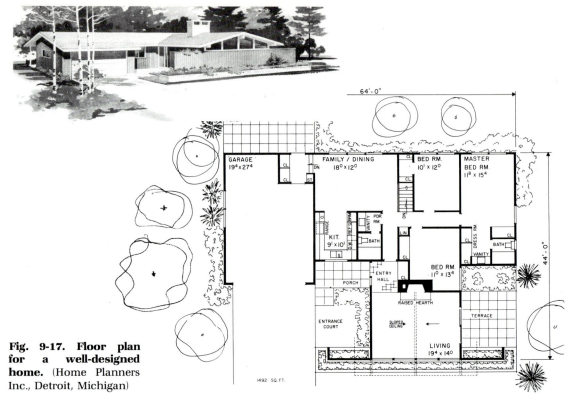

Fig. 9-17. Floor plan for a well-designed home. (Home Planners Inc., Detroit, Michigan)

the structure will look like from the front, back, and sides. Plans must also have a written list of the materials needed. These include lumber, panel stock, built-in appliances, cabinets, and everything else. These written lists are called *specifications* or *specs*.

People Who Will Build the House

The builder or general contractor is much like the director of a play. The builder takes the plans and makes them work. As a director, the builder must make sure that each of the different people, such as the mason, carpenter, electrician, plumber, and painter, comes to the job at just the right time.

Clearing the Land and Excavating

Land must first be cleared by a bulldozer, which moves the topsoil to one side. The

builder usually makes a special effort to save trees and other interesting landscape features. Stakes are driven into the ground to show the location of the structure. Next, the builder chooses the kind of foundation needed. Homes without basements are built on a concrete slab or with a crawl space underneath. For homes with a partial or full basement, mechanical equipment is brought in to *excavate* (dig) the basement. The framework is then built for the concrete footings. Concrete is cement that is mixed with water, sand, and gravel. Concrete arrives in huge mixing trucks and is poured into the footings. The basement walls come next. Some basement walls are made of poured concrete; others are built of concrete block (Fig. 9-18).

Framing a House

Lumber for framing the house is delivered and stacked on the site. A metal or wood

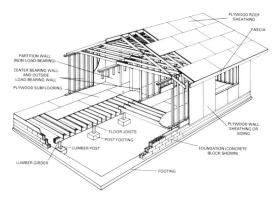

Fig. 9-18. Footings are poured, and the foundation is built of either concrete block or poured concrete. Girders may be steel or lumber.

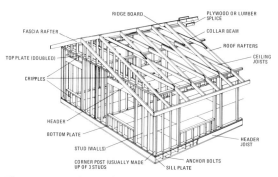

Fig. 9-19. Principal parts in framing the walls and ceiling of a home.

girder or beam, which is a horizontal load-supporting member, spans the distance between the foundation walls and the bearing posts. Holes are then drilled in the sole plate to go over the anchor bolts that are fastened in the basement walls. Next the carpenter puts down joists, which will support the floor on top of the foundation. Then rough floorboards or plywood sheets are laid for the subflooring. Now the carpenter is ready to start the framework (Fig. 9-19).

First the carpenters study the plans that show what sizes and lengths of lumber to use. These are measured and cut. The wall framing includes the sill plates, studs, headers, top plates, and fire stops. Usually, the walls are assembled on the subfloor

(Fig. 9-20) and then raised and nailed into place. The area for the doors and windows must be properly framed. Partitions are built for the rooms and closets according to the floor plan. Ceiling joists are nailed to the top plates. Sheathing is added to the exterior of the home. Now it is time to build the roof.

Roofs are built in one of two ways. The carpenter may nail joists across the top of the frame, put planks on top of the joists, and then stand on the planks while putting up the rafters. (The rafters are the sloped framing pieces that form the roof.) Carpenters today, however, are more likely to use ready-made joist-and-rafter units called prefabricated trusses. These look like triangles with braces (Fig. 9-21).

Fig. 9-20. Assembling wall panel framework.

Fig. 9-21. A load of trusses ready for delivery to a home site. (Panel Clip Company)

Sheathing is then nailed to the roof to close in the house. After the sheathing has been added, a cover of paper is tacked into place by the roofers. It must hold snow and ice all winter without leaking. The roofers must make it weather-tight. They put waterproof roofing felt on the roof to seal all cracks before installing some kind of wood or plastic shingles. Then they hang the *gutter* (trough), which catches rainwater, and the *leaders* (downspouts), which direct it away from the house.

If the structure has a chimney, fireplace, or brick walls, a mason builds them at this time. The outside is finished by installing the windows, doors, siding, shingles, brick, stone, or other materials.

Installing Electricity, Gas, and Other Utilities

While the carpenter and sheet-metal workers are completing the exterior of the house, the electrician, following the architectural drawing, installs wires for all electrical outlets and fixtures. Also the plumber installs all pipes that will bring gas and water into the house. The heating and air-conditioning workers install the furnace and air-conditioning equipment. Duct work necessary to carry the heat and air to all parts of the house is also done.

Inside walls can be finished in many ways. They may be covered with prefinished plywood, hardboard, or other material. Following this, the carpenters return to install the flooring and the interior trim such as floor molding and the molding around doors, windows, and ceilings. Also, tile workers put in any ceramic tiles or plastic tiles used in the bathroom, kitchen, and other parts of the home.

Other Jobs

Masons build the chimney and fireplace (Fig. 9-22). Electricians come back to add all switches and to connect appliances

Fig. 9-22. Masons must be skilled in using brick and other ceramic materials. (*Manpower and Vocational Education Weekly*)

and light fixtures. The plumber returns to connect many different kinds of fixtures and appliances for water, sewer, and gas service. The painter paints the outside and the inside of the house. Any other surfaces that are not covered by some prefinished or ceramic material are also painted. A paperhanger puts up wallpaper. The mason then comes back again to do several jobs—such as the outside steps, sidewalk, patio, and driveway. Finally, the landscapers smooth the ground around the house. They make sure that the ground slopes away from the house so that water will drain properly. They may also add trees, bushes, or plants as needed.

Your Place to Live

In purchasing or building a house, apartment, or mobile home, you should check the following:

1. Is the building architecturally pleasing?

2. Does the building cost no more than 2 to 2½ times your yearly income?

3. Do you like the neighborhood?

4. Are there enough rooms now, or is it possible to add more later on? Since many families live in four or five places over the years, there is a need to think about selling the house and moving elsewhere.

5. Are there enough closets?

6. Are bathrooms well placed and easy to get to?

7. Are there enough work surfaces and cabinets in the kitchen?

8. Is there a good place to eat?

9. Is the building well insulated for keeping down fuel costs?

10. Do windows and doors open easily?

11. Is the property termite-proofed?

Annual Checkup

After a person has bought a place to live, an annual checkup of the exterior and interior should be made as follows:

1. Decide if the exterior needs putty, caulking, or exterior paints.

2. Inspect glass and screens. Change storm windows or screens.

3. Check the roof for any damage.

4. Check for signs of termites.

5. Check interior paint and redecorate as needed.

6. Seed and fertilize the lawn. Prune shrubbery.

7. Shut off water hose connections in the fall to help prevent pipes from freezing.

8. Keep the driveway and walks free of ice and snow.

9. Have the heating system cleaned and repaired as necessary.

10. Check the cooling system.

11. Oil motors and appliances, following instructions. Check the humidifier, turning it on or off as necessary.

12. Check cords, plugs, and all electrical appliances.

Glossary

Batter board. A temporary framework used to assist in locating corners when laying out a foundation.

Blocking. Small wood pieces used between structural members to support panel edges.

Bottom plate (sill plate). The lowest horizontal member of a wall or partition that rests on the subflooring. Wall studs are nailed to the bottom plate.

Chalk line (snap line). A long spool-wound cord encased in a container filled with chalk. Chalk-covered string is pulled from the case, pulled taut across a surface, lifted, and snapped directly downward so that it leaves a long, straight chalk mark.

Course. A continuous level row of construction units, as a layer of foundation block, shingles, or plywood panels, as in subflooring or roof sheathing.

Cripple. Any part of a frame that is cut less than full length, as in cripple studs under a window opening.

Footing. The concrete (usually) base for foundation walls, posts, chimneys, etc. The footing is wider than the member it supports, and it distributes the weight to the ground over a larger area to prevent settling.

Gable. The triangular portion of the end wall of a house with a pitched roof.

Gusset. A small piece of wood, plywood, or metal attached to corners or intersections of a frame to add stiffness and strength.

Header. One or more pieces of framing lumber used around openings to support free ends of floor joists, studs, or rafters.

In-line joint. A connection made by butting two pieces of lumber, such as floor joists, end-to-end and fastening them together by using an additional splice piece nailed on both sides of the joint.

Joist. One of a series of parallel framing members used to support floor or ceiling loads, and supported in turn by larger beams, girders or bearing walls, or foundation

Kiln dried. Wood seasoned in a humidity- and temperature-controlled oven to minimize shrinkage and warping.

On center (o.c.). A method of indicating

the spacing of framing members by stating the measurement from the center of one member to the center of the next.

Plumb bob. A weight attached to a line for testing perpendicular surfaces for trueness.

Rafter. One of a series of structural members of a roof, designed to support roof loads.

Ridge board. Central framing member at the peak, or ridge, of a roof. The roof rafters frame into it from each side.

Sill (mudsill, sill plate). The lowest framing member of a structure, resting on the foundation and supporting the floor system and the uprights of the frame.

Soffit. Underside of a roof overhang.

Studs (wall). Vertical members (usually two-by-fours) making up the main framing of a wall.

Subflooring. Bottom layer of plywood in a two-layer floor.

Top plate. The uppermost horizontal member nailed to the wall or partition studs. Top plate is usually doubled with end joints offset.

Unit 10

Manufacturing Technology

Manufacturing is the system designed by *people* to convert *raw materials* from our planet, first into *standard stock*, and then into finished industrial and consumer products (Fig. 10-1 a and b). In this unit, you will learn about material processing and production activities and get an overall view of manufacturing. When you begin the study of wood, metal, and plastic technology, you will learn much more about how products are produced.

It is important to know about this industry. And to know about industry, you must know how a large industrial corporation works. These are the basic parts of all manufacturing *concerns* (businesses):

1. Product. A company is set up to produce products that people need. These

Fig. 10-1 a. **This steam iron is a typical consumer product.** (GE)

Fig. 10-1 b. **This arc welder is a standard industrial product.** (Hobart Brothers Co.)

Fig. 10-2. Planes built in the United States are used throughout the world. (McDonnell Douglas Company)

Fig. 10-3. Management is concerned with problems of the company, including sources of raw materials, sale of products, and distribution centers.

can be anything from paper clips and bicycles, TV sets and cars, to household appliances and furniture. Large industrial companies produce and distribute goods that are wanted and needed by millions of people and industries in many countries (Fig. 10-2). Most of you will someday work either directly or indirectly for a company that makes some kind of goods. Goods can be textbooks like this one, clothing to wear, or a car to drive. The tools and the machines you are using are also products of industry.

2. *Management.* The most common way of *organizing* (setting up) a business is to create a *corporation.* People buy a share of the corporation by buying stock. Stock must be sold to raise money (called *capital*). The stockholders can vote to choose a board of directors to lead the corporation. This board of directors then hires a management group. Management includes the president, vice president, managers, and other officers (Fig. 10-3). This group works to organize and operate the company so that it will make a profit (earn money). Just how a corporation is organized depends on how big it is and what kinds of goods it will make (Figs. 10-4 and 10-5). Management is responsible for making important decisions. With good management, it is

possible for a company to work well and earn a profit. Poor management can result in the loss of money, and the corporation may even go *bankrupt* (broke).

3. *Capital.* Money that is raised by the sale of stocks (and by borrowing money from banks, too) is used to build buildings

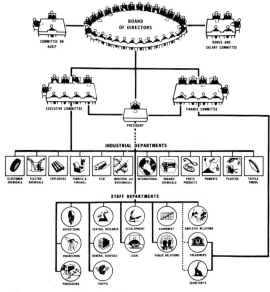

Fig. 10-4. This is the organization chart of a large corporation. Each division of this corporation operates much like a separate company.

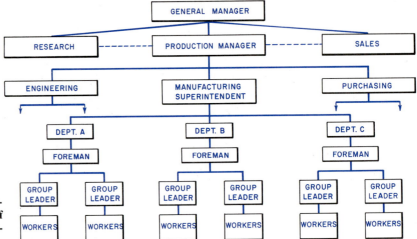

Fig. 10-5. This is a sample organization chart of a division of a larger corporation.

and to buy machines, tools, and materials. It also provides for all other things needed to produce goods (Figs. 10-6 and 10-7).

4. *Personnel.* People with many kinds of skills are needed to make up the company's work force. People do the work that makes the products. People other than the management group are often called the *labor force.* This group includes service workers, laborers, machine operators, office and clerical workers, the sales force, technicians, and professional workers such as engineers and researchers (Figs. 10-8 through 10-11).

5. *Raw materials.* Raw materials are natural resources. They are needed to produce goods or products. They can be classified as *mineral*, *vegetable*, or *animal*. These raw materials come from all over the world. Minerals can be divided into two types, metallic and nonmetallic. *Metallic* substances are materials such as iron ore, bauxite, lead, tin, and copper. *Nonmetallic* materials are those such as clay, limestone, and coal. *Vegetable* materials include those which come from trees and plants. A few examples are lumber, cotton, and rubber. *Animal* materials include hides for leather goods, wool for textiles, and bones for glue. These raw materials are reworked into standard or *semi-*

Fig. 10-6. Millions of dollars are invested in huge plants such as this one. (Butler Manufacturing Company)

Fig. 10-7. This automated metal machinery produces auto parts. (Allis-Chalmers)

Fig. 10-8. Employment personnel in large companies are well prepared for their jobs. You will need to learn how to apply for a job. (Western Electric Company)

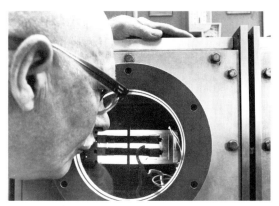

Fig. 10-9. This scientist is working in an engineering research center. (Western Electric Company)

Fig. 10-10. This skilled technician is checking an electron gun. (RCA)

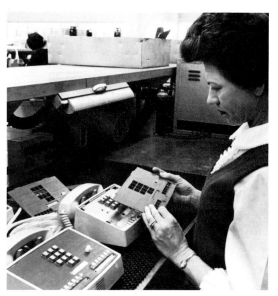

Fig. 10-11. This production worker is on the final assembly line for a telephone manufacturer. (Western Electric Company)

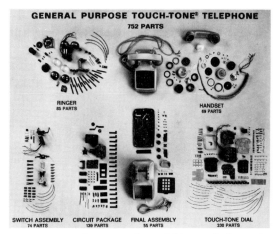

Fig. 10-12. There are lots of bits and pieces in the modern telephone system. Here the parts are displayed by function: ringer (to inform you someone is calling), handset (to convert your voice into electrical patterns and to do the opposite for the person you are listening to), switch assembly (to answer a call and hang up), touch-tone dial (to create tones that start the switching equipment and establish connection with another phone), circuit package (for electrical purposes, known technically as a *network*), and final assembly (the remainder of the phone). Where do all of these raw materials come from? (Western Electric Company)

processed materials. Semiprocessed materials include products such as plywood, steel, aluminum, copper, cement, textiles, and chemicals. These are in turn used in the manufacture of many different kinds of durable and nondurable goods for construction. Many raw materials are also used to manufacture components that are then used in the assembly of a finished product, as shown in Fig. 10-12.

6. *Research and development.* Research is done to find new knowledge and new materials. Both of these are needed for industrial progress. Not every manufacturing company does research. But most of the large corporations hire many professional and technical people to do this kind of work. Development people try to suggest how the new knowledge and materials can be used for new and better products.

Research and development are also carried on all the time to make the products we now use even better. Each year the telephone and the automobile become more efficient and useful (Figs. 10-13 through 10-16). Most of us take for granted new car models. We do not really know how much research, development, planning, and production are needed before we are able to see or buy these products (Fig. 10-17). The *customer* must always be happy with what is bought. Therefore, *market research* is always carried out to

find out what customers want. In the case of a car, the manufacturer has to think ahead for at least three years to know what

Fig. 10-14. These are some of the telephone models that are available for home and business use. (Western Electric Company)

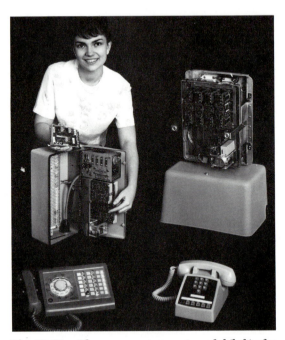

Fig. 10-15. The newer compact model (left) of a pushbutton phone, with its older equivalent on the right. (Western Electric Company)

Fig. 10-13. One of the earliest telephones developed by Alexander Graham Bell. (Western Electric Company)

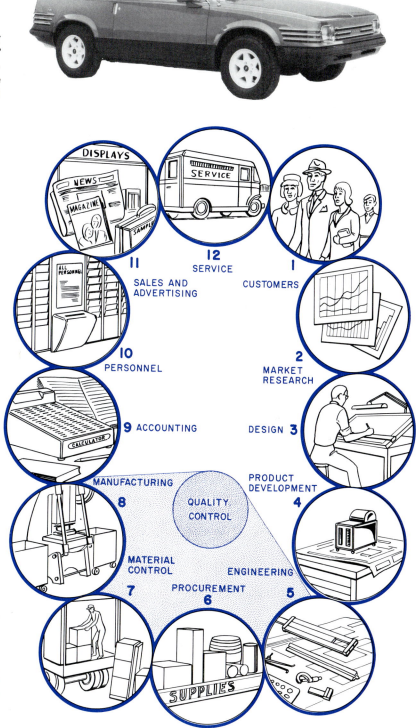

Fig. 10-16. A design of an agile, small "urban" car conceived for shopping and for use where parking is difficult. (Ford Motor Company)

DISPLAYS

NEWS

MAGAZINE

SAMPLE

SERVICE

ALL PERSONNEL

11

12 SERVICE

1

SALES AND ADVERTISING

CUSTOMERS

10 PERSONNEL

2 MARKET RESEARCH

CALCULATOR

9 ACCOUNTING

DESIGN **3**

MANUFACTURING

QUALITY CONTROL

PRODUCT DEVELOPMENT

8

4

MATERIAL CONTROL

ENGINEERING

7

PROCUREMENT

6

5

SUPPLIES

Fig. 10-17. This circle illustrates the general steps in producing and marketing a product.

Fig. 10-18. Student designers make a clay model of a compact two-door coupe. (Ford Motor Company)

Fig. 10-20. Using a wooden frame for the base, a full-scale model is built by the student designers. (Ford Motor Company)

the public will buy. About three years before a car is ready to be bought, it is designed as a new model. Designers, engineers, and drafters work on *design* and *product development.* They do all of the needed first-step research and development. They also make a full-size clay model of the new car (Figs. 10-18 through 10-22).

About 2 to 2½ years before production, engineering details must be worked out. Details must be known for the engine, transmission, and other subassemblies. Engineers must then build *prototypes* (models) of these *components* (parts that fit together). About two years before pro-

Fig. 10-21. The wooden frame is covered with clay and then smoothed. (Ford Motor Company)

Fig. 10-19. A full-size mock-up makes possible a study of the physical size of the car. (Ford Motor Company)

Fig. 10-22. The completed full-scale clay model gives these student designers an opportunity to study the three-dimensional effect. The design problem was to create a *congestion fighter.* (Ford Motor Company)

duction, the basic body style is fixed. Some style refinements or changes may be made at that time. The instrument panel and steering column have been designed and improved. The exterior of the car has been shaped and fixed in sheet metal.

Between 1 and 1½ years before production, testing is done on each of the parts that goes into the new model. At this time, the people in manufacturing engineering or production planning must begin to make needed changes in the assembly lines. They have already made the new tools, dies, and fixtures.

About six months to a year before the new car is produced, real models must be built and tested. Then, about five months before the model is sold to customers, the engineers approve the final production details. From 3 to 3½ months before the car is manufactured, production tooling, including the new dies, fixtures, and special tools, is ready for the assembly plant for final inspection, approval, and installation. At this time, the actual parts of the car are being made.

As you have learned, mass production includes two basic steps:

a. The production of interchangeable parts in large numbers

b. The assembly of these parts on a production, or assembly, line

An automobile manufacturer with many different models may have to make as many as 30 000 new and different kinds of parts for each model year. The number of each part produced will depend on how many cars are really manufactured and how many extra parts will be needed as replacements in damaged cars. For example, automobile manufacturers have found that they need about twice as many right-front fenders for replacement parts as left-front fenders. A whole new design of a car, as you can see, starts about 3 to 3½ years before car models are ready to be sold.

7. *Manufacturing engineering, or production control.* Manufacturing engineering, or production control, is concerned with all the problems that come up during the making of a product. People in this field must decide which processes can be done on which machines. They must also decide what new tools and machines are needed, what changes in plant layout must be made, and how the industrial processes are to be carried out. This group is also concerned with *material control* and *quality control*, including testing and standards.

In addition, this group must deal with such problems as deciding on the workload for all production workers. Production planning must also determine if the manufacturer will make all the parts or if the making of the parts will be "farmed out." When parts are farmed out, they are made most often by smaller manufacturers. A manufacturer may also buy parts from some other company. In manufacturing engineering, people must lay out the production lines and the assembly lines. In large manufacturing companies, much of the information is put on computers. Computer information can indicate whether there are enough parts ready for the final assembly.

The quality-control depatment must set up standards for testing the new materials, processes, and techniques. These people must make sure that products meet the standards of the manufacturer.

8. *Manufacturing.* Production deals with the manufacture of the individual parts. It also deals with the assembly of those parts into a finished product (Figs. 10-23 through 10-30). All manufacturing operations can be divided into four major areas:

a. *Cutting.* All materials must first be cut to size. This can be done in several ways, such as sawing, shearing, and stamping out.

b. *Forming.* Forming is the way materials are reshaped. Forming operations used most often are casting, bending, machining, and forging.

CUTTING	FORMING	ASSEMBLY	FINISHING
SAW	BEND	ADHESION	STAIN
SHEAR	FORGE	COHESION	DYE
FLAME CUT	CAST	SOLDER	BRUSH PAINT
SCORE	MOLD	WELD	ROLL PAINT
CHIP	BLOW-MOLD	FUSION	DIP PAINT
WEDGE	DRAW	MECHANICAL	SPRAY LACQUER
CHIP & WEDGE	ROLL	FASTENINGS	WAX
SHAPE	SPIN	ETC.	BUFF
GRIND	EXTRUDE		POLISH
TURN	PRESS		ANODIZE
ROUT	DIE FORM		ELECTROPLATE
PIERCE	ETC.		GLAZE
DRILL			ANTIQUE FINISH
ETC.			HARDEN-TEMPER
			ETCH
			SAND BLAST
			ETC.

Fig. 10-23. The common manufacturing processes.

c. *Assembling.* Assembling includes the techniques of putting parts together to make the finished product. Parts can be assembled by using mechanical fasteners, such as bolts or rivets, by using adhesives, by soldering or welding, or by using other methods.

d. *Finishing.* Finishing includes all the processes used to improve the appearance and protect the product parts.

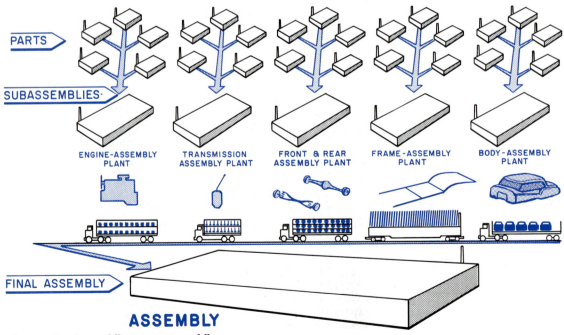

Fig. 10-24. Assembling an automobile.

Fig. 10-25. Drilling is one kind or form of cutting.

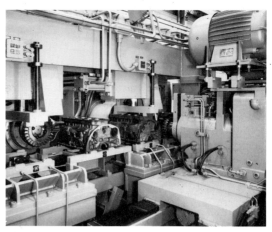

Fig. 10-26. Mass-producing metal parts for automobiles.

Fig. 10-27. (*a*) Exploded view showing the many parts of a bicycle coaster brake. (*b*) This is a subassembly of the brake parts onto the rear wheel and an exploded view of all the parts for a bicycle. (*c*) This is the assembled bicycle. (Schwinn Bicycle Company)

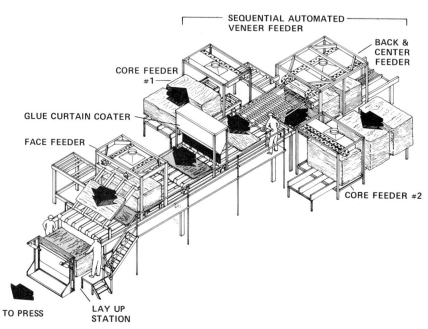

SEQUENTIAL AUTOMATED
VENEER FEEDER

BACK &
CENTER
FEEDER

CORE FEEDER
#1

GLUE CURTAIN COATER

FACE FEEDER

CORE FEEDER #2

**Fig. 10-28.
Mass-producing
plywood.**

TO PRESS

LAY UP
STATION

**Fig. 10-29. Testing the
final product.**

**Fig. 10-30 a.   Mass-produced telephones ready
for shipment.** (Western Electric Company)

**Fig. 10-30 b.   These manufactured products are
being stored ready for shipment.** (Western Electric Company)

All the major manufacturing processes can be done either by hand or by machine. Most of your experiences in this class will involve learning how to perform these processes.

The processes include painting, electroplating, and polishing.

9. *Marketing.* Marketing deals with *selling, advertising, distributing,* and *servicing* products. Most articles that we use are made many miles from where we live. Marketing brings these goods and services to us at the right time and at the right price.

10. *Finance.* A large part of any corporation's personnel must deal with *financial* (money) matters. These include account-

ing, purchasing materials, paying labor, and keeping records (Fig 10-31).

Fig. 10-31. Financial responsibilities are important to every company. (Du Pont Corp.)

Unit 11

Mass-production Technology

Most of the *products* (project activities) you will build are made in about the same way as they were by the custom (handicraft) method used in industry or by the artists and craftworkers of today (Fig. 11-1 a and b). Making a project activity by yourself gives you a chance to learn basic skills and gain knowledge. It also gives you a chance to be creative and to do problem solving in the use of tools, materials, and processes. However, you and your classmates may also have a chance to mass-produce a product. You may do this in a way similar to the methods of most modern industries. In order to organize for mass production, some of the things your class may have to do are as follows:

1. *Establish a manufacturing company.* The class can decide if it is necessary to organize a typical corporation. This is not necessary for mass-producing a product,

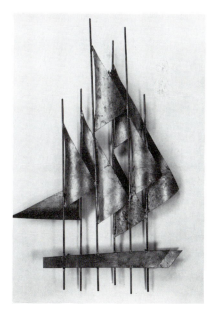

Fig. 11-1 a. This sculptured wall design is typical of most products made by custom (handicraft) methods. (The Sculpture Studio, Inc.)

	CUSTOM METHOD	MASS-PRODUCTION METHOD
Capital	None, except money for purchase of materials	Monies raised by *stock* for materials, labor, equipment, etc.
Location	Laboratory or shop	Shop organized as a manufacturing plant
Product selection	A design selected from a book, or you design your own	Market research is done to determine *what will sell*
Product development	Use the existing drawing or make your own sketch or drawing	After selection, the engineering department makes the drawings

Product planning	Making a *plan sheet:*
	  a. Bill of materials
	  b. Tools and machines
	  c. Order of procedure
		
		  a. Procurement of materials
		  b. Production line machines
		  c. Operation process charts

People	You are the individual artist-craftsperson	Division of labor force:
		  Management
		  Designers-engineers
		  Skilled workers, etc.

Producing the product	Individual parts built to fit only your project Put together as the parts are made	Standard interchangeable parts put together on an assembly line
Quality control and inspection	Quality depends on your skills	Careful quality control to make sure all parts and the complete product meet standards
Distribution	One "take-home" project	Identical products available for each member of the class, for sale, or for gifts

Fig. 11-1 b.   A comparison between the custom method of producing a product and the mass-production method.

especially if it is for class use alone. However, if a corporation is organized, experience will be gained in such areas as business, management, advertising, sales,

Fig. 11-2.   Experiences in managing a company are important. Monthly board of directors meetings held by each company (the entire class is the board) are used to formulate company policy and discuss problems.

and distribution. Your class must organize the company, sell stock, elect a board of directors, and hire the president and other management personnel (Fig. 11-2). Perhaps the first class experience will be concerned with manufacturing only enough of one product to allow each student to take one home. A formal company will not be needed for this.

2. *Market research.* If the class decides to mass-produce an article for sale, market research must be done to find out the kinds of products that will sell best (Fig. 11-3). A survey of your school or community will tell you what kinds of things people are willing to buy. Products such as drawing boards, lamps, bookshelves, picture frames, wall plaques, or small accessories would be good selections (Fig. 11-4 a). Your class may decide to produce products to donate to the community (Fig.

Fig. 11-3. This student has designed a metal guard for the rubber head and blade for small sailboats. It might be the kind of product that would sell well in areas where boating is popular. (Zinc Institute, Inc.)

Fig. 11-4 b. A metals technology class produced these identification signs to donate to a city so that the fire fighters could find the hydrants in winter months.

Fig. 11-4 a. Small accessories for kitchens and bathrooms may be good mass-produced products to sell.

Fig. 11-5 a. Some kinds of desk accessories may sell well. (MacDonald Products Company)

Fig. 11-5 b. Here are small desk accessories made of fine hardwood. (MacDonald Products Company)

11-4 b) or to needy children. These may include such things as games or toys. If a product is to be sold, it is important first to find out *what kind* and *how many* will be produced for sale. Some suggested projects are shown in Figs. 11-5 a and b and 11-6.

Fig. 11-6. These are some prize winning mass-produced products made in wood technology classes. (Stanley Tools)

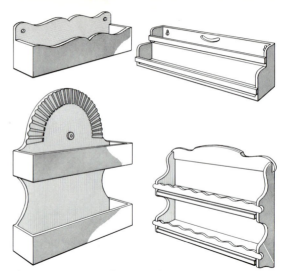

Fig. 11-7. Several possible designs for spice racks.

3. *Product development and engineering.* After the product has been chosen, it is necessary to do the development work. This includes designing it, producing the drawings, and making a pilot model. Suppose, for example, the class decides to produce a spice rack. The first thing to decide is what it will look like, or what the design will be. Class members can suggest several different ideas (Fig. 11-7). From these, one can be selected. Suppose the design in Fig. 11-8 is selected. The drafting class, operating as an engineering department, will make working drawings of the rack, including *detail* and *assembly* drawings (Fig. 11-9). Materials lists and procedure steps must also be made (Figs. 11-10 and 11-11). After the drawings are approved, *pilot models* (prototype units) must be built (Fig. 11-12). Pilot models are needed for the following reasons:

a. To discover any problems in construction

b. To develop a flowchart, or operation process chart, for manufacturing

c. To build the necessary jigs, fixtures, and gages (Fig. 11-13)

These steps take a great deal of "lead time." Therefore, your instructor may already have a product design for mass pro-

Fig. 11-8. The design selected.

duction, including all the necessary jigs, fixtures, and gages. As you learned in Unit 10, "Manufacturing Technology," automobile manufacturers must have at least three years of "lead time" before manufacturing can actually begin on a new model.

4. *Manufacturing engineering, or "tooling up."* Before manufacturing can begin, a good many things must be done. This is called "tooling up" for production:

a. *Plant layout.* In large companies, equipment is moved around and new plant layouts are made for more efficient

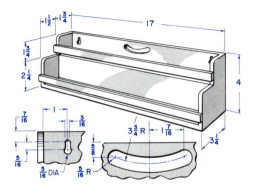

in	mm
$3/16$	5
$5/16$	8
$7/16$	11
$5/8$	16
1	25
$1\,7/16$	36
$1\,1/2$	38
$1\,3/4$	44
$2\,1/4$	57
$3\,1/4$	82
$3\,3/4$	95
4	102
17	432

Fig. 11-9. Detail drawings for each part should be made from these drawings. Use the conversion chart to make the project metric.

ORDER OF PROCEDURE

1. Lay out, saw, and jigsaw the two end pieces to size.

2. Cut the back piece, and lay out the handle and hanger holes.

3. Bore a $5/8$-inch hole at each end of the handle and saw out the remaining stock with the jigsaw. Another method – drill a small starting hole and jigsaw the entire opening.

4. Make hanger holes by drilling or boring upper holes first, following with the $5/16$-inch hole below and shaping with a rattail file or jigsaw.

5. Cut out shelves and bottom shelf support.

6. Cut guard rails.

7. Assemble with brads and glue.

8. Finish with paint or antique pine finish.

BILL OF MATERIALS

IMPORTANT: All dimensions listed below are *FINISHED* size.

No. of pieces	Part name	Thick-ness	Width	Length
1	Back	$1/4"$	$4"$	$16 1/2"$
2	Ends	$1/4"$	$3 1/4"$	$4"$
1	Top shelf	$1/4"$	$1 1/2"$	$16 1/2"$
1	Bottom shelf	$1/4"$	$1 1/2"$	$16 1/2"$
1	Top shelf support		$2"$	$16 1/2"$
2	Rails	$1/4"$	$1/2"$	$17"$

Fig. 11-10. The order-of-procedure steps and the bill of materials for making the spice rack. This is the method for producing one rack. From this, flowcharts must be made so that the parts can be mass-produced. Perhaps a separate "route sheet" listing the steps needed to make each part should also be made.

Fig. 11-11. Selection of materials for making the product is important.

Fig. 11-12. Checking the pilot model to make sure it meets customer needs. (Junior Achievement, Inc.)

Fig. 11-13. A jig for bending the rod when making the fire hydrant signs.

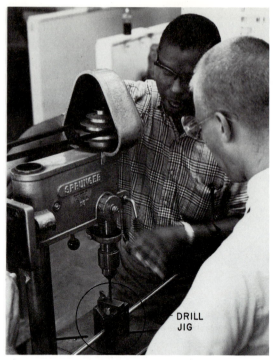

Fig. 11-14. A drill jig is used to hold the head of a mallet as the hole is being drilled.

production. In your shop or laboratory, you will want to plan the most orderly arrangement for manufacturing the product.

b. *Equipment.* Sometimes new machines and transfer systems are purchased for production. However, your class will have to use the tools and machines now on hand. It will be necessary, however, to develop the production devices, including any jigs, fixtures, and dies you will need (Figs. 11-14 and 11-15).

5. *Procurement of materials.* By now you have decided how many items will be produced. You have made a bill of materials for the product. Now the class must make sure that all these materials are available. It may be necessary to buy certain parts from other companies. For example, if desk sets with pens and penholders are to be made, the pens have

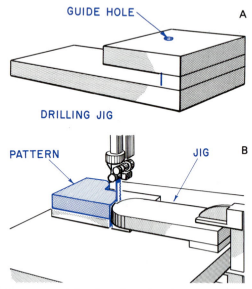

Fig. 11-15. (*a*) A drawing showing the design of a drill jig used to drill the holes in the back of the spice rack. (*b*) A drawing showing the design of a fixture used to hold the ends of the spice rack on a band saw.

to be bought. It may also be necessary to buy fastening devices, such as screws and nails.

6. *Personnel.* a basic idea in mass production is the *division of labor* among workers. You will need management personnel and production workers. Management personnel will include such persons as:

a. President, vice president, secretary, and treasurer

b. Director or manager of production, sales manager, personnel director, and others

Production workers will include:

a. A shop supervisor for each division.

b. An assistant shop supervisor.

c. Enough production workers for each work station to keep the production line going. For example, you may need three or four workers at each sanding station. However, only one may be needed at the drill press. Each person must know the job and be given the proper training for it (Fig. 11-16).

7. *Quality control.* All the parts and the final product must be made to certain standards. Poor parts are useless. Industry calls these *rejects.* To solve this problem, a quality-control program must be set up. This means that the parts will be checked for such things as size, finish, and accuracy at each step as they are produced. The completed product must be checked during assembly and finishing (Fig. 11-17 a and b).

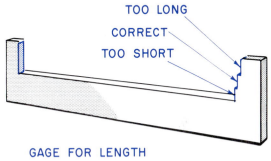

GAGE FOR LENGTH

Fig. 11-17 a. A drawing showing a simple go–no-go gage that can be used to check the length of the front supports of the spice rack.

Fig. 11-16. Each student worker must be well acquainted with his or her job and know how to perform it.

Fig. 11-17 b. A gage to check the angle of a twist drill.

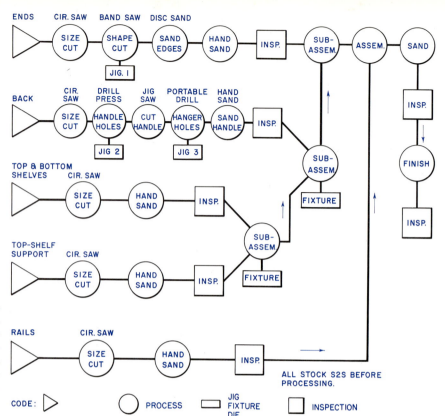

Fig. 11-18 a. A flow-chart, or flow process chart, showing how the spice rack can be manufactured: (*a*) storage, (*b*) process, (*c*) jig, fixture, or die, (*d*) inspection.

8. *Production control.* When you completed the plan sheet for making your own project activity, you had to list the *steps* in making the project. In industry this is called *production control.* This involves making flowcharts and operation process charts for such things as materials, manufacturing, and inspection (Fig. 11-18 a and b).

9. *Manufacturing.* You are now ready to manufacture the product. The plant has been organized. The equipment is available. The materials are ready, the workers have been trained, and the production devices have been made. The gages for quality control are complete. The flowcharts have been made. Now the materials must move from the storage area to manufacturing. Then they go on to finishing and assembly and finally to inspection and packaging (Fig. 11-19).

a. *Material control and movement.* For efficient manufacturing, a method must be set up for storing and moving materials and partly finished parts. In industry there are a number of methods. Small industries use *stock carts* and *pallets.* Larger industries use automatic conveyor systems. In your school, shop carts will probably be best for moving the parts from one place to another.

b. *Manufacturing processes.* All manufacturing processes can be divided into four basic types: *cutting, forming, assembling,* and *finishing.* All these processes are talked about in this book. Processes may be done with hand tools or machine tools.

c. *Inspection.* As the parts of the products are made, they must be inspected. They may be given a visual inspection

Product Name			Flow Begins	Flow Ends	Date
Prepared By:		Section:		Approved By:	

Process Symbols And No. Used	◯ Operations_____ ☐ Inspections_____ ⬠ Transportations
	D Delays_____ ▽ Storages_____

Task No.	Process Symbols	Description of Task	Machine Required	Tooling Required
	◯⬠☐D▽			
	◯⬠☐D▽			
	◯⬠☐D▽			
	◯⬠☐D▽			
	◯⬠☐D▽			
	◯⬠☐D▽			
	◯⬠☐D▽			
	◯⬠☐D▽			
	◯⬠☐D▽			

Fig. 11-18 b. A different kind of flow process chart that can be used in mass producing products.

and also may be inspected with gages and measuring devices. Visual inspection may be needed for seeing the quality of the finish and general appearance. Gages and measuring devices are used to see if the parts are made to the correct size. This is part of the quality-control program.

Fig. 11-19. Packaging the hydrant signs to make them ready for delivery.

10. *Business activity.* Business activities are an important part of any manufacturing industry (Fig. 11-20). Some of these activities include the following:

   a. *Accounting.* Before, during, and after manufacture, it is necessary to keep careful cost records of the items produced. In industry there are four major items: *materials, labor, overhead,* and *profit.* Cost of materials is quite easy to find out. This is done the same way as in making a *bill of materials* for an individual project activity. *Labor costs* are all the amounts paid to individuals to manage, manufacture, advertise, and sell the product. *Overhead* includes

Fig. 11-20. **These students are doing the record keeping for the mass production of products.**

such things as the cost of the building, taxes, rents, equipment replacement, electricity, and heat and light. Usually,

the school does not have to worry about overhead. Manufacturing concerns cannot stay in business very long unless they make a *profit*. It is not necessary for your class to make a profit on your production unless you have organized a corporation, sold stock, and hope to return a profit on each stockholder's investment.

b. *Advertising and sales.* If the mass-produced product is to be sold to the public, it is necessary to advertise and have a sales force.

c. *Service.* Although your group may not be concerned with servicing the items you manufacture, this is important in industry. Automobiles, lawn mowers, appliances—all must be serviced either by the manufacturer or by the dealers who sell the products.

Unit 12

Leadership Development Through AIASA

Leadership "is the ability of a person to work with other people to inspire and influence them to make desired decisions for definite actions." The dictionary also defines a leader as "a person who leads, or a principal performer of a group." There is always a small segment (group) of people in a community who actively guide decision-making policies.

In every large or small city, town, or community, there are organizations working actively for the overall benefit of the citizens. A few of these are the United Fund, Red Cross, Boy Scouts, Girl Scouts, Cub Scouts, Boys' Club, Girls' Club, YMCA, and YWCA. There are also church-oriented, civic, and service club groups. School-related clubs, especially the *American Industrial Arts Student Association (AIASA)* and others such as the Vocational-

Industrial Clubs of America (VICA), Future Farmers of America (FFA), 4-H, and Junior Achievement, offer people leadership roles.

People who assume a leadership role are usually among the most important people in the school community. They nearly always are among the busiest in terms of activities. They are respected by others as people who can organize and guide groups to work harmoniously in furthering a cause and who can achieve a definite objective (goal) set forth by the group. Leadership and club organizational practice and experience can be very helpful in selecting a career.

Leadership Qualities

Most students are potential leaders (Fig. 12-1). In assuming the responsibility of

Fig. 12-1. Student leaders preside at a flag-raising ceremony at a state AIASA convention. (Courtesy Art McNichol, AIASA photographer)

leadership, the student has the opportunity to learn and develop many important leadership qualities while participating in daily school-life activities. A good leader must:

1. Assume responsibility to work as a member of a group.

2. Be able to participate (work with) other people in a congenial (friendly) manner.

3. Be straightforward, pleasant, and industrious.

4. Give praise when it is due to a person who has contributed to the benefit of the group.

5. Be sensitive to the mood and thoughts of the group.

6. Have the ability to communicate thoughts and ideas both in writing and verbally.

7. Be willing to assume responsibility and work as a member of the group.

8. Respect and understand the rights of others, realizing that everyone can make worthwhile suggestions if given the opportunity to express an opinion.

9. Be well informed on the topics or matters on which the group is asked to respond. Worthwhile decisions can be arrived at only after pertinent information has been secured, studied, and voted on by the group.

10. Have self-confidence and have the integrity to show trust in other group members and their decisions.

11. Be positive in terms of enthusiasm and optimism, believing that group action can solve problems and achieve sought-after objectives (Fig. 12-2).

12. Have an open mind toward suggestions by others. Leaders cannot assume that they have the only answers; they must be willing to listen to the group.

13. Have the strength and conviction to support a position, even though it may be opposed by others. However, one must be willing to be convinced when better alternatives (plans) are presented.

Fig. 12-2. A student leader discusses problem-solving methods with a large group of AIASA members at a national convention. Several student chapter counselors are at the speaker's table to lend assistance when necessary. (Courtesy Art McNichol, AIASA photographer)

14. Be able to plan a procedure which will ultimately be respected by the group, thereby gaining their approval.

15. Become familiar with and understand the mechanics of group guidance. This includes study of *Roberts' Rules of Order* or other types of parliamentary procedure used to direct a meeting.

16. Be able to tactfully delegate or assign responsibilities to members of the group.

17. Know the duties and responsibilities of all officers serving in the organization. This includes studying the purpose for which meetings are scheduled and knowing how to plan a workable, realistic means of achieving (obtaining) the stated goals.

Leadership Personnel (Officers)

The many organizations and clubs referred to earlier in this unit require organization and management. There is no "one-person" leadership. Practically all organized groups require a slate of officers to manage basic organization and policies. Elements of leadership development emerge from serving in the capacity of any one of several offices. These offices are necessary for the management and success of an organization.

The American Industrial Arts Student Association (AIASA) is an example of how the general management and functioning of a group operates to achieve set goals. It includes a president, vice president, secretary, treasurer, reporter, sergeant at arms, and usually a parliamentarian. Each officer has specified (listed) duties. They are usually set forth in the group's constitution and bylaws.

The *president* presides at (conducts) the meeting. He or she observes the bylaws and constitution of the organization, based on parliamentary procedure. As the presiding officer, he or she keeps the membership informed of the subject to be discussed or acted on. The time allotted to the meeting is also recognized. The president coordinates the activities of the group by being advised of questions that may come to the attention of the members and by guiding the discussion. He or she keeps in touch with the other officers.

The president is the leader who sees that the organization of the club or chapter is in tune with designated plans. Basic knowledge of parliamentary law (procedure) is necessary; however, the parliamentarian can be called on to assist the president if technical questions of procedure arise.

The *vice president* is the first assistant to the president. This person helps the president perform his or her duties and presides at meetings when the president is absent. The vice president is in charge of the work of committees and monitors (checks on) their programs. The vice president serves as a member of all committees. Since the vice president is second in rank, he or she usually is in line to be elected the next president.

The *secretary* keeps a record of chapter business and other activities. These are the official minutes of the meetings and are available to the president for each meeting. The secretary takes care of correspondence, keeps a list of all committees, and files the committee reports. Minutes of the previous meeting are read to members and officers.

The *treasurer* has the responsibility of caring for all money matters and funds. The treasurer collects dues, keeps a record of disbursements (payments), and pays authorized bills. The treasurer keeps accurate financial books and is subject to an audit at the end of the fiscal year by an auditing committee that is appointed by the president.

The *reporter* is essentially the publicity director. Chapter plans and activities, both individual and group, that are worthy of publicity are written up. They are made available to school and local newspapers and to similar chapters in other schools. The reporter keeps a notebook of newspaper clippings and notices. He or she can also include pictures of chapter members and, possibly, photographs of club projects and gatherings. This calls the attention of others to the progress made by the group.

The *sergeant at arms* is responsible for the physical arrangement of the meeting room. This includes the placing of chairs for members, program participants, and possible guests. This officer must obtain the equipment required by the program committee or officers. Such items can include visual projection equipment, screens, microphones, and information about the location of power sources.

A *parliamentarian* assists the president to ensure a smooth and orderly meeting (order of business). This individual must be familiar with *Roberts' Rules of Order* and must see that the discussion follows accepted rules.

Leadership Activities

To establish and promote leadership qualities, there must be a variety of activities appropriate for the members of the club. Personal development grows as opportunities are offered and acted on. The activities planned should be suitable for students in industrial arts courses.

The following listed activities are but a few in which the student can participate as a member or in a leadership role:

1. Visiting *local industrial or business firms.* Industrial arts students can gain an appreciation and understanding of how industries and businesses operate. They can see how people work in these organizations and appreciate the fact that the skills and processes are sometimes the same as those which are experienced in the school industrial laboratory.

2. Inviting *industrial representatives to talk to the class, club, or group.* People from industry are most cooperative in talking with student groups. They can give valuable insight and information about how their particular industries operate and can name some qualifications for employment. This experience gives the industrial person a chance to visit the school. He or she sees the students' classroom or industrial laboratory and can gain an insight into the overall educational preparation.

3. Organizing *and publicizing (promoting) school exhibits.* Many school systems sponsor an annual exhibits day or night. Parents and friends visit to see the accomplishments of their sons, daughters, and friends in a school setting. This provides an opportunity to "sell" (acquaint) parents on student activities and accomplishments. Parents have an opportunity to become better acquainted with the faculty and learn why students study industrial technology.

4. Using *visual instruction, such as films, to show industrial processes in class.* Many educational industrial films (both sound and slides) are available from industries. Many come free of charge, while some require only a minimum shipping fee. Films with sound are usually in color. They have commentary that enlarges the students' educational and industrial knowledge.

5. Participating *in student organizations.* Leadership qualities can be achieved through student involvement (activity and interest) in student organizations, particularly the AIASA. Your school may have this organization; if not, try to interest your teacher in forming a chapter.

Junior Achievement is another nationally sponsored organization. It teaches self-enterprise, mass production, and techniques of selling a product made by the

Fig. 12-3. Student members participate in a state-wide AIASA conference. (Courtesy Art McNichol, AIASA photographer)

group. This organization is of interest to students studying industrial technology.

6. *Attending state, regional, or national conferences* (Fig. 12-3). These conferences relate to student involvement. There are opportunities available, particularly if there is a local chapter. Going to and participating in these meetings offers a broader insight (understanding) of leadership and responsibilities.

7. *Arranging for a school assembly program.* This experience involves many leadership qualities. It acquaints other students and faculty members with the importance of the student's place and potential in our present-day industrial technological society.

8. *Following a student personnel class management chart.* The purpose of this system of operation is to give the teacher more time for teaching and personal contact with students. At the same time, the teacher can give the students an opportunity to act in roles of leadership, as set forth on the chart.

Many teachers use a student personnel class management chart that outlines student responsibilities in the operation of a laboratory-type class. This chart lists individual duties in class management. Each student has a definite (specific) assignment to carry out during the semester or year.

Titles vary according to the teacher's choice. They may include:

a. *Shop supervisor,* who checks to see that the assignments are being done.

b. *Class secretary,* who checks the role at the beginning of each class.

c. *Safety inspector,* who sees that safety practices are observed in handling materials and tools and operating equipment.

d. *Tool supervisor,* who checks the return of all tools to their proper locations.

Generally, the student personnel class management chart duties are changed frequently. This rotation of students is done by name or bench number. Assignments give each student a definite responsibility. The teacher may make the assignments week by week, monthly, or for the entire semester. There are specific duties for each student in the class.

Ideas for AIASA Meetings and Activities

1. Elect officers for the AIASA or any designated related-activity group to plan a chapter calendar of activities for the year.

2. Appoint committees for chapter activities dealing with specific aims and accomplishments.

3. Organize a personnel-responsibility system for each class. This will involve a rotational system of management of classroom and laboratory activities in which all students participate. Each student will assist the instructor with regular classroom and laboratories procedures dealing with planning, record keeping, and maintenance.

4. Develop a workshop-concept session to train new officers for the incoming AIASA year.

5. Promote AIASA and gain identification by creating interest through publicity items and listing the organization in the schedule of school-year activities.

6. Support organized classes or groups for product-enterprise production. Mass-produce one or two items to become acquainted with industrial production methods.

7. Study safe practices as they relate to the use of machinery by students, learn how they are observed in industrial organizations, and understand the importance of this protection.

8. Organize and distribute (sell) the items manufactured by students in the product-enterprise activity. Recognize the importance of record keeping, advertising, and product financing.

9. Form an industry-community resource group. The purpose is to invite representatives from the local industrial community to speak and make demonstrations in various areas. This relates to manufacturing and distribution.

10. Explore (study and discuss) career and program opportunities through cooperation with industrial representatives. Encourage intensive study of selected topics such as library facilities.

11. Sponsor an open-house day or evening in the industrial laboratory. The purpose is to get parents and community residents to become more aware of career and potential industrial development possibilities.

12. Be enthusiastic about promoting and developing an *AIASA awards program.* This is done to recognize an outstanding

Fig. 12-4. Student activity contests are an important part of local and state AIASA meetings. Shown is a first place award for an architectural rendering. (Courtesy Art McNichol, AIASA photographer)

member or members for specific achievements and goals. Include many categories (types) of student exhibitions of completed project activities for community and school viewing (Fig. 12-4).

Outline for a Typical Business Meeting

1. Call the meeting to order.

2. Appoint a substitute officer for the meeting if the regular one is absent.

3. Have the minutes of the previous meeting read and corrected.

4. Call for reports by officers and standing or special committees.

5. Make announcements.

6. Discuss unfinished business (if any).

7. Discuss new business.

8. Adjourn (close) the meeting.

Unit 13

Careers in a Changing Industrial Technology

Approximately half of the students who graduate from high school each year enroll in colleges or universities. Students who decide not to get a baccalaureate degree in a specific area can attend a variety of specialized technical schools to acquire definite vocational skills. The National Association of Trade and Technical Schools takes the position that a person with no skills will have a hard time building a worthwhile career.

The greatest demand for employment in the next decade will be in those areas that require and involve technical knowledge. For example, employment opportunities for computer specialists and for medical secretaries are projected to increase significantly.

High school graduates can go into such areas as secretarial work, building trades, car mechanics, construction crafts, food service, paraprofessional health care, dental hygiene, and clinical laboratory technology. These are only a few of the many meaningful career opportunities for high school graduates. Some of these positions may require attending a trade or technical school.

Competition among college and university students is increasing as the demand for technical jobs grows. Students specialize in accounting, sales, marketing, and other technical fields. Computer science and engineering are also attractive career possibilities. At the graduate level beyond the bachelor's degree, there is a demand for dental, medical, and business degrees. As jobs become increasingly technical, it is also important that the job applicant present a favorable personal image along with a polished résumé in order to achieve a successful job interview.

Choosing a career is often a difficult and serious decision. Each person must know his or her abilities, interests, and training and should know about employment opportunities. The *Dictionary of Occupational Titles*, published by the U.S. Department of Labor, lists more than 30 000 job titles. After one studies the areas of interest, consideration must be given to how much education and training are needed. Other things to consider are the personal enjoyment and rewards of each particular career.

The student must know the differences between the terms *job*, *occupation*, *career*, and *profession*. A *job* is the process of doing a piece of work. It is hired work for a given service or a period of time. An *occupation* is the chief business of a person's life. A *profession* requires specialized knowledge, often acquired after long and very careful preparation in a college or university. A *career* is a profession and a long-lasting calling; the term is a broad one and can refer to any of these classifications.

The *Occupational Outlook Handbook*, published by the U.S. Bureau of Labor Statistics, forecasts a labor force of over 125 million by the year 2000. About 120 million people will have civilian careers; 5 million will be in the armed forces. People in this large labor force must adjust to living in a number of places because occupational requirements change rapidly and because industries move employees to different places.

An overall treatment of career trends is presented in this unit. Later in the text, specific careers are discussed as they relate to the sections on technology.

The U.S. Department of Labor lists industries as producing either *goods*, or *services*. The major divisions or groupings are

1. manufacturing
2. trade (sales)
3. government
4. services
5. transportation and public utilities
6. agriculture
7. finance, insurance, and real estate
8. contract construction
9. mining

Goods-producing Industries

Goods-producing industries include manufacturing, contract construction, mining, and agriculture. These industries have grown as a result of automation and other technological development. It is projected that by the year 2000 there will be over 50 million people in these occupations. This represents a 10 percent growth over the last decade.

Manufacturing is the largest division of goods-producing industries. A 13 percent increase in manufacturing employment is projected for the year 2000. Growing manufacturing activities are in the areas of plastic and rubber products; furniture and fixtures; stone, clay, and glass items; and technical instruments. There will be less activity and fewer chances for employment in leather and textile mill products, tobacco, and petroleum refining.

Contract construction jobs are expected to increase 40 percent by the year 2000. There will be growth in the construction of homes, offices, stores, apartment buildings, highways, bridges, dams, and other physical facilities because of population growth. Government spending will increase this growth by means of urban (city) renewal and new highway systems.

Mining offers fewer employment opportunities because of labor-saving technological changes. The development of power sources other than coal has caused mining to decline.

Agricultural employment has dropped because of mechanization of nearly all types of farm activity, even though the trend has been toward larger farmed areas. The decline in work possibilities may be as much as 20 percent in the near future.

Service-producing Industries

It is projected that job demand in service-producing industries will increase significantly during the current decade (10 years). This area currently employs more than 70 million people in trade (sales), government, transportation and public utilities, finance, insurance, and real estate.

Trade (sales) is the largest area within the service-producing industries. Wholesale and retail outlets (stores) have grown in number and size. Labor-saving technologies such as electronic data-processing equipment and automated warehouse equipment have become more common.

Government employment continues to expand in all levels of local, state, and federal activity. This includes people working in agencies involved in health, sanitation, welfare, and protective services.

Service industries offer basic (needed) activities such as maintenance, repair, advertising, and home and health-care services. These industries require expansion of business activities, data processing, and maintenance of equipment.

Transportation and *public utilities* are areas with steady occupational opportunities. While employment in the railroad industry has declined, the airline industry has experienced increased growth. Employment opportunities related to water transportation, electric and gas industries, and sanitary services are predicted to grow.

Finance, *insurance*, and *real estate* have continued to grow in all aspects, and job opportunities will increase.

Occupational Requirements

Specific career information is included in the major sections of this text. Occupations become more complex as these areas grow. There has been a definite shift to more *white-collar jobs.* This classification includes the *professions, managerial, clerical*, and *sales.* These workers now outnumber the *blue-collar* workers, who are *semiskilled* and *laborers.* Job opportunities for both white-collar and blue-collar workers will increase significantly in the next decade. White-collar workers will be needed as research and development are stressed and as services to education and health expand.

Professional and *technical* personnel are highly trained as *teachers, engineers, scientists, technicians, industrial-management persons, dentists, doctors, accountants,* and *clergy.* The number of these workers will also increase 50 percent during this decade. Activities influencing career opportunities for these educated persons are progress in education, health, welfare, urban renewal, transportation, oceanic discoveries, and environmental and ecological improvement.

Clerical workers are those who operate computers, keep records, take dictation, and type. Job opportunities will increase. More people will be trained to operate electronic data-processing machines. They work with many mechanical devices that do *repetitive* work (the same job over and over).

Sales employees work in retail stores, wholesale firms, insurance companies, and real estate agencies. Employment opportunities will also increase significantly. Residential and commercial construction and urban renewal influence the increased need for real estate agents. New laws for injured workers and automobile liability regulations have influenced the growth of insurance specialty occupations.

Management personnel, business officials, and *proprietors* (owners of businesses) total about 8.5 million. Employment should increase 20 percent in the near future. Management specialists are required in the growing areas of industries, business organizations, athletics, and government agencies.

Craft workers include *carpenters, toolmakers* and *die makers, instrument makers, machinists, electricians,* and *typesetters.* Industrial growth will expand 20 percent in these areas in the next decade, but technological developments will tend to limit the expansion of occupational opportunities for craft workers.

Semiskilled workers (often termed operatives) are the largest occupational work group, numbering over 14 million. They assemble goods in factories, operate machinery, and drive buses, trucks, and taxis. Demand for semiskilled workers will probably increase 13 percent in the next ten years. Factors influencing this growth are ongoing production caused by a rising population, increased economic growth, and growth of highway transportation.

Laborers (not counting those in farming and mining) make up a force of about 4 million. The U.S. Department of Labor defines their work as "labor involved to move, lift, and carry materials and tools in the work place." No increase is expected because new kinds of technological equipment are replacing the need for manual labor.

Service workers include people who assist professionals such as nurses; others act as aides. The category includes barbers, beauticians, people who wait tables, chefs, and domestic help. About 12 million are employed in these and similar occupations. The projected demand will increase this force 40 percent in the next decade.

Emphasis on Education

Education is increasing in importance in terms of finding and holding a job, and enhancing a satisfactory career in any field. Thousands of occupational titles have been grouped into 15 job clusters by the U.S. Office of Education:

1. health
2. agribusiness and natural science
3. business and office
4. public service
5. environment
6. communications and media
7. hospitality and recreation

8. manufacturing
9. marketing and distribution
10. marine science
11. personal services

12. construction
13. transportation
14. consumer and homemaking education
15. fine arts and the humanities

Unit 14

Discussion Topics on Elements of Industrial Technology

1. Tell in your own words what you think *technology* is.

2. Explain how people first started to make the products needed in daily life.

3. What was the role of Gutenberg in the development of mass production? In what area of industrial arts would Gutenberg be interested?

4. Who was Eli Whitney? What did he do for mass production?

5. When was the first automobile assembly line used? Who was the first to use it?

6. What does the term *high technology* mean?

7. Name the five major parts of a computer system.

8. What is the chip? Tell how it is made.

9. Define CPU.

10. Name the two memory units and describe the purpose of each.

11. What is software?

12. Describe the career of a computer programmer.

13. Name the two major types of robots.

14. Describe the five major parts of a basic robot.

15. What is the purpose of the robot arm in the space shuttle?

16. Define in your own words the terms *recycle*, *environment*, and *ecology*.

17. What is fiber optics, and where is this material used?

18. What does the term *laser* mean?

19. Describe five major uses of the laser in manufacturing.

20. What is a hologram?

21. What are the two major parts of mass production?

22. What is the meaning of *GNP*?

23. What is meant by secondary-materials industries?

24. Where can you find a description of every type of job or occupation in the United States?

25. About what percent of homes will be factory-built in the years to come?

26. What basic new development is greatly changing machinery manufacturing?

27. List the three most important historical developments in the growth of technology in the United States.

28. What do *VTOL* and *STOL* refer to in the aerospace area?

29. List the three factors that must be dealt with before any new idea is put into practice.

VISUAL COMMUNICATIONS TECHNOLOGY

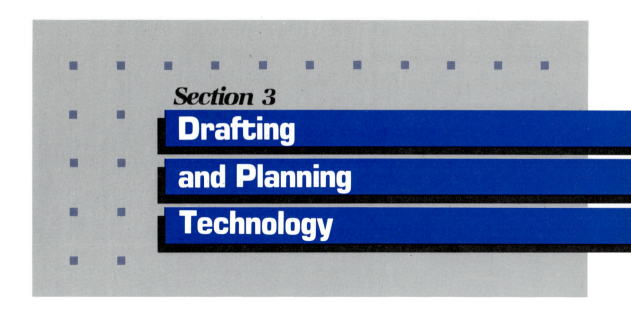

Section 3
Drafting
and Planning
Technology

Unit 15

Introduction to Drafting and Planning Technology

Imagine that it is your job to completely describe an oil refinery by using only words. You would have to describe the size and shape of every part and tell how all the parts fit together. If you could make an accurate drawing of the refinery, along with just a few words and numbers, your job would be much easier (Fig. 15-1). That is why drawing is so important to all industrial work.

Drafting begins with a good sketch of a project activity. Then a mechanical drawing is made, and dimensions (numbers that tell sizes) are added. When the drawing is complete, anyone can use it to build the project activity. A drawing is also helpful in making a list of materials to use.

No matter what language they speak, people around the world can understand a drawing. Learning to read and make drawings is important because drawings

Fig. 15-1. **Learning to read a drawing is as important as learning how to make one.** (Lockheed Aircraft Corp.)

are used often to describe the world around us.

If you become a designer, drafter, architect, or engineer, making and reading drawings will be especially important; you will be using drawings almost every day (Fig. 15-2 a and b).

Fig. 15-2 a.   Developing skills in drafting is essential. (Teledyne Post Company)

Fig. 15-2 b.   This architect is designing a dock system for lakefront vacation homes. The system is built of zinc extrusions with wood decking. (Zinc Institute, Inc.)

Unit 16

Drafting and Planning Technology-related Careers

Drafting is a language that is understood, used, and read throughout the world. It is needed by those who work in industry, manufacturing, and construction. Some of these workers are architects, building contractors, technicians, commercial artists, and industrial designers, to name only a few. Within each of these working lives, or careers, there are many different occupations and jobs. Most require education beyond high school, anywhere from two years to a college degree and sometimes more.

Drafters

Drafters are important to industries. Drafting deals with special areas, such as mechanical, electrical, aeronautical, structural, and architectural. Drafters *translate*, or explain, rough sketches and *specifications* (materials and measurements) (Fig. 16-1 a and b). They work with the carefully planned ideas of engineers, architects, and designers to make *working plans* or drawings.

These drawings are needed to make most of the things around you. Plans made by the drafter tell exactly what materials are to be used. They also tell exactly how something is to be made. See Fig. 16-2.

Drafting instruments, machines, and computers are used in making drawings. An advanced drafter also uses engineering and scientific handbooks to solve many problems. In the drafting area, the workers

Fig. 16-1 a. Many drafters are trained in vocational and technical schools. These students are participating in a national contest for young people who are members of the Vocational Industrial Clubs of America. (VICA)

Fig. 16-2. Engineers reviewing design plans. (Ford Motor Company)

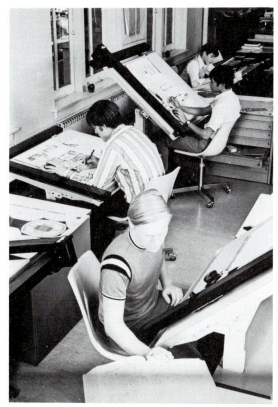

Fig. 16-1 b. Many different types of drafters are needed in industry. (Teledyne Post Company)

are called *senior drafter, drafting aide, detailer, checker,* and *tracer.*

A drafter may prepare for one of the many drafting careers by starting in courses like the one you are now taking. This person may specialize by taking advanced drafting courses in high school, technical institutes, and junior or community colleges. Home-study courses in drafting also help.

There is also on-the-job training for people who want to learn by working under skilled workers. The learner is called an *apprentice.* While learning under others, an apprentice may also take drafting and laboratory courses. Any well-trained drafter should be able to use and enjoy mathematics, physical science, and computers (Fig. 16-3).

There are over 1 million drafters in the world, but only one in twenty knows how to use computers. In the years ahead, all drafters will have to have this skill.

Industrial Designers

People who design new products are called industrial designers. They must have both artistic and drafting skills. De-

Fig. 16-3. Many drafters will need to know how to use computers in their work. (Eastman Kodak Company)

Fig. 16-4 b. A new design for an outdoor chair made of metal and plastics. (Zinc Institute, Inc.)

signers must know about materials that can be used in the final product. Designers may be self-employed or may work for design firms or manufacturing concerns (Fig. 16-4 a and b).

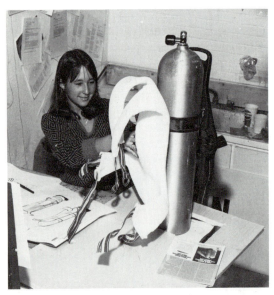

Fig. 16-4 a. This student designer is working on a scuba backpack. The designer is using a soft plastic material to hold the tank on the diver's back. (Zinc Institute, Inc.)

Technical Illustrators

The technical illustrator or illustrator-drafter must be able to produce all kinds of pictorial illustrations. These include everything from cartoons and pictorial technical illustrations to all kinds of visual presentations (Fig. 16-5 a, b, and c). Much of the work of technical illustrating can now be done with computers. The technical illustrator must have both artistic and technical skills in design, drawing, and drafting.

Engineers

Engineers offer a great deal to the welfare, technology, and defense of our nation. They design and plan industrial machin-

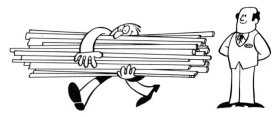

Fig. 16-5 a. The technical illustrator must also be able to draw cartoons.

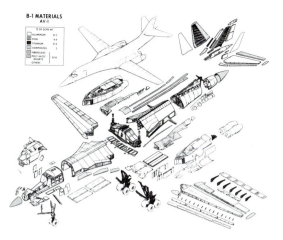

Fig. 16-5 b. This exploded view of an aircraft is typical of what a technical illustrator must be able to produce. (Rockwell International)

Fig. 16-6. These student engineers are checking the electronic circuits of communication equipment.

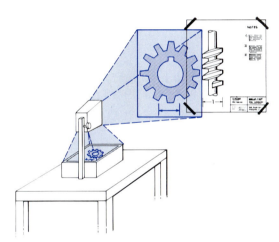

Fig. 16-5c. Technical illustrators must be able to make transparencies to use with an overhead projector.

Fig. 16-7. A chemical engineer making tests on petroleum products. (Gulf Oil Corporation)

ery and equipment. New designs or plans are needed to mass-produce products.

Engineers help make equipment to explore outer space. They design equipment that helps us find out about resources that are deep in the ocean. Highways and rapid-transit systems are also the result of engineers' work. And the research that made these new systems came after engineers used drafting and mathematics.

Other products that also result from engineers' planning include automobiles, refrigerators, and air-conditioning units, to name just three.

There are numerous special engineering areas. People may enter them by going to a college or university (Fig. 16-6) where engineering courses are offered. Some of the broad areas are *aerospace, agricultural, ar-*

chitectural, chemical, civil, electrical, geological, industrial, mechanical, metallurgical, mining, and *petroleum engineering* (Fig. 16-7). Young people considering careers in engineering should learn as much as they can about the field. They should study college catalogs and read information from professional associations.

In order to enter engineering jobs, a person usually needs a bachelor's degree in the field. Students must study drafting and have a talent and an interest in science, mathematics, and computers.

Many engineering jobs require education beyond a college degree. In engineering teaching and research, much more study is needed. And in areas such as nuclear engineering and atomic engineering, a master's degree or a doctorate is needed.

There are about 300 different engineering colleges, universities, and schools in the United States alone. In the first two years, the student will study basic sciences, mathematics, physical sciences, humanities, social studies, English, drafting, and computer science. The last two years are given to engineering science and areas that most interest the individual student.

Some institutions offer five- or six-year *cooperative* plans. Each student spends alternate *semesters* (terms) in the engineering school and in a job in industry.

Students often begin as trainees or assistants to professional engineers. Some large companies have special training programs. These programs introduce new engineers to the ways in which the companies work.

Technicians

Careers for technicians are growing rapidly because of the important work technicians do. Technicians work with engineers, scientists, and drafters. All these professional people should understand and enjoy the general field of drafting.

Technicians should understand and be able to use drafting and industrial processes. They must be able to use scientific and mathematical theories and have some talent for general mechanics. Technicians usually run experiments and tests. These workers set up and operate instruments and make calculations. They help engineers and scientists to plan and make new or experimental equipment and models.

A technician often makes drawings and sketches (Fig. 16-8). Next, a model may be built. Technicians work in jobs that are a part of production. They work out measurements and materials and manufacturing methods. Often, they carry out plans set up by engineers and production departments.

Technicians can prepare themselves for just about all the fields in which engineers work. Usually courses are taken for one or two years after high school. Technicians can attend technical institutes, junior and community colleges, and area vocational schools. These area schools are often operated along with programs in colleges

Fig. 16-8.   This student is planning to become a technician in the graphic arts industry. (VICA)

and universities that have engineering schools.

There are some four-year college programs in industrial technology. These schools offer more courses. The humanities, business, and certain types of technology courses in the engineering area are often required. Four-year programs lead to a bachelor's degree in engineering or industrial technology.

About six technicians are needed for every engineer and scientific person. More than 1 million new technicians will be needed in the near future.

Architects

Architecture is a challenging profession. Architects plan and design buildings and other structures to be safe, useful, and beautiful (Fig. 16-9).

Architects work with other professional people, such as engineers, city planners, landscape architects, and construction superintendents. An architect must know a great deal about buildings and building materials. An architect must also be able to take an idea and make it come true through careful planning and designing (Fig. 16-10).

There are about 40 000 registered architects in the United States. About 40 percent of them are self-employed or work with partners. The rest work for firms or companies of various sizes.

A person must have a license to practice architecture. In order to get a license, an architect must graduate from a university or college. This is usually followed by three years of on-the-job experience with a registered architect or an architectural firm. *Accredited* (approved) programs for architects in the United States have a five-year course of study leading to the bachelor of architecture degree.

A good job outlook in the near future is the result of new building in cities and suburbs, both in homes and in businesses. Other jobs related to architecture include *drafters*, *specification writers*, *printmakers*, and *office employees*. All of these jobs offer fine working conditions.

Fig. 16-9. Architects must be able to design all kinds of structures, including homes, commercial buildings, roads, dams, bridges, and many other units. (California Redwood Association)

Fig. 16-10. The architect has an interesting and challenging profession in designing buildings and other structures. (Inland Steel Company)

Metric Measurement and Standards

Various systems of measurements have been used throughout history. The system now used in the United States is called the *customary system* of measurement. For example, you can buy a *pound* of nails, a *quart* of paint, and several *feet* of lumber.

There was a time, however, when there was no accurate system of measurement. Instead, people used such things as three kernels of barleycorn and the distance from the elbow to the end of the hand as standards. The early "foot" was actually the length of a person's foot. Feet are many different sizes, of course, so you can see how measuring might have been a problem.

In the thirteenth century, King Edward I of England ordered a *standard* measuring stick made of iron. This was to be used as a *yardstick* for the entire kingdom. Over the years, a system of measurement based on inches and pounds was set up. It became known as the *U.S. customary system of measurement.* The United States has over 80 different units of weights and measures (Fig. 17-1).

In 1773, the French government adopted a new and different system of measurement. It was called the *metric system,* based on the *metre* (or meter). *Metre* is the official spelling in almost all the English-speaking countries using the SI metric system. SI stands for *Système International.* The United States is slowly becoming a metric country. But because no official policy has been formed by the U.S. government, the metric units *metre* and *litre* are usually spelled *meter* and *liter.* (See Fig. 17-2.)

The *meter* was supposed to be one ten-millionth (1/10 000 000) of the distance

SOME COMMON UNITS

LENGTH	MASS	VOLUME
	METRIC	
meter	kilogram	liter
	CUSTOMARY	
inch	ounce	fluid ounce
foot	pound	teaspoon
yard	ton	tablespoon
fathom	grain	cup
rod	dram	pint
mile		quart
furlong		gallon
		barrel
		peck
		bushel

Fig. 17-1. Comparison of the metric and some of the U.S. customary common units of measurement.

from the North Pole to the Equator. That distance was made by measuring on a straight line that runs along the surface of the Earth through Paris, France. But this measurement proved to be *inaccurate* (not exact). Now the meter is measured by an exact number of wavelengths of red-orange light given off by the element *krypton 86.*

The metric system is based on decimal units of 10. The metric system has seven basic units (Fig. 17-3). There are also many other units based on the basic units (Fig. 17-4). For everyday use, however, only three measurements are important to know: (1) the *meter,* for length, (2) the *kilogram,* for mass (weight), and (3) the *liter,* for volume.

It is fairly easy to remember that the meter is a little longer than the yard. The kilogram is a little more than 2 pounds. And

EVERYDAY UNITS

QUANTITY	UNIT	SYMBOL
Length	millimeter (one thousandth of a meter)	mm
	centimeter (one hundredth of a meter)	cm
	meter	m
	kilometer (1000 meters)	km
	international nautical mile (1852 meters)	n mile
Area	square centimeter	cm^2
	square meter	m^2
	hectare (ten thousand square meters)	ha
Volume	cubic centimeter	cm^3
	cubic meter	m^3
	milliliter (one thousandth of a liter)	ml
	centiliter (one hundredth of a liter)	cl
	deciliter (one tenth of a liter)	dl
	liter	l
	hectoliter (100 liters)	hl
Weight*	gram (one thousandth of a kilogram)	g
	kilogram	kg
	ton (1000 kilograms)	t
Time	second	s
	minute	min
	hour	h
Speed	meter per second	m/s
	kilometer per hour	km/h
	knot (international nautical mile per hour)	kn
Power	watt	W
	kilowatt (one thousand watts)	kW
Energy	kilowatt-hour	kWh
Electric potential difference	volt	V
Electric current	ampere	A
Electric resistance	ohm	Ω
Frequency	hertz	Hz
Temperature	degree Celsius†	°C

*Strictly, the gram, kilogram, and ton are units of mass. For most people and for ordinary trading purposes, the distinction between weight and mass is unimportant.
†This unit is often known in the United States as the "degree centigrade." To avoid confusion with a unit used in some other countries having the same name but used to denote fractions of a right angle, it has been agreed internationally that the name "degree centigrade" shall be replaced by "degree Celsius."

Fig. 17-2. Everyday metric units with symbols.

BASE UNIT	SYMBOL	DEFINITION
meter (length)	m	The measure equal to a certain number of wavelengths of light given off by the krypton-86 atom.
kilogram (mass)	kg	The measure equal to the mass of the standard kilogram artifact located at the International Bureau of Weights and Measures. The kilogram is also used to measure weight.
second (time)	s	The measure equal to a certain number of *oscillations* (back and forth movements) of the cesium atom in an atomic clock.
ampere (electric current)	A	The measure equal to the amount of current in two wires a certain distance apart that results in a specific force between the two wires.
candela (luminous intensity)	cd	The measure equal to the amount of light given off by platinum at its freezing point. At its freezing point, platinum is hot and glows.
mole (amount of substance)	mol	The measure equal to the number of particles contained in a certain amount of carbon. The mole is used only for very scientific measurements.
kelvin (temperature)	K	The measure of a certain fraction of the temperature of water at its triple point. The triple point of water is the temperature at which it exists as a solid, liquid, and vapor. The kelvin is used for special temperature measures. For practical purposes, the Celsius scale is used. The Celsius scale is 0 at the freezing point and 100 at the boiling point of water.

Fig. 17-3. Seven base units.

TABLE OF SI UNIT PREFIXES

MULTIPLE OR SUBMULTIPLE	PREFIX	SYMBOL	PRONUNCIATION*	MEANS
$1\ 000\ 000\ 000 = 10^9$	giga	G†	jig' a (a as in about)	One billion times
$1\ 000\ 000 = 10^6$	mega	M†	as in megaphone	One million times
$1000 = 10^3$	kilo	k†	as in kilowatt	One thousand times
$100 = 10^2$	hecto	h	heck' toe	One hundred times
$10 = 10^1$	deka	da	deck' a (a as in about)	Ten times
BASE UNIT $1 = 10^0$				
$0.1 = 10^{-1}$	deci	d	as in decimal	One tenth of
$0.01 = 10^{-2}$	centi	c†	as in centipede	One hundredth of
$0.001 = 10^{-3}$	milli	m†	as in military	One thousandth of
$0.000\ 001 = 10^{-6}$	micro	μ†	as in microphone	One millionth of
$0.000\ 000\ 001 = 10^{-9}$	nano	n†	nan' oh (an as in ant)	One billionth of

*The first syllable of every prefix is accented to make sure that the prefix will keep its identity. For example, the preferred pronunciation of kilometer places the accent on the first syllable, not the second.
†Most commonly used and preferred prefixes. Centimeter is used mainly for measuring the body, clothing, sporting goods, and some household articles.

Fig. 17-4. Table of prefixes.

COMMON CONVERSIONS

LENGTH

Customary to Metric (exact)

1 inch = 25.4 millimeters
1 inch = 2.54 centimeters
1 foot = 30.48 centimeters
1 foot = 0.3048 meter
1 yard = 91.44 centimeters
1 yard = 0.9144 meter
1 mile = 1.609 kilometers (approx.)

Metric to Customary

1 millimeter = 0.039 37 inch
1 centimeter = 0.3937 inch
1 meter = 39.37 inches
1 meter = 3.2808 feet
1 meter = 1.0936 yards
1 kilometer = 0.621 37 mile

VOLUME

Customary (U.S.) to Metric

1 fluid ounce = 29.57 milliliters
1 pint (liq.) = 473.18 milliliters
1 quart (liq.) = 0.9464 liter
1 gallon (liq.) = 3.7854 liters

Metric to Customary (U.S.)

1 milliliter = 0.0338 fluid ounce
1 liter = 2.1134 pint (liq.)
1 liter = 1.0567 quarts (liq.)
1 liter = 0.2642 gallon (liq.)

WEIGHT (MASS)

Customary to Metric

1 ounce (dry) = 28.35 grams
1 pound = 0.4536 kilograms
1 short ton (2000 lb) = 907.2 kilogram
1 short ton (2000 lb) = 0.9072 metric ton

Metric to Customary

1 gram = 0.035 27 ounce
1 kilogram = 2.2046 pounds
1 metric ton = 2204.6 pounds
1 metric ton = 1.102 tons (short)

AREA

Customary to Metric

1 square inch = 645.16 square millimeters
1 square inch = 6.5416 square centimeters
1 square foot = 929.03 square centimeters
1 square foot = 0.0929 square meter
1 square yard = 0.836 square meter
1 acre = 0.4047 square hectometer
1 acre = 0.4047 hectare
1 square mile = 2.59 square kilometers

Metric to Customary

1 square millimeter = 0.00155 square inch
1 square centimeter = 0.1550 square inch
1 square meter = 10.764 square feet
1 square meter = 1.196 square yards
1 square hectometer = 2.471 acres
1 hectare = 2.471 acres
1 square kilometer = 0.386 square mile

Fig. 17-5. Common conversions.

the liter is a little more than a quart. Figure 17-5 compares some common measurements.

In the metric system the prefixes giving the multiples of the unit are the same, whether they are the gram, liter, or meter: *kilo* = 1000, *hecto* = 100, *deka* = 10, *deci* = $^1/_{10}$, *centi* = $^1/_{100}$, and *milli* = $^1/_{1000}$ (Fig. 17-4). For everyday use, only two prefixes are generally needed: the kilo (1000) and the milli ($^1/_{1000}$). For example, 1 inch = 25.4 millimeters (the symbol is *mm*).

In woodworking, the 25.4 millimeters (mm) is usually rounded off to 25 millimeters. In drafting and metalworking, the millimeter measurements are most often carried to two places past the decimal point. The customary standard of measurement still used in the United States is based on the yard. Our standard is given in inches and feet. The inches divide into common fractions: $^1/_2$, $^1/_4$, $^1/_8$, $^1/_{32}$, and $^1/_{64}$. (See Fig. 17-6.)

The decimal-inch system of the cus-

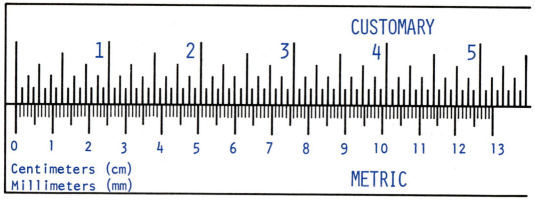

Fig. 17-6. Metric and U.S. customary equivalent measures of length.

tomary system is used in *precision* (highly accurate) measuring instruments. The micrometer divides the inch into tenths, hundredths, thousandths, and ten-thousandths. All precision measuring instruments, whether customary or metric, use the decimal system. For making measurements, you must learn to read a rule divided either into the customary or the metric system.

Reading a U.S. Customary Rule

A worthwhile drawing must be accurate. Therefore, you must know how to read a rule and make measurements. You probably have measured in feet (') and in inches ("). Now you will measure in parts of an inch. Most of the rules used in drawing are divided into *sixteenths* (¹/₁₆) of an inch.

Look at an inch and learn to read it (Fig. 17-7). Notice that (1) the longest line between the 0 and the 1-inch mark is the ¹/₂-inch mark, (2) the next longest lines are the ¹/₄-inch marks, (3) the next longest (getting shorter) lines are the ¹/₈-inch marks, and (4) the shortest lines are the ¹/₁₆-inch marks. Notice that ⁴/₁₆ equals ²/₈, or ¹/₄. If you want to measure a line, you can count the number of sixteenths (¹/₁₆). In Fig. 17-8 you see a measurement of 12 sixteenths (¹²/₁₆) past the 1-inch mark, or 1³/₄ inches. Read the rule in Fig. 17-9.

Reading a Metric Rule

A metric rule is usually available in meter, half-meter, 300-millimeter, or 150-millimeter lengths. Remember that a meter is divided into 100 centimeters and 1000 millimeters.

Usually the rule is marked with millimeters. The *divisions* (lines) between each of the numbered lines stand for millimeters (Fig. 17-10). You can see that 1 inch really equals 25.4 millimeters (or just about 25 millimeters). And 2 inches really equals 50.8 millimeters (or approximately 51 millimeters) as shown in Fig. 17-6.

Measure the lines shown in Fig. 17-11 in both U.S. customary measurements and in metric measurements. Remember that in the metric system, accurate measurements are always given in millimeters or meters. The *decimeter* and *centimeter* are not used very often.

Kinds of Metric Drawing

The three basic kinds of drawings used with the SI system are as follows:

Dual Dimensioned. In this kind of drawing, the size is shown in both the customary (inch) and the metric (millimeter) size for each dimension. If the design is in inches, the metric *equivalent* (equal dimension) is shown below the line or in

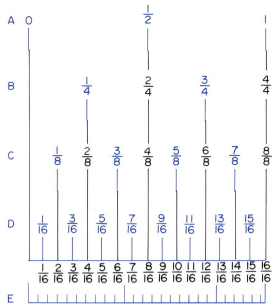

Fig. 17-7. The scale is divided into sixteenths of an inch. A millimeter scale is shown on line E.

Fig. 17-8. A dimension shows the length of a line.

brackets, [ ]. If the design is in metric, the inch equivalent is shown below the line or in brackets. With a dual-dimensioned drawing, a product can be made easily in either the customary or the metric system (Fig. 17-12).

Metric with Readout Chart. Most manufacturers are designing all new products in

the metric system. Only metric measurements are shown on the drawings themselves. A small *conversion* chart is often added to the drawing. This conversion chart shows only the dimensions on the drawing itself in millimeters and in inches. Anyone can easily produce the product in either the metric or the customary system (Fig. 17-13).

All-Metric. The goal in designing and planning will be an all-metric drawing, with only metric measurements.

Metric Standards

In industrial education, changing over to the metric system means a great deal more than changing the way people measure. Changing from inches to millimeters, pounds to kilograms, and quarts to liters is called a "soft" conversion. In other words, the real size of an object does not change. A board that is 1 inch is also 25.4 millimeters. A quart is equal to 0.948 liter.

SI is a measurement system only. Substituting metric equivalents for materials made in customary units is different from using metric standards for all materials. Getting people and industries to use metric standards is the job of a worldwide organization called ISO (International Standards Organization). It is also the job of ANSI (American National Standards Institute), an American organization. These organizations are planning standards for materials that will be designed and produced in metric measurements, not just converted from customary units. This is

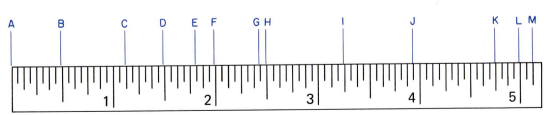

Fig. 17-9. Read the distance between the letters on this U.S. customary ruler.

READING A METRIC SCALE

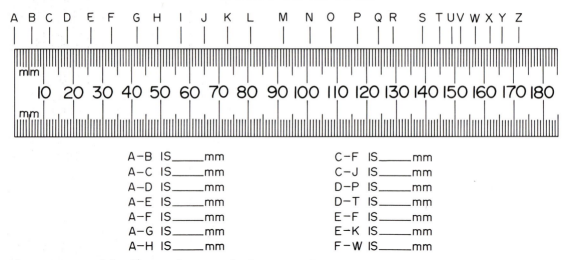

A–B IS＿＿＿mm
A–C IS＿＿＿mm
A–D IS＿＿＿mm
A–E IS＿＿＿mm
A–F IS＿＿＿mm
A–G IS＿＿＿mm
A–H IS＿＿＿mm

C–F IS＿＿＿mm
C–J IS＿＿＿mm
D–P IS＿＿＿mm
D–T IS＿＿＿mm
E–F IS＿＿＿mm
E–K IS＿＿＿mm
F–W IS＿＿＿mm

Fig. 17-10.   Read the distance between the letters on this metric ruler.

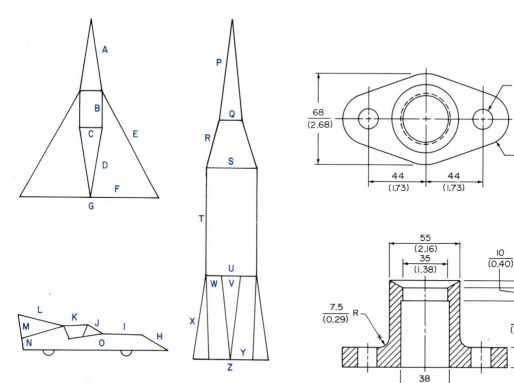

Fig. 17-11.   Measure the length of each line with U.S. customary measurements to one-sixteenth of an inch and the metric measurement to 1 millimeter.

Fig. 17-12.   Typical dual-dimensioned drawing.

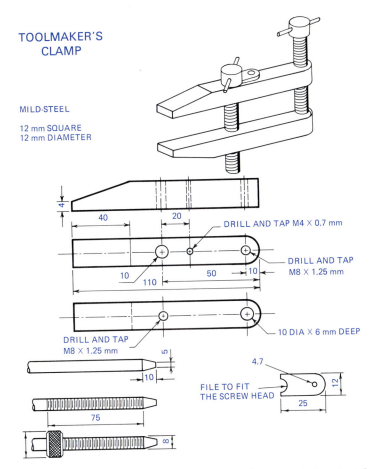

TOOLMAKER'S CLAMP

MILD-STEEL

12 mm SQUARE
12 mm DIAMETER

DRILL AND TAP M4 × 0.7 mm

DRILL AND TAP M8 × 1.25 mm

DRILL AND TAP M8 × 1.25 mm

10 DIA × 6 mm DEEP

FILE TO FIT THE SCREW HEAD

METRIC (mm)	CUSTOMARY ('')
4	0.16
4.7	0.19
5	0.20
6	0.24
8	0.32
10	0.40
12	0.48 (1/2'')
20	0.79
25	0.99 (1'')
40	1.58 (1 1/2'')
50	1.97 (2'')
75	2.95 (3'')
110	4.34
M4 × 0.7 mm	6-32 NC
M8 × 1.25 mm	5/16

Fig. 17-13. A metric drawing with a readout chart showing the equivalents of the metric dimensions in inches.

called "hard" conversion. An example of a hard conversion would be to measure and cut a board 100 millimeters wide and 510 millimeters long instead of 4 inches wide and 20 inches long.

In this book, the units on metrics show some of these metric standards. The ISO paper sizes used in drafting and the ISO thread standards used in metalworking are just two examples. When metric stand-

ards are given to some materials, the real sizes may change. For example, A4 paper (210 × 297 millimeters) is a little narrower and longer than the paper you are used to seeing, which is 8½ × 11 inches (216 × 280 millimeters). And while an M12 × 1.75 metric bolt seems to be about the same size as a ½-13 UNC bolt, the two cannot be used *interchangeably* (one for another).

Lines—Their Meanings and Uses

Each line on a drawing has a special meaning (Fig. 18-1). Lines are combined to make drawings in the same way that letters are combined to make words. It is important that you understand the *alphabet of lines* so that others can understand your drawings.

Construction Line

A construction line is a lightly drawn solid line. It shows the shape of the object. The line should be made so light that little or no erasing is needed. Use a 3H, 4H, or 5H pencil. Construction lines are also used as light guidelines for lettering.

Border Line

A border line is a very heavy solid line. It is used as a border or a frame for a drawing. Use an H pencil.

Visible, Outline, or Object Line

This solid heavy line shows all the edges and surfaces that can be seen from the outside of the object. It should be a clear, sharp line made with an H or 2H pencil.

Invisible, or Hidden, Line

Invisible lines are made up of short dashes about ¹/₁₆ inch (3 millimeters) long with spaces about ¹/₁₆ inch (1.5 millimeters) between. They are slightly lighter than visible lines and show all the *invisible* (unseen) object lines, edges, or surfaces. The first dash should start with the visible, or ob-

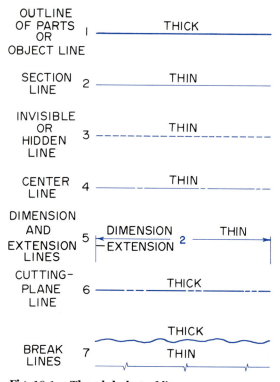

Fig. 18-1. The alphabet of lines.

ject, line. The last dash should also touch the visible, or object, line.

Center Line

This medium-light line is made up of a long dash about ³/₄ to 1 inch (19 to 25 millimeters) long, a short space about ¹/₁₆ inch (1.5 millimeters), a short dash about ¹/₈ inch (3 millimeters), and another short space. It is used to divide a drawing into equal, or *symmetrical*, parts. It is also used to find the centers of arcs and circles. For the centers of arcs, two short *intersecting* (crossing) dashes may be used.

Extension Line

An extension line is a thin, light, solid line. It makes longer, or *extends*, the object line. It should start about $\frac{1}{16}$ inch (1.5 millimeters) before the object line, and it goes beyond the last dimension line about $\frac{1}{8}$ inch (3 millimeters). The extension line is used with dimension lines to show the dimensions of the object.

Dimension Line

A dimension line is a thin, light, solid line with a space in the center for placing the dimensions. There is always an arrowhead at one or both ends of this line. A dimension line should start at about $\frac{3}{8}$ to $\frac{1}{2}$ inch (10 to 13 millimeters) away from the object line. There should be about $\frac{3}{8}$ inch (10 millimeters) between the dimension lines.

Section Line

A section line is a very light solid line. It is drawn as a slant line at an angle of about 45° right or left. It is used to show a cut section on a sectional view.

Cutting-plane Line

This heavy line is made up of a series of long dashes separated by two short dashes. An arrowhead is placed on either end to show the direction of the section that is shown. This line is used to show the cutting plane in making a sectional view.

Long Break Line

This medium-weight ruled line has freehand zigzags. It is used to show a break in an object that is too big to be put on the page. The long break line is used to shorten the length of the object.

Short Break Line

This medium-heavy wavy line is used to show that part of the object has not been drawn. It also shows that an outer surface has been cut away. Short break lines are always drawn freehand.

Figure 18-2 shows the weights of the lines used in sketching. Keep them in mind when you draw.

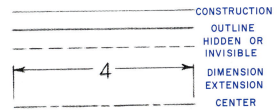

Fig. 18-2. The lines of freehand sketching are the same as in mechanical drawing. However, sketched lines do not have to be so accurately made.

Unit 19

Sketching and Freehand Drawing

A type of drawing that you should learn to do well is called *sketching*. A sketch is a way to communicate with people by making simple freehand drawings (Fig. 19-1).

Often you can explain ideas better by using a sketch than you can by using words, especially if you are going to work with materials and mechanical devices.

The only tools you need for sketching are a pencil and paper (Fig. 19-2).

A sketching pencil is usually 2H or H. Hold the pencil rather loosely in your right hand about 1 inch from the point (Fig. 19-3). Pull the pencil toward you as you sketch. Never push it. Use arm movement, not wrist and finger movements. Draw a light wavy line and put breaks in it (Fig. 19-4). Do not try to draw a continuous line.

Horizontal Lines

Place points for the beginning and the end of the line. Try to look at the whole line to keep it straight. When you get to the end of the line, go back over it to darken it (Fig. 19-5).

Vertical Lines

Draw from the top down. Pull the pencil toward you. Get used to making all similar lines on the paper in the same position (Fig. 19-6).

Sloping, or Inclined, Lines

Draw these lines from the top down. Draw them either from the upper right to the lower left or from the upper left to the lower right. (See Fig. 19-5.)

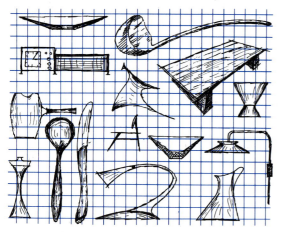

Fig. 19-1.　Freehand sketches of project activity designs.

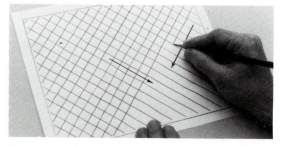

Fig. 19-3.　How to hold a pencil for sketching.

Fig. 19-4.　A sketched line.

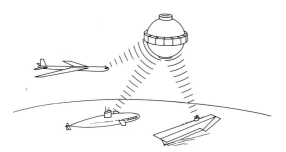

Fig. 19-2.　A sketch of a navigation satellite for planes, ships, and submarines. The satellite is like a small planet revolving around the earth. It can be used as an aid to navigation because its position at any time can be known.

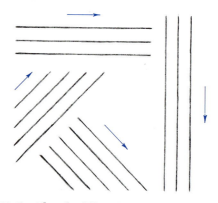

Fig. 19-5.　Sketched lines in different directions.

Square or Rectangle

Draw light *intersecting* (crossing) center lines. Then mark off the width and the height of the rectangle on these lines. Lightly sketch vertical and horizontal lines to intersect at the corners. Then darken with a heavier line (Fig. 19-7).

Angles

Draw a right angle from one point. Draw one line to the right and the other line down. To draw a 30° or 60° angle, divide the right angle into three equal parts (Fig. 19-8). Divide the right angle in half to make a 45° angle.

Triangles

Sketch light vertical and horizontal lines that intersect at a corner. Then mark off the height of the triangle on the vertical line. Mark the base of the triangle on the horizontal line. Darken the three sides with a heavy line.

Circles

Draw light center lines and a light square. Then start at each center line and work outward. Complete all four arcs. Darken to complete the circle (Fig. 19-9). A trick of the trade is to use a pencil and a string as a compass (Fig. 19-10). You may also use your finger and a pencil as a compass (Fig. 19-11).

Fig. 19-7. A sketched square.

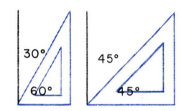

Fig. 19-8. Common angles that have been sketched.

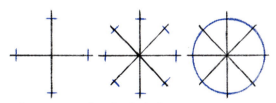

Fig. 19-9. A sketched circle.

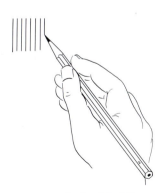

Fig. 19-6. One method of holding a pencil for sketching vertical lines.

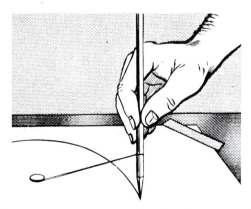

Fig. 19-10. Using a pencil and string as a substitute for a compass.

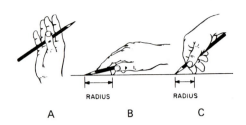

RADIUS   RADIUS

A   B   C

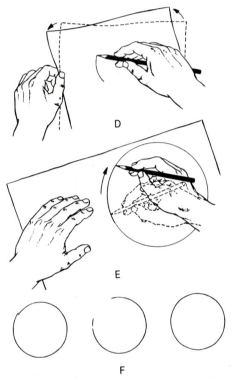

D

E

F

Fig. 19-11.   Suggestions for sketching a circle or arc. (*a*) Position of pencil in hand. (*b*) For a long radius, turn on middle finger. (*c*) For a short radius, turn on index finger. (*d*) Hold hand steady and rotate paper. (*e*) Move hand in a long stroke. (*f*) Samples of sketched circles.

Oval or Ellipse

Draw light center lines. Mark off the size of the oval and lightly sketch a box. Draw from each center line outward. Draw lines on the long side of the oval that are longer than those on the short side. Then darken the figure (Fig. 19-12).

Octagon (Eight Sides)

Draw a center line. Mark off the size of the octagon. Lightly draw a box. Use light lines to divide each side of the center line vertically and horizontally into equal parts. Draw eight sides like those shown in Fig. 19-13.

Hexagon (Six Sides)

Draw a center line. Draw inclined lines of 30° and 60°. Sketch a light circle for the inside of the hexagon. Connect every other line to form the hexagon (Fig. 19-14).

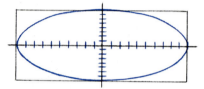

Fig. 19-12.   A sketched ellipse.

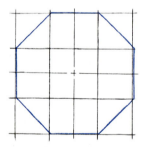

Fig. 19-13.   Method of sketching an octagon in a square.

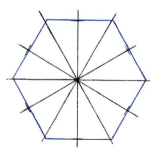

Fig. 19-14.   A sketched hexagon.

Isometric Drawings

Figure 19-15 shows the way to sketch an isometric rectangle.

Cabinet Drawings

Figure 19-16 shows a jewel box drawn as a cabinet sketch.

Customary Shop Drawings

Drawings using inches are easier to make and more exact if they are done on cross-section paper that has eight squares to the inch (Fig. 19-17 a and b). Each square measures $1/8 \times 1/8$ inch. Eight squares therefore equal 1 inch. Drawings done to this scale are full-scale drawings when each $1/8$-inch square equals $1/8$ inch of the project. Many project activities can be drawn full scale (Fig. 19-18).

Large projects or objects, such as automobiles and furniture, cannot be drawn full scale. To make reduced-scale drawings on cross-section paper, let each $1/8$-inch square stand for a larger dimension (Fig. 19-19). For example, if you want a one-half-scale drawing, let each square on the paper stand for $1/4$ inch. Then four squares will equal 1 inch.

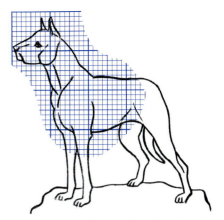

Fig. 19-17 a. Freehand sketch of a dog on squared paper.

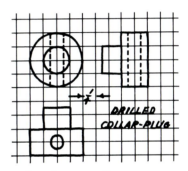

Fig. 19-17 b. Cross-section paper saves time when making a sketch.

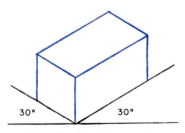

Fig. 19-15. Sketch of an isometric rectangle.

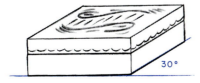

Fig. 19-16. A cabinet sketch of a jewel box.

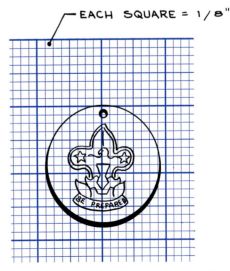

Fig. 19-18. Full-scale drawing of a medal on squared paper.

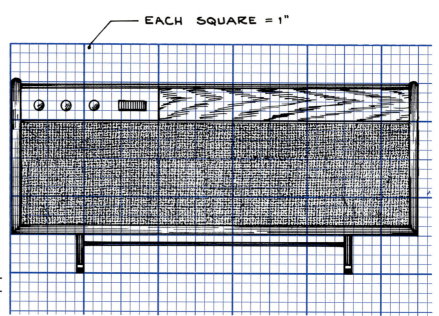

EACH SQUARE = 1"

Fig. 19-19. Reduced-scale drawing of furniture on squared paper.

In one-quarter-scale drawing, each ⅛-inch square equals ½ inch. Two squares will then equal 1 inch. For a one-eighth-scale drawing, let each ⅛-inch square equal 1 inch. Then eight squares will equal 8 inches.

Smaller-scale drawings, such as those needed for buildings and bridges, may be done by making each square stand for a certain number of feet (Fig. 19-20). If each ⅛-inch square equals 1 foot, then ⅛ inch equals 1 foot. If two squares equal 1 foot, then ¼ inch equals 1 foot.

Sometimes smaller objects, such as watch parts and jewelry, must be drawn larger than full scale. In this case, each ⅛-inch square should equal a lesser dimension. For example, in a double-sized drawing, each ⅛-inch square equals ¹⁄₁₆ inch. Then 16 squares (or 2 inches) will equal 1 inch. Larger drawings may also be made by letting each ⅛-inch square stand for a smaller dimension, such as ¹⁄₃₂ inch, ¹⁄₆₄ inch, and so on.

Circles and parts of circles can be sketched freehand. They may also be drawn with either a pencil compass or a circle template (Fig. 19-21). Isometric cross

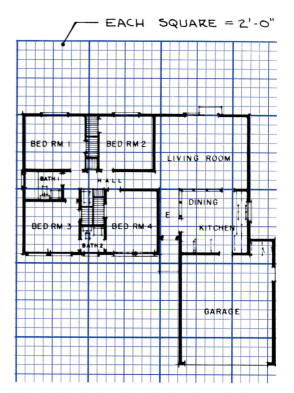

EACH SQUARE = 2'-0"

Fig. 19-20. Reduced-scale drawing of a house plan on squared paper.

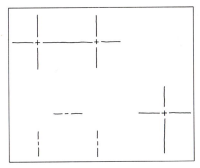

1 Sketch centerlines lightly for all views

2 Extend projectors and block in views

3 Establish points and draw arcs

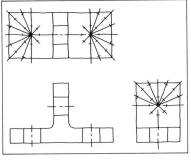

4 Darken outlines and draw hidden lines

Fig. 19-21. Sketching a simple object.

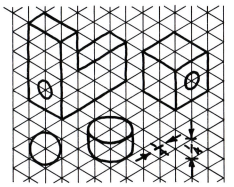

Fig. 19-22. Sketching with isometric cross-section paper.

section paper is also available for sketching (Fig. 19-22).

Metric Shop Drawings

In the metric system, cross-section paper has small squares measuring 1 millimeter by 1 millimeter (1 mm × 1 mm). Slightly heavier lines stand for squares that are 10 millimeters by 10 millimeters (Fig. 19-23). Designs may be drawn to large or small scales on metric paper. You use metric cross-section paper in the same way as inch cross-section paper.

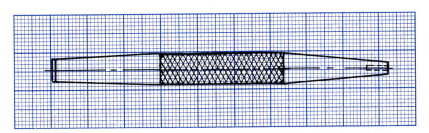

Fig. 19-23. Metric cross-section paper.

Unit 20

Lettering

The lines of sketches and drawings do not always explain your ideas completely. Words and letters are added to show information about what kinds of materials to use and how they should be assembled. Dimension numbers must be added to show sizes. The addition of letters, words, and numbers to a drawing is called *lettering*.

Good lettering has to be learned through practice. There are two types of letters: *single stroke Gothic vertical* letters and *single-stroke Gothic inclined* (slanted) letters. In single-stroke lettering, the letters are made with a single *stroke* (movement) of the pencil. Vertical letters are made at right angles to the horizontal. Inclined letters are made at an angle of about $67\frac{1}{2}°$ (Fig. 20-1).

Draw horizontal guidelines with a 4H or 5H pencil. These will help you keep the letters the same size. The lines will be straight, too. Guidelines may be left on the drawing.

To do a good job of lettering, you need a good sharp pencil about HB, H, or 2H. Hold the pencil as shown in Fig. 20-2. Sometimes light vertical guide lines are put in to help in spacing the letters. All lines should be made about the same way as in sketching.

All the letters of the alphabet may be placed in three groups as follows:

1. The group made up of *horizontal* and/or *vertical* lines includes E, F, H, I, L, and T.
2. The group made up of *vertical, hori-*

ABCDEFGHIJKLMNOPQRSTUVWXYZ

1234567890 $\frac{1}{2}$ $\frac{3}{4}$ $\frac{5}{8}$ $\frac{7}{16}$

ABCDEFGHIJKLMNOPQRSTUVWXYZ

1234567890 $\frac{1}{2}$ $\frac{3}{4}$ $\frac{5}{8}$ $\frac{7}{16}$

ABCDEFGHIJKLMNOPQRSTUVWXYZ

1234567890 $\frac{1}{2}$ $\frac{3}{4}$ $\frac{5}{8}$ $\frac{7}{16}$

ABCDEFGHIJKLMNOPQRSTUVWXYZ   abcdefghijklmnopqrstuvwxyz

.498 −.002   .498 $^{+.000}_{-.002}$ .498 .496 $\frac{1}{2}$ $\frac{3}{4}$ $\frac{5}{8}$ $\frac{7}{16}$

Fig. 20-1.   Inclined Gothic lettering.

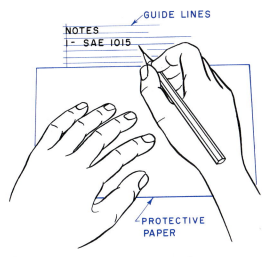

Fig. 20-2. Lettering with a pencil.

zontal and *slant* lines includes A, K, M, N, V, W, X, Y, and Z.

3. The group made up of *straight* and *curved* lines includes B, C, D, G, J, O, P, Q, R, S, and U. The basic letter in this group is O. It is necessary to form a good curved line to make all of these letters (Fig. 20-3).

You may find that a few letters in one group are harder to draw than the others. Practice each letter in each group until you master all of them. Study the shapes and sizes of the letters. All letters are made on the basis of six units in height. They may change in width from a single straight line, like the letter I, to the full six units in width in M and W (Fig. 20-4).

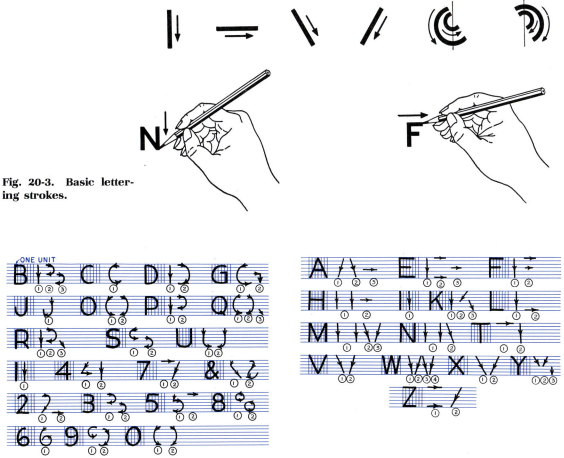

Fig. 20-3. Basic letter-ing strokes.

Fig. 20-4. Correct strokes in lettering single-stroke Gothic vertical letters.

In basic drawing, a simple way to form letters is to make all but four of them the same width. Make the J a little narrower than the others, the I just one line wide, and the W and M a little wider than the rest.

After you have learned to form each letter correctly, you can put them together into words and sentences. Letters are not equally spaced. If they were, some would appear farther apart than others. Closed-type letters, such as M and N, must be spaced farther apart than open-type letters, such as I, L, T, and J (Fig. 20-5). If all letters were spaced alike, it would look as if some letters used up too much space. As you form words, you will begin to understand the need for different spacing.

When you are lettering fractions, the fraction should be twice the height of the whole number (Fig. 20-6). The space between letters should be about one-fourth the height of the letter. The space between words should equal about one letter. The space between lines should equal about the height of the letter. Use a 1/8-inch (3-millimeter) letter for notes and for information in the title block.

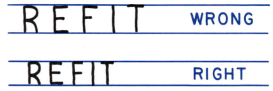

Fig. 20-5. The spacing of letters. Note that when the letters are equally spaced, the word does not appear as a unit.

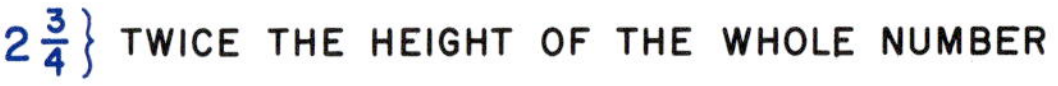

$2\frac{3}{4}$ } TWICE THE HEIGHT OF THE WHOLE NUMBER

Fig. 20-6. The method of lettering fractional numbers.

Unit 21

Dimensions, Conventions, and Symbols

A drawing or sketch is used to show the shape of an object, but the lines are not usually used to indicate exact sizes. Instead, *dimensions* are added that "call out" exact sizes. These are carefully followed when the drawing is used.

Conventions and *symbols* are used to help the drafter save time on drawings without losing information. It is important to know what the various conventions and symbols mean.

Dimensions

There are two kinds of dimensions. *Size dimensions* show the total height, width, and length of the object or the size of some detail. *Location*, or *position*, *dimensions* show where the details are located (Fig. 21-1). Follow these dimensioning procedures:

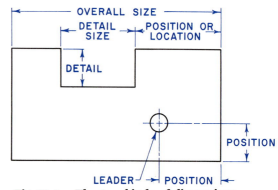

Fig. 21-1. The two kinds of dimensions.

1. Place dimensions so that they can be read either from the bottom (*unidirectional dimensions*) or from the bottom and the right side (*aligned dimensions*) (Fig. 21-2).

2. Place dimensions between views on simple drawings.

3. Start dimension lines 3/8-inch (10 millimeters) from the view. Leave about 1/4 inch (6 millimeters) between the dimension line.

4. Dimension the view that best shows the shape (Fig. 21-3 a and b).

5. Place the smaller detailed dimension inside the overall dimension.

6. Dimension a hole or a circle by giving its *diameter* (distance across) (Fig. 21-4). A circle may be dimensioned by giving its diameter in the rectangular view. Often a *leader* (a line with an arrowhead at the end) is used to dimension holes or openings.

7. Dimension the *radius* (half the diameter) of arcs (Fig. 21-5).

8. Write down dimensions in places where they can be read easily.

9. Do not repeat dimensions. Each dimension should appear only once.

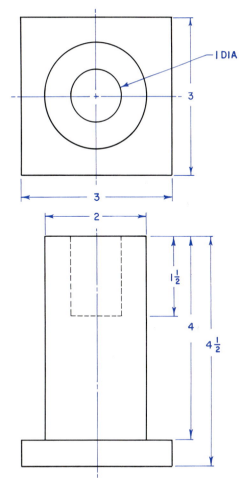

Fig. 21-3 a. **Sufficient dimensions must be placed on the drawing so that the object can be constructed.**

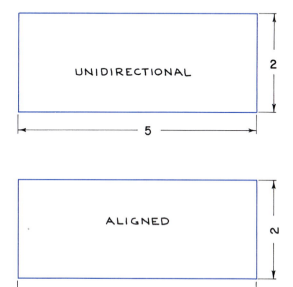

Fig. 21-2. **Systems of placing dimensions.**

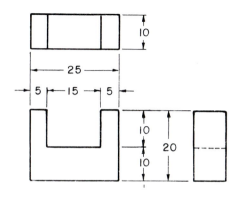

ALL DIMENSIONS IN mm

Fig. 21-3 b. **A simple metric drawing.**

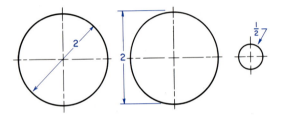

Fig. 21-4. **Ways of dimensioning a hole or a circle.**

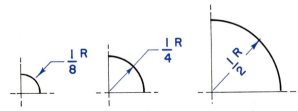

Fig. 21-5. **Methods of dimensioning radii.**

10. Never let dimension lines cross extension lines.

11. Dimension in inches, up to and including 72 inches. Use feet and inches above this size. If all dimensions are in inches, the inch mark (") is usually not used.

In the metric system, most dimensions are given in millimeters. Very large items are dimensioned in meters. On metric drawings, the symbol for the circle, $\varnothing$, and the square, $\square$, are commonly used (Fig. 21-6).

12. To dimension small spaces, place the arrowheads outside the extension lines pointing in, or use any of the methods shown in Fig. 21-7.

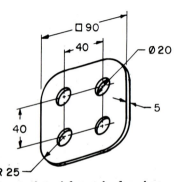

Fig. 21-6. **A pictorial metric drawing.**

13. Show angles by degrees or by two dimensions (Fig. 21-8).

14. Draw arrowheads as shown in Fig. 21-9. Arrowheads are drawn at one or both ends of dimension lines. Use only one arrowhead on a dimension line for the ra-

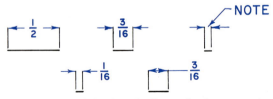

Fig. 21-7. **Methods of dimensioning narrow spaces.**

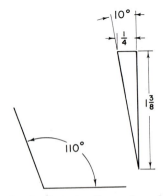

Fig. 21-8. **Angles are dimensions in degrees or a combination of degrees and linear measurements.**

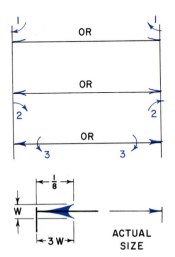

Fig. 21-9. **The correct way to draw an arrowhead.**

dius. The arrowheads should be neat and not too dark. They should be about three times as long as they are wide.

Conventions and Symbols

It would take a long time to draw many mechanical, architectural, and electrical parts as they really appear. For example, an external screw thread would truly look like Fig. 21-10. An electric-bell circuit would look like Fig. 21-11. To help the drafter, a system of *conventions* and *symbols* has been worked out.

A schematic of an electric-bell circuit would look like Fig. 21-12. Using symbols, an external screw thread would look like Fig. 21-13. In Fig. 21-13, you see the two kinds of external thread symbols—*regular* and *simplified*. Another symbol is used for an internal thread (Fig. 21-14).

A note on the inch drawing shows the diameter, the number of threads per inch, and whether it is National Fine (NF) or National Coarse (NC) series. A machinist looking at Fig. 21-14, for example, would know that this is a tapped hole. It must be made with a ½-inch National Coarse tap with 13 threads to the inch.

A note for a metric thread such as M6 × 1 means the following: M shows that it is an ISO metric thread; 6 is the diameter in millimeters; 1 is the pitch in millimeters (*pitch* is the distance from the top of one thread to the top of the next) (Fig 21-15).

There are electricity symbols for standard electrical items (Fig. 21-16). These are used to make layouts of electric circuits. The layouts are called *schematic drawings*. Figure 21-11 is a pictorial drawing of a dry cell, a wire, and a small push button and bell. This looks quite different as a schematic drawing. See Fig. 21-12. The dia-

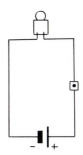

Fig. 21-12.   A schematic drawing of a simple bell circuit showing the use of electrical symbols.

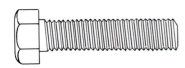

Fig. 21-10.   A drawing of a thread on a bolt.

Fig. 21-11.   Pictorial drawing of a bell circuit.

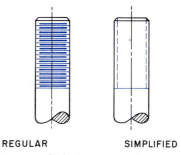

REGULAR   SIMPLIFIED

Fig. 21-13.   Symbols for external threads.

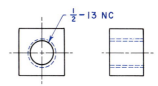

$\frac{1}{2}$ – 13 NC

Fig. 21-14.   Symbols for internal inch threads that show the size of the tapped hole, the number of threads per inch, and the type of thread.

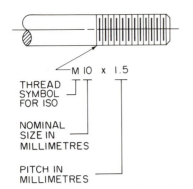

M 10 x 1.5

THREAD SYMBOL FOR ISO

NOMINAL SIZE IN MILLIMETRES

PITCH IN MILLIMETRES

Fig. 21-15. Symbols for external metric threads.

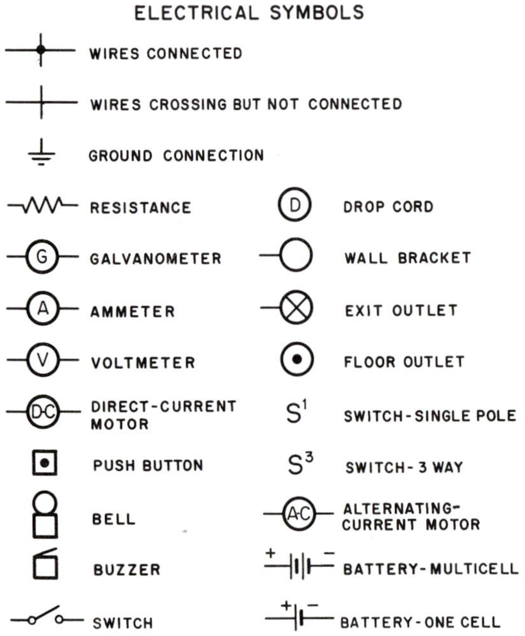

ELECTRICAL SYMBOLS

—•— WIRES CONNECTED

—|— WIRES CROSSING BUT NOT CONNECTED

⏚ GROUND CONNECTION

—⋎⋎⋎— RESISTANCE

—(G)— GALVANOMETER

—(A)— AMMETER

—(V)— VOLTMETER

—(DC)— DIRECT-CURRENT MOTOR

▣ PUSH BUTTON

⎕ BELL

⎕ BUZZER

—o o— SWITCH

(D) DROP CORD

—(O) WALL BRACKET

—(X) EXIT OUTLET

—(•) FLOOR OUTLET

S¹ SWITCH-SINGLE POLE

S³ SWITCH-3 WAY

—(AC)— ALTERNATING-CURRENT MOTOR

—|⫶|⫶|— BATTERY-MULTICELL

—|⫶|— BATTERY-ONE CELL

Fig. 21-16. Symbols for standard electrical items.

grams of all electric wiring for homes, radio, and television have these symbols.

You will also find conventions and symbols for architectural details. Conventions and symbols are used for identifying materials. For instance, in Fig. 21-17 it is easy to see the difference between wood and cast iron.

Conventional *breaks* are used to shorten a part that is too long for the page (Fig. 21-18). In almost every area, whether it is mapmaking, electrical, welding, gears, plumbing, architecture, or aircraft, symbols and conventions are used to make drawings simple. The drafter must know how to use them.

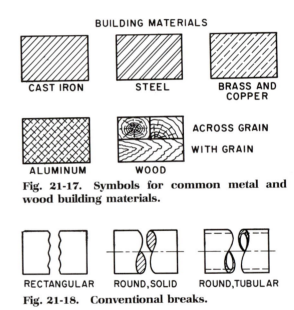

BUILDING MATERIALS

CAST IRON   STEEL   BRASS AND COPPER

ALUMINUM   WOOD   ACROSS GRAIN / WITH GRAIN

Fig. 21-17. Symbols for common metal and wood building materials.

RECTANGULAR   ROUND, SOLID   ROUND, TUBULAR

Fig. 21-18. Conventional breaks.

Unit 22

Drawing Equipment

No matter what your job will be, tools of some kind will be an important and useful part of your work. Drafters, architects, and designers use special drawing tools to help them make a neat and accurate drawing (Fig. 22-1). With proper care, these tools can last a long time and help you with many drawings.

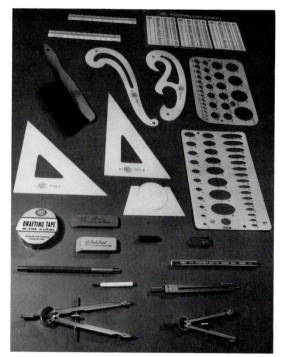

Fig. 22-1. **Drafting equipment.** (Teledyne Post)

Drawing Board

A drawing board should be made of a soft wood, such as bass or pine. It should have a smooth surface and a straight working edge (Fig. 22-2).

Drawing boards are made in many different sizes. However, the 16 × 22 inch (406 × 560 millimeter) or the 18 × 24 inch (457 × 610 millimeter) size is best for most work. Keep the surface of the drawing board free of nicks and scratches. Sometimes a heavy piece of paper is placed over the working surface to protect it.

T Square

The T square has two parts. The *head* and the *blade* are fastened together at right angles (Fig. 22-3). Simple T squares are all wood, but better-quality ones have a plastic edge on either side of the blade. To be useful, the head and the blade must be exactly 90° to each other.

This tool must be handled with great care. Never drop it or use it as a hammer. A T square is used to draw all horizontal lines. It is also a guide for other tools, such as the triangle.

Note: If you are right-handed, hold the head of the T square against the left edge of the drawing board. If you are left-handed, hold the head of the T square against the right edge of the drawing board.

Triangles

Triangles are used to draw vertical and inclined lines. You need two triangles: (1) a 45° triangle, which has a 90° angle and two 45° angles (Fig. 22-4), (2) a 30-60° triangle, which has a 90° angle, a 30° angle, and a 60° angle (Fig. 22-5). Use the 30-60° triangle to draw angles of 30° and 60°. Use the 45° triangle to draw an angle of 45° (Fig. 22-6). The triangles used together will make angles of 15° and 75° (Fig. 22-7).

Fig. 22-2. **A metal-edge drawing board.** (Teledyne Post)

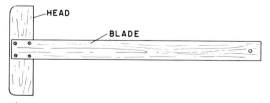

Fig. 22-3. **An example of a T square.**

■ 143

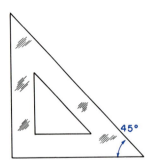

Fig. 22-4. A 45° triangle.

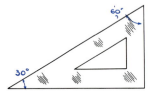

Fig. 22-5. A 30-60° triangle.

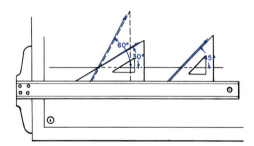

Fig. 22-6. Layout of a 60°, a 30°, and a 45° angle.

Scales

A scale is a measuring tool that helps you draw objects at larger or smaller sizes than they actually are (Fig. 22-8).

The three common kinds of scales are the *architect's*, the *metric*, and the *engineer's* (Fig. 22-9). The architect's scale is used for making customary (inch) drawings. The metric scale is used for metric drawings.

A comparison of the common metric and customary equivalents is shown in Fig. 22-10 and Fig. 22-11 shows an architect's scale used to draw an object one-quarter size, or 3 inches = 1 foot. The dis-

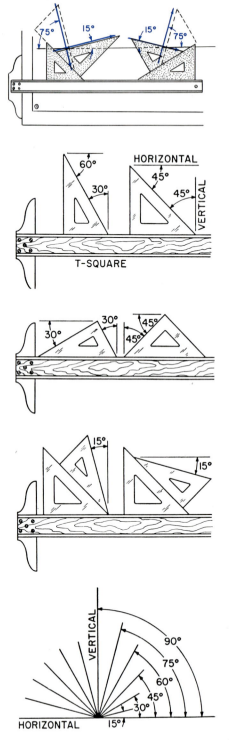

Fig. 22-7. Method of using two triangles to lay out a 15° or a 75° angle.

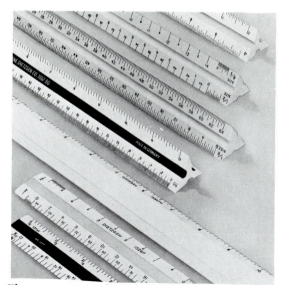

Fig. 22-8. Several kinds of scales useful to career drafters. (Eugene Dietzgen Company)

tance from 0 to the left end of the scale *represents* (stands for) 1 foot. The small numbers—3, 6, and 9—represent inches. Longer divisions start from 0 and go to the right end. They represent feet. For example, the distance of 15 inches—or 1 foot and 3 inches—is from the 1-foot mark past 0 to the 3-inch mark (reading from right to left). In the metric system, the scale 1:5 is similar to but not exactly the same as this customary scale (Fig. 22-12).

Pencils

Pencils come in many grades of hardness, from 6B, which is the softest, through grades 5B, 4B, 3B, 2B, B, HB, F, H, 2H, 3H, 4H, 5H, 6H, 7H, 8H, and 9H, which is the hardest. Usually a 3H or 4H pencil is used

Fig. 22-9. The three common types of scales. Top: full-size architect's scale. Middle: metric scale. Bottom: engineer's scale.

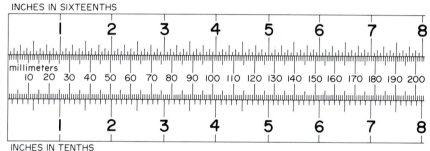

for making light lines. 2H or H pencils are used for most drawing. HB pencils are used for lettering or sketching. A standard pencil sharpener can be used, but it is better to have a drafter's pencil sharpener. This cuts off the wood but does not sharpen the lead to a point.

METRIC SCALE	CUSTOMARY EQUIVALENT
1:1 (FULL SIZE)	1"=1" or 12"=12"(FULL SIZE)
1:2 (HALF SIZE)	$\frac{1}{2}$" = 1" or 6" = 12"(HALF SIZE)
1:3 (THIRD SIZE)	$\frac{3}{8}$" = 1" or (THREE-EIGHTH SIZE)
1:5 (FIFTH SIZE)	$\frac{1}{4}$" = 1" or 3" = 12"(QUARTER SIZE)
1:10(TENTH SIZE)	$\frac{1}{8}$" = 1" or (EIGHT SIZE)

■ APPROXIMATE EQUIVALENT

Fig. 22-10. A comparison of the common metric and customary reduction scales.

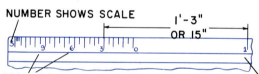

NUMBER SHOWS SCALE   1'-3"
OR 15"

NUMBERS INDICATE INCHES   FEET FROM ZERO

Fig. 22-11. Fifteen inches on the architect's scale.

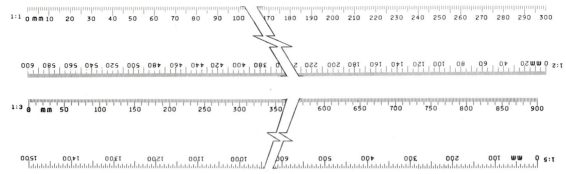

Fig. 22-12. The four most common scales used in making a metric drawing.

To sharpen drawing pencils, shape the lead to a long, cone-shaped point with a sandpaper pad or a file (Figs. 22-13 a and b and 22-14). Always clean the point on a piece of scrap paper before drawing. Remember to sharpen the end opposite the grade stamp.

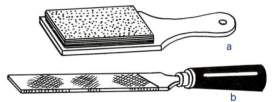

Fig. 22-13. (*a*) Sandpaper pad. (*b*) File.

Protractor

Use a protractor to lay out an angle that cannot be measured with the triangles (Fig. 22-15). The protractor is made up of a semicircle divided into 180°. The outer edge starts at 0° on the right side and goes to 180° at the left side. The inner edge reads from 0 to 180° from left to right.

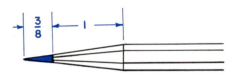

Fig. 22-14. A drawing pencil correctly sharpened to a long cone-shaped point.

Compass and Dividers

A compass is used to draw circles and arcs. The simplest kind is called a *bow compass* (Fig. 22-16). The point of a compass pencil should be sharpened to a wedge shape, as shown in Fig. 22-17. To adjust a compass, place a rule or a scale on a piece of paper. Then place the point of the compass on the 0 or the 1-inch mark. Open the compass until the pencil point is at the correct radius of the arc or the circle you want.

The compass pencil should be the same grade as or one grade softer than the pencil used for drawing. To use a compass,

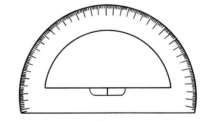

Fig. 22-15. A protractor.

Fig. 22-16. A bow compass with pencil and inking attachments.

place the point on the center of the circle or arc. Start a circle by tipping the compass slightly and turning it from left to right (Fig. 22-18). Be careful not to let the needle point slip from its position or the circle will not meet.

Dividers have a needle point at both ends. They are used to divide lines or transfer distances (Fig. 22-19).

Erasers and Eraser Shields

You need a good eraser for erasing lines (Fig. 22-20). An *Artgum eraser* is used for cleaning the drawing. Use a metal eraser shield as shown in Fig. 22-21

French, or Irregular, Curve

This tool is used to draw irregular curves (Fig. 22-22). An irregular curve is one that cannot be drawn with a compass because

Fig. 22-19.   **Transferring distances with dividers.**

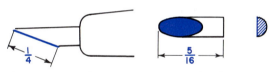

Fig. 22-17.   **A sharpened compass lead.**

Fig. 22-18.   **Using a compass.**

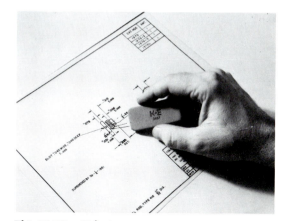

Fig. 22-20.   **Using an eraser.**

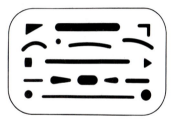

Fig. 22-21.   **An eraser shield.**

■ **147**

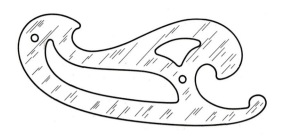

the curved line is not always the same distance from the center point. Locate several points on the curve to follow. Line up at least three points with the irregular curve. Draw a line about half the distance between the outside points. Then line up three more points. By continuing the same way, you will get a nice smooth curve.

Drafting Machines

Many drafters use a drafting machine. It does away with the need for a T square, triangle, and protractor (Fig. 22-23 a and b).

Drafting Tape

Always use drafting or masking tape to hold drawing paper to the board. Use a piece to tape down each corner of the paper.

Templates

Templates are plastic devices made with different cutouts. They are made in either customary (inch) or metric styles (Fig. 22-24). Templates make drawing faster and easier to do.

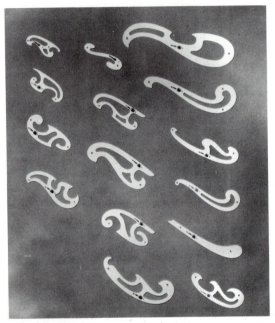

Fig. 22-22. French, or irregular, curves are made in many shapes.

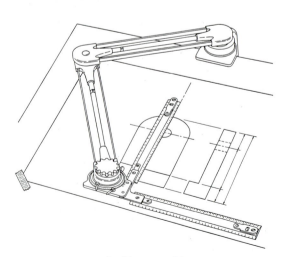

Fig. 22-23 a.   A drafting machine.

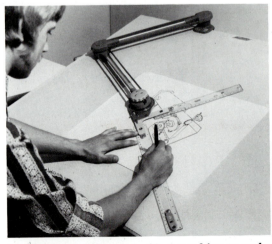

Fig. 22-23 b.   Using a drafting machine to make a technical illustration.

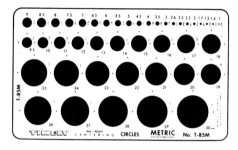

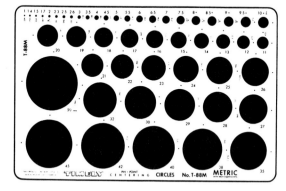

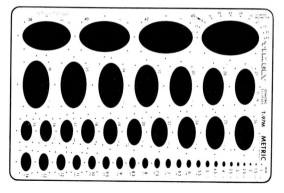

Paper

There are several kinds and colors of drawing paper. Paper can be plain white, buff, cream, or green. For a print, the drawing must be made on tracing paper. For sketching, squared paper or paper with equally spaced dots is used. The common sizes of paper are $8\frac{1}{2} \times 11$ inches or 9×12 inches, and 11×17 inches or 12×18 inches. You will probably use A size paper that is $8\frac{1}{2} \times 11$ inches (or 9×12 inches) or A4 metric sized paper for most of the drawings needed for your project activities (Fig. 22-25).

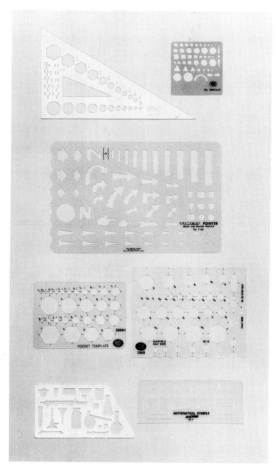

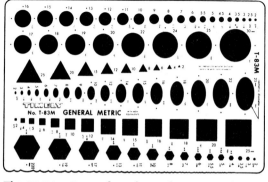

Fig. 22-24. Examples of templates. (Timely Products Company)

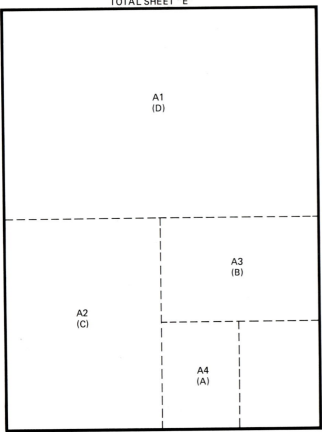

TOTAL SHEET "A0"
TOTAL SHEET "E"

A1
(D)

A3
(B)

A2
(C)

A4
(A)

STANDARD DRAWING SIZE
ISO STANDARD

SIZE	MILLIMETERS	INCHES
A0	841 X 1189	33.11 X 46.81
A1	594 X 841	23.39 X 33.11
A2	420 X 594	16.54 X 23.39
A3	297 X 420	11.69 X 16.54
A4	210 X 297	8.27 X 11.69

AMERICAN STANDARD

SIZE	INCHES	MILLIMETERS
E	34 X 44	863.6 X 1117.6
D	22 X 34	558.8 X 863.6
C	17 X 22	431.8 X 558.8
B	11 X 17	279.4 X 431.8
A	8½ X 11	215.9 X 279.4
E	36 X 48	914.4 X 1219.2
D	24 X 36	609.6 X 914.4
C	18 X 24	457.2 X 609.6
B	12 X 18	304.8 X 457.2
A	9 X 12	228.6 X 304.8

Fig. 22-25.   A comparison of ISO and American standard paper sizes.

Unit 23

Making a One-view Drawing

A one-view drawing is often enough to show an object, and many of your project activities can be represented this way. For example, the design for a cutting board or the pattern for a plastic letter opener needs only one view. The floor plan for a house is also a one-view drawing.

Making a One-view Drawing

In order to make a one-view drawing, you must know how to draw several kinds of lines. Practice drawing these lines:

1. Place the T square with the head against the left (if you are right-handed) or the right edge. The blade should be about three-fourths of the way down on the board. Place the paper on the board about 2 or 3 inches (50 or 75 millimeters) from the working edge. Place the T square on top of the paper. Line up the top edge of the paper with the top edge of the T square. Hold the paper in place. Now carefully slide the T square down the board far enough to see the corners of the paper.

Tear off two pieces of masking tape. Place them across the upper corners to fasten the paper to the board. Keep the paper smooth and tape the lower corners of the paper to the board (Fig. 23-1). You are now ready to begin drawing lines (Fig. 23-2).

2. Hold the T square firmly against the working edge of the drawing board with one hand. Tilt the pencil so that the point is in the corner made by the blade and the paper. Use even pressure and pull the pencil along, turning it slightly as you make the line (Fig. 23-3). Try to get a sharp, clear, equal-weight line. Do not press too hard or too lightly. Try several horizontal lines, moving the T square after each line has been drawn.

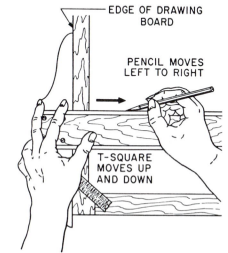

Fig. 23-3. **Drawing a horizontal line along a T square.**

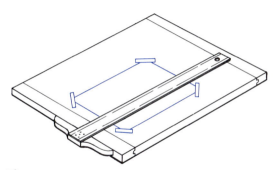

Fig. 23-1. **Paper correctly fastened to the drawing board.**

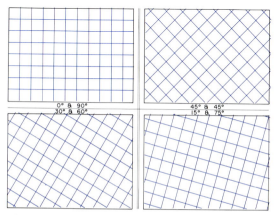

Fig. 23-2. **Straight lines drawn at different angles.**

3. Place a triangle against the upper edge of the blade of the T square. Hold it firmly against the blade. The thumb and little finger should put pressure sideways to the T square. The other three fingers should hold the triangle against the blade. Draw vertical lines from the bottom to the top (Fig. 23-4).

4. A slanted, or inclined, line is a line drawn at an angle other than 90°. Inclined lines at the common angles may be drawn with one or both of the triangles. Draw lines at angles of 30°, 45°, and 60°. Then draw lines at 15° and 75°. You will need to use both triangles to draw lines at these angles.

5. Draw circles and arcs. Divide another sheet of paper into four equal parts. Mark

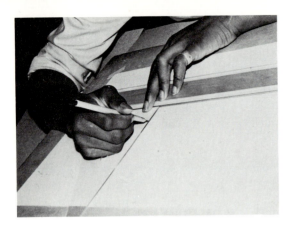

1. Fasten another piece of 8½ × 11 inch (216 × 280 millimeter) paper to your board.

2. Measure ¼ inch (6 millimeters) in from all edges. draw a heavy border line. This line will serve as the frame for your drawing.

3. Then measure ½ inch (13 millimeters) up from the lower border line. Draw another heavy line. This space may be divided into three or four parts for your title block. The title block may have the name of your school, the name of the drawing, the scale, your own name, and the number of the drawing.

In the center of the title block, draw light guidelines to be used in lettering the information. Usually ⅛-inch (3-millimeter) letters are used in the title block.

4. You may want to lay out the cutting board shown in Fig 23-6. The overall height is 6 inches (152.5 millimeters). The overall length is 8¾ inches (222 millimeters). You have a space of 7½ × 10½ inches (191 × 267 millimeters) inside the border and the title block. Measure up half this distance, about 3¾ inches (95 millimeters). Draw a horizontal center line. This will be the center of the drawing.

5. Measure over from the border 3⅞ inches (98 millimeters) (Fig. 23-7). Then draw the first vertical center line.

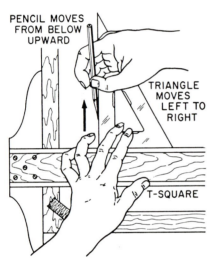

Fig. 23-4. Drawing a vertical line along a triangle.

them *1, 2, 3,* and *4.* In part 1 draw several circles. In part 2 draw several semicircles and arcs. In part 3 draw both arcs and straight lines. In part 4 draw circles, arcs, and straight lines (Fig. 23-5).

When you can draw these lines correctly and easily, you are ready to make a one-view drawing.

One-view Drawing

Follow these steps to make a one-view drawing:

Fig. 23-5. An exercise in drawing circles and arcs.

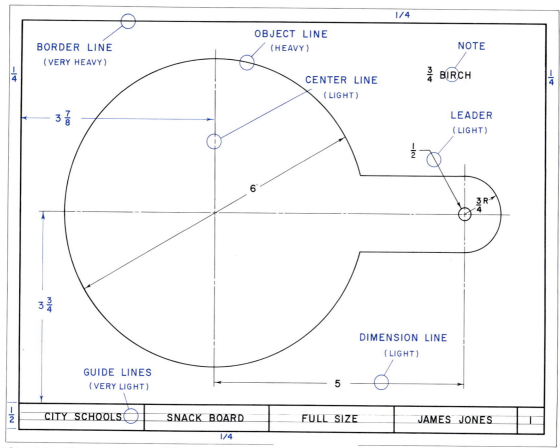

BORDER LINE
(VERY HEAVY)

OBJECT LINE
(HEAVY)

NOTE

$\frac{3}{4}$ BIRCH

CENTER LINE
(LIGHT)

LEADER
(LIGHT)

$\frac{1}{2}$

$3\frac{7}{8}$

$\frac{1}{4}$

$\frac{1}{4}$

6

$\frac{3}{4}$ R

$3\frac{3}{4}$

DIMENSION LINE
(LIGHT)

GUIDE LINES
(VERY LIGHT)

5

$\frac{1}{2}$

CITY SCHOOLS | SNACK BOARD | FULL SIZE | JAMES JONES | I

1/4

Fig. 23-6.   The correct use of lines in making a one-view drawing.

6. Adjust your compass to a 3-inch (76 millimeter) radius. Draw a light circle. Adjust the compass to ³/₄ inch (19 millimeters). Draw a light semicircle at the other end.

7. Measure up ³/₄ inch (19 millimeters) and down ³/₄ inch (19 millimeters) from the horizontal center line. Draw a light construction line to complete the handle.

8. Now darken the circle, semicircle, and straight lines to complete the outline of the cutting board.

9. Adjust the compass to ¹/₄ inch (6.5 millimeters) and draw the hole for the handle.

10. Add the dimensions as shown in Fig. 23-6.

Fig. 23-7.   Using the architect's scale and the T square for layout work.

11. Add a note indicating the thickness and kind of material.

12. Check all your lines. Make sure they are the correct type and weight.

Figures 23-8 and 23-9 show additional examples of one-view drawings with metric dimensions.

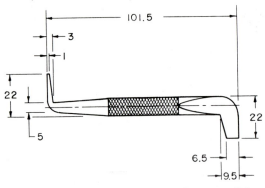

Fig. 23-8. A one-view drawing of a screwdriver using metric dimensions.

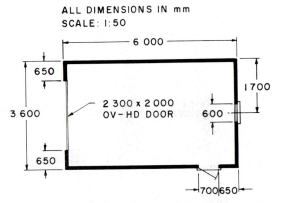

Fig. 23-9. A one-view drawing of a small garage. This floor plan uses metric dimensions.

Enlarging a Design

An irregularly shaped design usually has to be laid out on paper first. Then it must be transferred to the material. You will see designs in books or magazines that are smaller than full size. These must be enlarged. The design may be covered with squares that are smaller than full size. If not, cover the design with squares of a *fractional* size (Fig. 23-10).

Suppose the design is one-fourth full size. First cover the design with ¼-inch

squares. Then draw 1-inch squares on a piece of layout paper. Letter across the bottom of both the original and the full-sized paper. Number up the sides of each sheet. Then find a point on the original drawing. Transfer it to the full-sized pattern. Do this until you have found enough points to draw the enlarged design. Connect all the points with freehand sketching or with a French curve. If metric cross-section paper is used, the 1-mm squares can stand for 2, 3, or 4 millimeters, and so on, to get the correct enlargement.

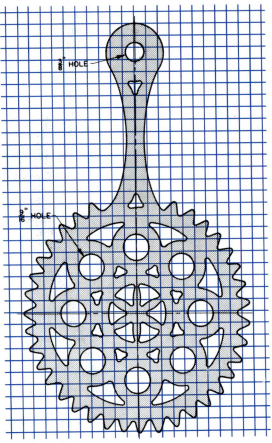

Fig. 23-10. Can you make a full-size pattern for this trivet?

Geometric Construction

What does a light fixture have in common with a racetrack or a coffee table? The shape of each one is basically geometric (Fig. 24-1). It is important for a drafter or designer to learn how to draw geometric shapes and angles accurately because so many objects are based on them.

Dividing a Line into Two Equal Parts

To divide the line *AB* (Fig. 24-2), set a compass to a radius greater than one-half *AB*. Then, with points *A* and *B* as centers, draw arcs crossing at *C* and *D*. Draw line *CD*. Line *CD* will divide line *AB* into two equal parts.

Bisecting an Angle

Suppose *BAC* is the given angle. Open the compass to about a 1-inch (approximately 25-millimeters) radius. Draw an arc intersecting *AB* at *D* and *AC* at *E*. Use *E* and *D* as centers and a radius of slightly more than half *ED*. Now draw arcs intersecting at *H*.

Fig. 24-1. How many geometric shapes can you find in this room? (Georgia-Pacific Corp.)

Line *AF* will divide the angle into equal parts (Fig. 24-3).

Dividing a Line into Several Equal Parts

Suppose you want to divide *AB* into nine equal parts. Start at point *B*. Draw a line, *BC*, that is *perpendicular* (at a 90° angle) to *AB*. Line *BC* may be any length. Place a rule at point *A*. Move it across line *BC* until a measurement is reached that can be easily divided by nine. Mark off nine spaces on line *AC*. Then draw vertical lines from these points. The vertical lines will divide line *AB* into equal parts (Fig. 24-4).

Drawing a Hexagon

Open a compass to an amount equal to one side of the hexagon. Using this setting,

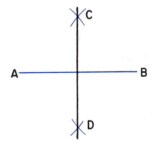

Fig. 24-2.   A line divided into two equal parts.

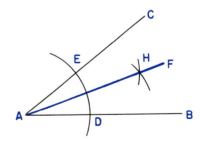

Fig. 24-3.   A bisected angle.

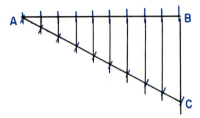

Fig. 24-4. A line divided into several equal parts.

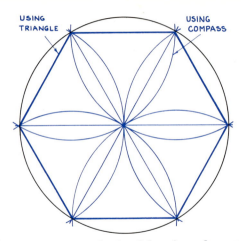

USING
TRIANGLE

USING
COMPASS

Fig. 24-5. Two methods of drawing a hexagon.

draw a circle. With the same setting, start at any point on the circle. Swing arcs to divide the circumference into six parts. Or divide the circle into six equal parts with a triangle. Then draw the six straight lines that will form the sides (Fig. 24-5).

Drawing an Octagon

Lay out a square that is the same width and height as the octagon you wish to draw. Draw two diagonals across the square. Set a compass equal to one-half the diagonal measurement. Place the point of the compass at each of the corners. Swing an arc that will intersect the sides of the square. Connect these intersections with eight straight lines to form the octagon (Fig. 24-6).

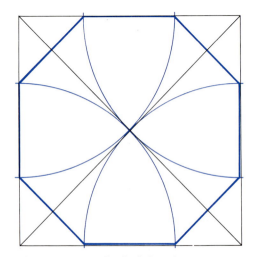

Fig. 24-6. One method of drawing an octagon.

Drawing an Ellipse

Lay out two lines intersecting at 90°. Measure the width of the ellipse along the vertical line AB. Measure the length across the horizontal line CD. Half will be on either side. Open a compass to a distance equal to one-half the horizontal lines, or ZC. Place the point of the compass at B and intersect line CD at X and Y. Then place pins or thumbtacks at points X, Y, and A. Tie a string around them. Remove the pin or the thumbtack at point A. Put a sharp pencil in its place. Hold the pencil against the string and draw the ellipse (Fig. 24-7).

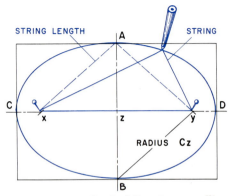

STRING LENGTH   A   STRING

C   D

x   z   ẏ

RADIUS   Cz

B

Fig. 24-7. One method of drawing an ellipse.

Multiview Drawings

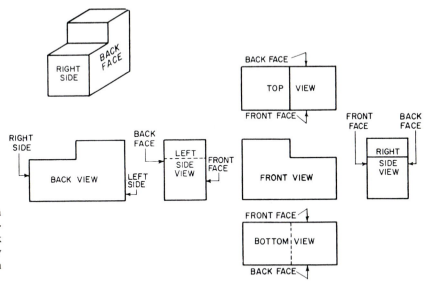

Fig. 25-1. Six views of a simple block. Try to follow the front and back face arrows to see how each view is shown in its different position.

A one-view drawing is not always enough to show an object accurately. For example, if you were shown only the top view of the block in Fig. 25-1, you would probably not be able to tell what the rest of the block looked like. Notice how the other views help you see its true shape. Together, these views are called a *multiview* drawing, and you will see this type of drawing often in this book.

Multiview drawing is also called *orthographic projection*. The word *orthographic* means "straight line," and *projection* means "representing the parts of an object on a flat surface."

Orthographic Projection

In orthographic projection, you can draw lines from one view to another to help complete the drawing. For most objects three views are needed to show the front, the top, and the right side (Fig. 25-2 a, b,

and c). Two views are usually enough for cylindrical objects (Fig. 25-3 a and b). Sometimes more than two or three regular views are needed, as shown in the drawing of the cold chisel (Fig. 25-4).

The way to construct multiview drawings is shown in Fig. 25-5. Note that when

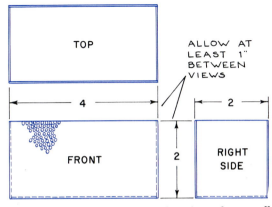

Fig. 25-2 a. A three-view drawing of a small planter with customary dimensions.

METRIC DRAWING—WOODWORKING
(VENEERED BOX)

CONVERSION CHART		
mm	decimal inch	inch to nearest 16th
30	1.181	1 3/16
75	2.953	2 15/16
91	3.583	3 9/16
125	4.921	4 15/16
150	5.905	5 7/8
250	9.842	9 13/16

NOTE :
16 mm BASSWOOD USED
HARDWARE TO SUIT

THIRD ANGLE PROJECTION

RABBET JOINT
4 PLACES

1 mm VENEER

250

125

SAW CUT

30

91   75

150

Fig. 25-2 b. This is the most common kind of multiview drawing showing the front, top, and right-side views.

Fig. 25-2 c.   The three views of this helicopter are considered to be the front, top, and left-side views. (Textron Company)

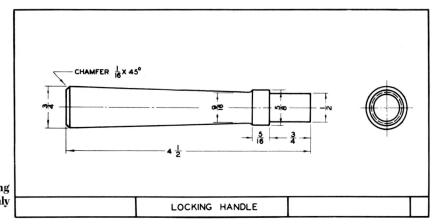

CHAMFER 1/16 X 45°

3/4

9/16   5/8   2

5/16   3/4

4 1/2

LOCKING HANDLE

Fig. 25-3 a. A locking handle requires only two views.

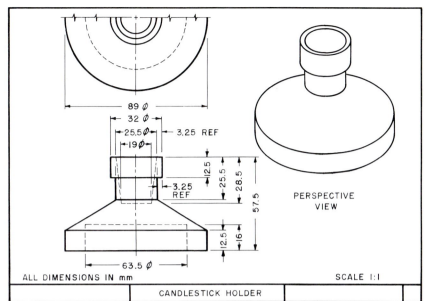

Fig. 25-3 b. This candlestick holder requires two views and is dimensioned in millimeters. The perspective view is not needed but is here to help you understand the two-view drawing.

PERSPECTIVE VIEW

ALL DIMENSIONS IN mm SCALE 1:1

CANDLESTICK HOLDER

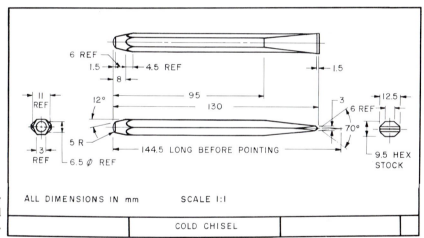

Fig. 25-4. A four-view drawing of a cold chisel with metric dimensions.

144.5 LONG BEFORE POINTING

9.5 HEX STOCK

ALL DIMENSIONS IN mm SCALE 1:1

COLD CHISEL

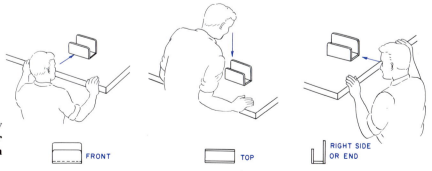

Fig. 25-5. A three-view drawing of a letter holder as you would see it from the three positions.

FRONT

TOP

RIGHT SIDE OR END

■ **159**

you look at the front of the object, you see what would be drawn on the paper as the front view. This front view should always be the best or most important view. When you look down on the object, the top view is seen. The top view is placed right above the front view. When you look at the right side of the object, the right-side view is seen.

Another way to see how this drawing is made is to imagine that an object is in a clear plastic box. The box is joined together as shown in Fig. 25-6 a. First sketch the front of the object on the front of the box. Next sketch the top of the object on the top of the box. Then sketch the right side of the object on the side of the box. When you have finished sketching, lift up the top of the box and swing out the end. You will see all three views in their correct positions on a flat surface. That shows just how you would draw them on a piece of paper.

In Fig. 25-1 there are six views that show all six sides of an object. However, the left-side, the back, and the bottom views are not really needed. But sometimes a left view is needed (Fig. 25-6 b). Notice how you can *project* (draw) from the front view to the right-side view to get the height of the right-side view. You can also project from the front view to the top view to get

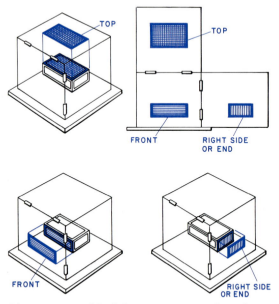

Fig. 25-6 a. A block in a transparent box showing the relationship of the views.

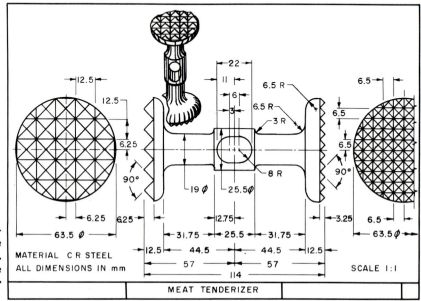

Fig. 25-6 b. This three-view drawing shows the front, right, and left views. Can you see why these are the best views for producing this product?

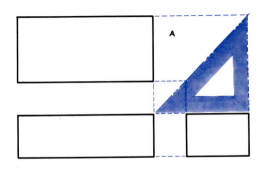

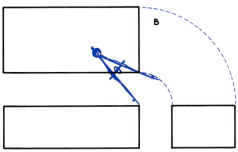

Fig. 25-7. Two ways of projecting from one view to another.

arcs and circles with your compass when it is needed. Make sure that all lines are the same quality. Do not make them so dark that they look rough or so light that they cannot be seen easily. Do not cross or round corners.

Add Dimensions

Draw in extension lines when they are needed. There should be a break of about $1/16$ inch (1.5 millimeters) between the outline of the object and the extension lines.

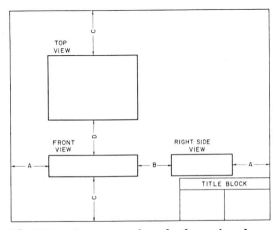

Fig. 25-8. Proper spacing of a three-view drawing. Distances A and C should be about equal. Distances B and D should be about equal. Sometimes D is made slightly larger than B for added dimensions.

the length of the top view. Figure 25-7 a and b shows two other ways of transferring the width from the top view to the right-side view.

Centering a Three-view Drawing

A three-view drawing should be put in the center of the paper. Enough space should be left between the views for dimensions. The drawing should appear to be well balanced. It is a simple job to center the object if you follow these suggestions:

1. Select the best view for the front view. Check the object to find its height, depth, and length.
2. Leave a space of about 1 inch (25 millimeters) between the views for dimensions. Figures 25-8 and 25-9 show drawings that have been centered.

Retrace Lines to Finish Drawing

Use an H or 2H pencil to darken the lines and make them clear and sharp. Darken

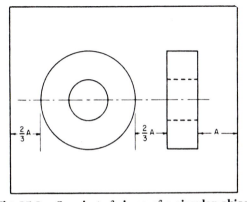

Fig. 25-9. Spacing of views of a circular object.

These lines should be very light. Draw in dimension lines between the extension lines.

An arrowhead should be placed at both ends of the dimension lines except when an arc is dimensioned. In this case, only one end should have an arrowhead. Always leave space in the line for the dimensions. If there are several dimension lines, one under the other, space them about $3/8$ inch (9.5 millimeters) apart. The breaks in the lines should be staggered for easy reading. Make the arrowheads neat and pointed. Letter in the dimensions using guidelines to make the numbers the same height.

Add All Lettering

Letter in notes that give information about building the object. Also letter in the title block, including all necessary information.

Points to Remember in Making a Multiview Drawing

1. Select the view that best shows the shape of the object as the front view.
2. Set up the views so that the object is in its natural position. For example, do not draw a pail upside down.
3. Arrange the views so that you will have the smallest number of invisible lines.
4. Set up the views so that they will look well balanced on the page.
5. Use only the views that you really need.

A Working Drawing

Drawings used in the manufacture of a product are called *working drawings*. Often a single-view drawing is all that is needed. For some types of products, a two- or three-view drawing is needed. In making wood products, an *isometric drawing* or a *cabinet drawing* is often used. These drawings are explained in Unit 27, "Pictorial Drawings."

There are two kinds of working drawings—*detail drawings* and *assembly drawings*. The detail drawing gives a complete picture of one part. It includes the shape, the dimensions, and information about the kind of material and the finish needed. For a complicated product such as an outboard motor, a detail drawing would be needed for each small part (Fig. 25-10).

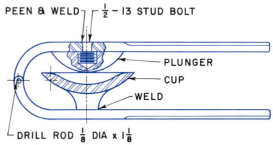

PEEN & WELD — $\frac{1}{2}$ – 13 STUD BOLT
PLUNGER
CUP
WELD
DRILL ROD $\frac{1}{8}$ DIA x $1\frac{1}{8}$

Fig. 25-10.   A detail drawing.

The assembly drawing shows how all the different parts go together. There are several kinds of assembly drawings. Sometimes they are made as multiview drawings (Fig. 25-11). Often, however, they are made as an *exploded pictorial drawing*. An exploded view easily shows how each part fits the next one (Fig. 25-12). Follow these suggestions to make a detail or assembly drawing:

1. Select views that will best show the object.
2. Use only the views that are needed. Sometimes one view is enough. Sometimes three views or more may be required.
3. Place the drawing on the page so that it is centered and balanced.
4. Add the dimensions so that they can be read easily.
5. Place both the detail and the assembly drawings on one page if a simple object with few parts must be drawn. Draw detail views on separate sheets if the design is complex.

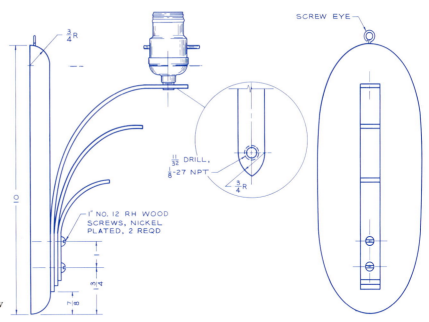

$\frac{3}{4}$R

10

$\frac{11}{32}$ DRILL, $\frac{1}{8}$-27 NPT

$\frac{3}{4}$R

1" NO. 12 RH WOOD SCREWS, NICKEL PLATED, 2 REQD

$\frac{3}{4}$

$\frac{7}{8}$

SCREW EYE

Fig. 25-11. A multiview assembly drawing.

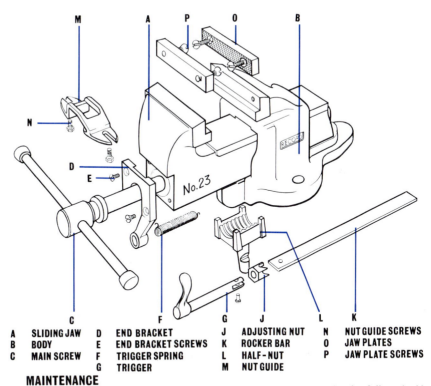

M   A   P   O   B

N

D

E

No.23

RECORD

C   F   G   J   L   K

A	SLIDING JAW	D	END BRACKET	J	ADJUSTING NUT	N	NUT GUIDE SCREWS

A   SLIDING JAW   D   END BRACKET   J   ADJUSTING NUT   N   NUT GUIDE SCREWS
B   BODY   E   END BRACKET SCREWS K   ROCKER BAR   O   JAW PLATES
C   MAIN SCREW   F   TRIGGER SPRING   L   HALF-NUT   P   JAW PLATE SCREWS
                     G   TRIGGER   M   NUT GUIDE

MAINTENANCE
All working parts should be oiled periodically. To do this, first open the vice fully and add a few drops of light oil to the following points:

(1) Behind main screw head   (2) Dual Action trigger mechanism (if fitted)

(3) Main screw and nut

Fig. 25-12. An exploded pictorial assembly drawing.

Unit 26

Sectional and Auxiliary Views

Figure 26-1 shows a cutaway photo of an alternator (used to generate electricity for your car) with the parts named. It would be difficult to make a drawing of this using hidden lines for the inside details. This is why sectional views are often used to show complex objects clearly.

Sectional Views

The most common types of sectional views are called *full sections* and *half sections*. A full section is drawn as if the object were cut in half with the front part taken away to show the inside features (Figs. 26-2 and 26-3). The location of the cut is shown by a cutting-plane line. Arrows at the ends of the line show the direction in which you are looking (Fig. 26-4). Sometimes the cutting-plane line does not go straight through an object. It

may move over to include another feature. It is then called an *offset cutting-plane line* (Fig. 26-5). *Section-lining symbols*, such as cross-hatching, show which parts are cut.

In a half section, only half of the front is cut away. This is used when you want to

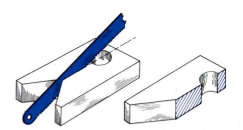

Fig. 26-2.   Full-section view of a steel casting.

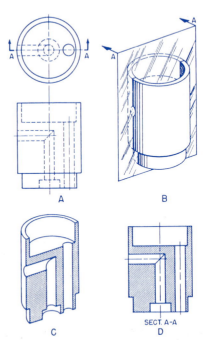

Fig. 26-3.   The inside of a part shows more clearly in a section view. Compare the views at A and D.

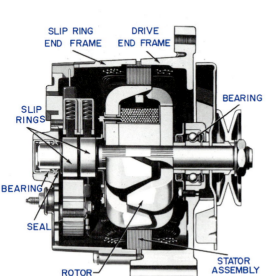

Fig. 26-1.   Cutaway view (sectional view) of an alternator. (Delco-Remy Company)

■ **164**

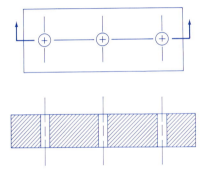

Fig. 26-4.   A cutting-plane line.

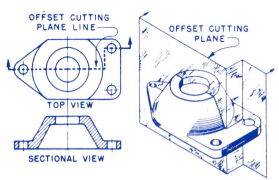

Fig. 26-5.   An example of an offset section.

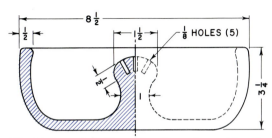

Fig. 26-6 a.   A half-section view of a nut bowl.

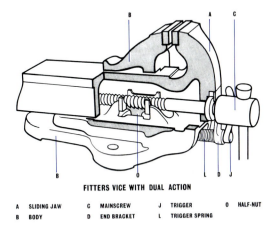

FITTERS VICE WITH DUAL ACTION

A	SLIDING JAW	C	MAINSCREW
B	BODY	D	END BRACKET

J   TRIGGER   O   HALF-NUT
L   TRIGGER SPRING

Fig. 26-6 b.   A half-section view showing the major parts.

show both the inside and the outside on the same view (Fig. 26-6 a and b).

Section lines on most drawings should be medium-weight, sharp lines. They are drawn at an angle of 45°. A good way to space section lines is to scratch a sharp line parallel to the *hypotenuse* (long side) on a 45° triangle, about 1/16 inch (1.5 millimeters) from the edge. Then, after you have drawn the first section line, you can move the triangle so that the section line is over the first line of the drawing. This will produce evenly spaced section lines.

Sometimes other section-lining symbols are used to show other kinds of materials (Fig. 26-7).

Auxiliary Views

An *auxiliary view* is an extra view. This view is needed when one of the main surfaces

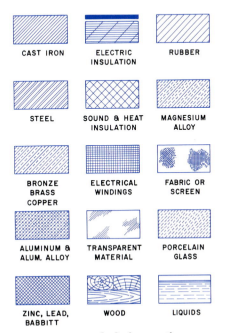

Fig. 26-7.   Some symbols for sections.

CAST IRON   ELECTRIC INSULATION   RUBBER

STEEL   SOUND & HEAT INSULATION   MAGNESIUM ALLOY

BRONZE BRASS COPPER   ELECTRICAL WINDINGS   FABRIC OR SCREEN

ALUMINUM & ALUM. ALLOY   TRANSPARENT MATERIAL   PORCELAIN GLASS

ZINC, LEAD, BABBITT   WOOD   LIQUIDS

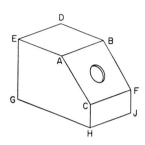

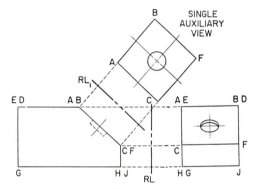

Fig. 26-8. An auxiliary view.

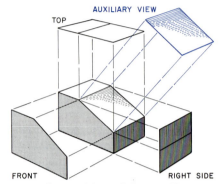

Fig. 26-9. The position of the auxiliary view.

(front, top, or side) is not at a right angle to the other surfaces (Fig. 26-8).

Auxiliary views are used mostly in machine drawings. The views show where a bracket or metal part is bent at an angle other than 90°. The auxiliary view shows these parts in true size and shape. It is always drawn at right angles to the inclined surface (Fig. 26-9). Usually the auxiliary view will have some dimensions on it.

Unit 27

Pictorial Drawings

If you drew an object exactly as it appears to you without including the hidden lines and surfaces, you would be making a *pictorial drawing*. There are three types of pictorial drawings: perspective, oblique, and isometric (Fig. 27-1 a, b, and c). They realistically represent scenes and objects just as a photograph does (Fig. 27-2).

Perspective Drawing

A perspective drawing gives the illusion of depth to an object or scene. Notice that the railroad tracks in Figs. 27-3 and 27-4 seem to come together at some faraway point. This point is called the *vanishing*

point. One or two vanishing points are used in drawing a perspective.

The least-used perspective has only one vanishing point and is called *parallel perspective* (Fig. 27-5). It looks something like a cabinet drawing. However, most perspectives are made with two or more vanishing points and are called *angular perspectives* (Fig. 27-6). These look something like an isometric drawing.

Perspective drawings are used by architects and drafters. The drawings give a picture of a building or a product before it is built. You have probably seen perspective drawings of homes and office buildings that look like photographs.

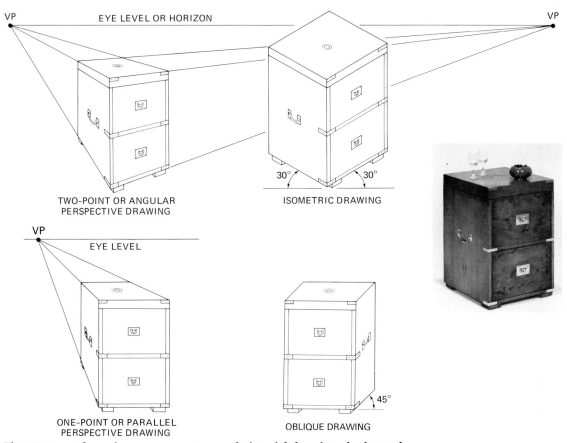

VP

EYE LEVEL OR HORIZON

VP

30° 30°

TWO-POINT OR ANGULAR
PERSPECTIVE DRAWING

ISOMETRIC DRAWING

VP

EYE LEVEL

45°

ONE-POINT OR PARALLEL
PERSPECTIVE DRAWING

OBLIQUE DRAWING

Fig. 27-1 a.   These four common types of pictorial drawings look much like the photograph.

Fig. 27-1 b.   A perspective drawing of a remodeled garage. Can you tell which type of perspective it is?

Fig. 27-1 c.   A perspective drawing of a utility room including washer, dryer, and cabinets. Is this the same perspective shown in Fig. 27-1 b?

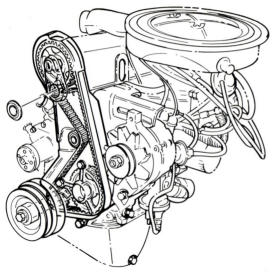

Fig. 27-2. A pictorial drawing of an engine. (Ford Motor Company)

Fig. 27-3. Photograph of railroad tracks. (Association of American Railroads)

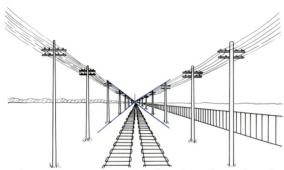

Fig. 27-4. A perspective drawing of a railroad track.

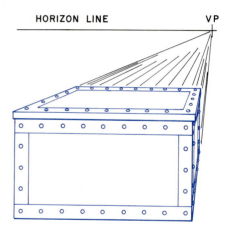

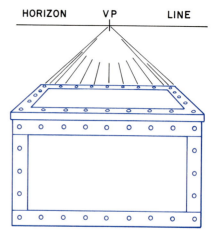

Fig. 27-5. A one-point, or parallel, perspective drawing of a box.

Oblique Drawing

Oblique means "slanting." An *oblique drawing* is a simple pictorial drawing where only the front surface shows the true size and shape. The sides and the top slant back at an angle of 30°, 45°, or 60°. There are two kinds of oblique drawings: *cavalier* and *cabinet* (Fig. 27-7). When the slanted lines for the top and the side are made true length, the drawing is called *cavalier oblique*. However, this type of drawing makes an object look unreal. To correct this, the depth is made half its true length. This is called *cabinet oblique*, or a *cabinet drawing*. Most furniture that you

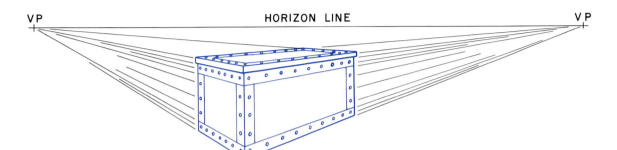

Fig. 27-6. A two-point, or angular, perspective drawing of a box.

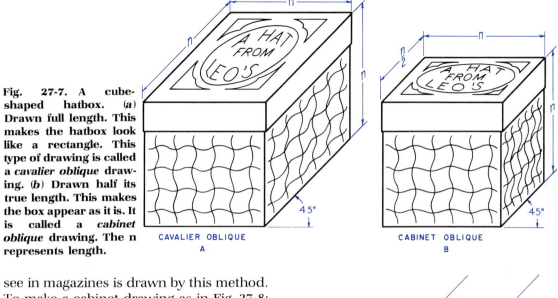

Fig. 27-7. A cube-shaped hatbox. (*a*) Drawn full length. This makes the hatbox look like a rectangle. This type of drawing is called a *cavalier oblique* drawing. (*b*) Drawn half its true length. This makes the box appear as it is. It is called a *cabinet oblique* drawing. The n represents length.

see in magazines is drawn by this method. To make a cabinet drawing as in Fig. 27-8:

1. Select the view that will show the best surface. If the object contains circles or arcs, select the view showing them as the front view.

2. Draw the front view to the true size and shape.

3. Draw slanted lines to the right or the left to form the top and right sides. Measure along these lines half the true length. Complete the cabinet drawing.

Points to Remember

1. Circles and arcs on the top or the side view are shown as ellipses.

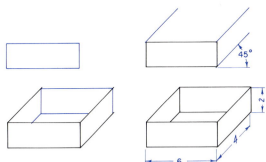

Fig. 27-8. The steps in making a cabinet drawing begin with the front view.

2. If a long object is to be drawn, always make the long side the front view.

3. Cabinet drawings can be made to appear above eye level by drawing the slanted lines down at an angle of 30°, 45°, or 60°.

4. When dimensioning a cabinet drawing, always put the true dimensions on it even though the lines are not true length.

5. Hidden lines are not usually shown in a cabinet drawing unless they are absolutely needed.

6. Place dimensions on a cabinet drawing, as shown in Fig. 27-8.

Isometric Drawing

Isometric means "equal measurement." An isometric drawing looks much like an angular perspective. In this drawing one corner of the object appears nearest the viewer. It is best used for drawing rectangular-shaped objects. An isometric drawing is constructed around three lines. All the lines are exactly 120° apart. One line is drawn vertical. The other two lines are drawn at 30° to the horizontal line. To construct an isometric drawing as shown in Fig. 27-9 a and b:

1. Draw a light, horizontal construction line. Then lay out a vertical line and two inclined lines at 30° to this horizontal. These three lines will form the base of the object.

2. Lay out the true length along these three lines and then mark off the width, the length, and the height.

3. Draw the parallel lines to complete the view.

4. Add dimensions, as shown in Fig. 27-10. Hidden or invisible lines are not used very often.

Lines drawn at an angle of 30° or 90° from the horizontal plane are called *isometric lines* (Fig. 27-11). All other lines are *nonisometric lines*. To draw nonisometric lines, you must find the points of intersection along the isometric lines. Then join the two points. Nonisometric lines are always longer or shorter than true length.

Circles and arcs must be drawn in an isometric square, as shown in Figs. 27-12

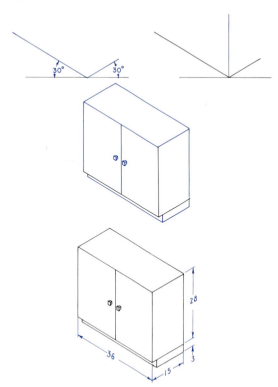

Fig. 27-9 a.   Steps in making an isometric drawing.

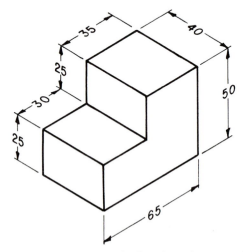

Fig. 27-9 b.   An isometric drawing of a step cube with metric measurements.

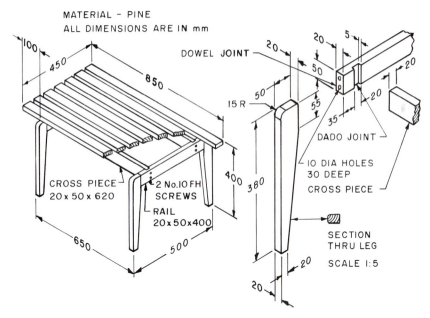

MATERIAL – PINE
ALL DIMENSIONS ARE IN mm

DOWEL JOINT

DADO JOINT

15 R

400   380

650   500

CROSS PIECE
20 x 50 x 620

2 No.10 FH
SCREWS

RAIL
20 x 50 x 400

10 DIA HOLES
30 DEEP
CROSS PIECE

SECTION
THRU LEG

SCALE 1:5

Fig. 27-10. An isometric drawing of a patio table with dimensions in millimeters.

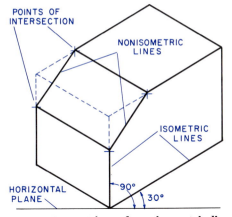

POINTS OF
INTERSECTION

NONISOMETRIC
LINES

ISOMETRIC
LINES

HORIZONTAL
PLANE   90°   30°

Fig. 27-11. Isometric and nonisometric lines.

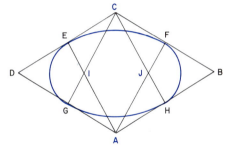

Fig. 27-12. A circle drawn on a top isometric plane.

and 27-13. To draw a circle in an isometric square, do the following:

1. Divide the sides of the square equally.
2. Draw construction lines from the opposite corners to the opposite sides.
3. Open a compass to the long radius *AF* and draw two points of the oval. These will be arcs *EF* and *GH*.
4. To complete the circle, adjust the compass to the short radius *JF*. Draw the other two arcs, *FH* and *EG*.

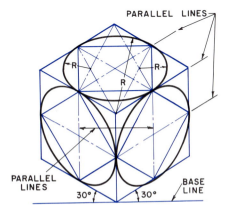

PARALLEL LINES

R   R
R

PARALLEL
LINES   30°   30°

BASE
LINE

Fig. 27-13. Circles drawn on all isometric planes.

■ 171

Unit 28

Pattern Development

Sheet metal can be cut, rolled, bent, crimped, riveted, soldered, and formed into such shapes as boxes, scoops, funnels, and fireplaces (Fig. 28-1). Before you can begin, you must first draw a plan on the metal itself. This is called a *sheet-metal development* or *pattern development*. Figure 28-2 illustrates the four basic kinds of pattern development.

Direct Development

Flat patterns, such as the box shown in Fig. 28-3 a, are simple. They are usually laid out directly on the metal. The direct method can be used only when there are no curved or intersecting surfaces on the pattern.

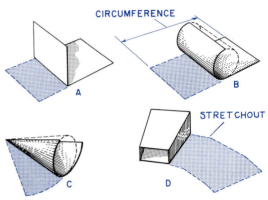

Fig. 28-2. Four types of pattern development: (*a*) direct, (*b*) parallel, (*c*) radial, (*d*) triangular.

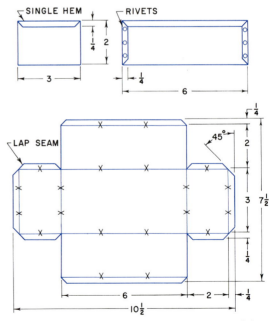

Fig. 28-3 a. The layout for a sheet-metal box with inch dimensions.

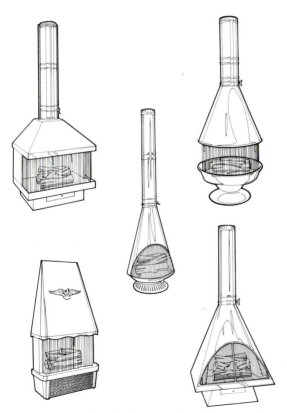

Fig. 28-1. Each of these unique fireplace designs requires a different pattern development.

Parallel-line Development

Figure 28-3 a and b shows the layout of two rectangular project activities. One uses customary and the other uses metric dimensions. A simple cylinder is laid out as a rectangle (Fig. 28-4). The length of the rectangle is the same as the circumference of the cylinder. It is determined by the formula $C = \pi D$. Leave extra metal for making the seam. To cut off the end of the metal for a scoop shape, make a pattern development as follows:

1. Make a two-view drawing of the object. Show the front and the top views.
2. Divide the top view into equal parts, such as 12.
3. Draw parallel lines from the division points to the front view.
4. Lay out a base line to the right of the front view. This is the circumference of the scoop. Divide it into as many equal spaces as you have divided the circle or the top view.
5. Intersect the lines on the flat-pattern layout. Use the same number of lines as in the front and the top views.
6. Lay out the shape of the end of the scoop with a French, or irregular, curve.

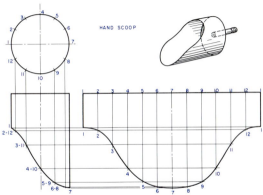

Fig. 28-4. A pattern developmet of a scoop showing the projection of the stretch-out.

Radial-line Development

This method is used for many cone-shaped objects. A funnel is a good example (Fig. 28-5). To make a development of a funnel:

1. Use a piece of paper large enough to make the flat-pattern layout.
2. Draw a front view of the funnel to full size. Continue the lines for the tapered sides of the body and spout until they intersect at points *a* and *a'*.
3. Draw semicircle *bc* at the large end of the body.
4. Divide semicircle *bc* into an even number of equal parts, perhaps eight. Of course, any number may be used.
5. Make a flat-pattern layout for the body of the funnel as follows:

    a. With the distance *ab* as a radius, draw an arc that will be the outside edge of the flat-pattern layout.
    b. With the distance *ad*, draw another arc using the same center.
    c. Along the outer edge, space off twice the number of segments that are on semicircle *bc*. Draw a line to the center at either end of the arc to form the flat-pattern *fghi*.

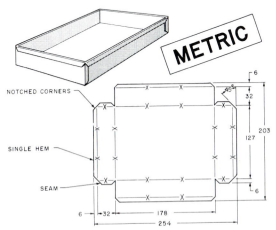

Fig. 28-3 b. The layout for a sheet-metal box with metric dimensions.

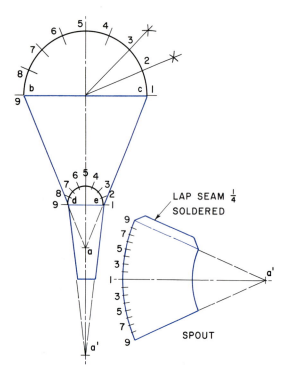

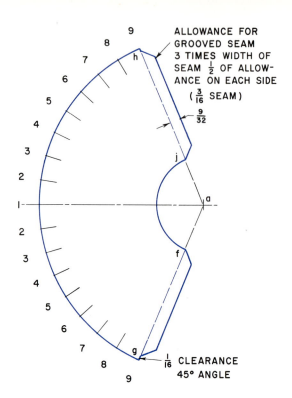

Fig. 28-5.   A pattern development of a funnel.

d. For making the wired edge, add an extra amount, or allowance. This amount should be 2¹/₂ times the diameter of the wire along the outer edge.

e. For making a grooved seam, add an allowance on each side of the pattern.

6. Make a flat-pattern layout for the spout in the same way. Add enough for a lap seam on one side.

Triangular Development

Triangular development is also called *transition*. It is done when shapes must be changed from square to round. It must be done when stretch-outs are made from irregular geometric surfaces. Triangular development is also used when surfaces are parallel geometric surfaces. In this case, triangular development is used in place of parallel development.

Unit 29

Reproducing Shop Drawings

In industry, an original mechanical drawing is never used at the job site or in the lab because of the danger of damaging it. Instead, the drafter makes a *tracing* on thin paper or cloth that can be seen through. Copies, or *prints*, are then made from the tracing and are used by whoever needs them. The original tracing is stored in a safe place and is used only if more prints are needed or if a change needs to be made in the design.

There are many methods of making prints and copies of tracings. Most are made as follows:

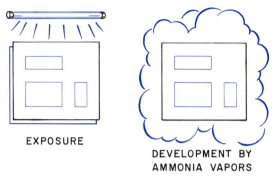

EXPOSURE

DEVELOPMENT BY AMMONIA VAPORS

Fig. 29-1.   The steps in duplicating a drawing by the blueprint method.

EXPOSURE

WATER WASH

POTASSIUM DICHROMATE BATH

WATER RINSE

PRINT DRYING

Fig. 29-2.   The dry diazo process of duplicating a drawing.

The tracing is placed over a piece of chemically treated paper. These papers are placed under a strong light. The light produces a change on the chemically treated paper. The areas under the lines of the tracing are protected from the light. They do not undergo a chemical change. Then the tracing is removed. The print paper is developed like a photograph in a chemical solution.

One of the most common ways of making copies is the *blueprint* process used in the building industry (Fig. 29-1). This process produces a white line on a blue background. In the *dry diazo*, or Ozalid, process (Fig. 29-2), a developer of potassium dichromate is used. This developer makes a print with a white background and colored lines. The color of the lines depends on the kind of paper used. A third way that is often used is called the *moist diazo*, or Bruning, process (Fig. 29-3). The liquid developer used produces brown or black lines on a white background.

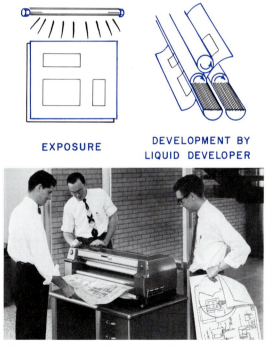

EXPOSURE

DEVELOPMENT BY LIQUID DEVELOPER

Fig. 29-3.   Using the moist diazo process to make a print. (Multigraph Corp.)

Discussion Topics and Suggested Problems on Drafting and Planning Technology

Discussion Topics

1. Why is it easier to explain a project with a drawing or a sketch than it is with words?

2. List 10 careers in which drafting is needed to show ideas to others.

3. Name 10 industries in which engineering can be a career.

4. What are the differences in educational courses needed to become an engineer, a technician, and a craftsperson?

5. How many technicians are needed for every engineer or scientific person?

6. What education is needed to become a professional architect?

7. Select and describe the education needed and the duties of the three careers in this unit that interest you most.

8. Why is it impossible to make a drawing without knowing how to measure?

9. What is meant by the alphabet of lines?

10. Name and describe the tools and materials needed for sketching.

11. Name the three groups of letters.

12. Why are letters drawn with unequal spaces between them?

13. What is the purpose of a dimension?

14. List the drawing equipment needed to make a mechanical drawing.

15. What kind of drawing would you make for the layout of a football field?

16. Explain how to make a multiview drawing.

17. What is meant by scale drawing?

18. What is the best kind of drawing for construction purposes?

19. Name the three kinds of pictorial drawings talked about in this section.

20. Tell why geometric construction is important in drawing.

21. Name the four kinds of pattern development.

22. Tell how a print is made.

Suggested Problems

1. Divide a piece of paper into four equal parts and sketch horizontal lines, vertical lines, inclined lines, and circles.

2. Use a sheet layout as shown in Fig. 23-2 and sketch lines at different angles.

3. Use a sheet layout as shown in Fig. 23-5 and sketch circles and arcs.

4. Divide a sheet into four equal parts. Sketch a hexagon, an octagon, an isometric box, and a cabinet box.

5. Make an enlargement for a wall plaque of a butterfly (Fig. 30-1).

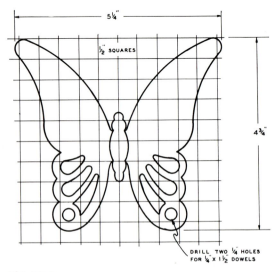

Fig. 30-1.

6. Make an enlargement to 6 inches (150 millimeters) in height for the elephant (Fig. 30-2).

7. Make a lettering exercise sheet and copy each letter of the alphabet five times.

8. Follow the layout sheet as shown in Fig. 23-2 and use triangles to lay out lines at the correct angles.

9. Do the following geometric constructions:

a. Divide a 3¹/₄-inch line into two equal parts.

b. Bisect a 38° angle.

c. Divide any line into 13 equal parts.

d. Draw an octagon in an 80-millimeter square.

10. Make a one-view layout of a checkerboard: 64 squares, 8 on each side. The side of each square should measure 1¹/₂ inches (40 millimeters). Use the scale 1:2 (6 inches = 1 foot).

11. Make a one-view layout drawing of a volleyball court.

12. Make a one-view layout drawing of a softball field.

13. Make a two-view full-size drawing of a hockey puck. Use these dimensions: 1 inch (25 millimeters) thick and 3 inches (75 millimeters) in diameter.

14. Make a three-view drawing of one or more of the following:

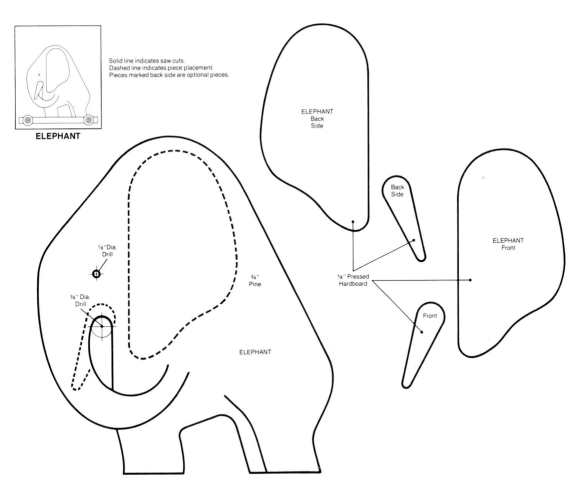

Fig. 30-2.

a. A footstool of ¾-inch material; each square is equal to 1 inch (Fig. 30-3 a).
b. A knife rack (Fig. 30-3 b).
c. A pipe holder (Fig. 30-3 c).
d. A spice holder (Fig. 30-3 d).

No. of Pieces	Part Name	Thickness	Width	Length
1	Base	½"	3"	6¾"
1	Bowl Rest	½"	2"	6¼"
1	Stem Support	¼"	1¼"	6¼"
1	Back	¼"	$6\frac{7}{16}$"	7"

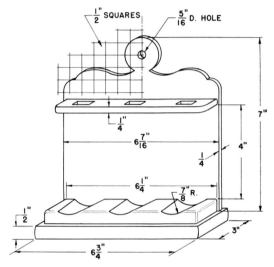

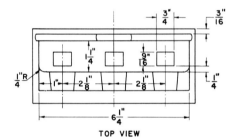

TOP VIEW

Fig. 30-3 c.

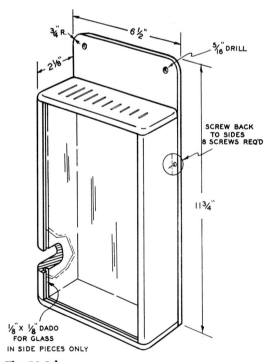

Fig. 30-3 b.

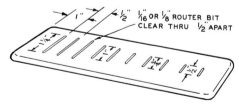

No. of Pieces	Part Name	Thickness	Width	Length	Stock
1	Back	¼"	6½"	11¾"	Pine
2	Top and Bottom	⅜"	2⅛"	6½"	Pine
2	Sides	½"	2⅛"	8½"	Pine
1	Front	Double Strength	5¾"	8½"	Glass
8	¾"-6 Flat-Head Wood Screws.				
	¾" Brads				

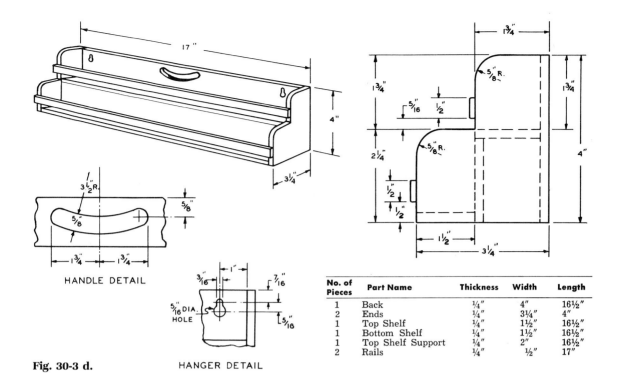

No. of Pieces	Part Name	Thickness	Width	Length
1	Back	¼"	4"	16½"
2	Ends	¼"	3¼"	4"
1	Top Shelf	¼"	1½"	16½"
1	Bottom Shelf	¼"	1½"	16½"
1	Top Shelf Support	¼"	2"	16½"
2	Rails	¼"	½"	17"

HANDLE DETAIL

HANGER DETAIL

Fig. 30-3 d.

15. Make a cavalier oblique drawing of the peg puzzle (Fig. 30-4).

16. Make a cabinet drawing of Fig. 30-3 a or b.

17. Make a freehand sketch of an electric circuit consisting of three cells, a push button, and three bells hooked in series. Make another sketch with three bells hooked in a parallel circuit. (See Section 9.)

18. Make a full-size layout of a funnel (Fig. 28-5). The body measures 5 inches on top and 1½ inches on the bottom. The spout measures ¾ inch on the small end. The body height is 6 inches, and the spout height is 3 inches.

19. Make a drawing or a shop sketch for each of the project activities you will build during this term. Add dimensions in millimeters or inches.

20. Make a full-section drawing of the hollow punch shown in Fig. 30-5.

21. Make a pattern development for one of the projects shown in Fig. 30-6.

22. Make a working drawing (two or three views) of the metal project activities shown in Fig. 30-7. Convert the dimensions to millimeters.

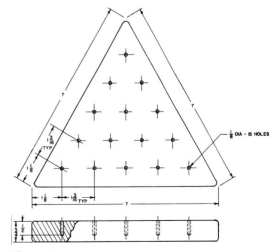

Fig. 30-4.

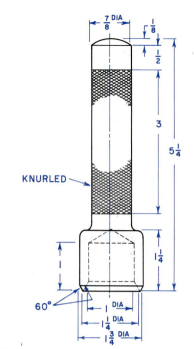

$\frac{7}{8}$ DIA $\frac{1}{8}$ $\frac{1}{2}$ 3 $5\frac{1}{4}$ KNURLED $1\frac{1}{4}$ 60° 1 DIA $1\frac{1}{4}$ DIA $1\frac{3}{4}$ DIA

Fig. 30-5.

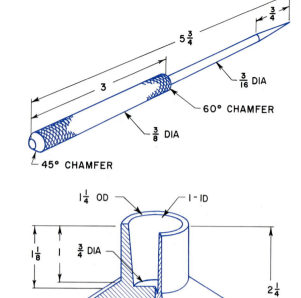

$5\frac{3}{4}$ $\frac{3}{4}$ 3 $\frac{3}{16}$ DIA 60° CHAMFER $\frac{3}{8}$ DIA 45° CHAMFER

$1\frac{1}{4}$ OD 1 - ID $1\frac{1}{8}$ 1 $\frac{3}{4}$ DIA $2\frac{1}{4}$ $\frac{1}{2}$ $2\frac{1}{2}$ DIA x $\frac{5}{8}$ HIGH $3\frac{1}{2}$ DIA

Fig. 30-7.

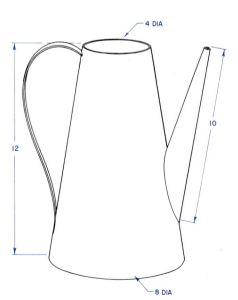

4 DIA 10 12 8 DIA

Fig. 30-6.

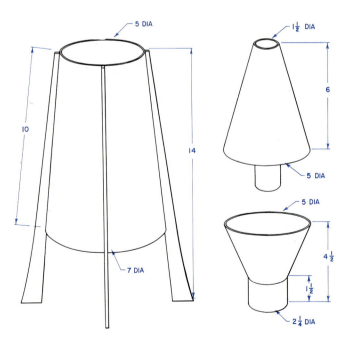

5 DIA 10 14 7 DIA $1\frac{1}{2}$ DIA 6 5 DIA 5 DIA $4\frac{1}{2}$ $1\frac{1}{2}$ $2\frac{1}{4}$ DIA

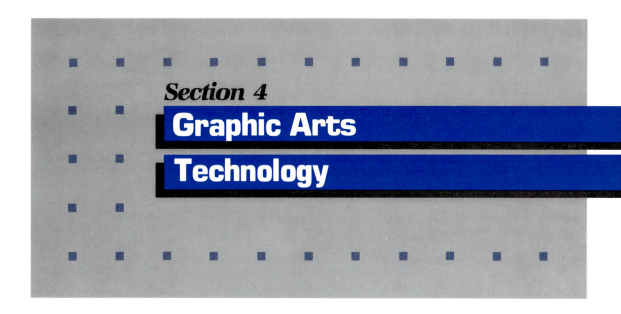

Section 4

Graphic Arts

Technology

Unit 31

Introduction to Graphic Arts Technology

Graphic arts is a form of visual communication that informs, educates, and serves all aspects of modern industry and personal life. It places an image (something that can be seen) on a solid material. Paper is the medium used most in this process. Other materials are printing ink and photographic media. Where we make use of these materials, a message is presented. Several large printing presses are shown in Fig. 31-1. These produce printed images.

The graphics arts industry is among the 10 largest manufacturing industries in the United States. The yearly sales total is over $12 billion. There are more than 850 000 workers in this industry, and this number would easily increase to well over 1 million workers if the allied industries were counted.

Graphic arts is a very old craft. It dates back several thousand years, when the an-

cient Chinese printed from wooden blocks. In 1441, Johann Gutenberg made the greatest single discovery in this ancient craft. He designed and made mov-

Fig. 31-1.   A pressroom in a large printing plant. (Brown & Bigelow)

able type. Graphic arts is the oldest mass-production industry in the world.

A typical eighteenth-century pressroom is shown in Fig. 31-2. The puller and beater are shown in two stages of operation. On the left, a sheet of paper is placed on the cylinder by the puller (worker), and the type is inked by the beater (another worker). On the right, the puller is printing an impression on the paper, and the beater is distributing (spreading) ink on the stocks while he inspects the previous pull.

Fig. 31-2. An eighteenth-century printer's pressroom. (Colonial Williamsburg)

Unit 32

Metrics in Graphic Arts Technology

People who work in the graphic arts industry must know metric measurements, especially in the area of photography (Fig. 32-1). Film sizes are given in 8, 16, and 35 millimeters, not in inches (such as 0.315, 0.63, or 1.38).

Measuring Systems in Graphic Arts

Three different measuring systems are used most frequently in graphic arts. These are the *customary* (in inches, or in), the *metric* (in millimeters, or mm), and the familiar *printers' measurement* of *points* and *picas* (Fig. 32-2). A pica is 1/6 of an inch, that is there are 6 picas per inch. There are 12 points in each pica, or 72 points per inch. These measurements will be discussed later. The point system is talked about in Unit 36, "Hand Composition."

Paper Sizes

The International Standards Organization (ISO) paper sizes are based on a unit of 1 square meter (1 m²) as the basic size. It has size. It has a proportion of 1:$\sqrt{2}$ (Fig. 32-3).

Each smaller size is exactly half of the next larger size. The short side of the larger size becomes the long side of the next smaller size. The common metric stationery size is called *A4* and is 210 × 297 millimeters. (Fig. 32-4). The A4 size is slightly narrower and somewhat longer than the customary 8½ × 11 inch paper. However, the A4 size does not fit well into American-made file drawers or copying machines. Most people in European counties use ISO paper sizes for both personal and business letters.

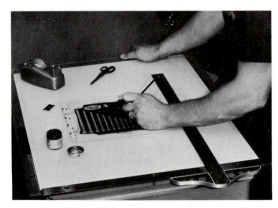

Fig. 32-1. The lithography artist must know and use the metric system.

PRINTER'S UNITS
CUSTOMARY AND METRIC EQUIVALENTS

| PRINTER'S | | CUSTOMARY (inches) | | METRIC |
Picas	Points	Approximate Fraction	Decimal	Millimeters
	1	$1/64$	0.014	0.35
	2	$1/32$	0.028	0.70
	3	$3/64$	0.042	1.05
	4	$7/128$	0.055	1.40
	5	$1/16$	0.069	1.75
	6	$5/64$	0.083	2.10
	7	$3/32$	0.097	2.45
	8	$7/64$	0.111	2.80
	9	$1/8$	0.125	3.15
	10	$9/64$	0.138	3.50
1	12	$21/128$	0.166	4.20
	14	$25/128$	0.194	4.90
	18	$1/4$	0.249	6.30
2	24	$21/64$	0.332	8.40
	30	$53/128$	0.414	10.50
3	36	$1/2$	0.498	12.60
	42	$37/64$	0.581	14.70
4	48	$85/128$	0.664	16.80
5	60	$53/64$	0.828	21.00
6	72	1	0.996	25.20

Fig. 32-2.   Printer's units with the customary and metric equivalents.

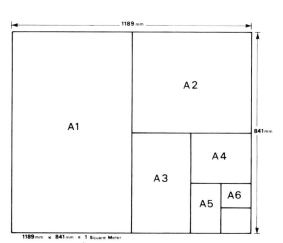

Fig. 32-3.   The ISO A size paper. With this system, many of the unusual sizes can be eliminated.

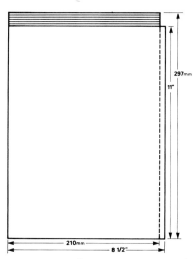

Fig. 32-4.   Comparing a customary $8^{1}/_{2} \times 11$ inch sheet and an ISO 210×280 millimeter sheet to A4 size paper. The ISO sheet is as wide as the A4 and is 11 inches long.

The ISO has approved a sheet size that is 210 × 280 millimeters or about 8¼ × 11 inches. The width is slightly less than that of the customary size. The length, however, is exactly the same. The width is changed because there is a C series of ISO paper sizes (plus a DL) designed to be used for postcards and envelopes (Fig. 32-5). Notice that with these envelopes the width of the paper is the most important measurement. Either the A4 or the ISO sheet will fit nicely (Figs. 32-6 and 32-7). The weight of the paper is also given in the metric system, in grams per square meter (g/m²) (Fig. 32-8).

Cold-type Sizes

There is no standard for cold-type sizes. But the most common standard now suggested is known as *Heden's system*. In Heden's system, the millimeter is the base unit. The proposed unit of measurement for the Heden system is called *d*. It is one-tenth of a millimeter (0.1 mm). The *d* unit specifies body size, type size, margin widths, and line spacing.

It will be helpful to review the general information concerning the metric system of measurement in Unit 17, "Metric Measurement and Standards."

COMPARISON CHART: C SIZE ENVELOPE		
ISO (size)	Metric (millimeters)	Customary (inches)
C4	229 × 324	9.02 × 12.76
C5	162 × 229	6.38 × 9.02
C6	114 × 162	4.49 × 6.38
DL	110 × 220	4.33 × 8.66

Fig. 32-5. The C size envelopes will be used for all international mailings.

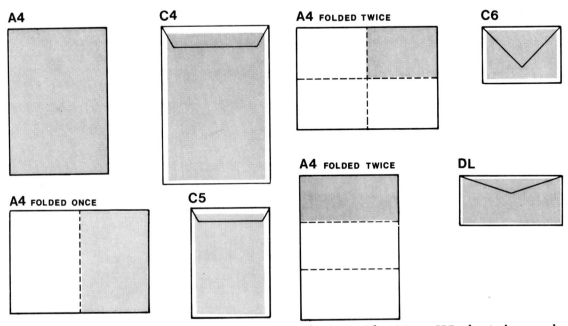

A4

C4

A4 FOLDED TWICE

C6

A4 FOLDED ONCE

C5

A4 FOLDED TWICE

DL

Fig. 32-6. Note how the A4, or ISO sheet, will fit into either the C4 or the C5 envelope.

Fig. 32-7. The A4, or ISO sheet size can be folded to fit into the C6 or DL envelope.

R20 SERIES OF PAPER WEIGHTS AND EQUIVALENT WEIGHTS

R SERIES g/m²	BOND 17 × 22 432 × 559 mm lb/ream	COVER 20 × 26 508 × 660 mm lb/ream	INDEX 25½ × 30½ 648 × 775 mm lb/ream	NEWSPRINT 24 × 36 610 × 914 mm lb/ream	BOOK 25 × 38 635 × 965 mm lb/ream
20.0	5.32	7.39	11.00	12.29	13.51
22.4	5.95	8.28	12.39	13.77	15.13
25.0	6.65	9.24	13.83	15.36	16.89
28.0	7.44	10.35	15.49	17.21	18.92
31.5	8.37	11.65	17.43	19.36	21.28
45.0	11.97	16.64	24.89	27.66	30.40
50.0	13.30	18.49	27.66	30.73	33.78
56.0	14.89	20.71	30.98	34.42	37.84
63.0	16.75	23.30	34.85	38.72	42.57
71.0	18.88	26.26	39.28	43.63	47.97
85.0	22.61	31.45	46.90	52.27	57.46
100.0	26.60	36.98	55.32	61.46	67.57
112.0	29.79	41.42	61.96	68.83	75.68
140.0	37.24	51.78	77.45	86.04	94.60
180.0	47.88	66.57	99.58	110.62	121.63
200.0	53.20	73.97	110.64	122.91	135.14
250.0	66.50	92.46	138.30	153.64	168.93
400.0	106.41	147.95	221.29	245.83	270.29

Fig. 32-8. The weight of paper in the metric system is in grams per square meter (g/m²). This standard applies to all grades of paper.

Unit 33

Graphic Arts Technology-related Careers

There are many careers in graphic arts. These will increase as the population grows. Occupations within this field form "clusters," including those involved in printing: composing room workers, photoengravers, electrotypers and stereotypers, press operators (Fig. 33-1), lithographic specialists, phototypesetting operators (Figs. 33-2 through 33-4), and bookbinding and related workers.

Newspaper reporters and technical writers form the "cluster" in journalism. Photography and many photographic laboratory occupations make up a small but important group in graphic arts.

Printing

Printing is a means of communication. The careers related directly to printing, publishing, and allied industries also have a direct effect on businesses that depend on graphic arts. Among these are banks, insurance companies, and manufacturers of paper products.

Fig. 33-1. **Skill is needed to operate the platen press.** (Heidelberg Eastern, Incorporated)

Fig. 33-2. Phototypesetting operators. (York Graphic Services)

Fig. 33-3. The phototypesetting operator has a very interesting craft. (York Graphic Services)

The largest division of graphic arts careers is newspaper printing and publishing. Practically every community in the nation has people working in daily or weekly newspaper production. Commercial, or job, printing is the second largest *segment* (part). These companies produce such materials as advertising matter, letterheads, business cards, calendars, catalogs, labels, maps, pamphlets, books, and magazines.

Some of the numerous printing methods for reproduction are described in this section. Knowledge and experience in those areas can prepare one to become a *hand compositer, typesetting-machine operator, phototypesetting operator, make-up arranger, teletype setter, dummier* (Fig. 33-5), or *proofreader* (Fig. 33-6), to name but a few. You may also be interested in becoming a *printing press operator* (Fig. 33-7), *camera operator, artist, stripper* (Fig. 33-8), *platemaker, lithographic press operator, bookbinder, electrotyper,* or *maintenance machinist.*

With additional education and training, there are opportunities for *executives, salespersons, teachers* (Fig. 33-9), *accountants, engineers, stenographers,* and *clerks.* The journalistic aspect (consideration) may interest a person in becoming a *reporter, writer,* or *editor.*

Fig. 33-4. Phototypesetting is a graphic arts area open to all. (Black Dot, Incorporated)

Fig. 33-5. The dummier has a very fascinating occupation. Dummying requires originality. (Black Dot, Incorporated)

Fig. 33-6. The proofreader is responsible for make-up work and for finding errors. (Black Dot, Incorporated)

Fig. 33-7. Pressroom supervisor discusses problems with printing press operators. (J. Paul Kirouac)

Fig. 33-8. The stripper has the responsibility of cutting phototypesetting galleys to make up pages. (Black Dot, Incorporated)

Fig. 33-9. Teaching graphic arts is a challenging and rewarding profession. (Heidelberg Eastern, Incorporated)

Most careers in the graphic arts industries, especially those involving crafts, require an apprenticeship (Fig. 33-10). The educational experiences acquired and developed in school create a better selection of career possibilities. Students can determine whether they wish to enter the crafts areas or go into the professional activities that require higher formal education.

Apprenticeship usually requires four to six years, depending on the occupation

Fig. 33-10. Apprenticeship is the usual method of entry into the graphic arts field. (Heidelberg Eastern, Incorporated)

Fig. 33-11. Reporting political activities embodies knowledge of political science, psychology, and journalism. (*Houston Chronicle*)

chosen. Approximately 4000 high schools, vocational schools, technical institutes, colleges, and universities offer courses pertaining to graphic arts.

Technological changes in graphic arts production will create many new career opportunities. *Lithography* (offset printing) is one of the areas that has a good growth pattern. This is because lithography is one of the more popular methods of graphic reproduction.

Writing

Newspaper reporting and *technical writing* are two basic areas related directly to the graphic arts (Figs. 33-11 and 33-12). The many career possibilities in either area require a person who is creative, clear-thinking, interested, and dedicated. There are many areas of newspaper reporting. It is important that a person know about the specialty area on which he or she will report. Some of these specialty areas are medicine, politics, science, education, business, labor, religion, society, and sports. Reporters often become editors (Fig. 33-13).

Many writers, reporters, and editors start with a minimum of professional

Fig. 33-12. A technical writer must have a good background in journalism and in the technical area on which he or she is reporting. (*Houston Chronicle*)

preparation. It is suggested that one prepare through professional studies leading to a bachelor's degree in journalism. There are about 150 colleges and universities in the United States that offer this degree.

There are employment opportunities for people with journalistic hopes in newspaper, magazine, and book publishing. People who have real talent and training and who are curious, interested, and clever often seek careers in these fields.

Fig. 33-13. **Experience in reporting on various fields is an excellent background for becoming a city editor.** (*Houston Chronicle*)

Fig. 33-14. **Knowledge of the use of lighting is necessary for a career in photography.** (North American Rockwell)

Photography

Photography is an artistic and technical occupation involving much more than working a camera. The work of photographers varies greatly, depending on the area of specialization. A photographer must be well informed about photographic equipment and materials. One needs a knowledge of art and design and must understand the technical details of how the equipment works (Fig. 33-14). Also study Unit 48 on photographic careers.

Preparation for photography requires education and training. This can be gotten in graphic arts courses in high school and technical schools or college.

The number of employment opportunities in the photographic field is rising quickly, especially for industrial photographers. More and more of these specialists will be needed for research and development. They will also be needed in the widespread production of audiovisual aids for business, industry, schools, and government. Opportunities are expected to be very good for photographers working in scientific and engineering photography, illustrative photography, photojournalism,

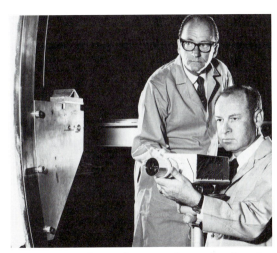

Fig. 33-15. **These photographers check out a new color (TV) camera that can function under lunar (moon) lighting conditions.** (Radio Corporation of America)

microfilming, and other highly specialized areas (Fig. 33-15).

The *Dictionary of Occupational Titles* (DOT) lists many photographic laboratory occupations on the technician level. These are very important for the graphic arts. *Technicians* work in color, developing, retouching, and mixing chemicals. They may also be classified as *photo checkers* and *slide mounters*, or they may be classified on a semiskilled level.

Paper and Allied Products Industries

Paper and allied products industries need professional and technical people such as *engineers*, *chemists*, and others with technical training. There is also a need for *scientists*. Some engineering opportunities are found in mechanical, chemical, electrical, and packaging engineering. There are also opportunities for trained *foresters*. Other areas include specialists in the production of pulp, paper, and paperboard (Fig. 33-16).

The growth in graphic arts technology means that more *systems analysts* and *computer programmers* are needed. Paper and allied products industries also need people with managerial and administrative abilities.

The careers relating to professional occupations require college preparation leading to a bachelor's degree. The technician and semiskilled fields often do not require more than fair mechanical preparation and a general knowledge about the exact job for which one is preparing.

Fig. 33-16. **The production of paper is basic to the graphic arts. Many excellent careers are available in the pulp, paper, and paperboard industries.** (St. Regis Paper Company)

Unit 34

Methods of Printing

Most printing is done by four major processes: (1) letterpress, (2) lithography, (3) gravure, and (4) screen.

Letterpress

This is the oldest printing method. Both the wooden block used by the Chinese and the movable type used by Gutenberg were letterpress printing methods. Letterpress printing is done from a raised surface (Fig. 34-1).

Single letters, entire lines of type, and plates are used to print daily newspapers

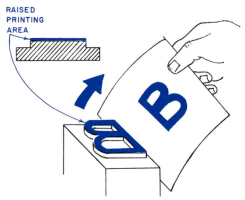

Fig. 34-1. **Printing from a raised surface (letterpress printing).**

and school textbooks. Many large letterpress printing presses print both sides of a continuous sheet of paper as it passes through the press (Fig. 34-2).

Lithography

Aloys Senefelder discovered this printing process by accident. He used a stone for his printing surface. Today metal and special paper printing surfaces are used.

Lithography (sometimes called *offset*) prints from a smooth surface (Fig. 34-3). The chemical principle that oil and water do not mix makes the offset-lithography process possible.

Advertising brochures and many food packages are good examples of this printing method. The operating principle of a commercial offset-lithography printing press is shown in Fig. 34-4.

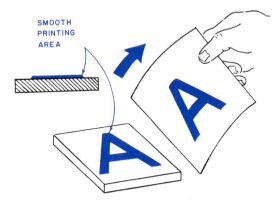

Fig. 34-3. **Printing from a smooth surface (lithography printing).**

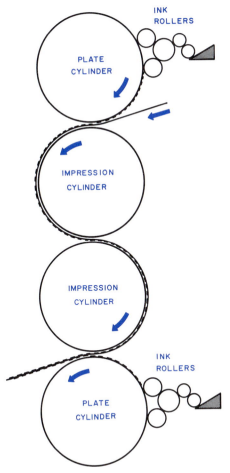

Fig. 34-2. **The way a letterpress works to print on both sides of the paper.**

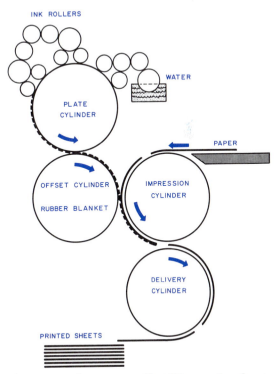

Fig. 34-4. **The way an offset-lithography sheet-fed press works.**

Gravure

Gravure is exactly the opposite of letterpress. In this process, the image is etched below the plate surface (Fig. 34-5). The entire metal printing plate runs in a vat of ink (Fig. 34-6). A sharp metal blade (*doctor blade*) scrapes the ink from the nonprinting surface. This leaves ink in the etched-out letters and illustrations. The paper then passes against the printing plate and draws the ink from the etched image. Sunday newspaper supplements and paper money are good examples of gravure (sometimes called *intaglio*) printing.

Screen

The process of screen printing is very different from the other three methods. Ink is forced through a cutout area somewhat like a stencil (Fig. 34-7). Screen printing uses a silk or metal screen material stretched over a frame.

A screen printing press is shown in Fig. 34-8. A cutout stencil is attached to the

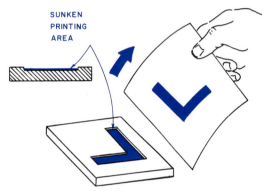

Fig. 34-5. Printing from a sunken surface (gravure printing).

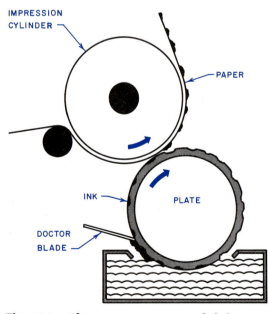

Fig. 34-6. The way a gravure web-fed press works.

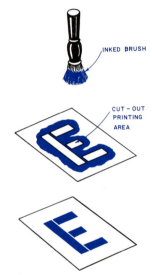

Fig. 34-7. Printing through a cutout area (stencil printing).

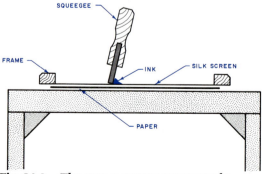

Fig. 34-8. The way a screen press works.

■ 192

screen. Ink is poured onto it, and a squeegee (Fig. 34-8) is pulled across the image area. This forces the ink through the stencil openings, causing it to print. Felt pennants, coasters, bumper stickers, cookware decorations, signs, wallpaper, fabrics, circuit boards, and membrane switches are examples of screen printing.

Unit 35

Allied Processes and Materials

Two necessary materials for the graphic arts industry are paper and ink. Without paper (the image carrier) and ink (the image maker), there would be no graphic arts industry.

Paper

It has taken many centuries to develop the material called paper. A paper similar to that used today was first made by a Chinese government official named Ts' ai Lun in A.D. 105. This form of paper was made from the inner fibers of the mulberry tree. Before paper as we know it was made, papyrus had been used in ancient Egypt. *Papyrus* was made from reeds that grew along the Nile River. Two other early paper forms were *vellum* and *parchment*. Both of these were made from animal skins. The papermaking machine was not developed until 1804. Before this date, all paper had to be made by hand.

Some historians call the nineteenth century the *paper age*. Many developments that took place then laid the groundwork for twentieth-century progress. Chemistry, physics, and other sciences have played a large part in the growth of paper technology.

Wood is the most important ingredient in paper, but other materials are also used. Some of these are linen, flax, cotton, and recycled wastepaper. Paper is one of the most important goods in the modern business world. Figures 35-1 and 35-2 show paper-manufacturing machines.

Fig. 35-1. The wet end of the Fourdrinier paper-manufacturing machine. (Crown Zellerbach Corporation)

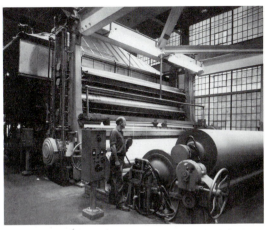

Fig. 35-2. The dry end of the Fourdrinier paper-manufacturing machine. (Crown Zellerbach Corporation)

Ink

Another element important to the printing process is ink. In fact, ink ranks as the second most important material used in printing. Printing ink is made in many colors, thicknesses, and compositions to serve many needs. As with paper, the Chinese were the first to experiment with printing ink. Their inks were made from plant substances mixed with various colored natural materials and soot, or lampblack. Printing inks are made up of the basic pigment (the coloring agent) and the vehicle, or carrier, for the pigment. Quality *synthetic* (not natural) compounds are also being made and used.

Millions of pounds of inks are used each year to print daily and weekly newspapers, magazines, books, advertisements, and other printed materials. Newspapers alone in the United States use approximately 600 000 pounds (265 800 kilograms) of ink daily. An industrial mill operator is shown adjusting an ink mill in Fig. 35-3.

Binding

Binding fastens sheets of paper into a usable form. A modern binding machine is shown in Fig. 35-4. It is *ejecting* (releasing) finished magazines ready for shipping.

Nearly 6000 years ago, the Egyptians practiced a form of bookbinding. They fastened papyrus sheets end to end, sometimes forming lengths over 18 feet (6 meters). These were made into rolls. Rolls of papyrus were called *scrolls*. Today we see scrolls only in museums.

The type of binding used to form this textbook was not developed until the art of printing on movable metal type was discovered.

Fig. 35-3. Adjusting an ink mill. (Sinclair and Valentine)

Fig. 35-4. Finished magazines coming off a modern high-speed binding machine. (Consolidated International Corporation)

Unit 36

Hand Composition

Some type is still *composed* (put together) by hand. This is *foundry type*, and it is similar to the type Gutenberg used over 500 years ago, when newspapers and books had to be composed by hand. Foundry type is used mostly for advertising and small business jobs, such as letterheads, envelopes, and calling cards.

The Point System

Type cannot be composed without knowing the printer's system of measurement. This measurement system is called the *point system.*

The point system developed because the customary system of inches and fractions was not easy enough for measuring type. A *point* is equal to one seventy-second ($^1/_{72}$) of an inch. Therefore, an inch contains 72 points.

Another measurement term in the point system is the *pica.* It is equal to 12 points. Therefore, there are 6 picas in 1 inch.

The *nonpareil,* another measurement term in the point system, is equal to 6 points. All hand composition is thought of in terms of the point system of measurement.

The printer's measuring device is called a *line gage* (Fig. 36-1). It is designed so that it may be easily hooked at the end of a line when it is being used. The line gage is divided into nonpareils and picas on the left side. The center is divided into *agates* (approximately 5$^1/_2$ points) and the right side into inches.

Type

The most important tool of the printer is *type.* A piece of foundry type is shown in Fig. 36-2. The major parts of a piece of type should be learned as quickly as possible. This will help you identify *typefaces* (surfaces) when you begin using different type styles.

Foundry type is measured by points. Type is measured from the nick side to the back side of the body. The more common sizes of type are from 6 point through 72 point (Fig. 36-3).

Spacing Material

Space is needed between words. Several different sizes of spacing are used (Fig. 36-4). Three points is the amount of space most often used between words, but other spaces are also used. The *em quad* is the basic spacing unit. The other six spacing materials come from it. Study Fig. 36-4 to learn how each space and quad is related to the basic unit of spacing, the em quad.

To get spacing between lines, thin strips of metal, called *leads* and *slugs* are used. The lead most often used is 2 points in thickness. The slug most often used is 6 points (Fig. 36-5). Leads and slugs can be cut to any pica lengths from 10 to 50.

The Type Case

Separate pieces of type are stored in a shallow drawer that is divided into many small *compartments* (sections). This drawer is called a *type case.* It contains compartments for all the letters (both small and capital), numbers, punctuation

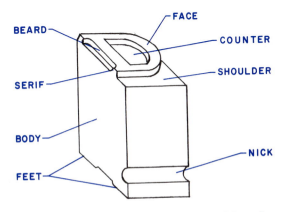

Fig. 36-2. The names of the parts of foundry type.

Fig. 36-1. A printer's line gage.

Shelby

120 pt. 3A 4a 3-1

Shelby

96 pt. 3A 4a 3-1

Pictured

84 pt. 3A 4a 3-1

Changed

72 pt. 3A 4a 3-1

THEprinter

60 pt. 4A 6a 3-1

WHILE apart

48 pt. 4A 7a 3-1

FINE quality coat

42 pt. 4A 8a 4-1

SOME xylophones

36 pt. 5A 9a 4-1   Lower case alphabet 404 pts.   Characters per pica .85

KNIGHTS usually are

Fig. 36-3. Common type sizes. (American Type Founders)

Spartan Medium
Series Number 680

Characters in complete font

A B C D E F G H I J K L M N
O P Q R S T U V W X Y Z &
$ 1 2 3 4 5 6 7 8 9 0 * ¢ %
a b c d e f g h i j k l m n o p q
r s t u v w x y z . , - : ; ! ? '´"" ( )

Ligatures are included in fonts of 6 to 18 point
sizes, and are obtainable in 24 to 120 point
sizes in foundry lines.

fi ff fl ffi ffl

---

30 pt. 6A 11a 5-1   Lower case alphabet 328 pts.   Characters per pica 1.0

JADE varies into

24 pt. 9A 15a 7-1   Lower case alphabet 246 pts.   Characters per pica 1.4

BRAZIL and countries

18 pt. 10A 21a 8-1   Lower case alphabet 205 pts.   Characters per pica 1.7

THE EARLY printers came

16 pt. 13A 24a 9-1   Lower case alphabet 178 pts.   Characters per pica 1.9

THE EARLY printers cast their

14 pt. 18A 32a 10-1   Lower case alphabet 158 pts.   Characters per pica 2.2

THE EARLY PRINTERS CAST THE
They instructed some local black

12 pt. 21A 38a 12-1   Lower case alphabet 137 pts.   Characters per pica 2.5

THE EARLY PRINTER CAST THE TYPE
They instructed some local blacksmith

10 pt. 25A 45a 13-1   Lower case alphabet 117 pts.   Characters per pica 2.9

THE EARLY PRINTERS CAST THEIR TYPES
They instructed the local blacksmith to make

8 pt. 26A 48a 19-1   Lower case alphabet 93 pts.   Characters per pica 3.7

THE EARLY PRINTERS CAST THEIR OWN TYPES AND
They instructed some local blacksmith to make the iron

6 pt. 28A 51a 21-1   Lower case alphabet 89 pts.   Characters per pica 3.8

THE EARLY PRINTERS CAST THEIR OWN TYPES, MADE INK
They instructed some local blacksmith to make the iron frames
or chases in which the types are confined for printing, and

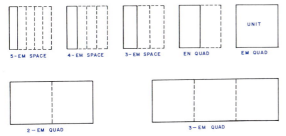

Fig. 36-4. The seven common sizes of word-spacing materials.

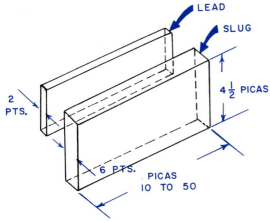

Fig. 36-5. Line-spacing materials: leads and slugs.

Fig. 36-6. Layout of the California job case.

marks, and word-spacing materials. The type case most often used is the *California job case* (Fig. 36-6).

Type cases are stored in a cabinet called a *bank* (Fig. 36-7). Banks also have space to store leads and slugs and other small materials. The working surface on the top of

Fig. 36-7. A bank providing storage for type cases, leads, and slugs. (Thompson Cabinet Company)

Fig. 36-8. A composing stick. (H.B. Rouse and Company)

the bank can be used to hold the job case as you are composing type.

Composing Type

Type is composed in a device called a *composing stick* (Fig. 36-8). Composing sticks can be adjusted to pica and non-pareil line lengths. The following steps will guide in composing type:

1. Adjust the composing stick to the desired line length. Raise the clamps and slide the *knee* along the *body* to the desired length. Press the clamp back down into position.
2. Stand in front of the type case and hold the composing stick in the left hand, as shown in Fig. 36-9. The composing stick should be held at a slight angle, with the thumb placed in the *throat* of the stick.
3. Place a slug into the composing stick. Type can then be placed into it. Each letter

Fig. 36-9. The proper method of holding a composing stick.

Fig. 36-11. Dumping the composing stick.

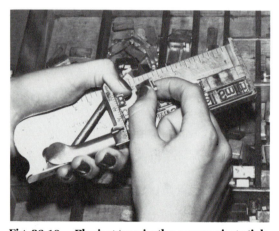

Fig. 36-10. Placing type in the composing stick.

of type is picked separately from the type case. Place it in the composing stick (Fig. 36-10). The nick of the type should always face toward the open side of the throat of the composing stick.

Note: Type is always set from left to right (in the same way that we read). Hold each piece of type in place with the thumb of the left hand.

After the words have been placed in each line, the line must be made the proper length. This is called *justification.* Spaces and quads are used to make each line exactly the same length.

4. Remove the *typeform* (several lines of type composed together) from the composing stick. This is called *dumping.* The proper method for dumping the composing stick is shown in Fig. 36-11.

5. Place the composing stick into a *galley* (a shallow three-sided metal tray). The open throat of the composing stick is always turned away from the *compositor* (the person doing the composing).

Grip the typeform by placing the thumbs against the bottom line and the index fingers of both hands against the top line. The typeform is then slid from the composing stick into the corner of the galley.

CAUTION: Be sure to *slide* the typeform on the galley. Do not pick up the typeform.

6. Tie (make secure) the lines of type with string after the type is removed from the composing stick. Figure 36-12 shows the right way of tying a typeform. A typeform should always be tied with the string going around it in a clockwise direction (to the right). Wrap the string around the typeform at least six or seven times. A *printer's*

Fig. 36-12. Tying the typeform.

knot is then tied into the corner of the typeform (Fig. 36-13). The typeform is now ready for *proofing* (making a sample copy).

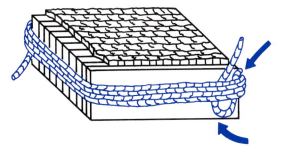

Fig. 36-13. A printer's knot.

Unit 37

Proofing Type

Skill and practice are required to make a good *proof* (sample copy). A proof reveals errors such as a spelling mistake, a piece of type upside down, type with a different face (image), and wrong or damaged (worn) type pieces.

The Proof Press

The most common proof press used in schools is the one shown in Fig. 37-1. It is designed for easy operation.

The typeform is inked by using the *brayer* (an ink roller with a handle). Inking is done before the *impression cylinder* is rolled over the type to make the *impression* (the print).

Proofing

1. Place the galley with the tied typeform on the bed of the proof press (Fig. 37-2). To get a good proof, place the typeform in a position so that the horizontal lines are at an angle of 90° to the proof-press impres-

sion cylinder. Be sure that the ends of the string are not under the typeform.

2. Ink the brayer and the ink plate. With the ink knife, place a small amount of ink

Fig. 37-1. A proof press used in many school graphic arts laboratories. (The Challenge Machinery Company)

Fig. 37-2. **The correct position of a galley and typeform in the bed of a proof press.**

Fig. 37-3. **Inking the ink plate on the proof press.**

on the ink plate. Spread it over the plate with the brayer (Fig. 37-3). Use only a small amount of ink.

3. Ink the typeform. Roll the brayer across the face of the typeform at least two or three times to get the proper ink coverage. Allow only the weight of the brayer to push against the typeface (Fig. 37-4).

4. Place a sheet of proof paper on top of the typeform. Pull the impression cylinder across it (Fig. 37-5). Using one sheet of paper, roll the impression cylinder over the typeform *only once* to make the proper proof.

5. Wipe away the ink from the face of the type after proofing a typeform. If the ink is not removed from the type immediately after printing, it will dry on the face and will fill the *counters* (centers) of the type (Fig. 37-6).

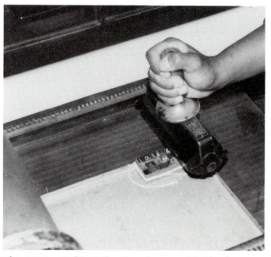

Fig. 37-4. **Inking the typeform with the brayer.**

Fig. 37-5. **Pulling a proof. Note how clean it is.**

6. Read the proof for errors or damaged type. If anything is found to be wrong, correct the typeform and reproof. After the typeform has been corrected, it is ready to be prepared for the printing press.

7. Return the type to the type case after the typeform has been printed on the proof press or on a regular press. Untie the typeform in the galley. Grasp the typeform with the fingers of both hands. Lift the form and place it in the left hand between the thumb and the middle finger (Fig. 37-7).

Place the index finger right under the bottom slug of the typeform. This gives added support (Fig. 37-7). Distribute each line of type. Begin with the right side of the line. Pick up an entire word between the thumb and the index finger of the right hand. Distribute each character into its proper compartment in the type case. Replace all spacing material in the correct locations.

Fig. 37-6. **Washing a typeform with a cloth dampened with solvent.**

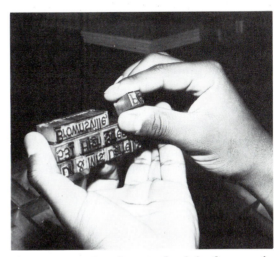

Fig. 37-7. **Putting the type back in the case is called distributing a typeform. Several lines of type can be firmly held in the left hand. Individual letters are put back with the right hand.**

Unit 38

Preparing Type for the Press

Lockup prepares a typeform for the press. It holds the typeform in place while it prints on the press.

Tools

The tools and materials most often used in the lockup procedure are shown in Fig. 38-1.

A. *Chase.* This is a metal frame in which typeforms are locked so that they may be held firmly in the press.

B. *Furniture.* These are pieces of metal or wood used in locking and making up typeforms. They come in different pica widths and lengths.

C. *Quoins.* These are mechanical devices that are made so that they can be ex-

Fig. 38-1. **The tools used in locking up a typeform.**

Fig. 38-2. **A stone, sometimes called an imposing table. Storage is provided for furniture, reglets, and galleys.** (Thompson Cabinet Company)

Fig. 38-3. **Furniture placed around the typeform.**

panded. In this way, they put added pressure against the typeform and the type.

D. *Quoin key.* This device tightens quoins. It is usually in the shape of a key.

E. *Reglet.* This is a thin wooden strip used to fill a narrow space in the chase. The most frequently used thicknesses are 6 and 12 point. Reglets come in all pica lengths from 10 to 60.

F. *Planer.* This is a block of hardwood. One face is level and smooth. It is used on the face of typeforms to push down all the letters and make the form level.

G. *Mallet.* This is a wooden, rawhide, or fiber-headed hammer used to tap the planer.

H. *Stone.* This is a flat-surfaced table on which the typeforms are locked. The surface of the table is either stone or metal. The stone is sometimes called an *imposing table* (Fig. 38-2).

Lockup Procedure

Follow these steps to lock up the typeform for the press properly:

1. Clean the surface of the stone.

2. Slide the typeform from the galley onto the stone.

3. Place the chase over the typeform.

4. Place the typeform in the proper position in the chase. The long side of the paper to be printed on should be parallel with the long side of the chase. The typeform heading should be to the left or to the bottom of the chase.

5. Place furniture around the typeform (Fig. 38-3). Furniture should surround the entire typeform. Each piece of furniture should *chase* (follow) the other pieces of furniture around the typeform.

6. Fill in the left side and bottom of the chase with furniture (Fig. 38-4).

7. Remove the string from the typeform.

CAUTION: Do not remove the string before this step.

8. Place the quoins to the top and the right side of the chase (Fig. 38-5).

9. Fill in the right side and the top of the chase with furniture (Fig. 38-6).

10. Lightly tighten the quoin with the quoin key. (See Fig. 38-8.)

11. Plane the typeform (Fig. 38-7). Hold the planer block level with the typeform and tap lightly with the mallet.

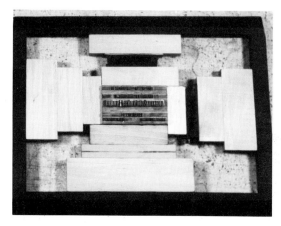

Fig. 38-6. The typeform, furniture, quoins, and a reglet properly positioned in the chase.

Fig. 38-4. The left side and bottom of the chase filled with furniture.

Fig. 38-7. A typeform being planed (each character made the same height).

Fig. 38-5. The quoins placed to the top and the right side of the chase.

Fig. 38-8. Tightening a quoin with a quoin key.

12. Tighten the quoins (Fig. 38-8). Each quoin should be tightened a small amount until the form is held firmly in place.

13. Check the form for *lift* (Fig. 38-9). Lift (raise) one end of the chase and slide a quoin key under it. Lightly tap each piece of type with a finger. If letters slide through the typeform, more spacing material must be placed into the line.

The typeform is now ready for the press.

CAUTION: Handle the locked chase *very* carefully.

Fig. 38-9.   Checking the typeform for lift (raise).

Unit 39

Letterpress Printing Presses

There are three types of letterpress printing machines in use. These are the (1) platen, (2) cylinder, and (3) rotary presses. The working principles of each are different. But each press uses typeforms that have raised images.

The Platen Press

The typeform for the platen press is held on a vertical bed. It is printed at one time (Fig. 39-1). Platen presses are either power-driven (Fig. 39-2) or hand-driven (Fig. 39-3). Both kinds are used in school graphic arts laboratories. Cards and stationery are often printed on the platen press.

The Cylinder Press

The typeform on a cylinder press is held on a large, flat bed that moves back and forth under the impression cylinder (Fig. 39-4). Sheets of paper are held to the cylin-

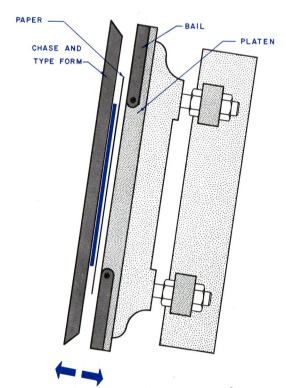

Fig. 39-1.   The way a platen press works.

Fig. 39-2. A power-driven, hand-fed platen press. (Chandler and Price)

Fig. 39-3. A platen press that works by hand (pilot press). (Chandler and Price)

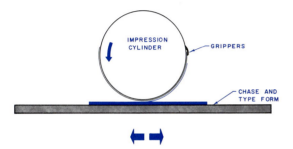

IMPRESSION CYLINDER

GRIPPERS

CHASE AND TYPE FORM

Fig. 39-4. The way a cylinder press works.

der by the grippers. Ink is transferred to the paper as it passes between the impression cylinder and the typeform. With this kind of press, only a small part of the typeform touches the paper at any one time.

Cylinder presses are of two designs: the flat-bed (Fig. 39-5) and the vertical-bed (Fig. 39-6). These presses are used to print business forms, advertising matter, and pamphlets of all kinds.

Fig. 39-5. A flat-bed cylinder press. (Mergenthaler Linotype Company)

Fig. 39-6. A vertical-bed (up-and-down) cylinder press. (Miehle-Goss-Dexter)

The Rotary Press

Typeforms called printing plates are shaped into half circles to fit the impression cylinder of a rotary press. To print the image, a continuous roll of paper is fed between the impression cylinder and the plate cylinder of this press. Two entire pages are printed with one *revolution* (turn) of the cylinder (Fig. 39-7).

Nearly all rotary presses print two or more colors at one time. These presses are large and are used for high-speed work, such as newspapers, magazines, catalogs, and advertising matter (Fig. 39-8).

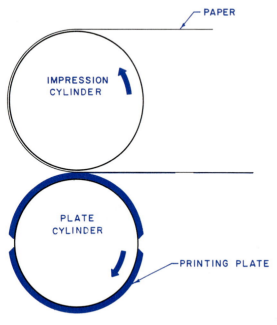

Fig. 39-7.   The way a rotary press works.

Fig. 39-8.   A five-color rotary magazine press. Paper passes between the cylinders at a rate of 2000 feet per minute. (The Goss Company)

Unit 40

Preparing and Printing with the Platen Press

The platen press is an important machine in the graphic arts industry. Many business forms are printed with this kind of press. Follow the steps listed for the best operation.

Dressing the Platen

1. Remove the old *dressing* (platen covering used before). Raise the *bails* (paper clamps) and pull out the dressing sheets. Save the gage pins and good paper.

2. Dress the platen. Use one tympan sheet (a *tympan* is the sheet of paper placed around the impression cylinder), three sheets of 60-pound (89-grams-per-square-meter) coated book paper (hanger sheets), and a pressboard (Fig. 40-1).

3. Place the tympan over the three sheets of coated book paper (hanger sheets). Extend these sheets about ³/₄ inch (20 millimeters) below the platen; clamp them to the bottom bail.

4. Insert the pressboard under these sheets. See Fig. 40-1. Clamp the top bail. *Note:* The pressboard must not be clamped under the bails.

Inking the Press

1. Place a small amount of ink on the lower left side of the ink disk (Fig. 40-2).

2. Turn on the press and allow the ink rollers to spread the ink.

3. Stop the press.

Fig. 40-1. Dressing the platen.

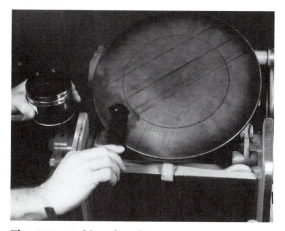

Fig. 40-2. Inking the platen press.

Positioning the Gage Pins

1. Place the chase (see Fig. 40-6) into the bed of the press (Fig. 40-3).

2. Take a trial impression by rolling the press over by hand. The type will print on the tympan (Fig. 40-4).

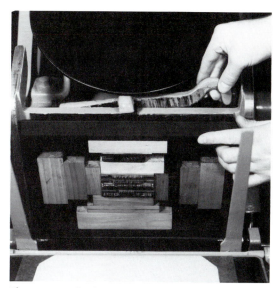

Fig. 40-3.   Placing the chase on the bed of the platen press.

Fig. 40-4.   The first impression with gage pins in place.

■ **207**

3. Measure and mark the paper margins you want on the tympan.

4. Place two gage pins at the bottom of the sheet and one gage pin on the left side of the sheet (Fig. 40-5). The sharp point of the pin should *penetrate* (go through) the tympan about 1/8 inch (3 millimeters) beyond the paper edge. Be sure the points come back through the tympan (Fig. 40-6).

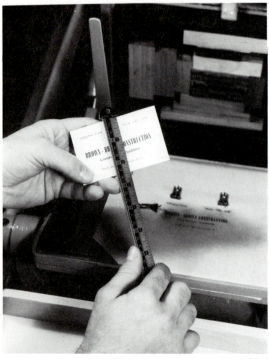

Fig. 40-5. Using the line gage to position the printing properly.

Fig. 40-6. A properly attached gage pin.

Taking the Trial Impression

Place a paper sheet against the pins and print a copy. Turn the press by hand. Adjust the pins for type position. (See Fig. 40-5.)

Set the gage pins by tapping the front edge with a quoin key. The position permission should now be gotten from your teacher.

Performing the Make-ready Operation

1. Poor (depressed) printing areas must be built up with added packing. On a printed sheet, circle the *faint* (light) printed areas.

2. Glue tissue paper inside these circled areas (Fig. 40-7).

3. Glue this *make-ready* (built-up) sheet on the top hanger sheet (Fig. 40-8). This sheet must be right under the printing area on the tympan. Caret marks (∧) will help mark the proper position.

4. Place the pressboard on top of this hanger and make-ready sheet. Replace the dressing sheets under the top bail. The ready-to-run permission should now be gotten.

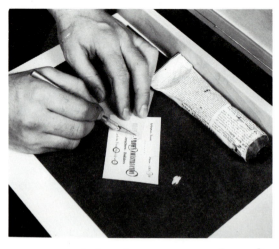

Fig. 40-7. Placing tissue paper on the make-ready sheet.

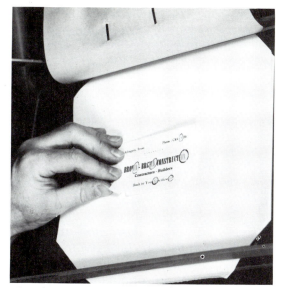

Fig. 40-8. Hanging in the make-ready sheet.

Fig. 40-9. Pulling the impression lever.

Feeding the Platen Press

1. Turn on the power. Give the flywheel a push to help the motor.
2. Place a sheet of paper against the pins with the right hand.
3. Pull the impression lever back to print (Fig. 40-9).
4. Remove the printed sheet with the left hand. While this is being done, the right hand should be placing a new sheet against the pins (Fig. 40-10).

Cleaning the Press

1. Remove the chase from the press.
2. Wash the ink from the type, the ink disk, and the rollers (Fig. 40-11). No ink should be allowed to dry on any of these surfaces.
3. Wipe down and oil the press at regular intervals.

Replacing All Tools and Supplies

1. Unlock the chase. Return the furniture and quoins to their proper places.
2. Retie the typeform. Distribute the type into the case as soon as possible.
3. Return all other tools and supplies that were used while printing to their proper places.

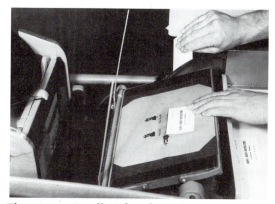

Fig. 40-10. Feeding the platen press.

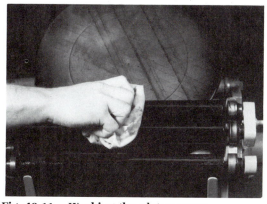

Fig. 40-11. Washing the platen press.

Tools and Equipment for the Lithographer

Basic tools and supplies for the lithographer (the person working with an offset-printing press) include type, layout equipment, camera, printing plates, and offset presses. These essentials for offset-lithography are discussed and explained in the following sections. Review Unit 34, "Methods of Printing."

Type

The two methods of type composition are *hot* and *cold*. Hot-type composition includes any typesetting done by methods that make use of type metal or *metal slugs* (entire lines of type). Figure 41-1 shows a high-speed slug-casting machine. It can be operated by hand or by perforated tape. It produces up to 15 lines of type per minute. Setting foundry type, which you learned about in Unit 36, "Hand Composition," is hot composition.

Cold-type composition is any typesetting produced by methods that do not use metal type characters. *Photographic, strike-on,* and *hand illustration* are three basic techniques of cold-type composition. Figure 41-2 shows a machine that produces type on photographic paper. Several different kinds and sizes of type can be composed with this machine.

One of the most frequently used cold-composition machines is a common *typewriter* (strike-on method). Hand illustration and lettering are also common techniques for producing copy.

Layout Equipment

A needed item for the *litho* (lithography) artist is a *light table.* The top of a light table

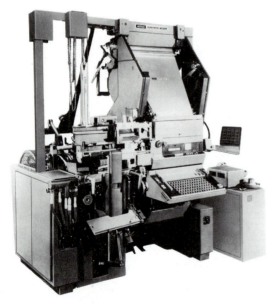

Fig. 41-1.   A high-speed slug-casting machine. It can be operated by punch tape or by the keyboard. (Mergenthaler Linotype Company)

Fig. 41-2.   A cold-type composition machine. It makes large type by the photographic method. (Visual Graphics Corporation)

is frosted glass. When lighted from below, it makes a working surface on which it is easy to read and work with photographic negatives. A typical light table is shown in Fig. 41-3.

Other pieces of layout equipment include a drafter's drawing set, a T square, and a triangle. These are used to prepare copy and to make the *flat* (assembled photographic negatives).

Camera

A good camera is one of the most important pieces of equipment for a lithographer. It is used to make *glossy prints* (pictures). It comes in many styles and makes that use different types of film.

The *process camera* is built just for the printer. Unlike the usual camera, it holds the copy to be photographed in a copyboard. These cameras can *reduce* (make smaller) the copy to one-third of the original size. They can *enlarge* (make larger) the copy to twice the original size. A horizontal process camera is shown in Fig. 41-4.

Plates

Basically, there are two types of offset-lithography printing plates: those produced by photographic means and those in which the image is placed right onto the plate. *Photographically produced* plates can be made of metal, plastic, or paper. They are designed for medium to long runs of 1000 or more copies.

Direct-image plates are made by typing or drawing right onto the plate. They are used for shorter runs of less than 1000 copies.

Presses

Several different makes, models, and sizes of offset-lithography printing presses are manufactured. Office presses print pieces of paper as small as 3 × 5 inches (76 × 127 millimeters) and up to 14 × 20 inches (335 × 508 millimeters). Two small offset presses used in business offices and in small commercial plants are shown in Figs. 41-5 and 41-6.

Offset presses designed for commercial use are manufactured in many sizes. Some

Fig. 41-3. A light table is used by the litho artist to make up a flat. (nuArc Company)

Fig. 41-4. A horizontal process camera used in making film negatives.

Fig. 41-5. An offset-lithographic printing press used in schools and offices. (Addressograph-Multigraph Corporation)

Fig. 41-6. A small offset-lithographic printing press used in small commercial plants. (Addressograph-Multigraph Corporation)

of these presses can be used to print multicolor (three or more colors) materials (Figs. 41-7 and 41-8). They are designed for perfect *register* (printing-position) control and for high-quality work. Large commercial printers use four- and five-color offset presses.

Fig. 41-7. A one-color commercial offset-lithographic printing press. (Mergenthaler Linotype Company)

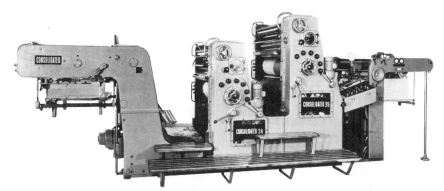

Fig. 41-8. A multicolor commercial offset-lithographic printing press used in large commercial plants. (Consolidated International Corporation)

Steps in the Offset-Lithography Process

Several steps are needed to get printed material in the offset-lithography process. The following four will help you learn the process: (1) composing and photographing copy, (2) stripping negatives into a flat, (3) making the plate, and (4) preparing and printing with the offset press.

Composing and Photographing Copy

Basically, there are two kinds of copy: line and halftone. *Line copy* is made up of type or illustrations that are either black or white. This means there is no *gradation* (shading) of color. *Halftones* are made from glossy pictures and are made up of hundreds of small dots. Examine the pictures in this book with a magnifying glass. You will see the dots. The basic kinds of copy are shown in Fig. 42-1.

Stripping Negatives into a Flat

After the copy has been photographed with the process camera, negatives are made. They must be placed and taped onto a special sheet of paper. Placing the negative or negatives (Fig. 42-2) onto this paper (*stripping paper*) is a very important process. This work (called *stripping*) is done on the light table (Fig. 42-2).

After being taped, the negatives and stripping paper are turned over. A razor blade or an artist's knife is used to cut out the printing area from the stripping paper. Small dots and unwanted printing areas are then blocked out by use of an *opaquing solution* (Fig. 42-3). When the negative has been *opaqued*, the nonprinting areas have been blocked out. The negative and stripping paper together in this form are called a *flat*.

Making the Plate

After the flat has been made, the next big step is to make the *plate*. The flat is laid on a light-sensitive metal plate. This is placed into a vacuum frame for exposure to bright light (Fig. 42-4). After the plate has been ex-

Fig. 42-1.   Kinds of copy: line type, line illustration, and halftones (pictures).

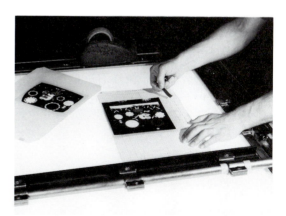

Fig. 42-2.   A litho artist taping a negative to stripping paper.

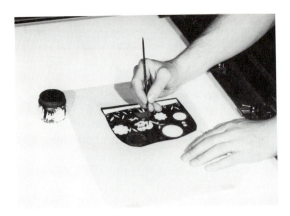

Fig. 42-3.   Opaquing a negative.

Fig. 42-4.   The flat and the metal plate in a vacuum frame ready for exposure.

adjustment for 8½ × 11 inch (216 × 279 millimeter) paper.

2. Make the correct adjustment on the register board. In that way, the paper will feed properly into the press. The different adjustments and settings are shown in Fig. 42-7 for a 8½ × 11 inch (216 × 279 millimeter) sheet.

3. Place ink in the *ink fountain* (or *ink reservoir*) (Fig. 42-8).

4. Prepare the *water fountain* (or *water reservoir*) and the rollers (Fig. 42-9). Allow the press to run for about 10 minutes to

Fig. 42-5.   Developing a metal offset-lithography plate.

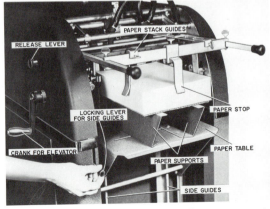

Fig. 42-6.   The proper setup for the paper magazine.

posed, it is *developed* (Fig. 42-5). For most plates, two basic chemicals are used in developing: an *etching solution* and a *developing solution*. The etching solution is spread over the plate first. The developing solution is then rubbed over the plate to bring out the image.

Preparing and Printing with the Offset Press

1. Adjust the paper *magazine* (feeding mechanism) for the size of paper that will be printed. Figure 42-6 shows the proper

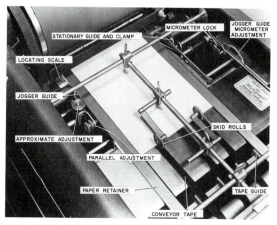

Fig. 42-7. The proper setup for the register board.

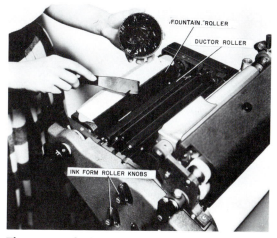

Fig. 42-8. Placing ink in the ink fountain.

Fig. 42-9. The water fountain and rollers.

moisten the water rollers thoroughly before printing.

5. Install (fasten) the plate on the press. Attach the *lead end* (top) of the plate to the plate cylinder, as shown in Fig. 42-10. The *trailing end* (bottom) of the plate is attached to the plate cylinder (Fig. 42-11). Be sure to lock the plate into place by tightening the *lock screw* on the clamp (Fig. 42-11).

6. Complete the printing operation:

> **a.** Start the press and contact (lower) the water roller to the plate.
>
> **b.** Contact the ink roller to the plate.
>
> **c.** Contact the plate cylinder to the blanket (printing) cylinder.

Fig. 42-10. Attaching the lead end of the printing plate to the plate cylinder.

Fig. 42-11. Attaching the trailing end of the printing plate to the plate cylinder.

d. After one or two revolutions (full turns), start the paper through the machine. Print the number of copies you want.

7. Clean the press:

a. Remove the plate from the plate cylinder. Prepare the plate for storage.

b. Remove the ink from the ink fountain (Fig. 42-12).

c. Attach the cleaning unit and apply solvent to the ink rollers (Fig. 42-13).

d. Clean the rollers and water fountain.

e. Wipe all ink, oil, and dirt from the press.

Fig. 42-12. Removing ink from the ink fountain.

Fig. 42-13. Applying cleaning solvent to the ink rollers.

Unit 43

Binding Equipment

Several pieces of equipment are needed to do basic binding. Almost all commercial printing plants use such equipment.

Paper Cutters

There are two kinds of paper cutters: *hand-operated* (Fig. 43-1) and *hydraulic-operated* (Fig. 43-2). Both of these machines can cut more than a 2-inch (50-millimeter) thickness of paper with hairline accuracy.

Paper Folders

Paper folders are manufactured in many sizes and designs. Small ones that sit on

Fig. 43-1. A hand-operated paper cutter. (Chandler and Price Company)

Fig. 43-2. A hydraulic-operated paper cutter. (Chandler and Price Company)

Fig. 43-4. A power-operated wire stitcher. (Interlake Steel Corporation)

Fig. 43-3. A high-speed paper folder that can make up to five folds in a sheet in one pass through the machine. (Consolidated International Corporation)

tables are designed for folding letterheads. Large paper folders, such as the one shown in Fig. 43-3, are designed for making *multiple* (many) folds.

Stitchers

Stitchers are designed to make staples from a continuous roll of wire and then to insert the staples into the paper. These machines produce the proper length of staple needed for the thickness of the paper being bound. This machine can staple paper two sheets thick up to 1 inch (25 millimeters) thick (Fig. 43-4).

Fig. 43-5. A power paper drill. (Chandler and Price Company)

Paper Drills

A paper drill machine *drills* (bores) holes in paper at a high speed. The center of the drill bit is hollow to allow the cut centers of the holes to be forced through the bit. This machine drills through a thickness of paper 1 inch or greater (Fig. 43-5).

Basic Methods of Binding

There are nine basic methods of binding sheets of paper together. Each has advantages and disadvantages. The method to use depends on the type of printed matter and its use. An illustration and brief description of each type of binding follows.

Saddle-wire Stitch

Saddle-wire-stitch booklets (Fig. 44-1) are the simplest and cheapest to bind. This type of binding cannot stand up to heavy use and is not considered long-lasting. Small advertising booklets, programs, catalogs, and other printed matter are bound by the saddle-wire-stitch method.

Side-wire Stitch

Side-wire-stitch booklets (Fig. 44-2) are simple and not costly to bind. Booklets of any thickness under 1 inch (25 millimeters) can be bound in this way. A disadvantage is that the book will not lie open for easy reading. Thick catalogs and magazines are often bound by the side-wire-stitch method.

Sewn Softcover Binding

Sewn softcover bindings (Fig. 44-3) cost more than either saddle- or side-wire bindings. In this long-lasting method, each *signature* (sheet of paper folded into several pages) is sewed to another signature to form the wanted thickness. The bound sheets (book) will lie flat when opened. Thick books are sometimes bound by this method.

Sewn-case Binding

The sewn-case binding method (Fig. 44-4) is used when a *permanent* (long-lasting) binding is wanted. This is the most costly of the nine types of binding. Sheets of any

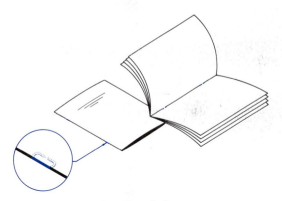

Fig. 44-1.   Saddle-wire stitch.

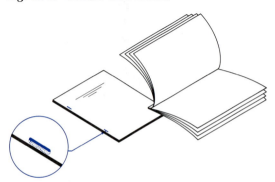

Fig. 44-2.   Side-wire stitch.

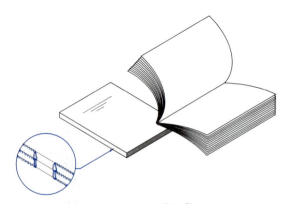

Fig. 44-3.   Sewn softcover binding.

size and number can be sewn-case bound. This method features hard and thick covers to protect the printed sheets. Many school textbooks and library books are bound in this style.

Binding Post

The binding-post method (Fig. 44-5) of binding several sheets of paper has some advantages. The boltlike binding element can be of any length. As many sheets of paper as wanted can be bound. Single sheets can be added or removed without affecting the strength of the binding. Photographic albums are bound this way.

Ring Binder

The ring-binder method (Fig. 44-6) has many uses and advantages. Punched sheets of paper are held together by rings. They can be opened easily for adding or removing one or all of the sheets from the binder. School notebooks and company catalogs are examples.

Plastic Cylinder

Plastic-cylinder binding (Fig. 44-7) holds single sheets of paper together with several plastic rings. The individual sheets of paper are punched with rectangular holes. The plastic-cylinder rings are inserted through these holes to hold the sheets. Plastic cylinders come in many colors. Labeling can be done on the wide part of the plastic cylinder. A few sheets or up to 250 can easily be bound by this method. Booklets, brochures, and company catalogs are well-known examples.

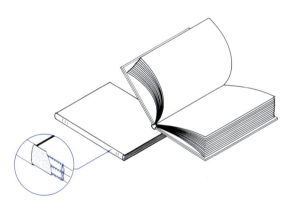

Fig. 44-4. Sewn-case binding.

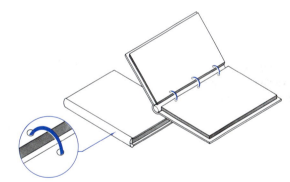

Fig. 44-6. Ring binder.

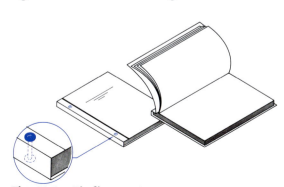

Fig. 44-5. Binding post.

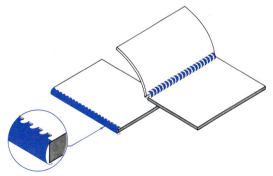

Fig. 44-7. Plastic-cylinder binding.

Spiral Wire

Spiral-wire binding (Fig. 44-8) holds the sheets of paper together with a wire coil. It runs through small cylindrical holes punched into the sheets. One good point about this kind of binding is that books, when opened, always lie flat for easy reading. This kind of binding is also economical. School notebooks are usually bound by this method.

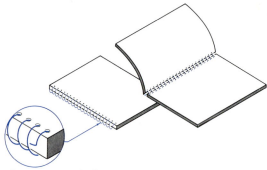

Fig. 44-8.   Spiral-wire binding.

Perfect (Padding) Method

Perfect binding, commonly called *padding* (Fig. 44-9), is one of the newest binding methods. A synthetic glue that stays rubberlike after it has dried is applied to the edge of a stack of paper. This binding is not permanent. With careful handling, however, it will last for a long time. Paperback books and scratch pads are often bound this way as are many hardcover books today.

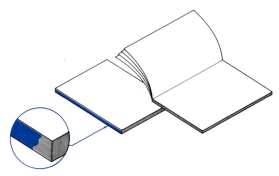

Fig. 44-9.   Perfect binding (padding).

Unit 45

Binding in the School Laboratory

Binding sheets of paper together is an interesting process. The steps listed under each of the following four methods give the information you will need for doing four different types of booklet bindings.

Saddle-wire Stitching

1. Fold several sheets of paper in half (Fig. 45-1).
2. Assemble the sheets and the front and back covers into the correct order.
3. Saddle-wire stitch the booklet together by using a wire stitcher (Fig. 45-2).
4. Trim the booklet on the three *unbound* (unstitched) sides (Fig. 45-3).

Printing companies use heavy production machines (Fig. 45-4) to do large numbers of saddle-wire booklets. Production machines of this type gather the signatures together in the correct order, stitch the booklet with two or more staples, and then trim it on three sides. Several hundred copies can be completed by this machine in a very short period of time.

Side-wire Stitching

1. Assemble the single sheets and the front and back covers into the correct order for the book.

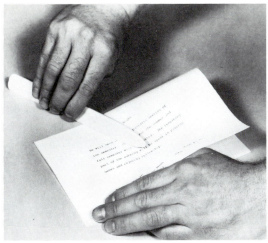

Fig. 45-1. Folding a sheet of paper with a folding bone.

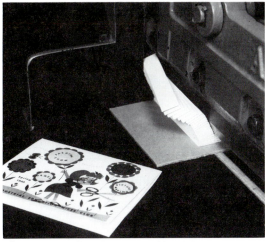

Fig. 45-3. Trimming several booklets with a paper trimmer.

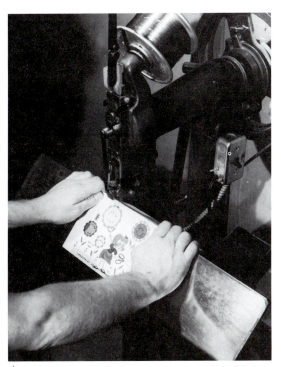

Fig. 45-2. Saddle-wire stitching of a booklet.

Fig. 45-4. A commercial machine that gathers, stitches, and trims saddle-wire-stitched booklets. (Consolidated International Incorporated)

2. Side-wire stitch the booklet with the stitching machine (Fig. 45-5).

3. Tape the *bound* (stitched) edge of the booklet with binding tape. This covers the staples and makes the booklet better looking (Fig. 45-6).

4. Trim the booklet with a paper cutter, as was done with the saddle-wire-stitched booklet. (See Fig. 45-3.)

To increase speed in binding side-wire-stitched booklets, large industrial printing companies use a multiple-head side-wire stitcher (Fig. 45-7). This machine places six staples into a booklet at one time.

Fig. 45-5. Side-wire stitching a booklet.

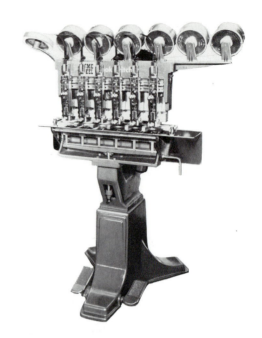

Fig. 45-7. A multiple-head side-wire stitcher. (Interlake Steel Corporation)

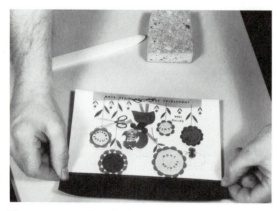

Fig. 45-6. Taping the bound edge of a side-wire-stitched booklet.

Fig. 45-8. Punching a series of rectangular holes for plastic-cylinder binding. (General Binding Corporation)

Plastic-cylinder Binding

1. Assemble the single sheets and the front and back covers into the correct order for the book.

2. Punch holes in the sheets (Fig. 45-8). Punch only four or five sheets of paper at one time.

3. Attach the plastic-cylinder binder (Fig. 45-9). No trimming is needed with this binding.

Automated equipment speeds this type of binding. It gathers, punches, and attaches plastic-cylinder binding at a high speed (Fig. 45-10).

Fig. 45-9. Attaching plastic-cylinder binder to the prepunched booklet. (General Binding Corporation)

Fig. 45-11. Applying padding cement to the edges of a stack of paper.

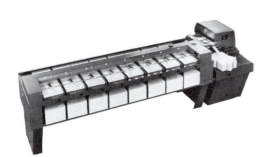

Fig. 45-10. A machine that gathers, punches, and attaches plastic-cylinder bindings. (General Binding Corporation)

Perfect (Padding) Binding

1. Assemble single sheets of paper into the correct order.

2. Place a *chip board* (stiff cardboard) on the bottom of each group (stack) of sheets.

3. *Jog* (straighten) the sheets and the pieces of chipboard to the binding edge.

4. Place the groups of sheets on the edge of a table with the binding edge facing out. Place a heavy weight on top of the paper near the binding edge. Special padding presses can also be used.

5. Apply the liquid padding cement to the binding edge of the sheets with a brush (Fig. 45-11). Two thin coats of padding cement have better holding power than one thick one. Allow 15 minutes for each coat to dry.

6. Separate each pad or group of sheets with a bookbinder's knife (Fig. 45-12).

7. Trim the three *unbound* (uncemented) edges of the pad in a paper cutter. (See Fig. 45-3.)

Power equipment has been developed to speed this process. The industrial machine shown in Fig. 45-13 puts padding cement on the binding edge of the booklet and attaches the covers in one process.

Fig. 45-12. Separating the pad (stacks of sheets) with a binder's knife.

Fig. 45-13. An automated machine that applies padding cement and attaches covers to the booklet simultaneously. (Consolidated International Corporation)

Unit 46

Discussion Topics on Graphic Arts Technology

1. What does *graphic arts* mean?

2. List and briefly tell about the four major printing processes.

3. What is most needed to make paper?

4. Name and tell about the printer's system of measurement.

5. What is a composing stick used for?

6. Tell about the basic steps in doing offset lithography.

7. Who is said to have made the single greatest discovery in printing? Name the discovery and tell when it happened.

8. List the three measuring systems used in graphic arts, and explain each.

9. Name four clusters, or groups, of graphic arts technology-related careers.

10. What is meant by *hand composition*?

11. What does *dumping the composing stick* mean?

12. What is a printer's knot used for?

13. What are the basic steps in the lockup procedure?

14. Name and describe three letterpress printing processes.

15. Describe four essential tools for the lithographer.

16. Define and describe four basic methods of binding sheets of paper together.

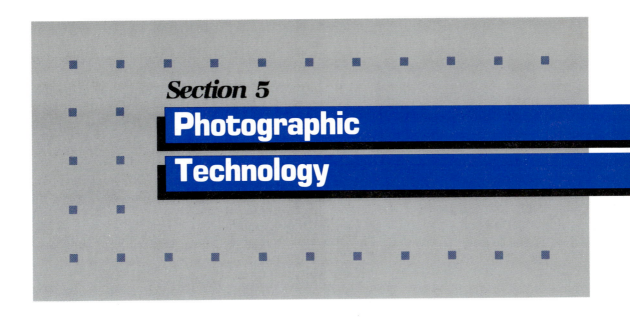

Section 5

Photographic
Technology

Unit 47

Introduction to Photographic Technology

Photography is "drawing with light." In this section, you will learn to use the camera as a tool much as an artist uses a brush to create a painting. You will learn to plan your photographs, develop film in light-tight tanks, and create black-and-white enlargements in the darkroom.

Photographs are used to illustrate books, magazines, and newspapers. Without photographs, communicating ideas would take thousands of words and hundreds of pages.

Photography allows you to record summer vacations, remember friends and relatives, and obtain pleasing photos for home decoration.

Photography is fascinating and personally rewarding. The photographic industry employs thousands of people in scientific, medical, and engineering applications. It can be an exciting career.

History

Camera Obscura. The history of modern photography began in the sixteenth century when Robert Boyle constructed a box type camera in which light passing through a small hole would form an image inside a darkened area. This forerunner of the later box camera (Fig. 47-1) was known as the camera obscura (dark chamber).

Permanent Photographs. During the nineteenth century, many scientists and inventors tried to make the image formed by the camera obscura permanent. It was discovered that when silver iodide coatings of metal were exposed to light, this would produce images if "fixed" with a hot salt solution after exposure. These first photographs were known as daguerreotypes (Fig. 47-2), named after their inventor, the French painter Louis Daguerre.

Fig. 47-1. Box camera.

Fig. 47-3. Eastman camera with adjustable lens.

New Processes. By the late nineteenth century, there were many new processes for making prints and many new types of cameras. The paper negative was invented by William Fox Talbot, making it possible to produce duplicate prints.

Richard Maddox, a British physician, invented the dry plate negative, which was made from gelatin and coated with silver nitrate. Modern film is made much the same way today.

Flexible Film. During the 1880s the "father of modern film and processing," George Eastman, invented flexible roll film. He introduced the 100-shot camera, the "Brownie" camera, and the pocket camera (Fig. 47-3).

Modern Trends. Modern cameras have been improved continuously. Special underwater cameras record objects at depths people cannot reach. Medical micro cameras probe the inner workings of the human body. Fast film speeds, instant photographs, auto-focus, and auto-exposure are just a few of the more recent developments in photography.

Fig. 47-2. Example of daguerreotype.

Photographic Technology-related Careers

Photography is an interesting hobby for many people. Taking pictures, developing film, and making prints are enjoyable and creative experiences for those who have mastered the basics of photography. Many people have turned this hobby or avocation into an exciting career. Opportunities in the photographic industries are always available for the well-prepared and creative individual.

Preparation for a career in the field of photography can originate in a college or university, a technical institution, military school, or on-the-job training with a working professional photographer. In many cases, a combination of apprenticeship and formal education is necessary to qualify for a photographic career.

One could define a professional photographer as any individual who makes a living in the photographic industry. There are numerous specialists in the field of photography. These can be grouped broadly into the following areas: portrait or wedding photographer, industrial or commercial photographer, photojournalist, photofinishing technician, photographic retail salesperson, and camera repairer.

Portrait or Wedding Photographer

Many individuals get into photography as a career by first taking pictures for family and friends. Almost everyone wants photographs of children as they are growing up and getting married and during special occasions. The advanced amateur photographer gains much experience by photographing these assignments and is thereby better able to decide whether this is what he or she wants to do as a career (Fig. 48-1).

The portrait photographer must enjoy working with people. Special lighting equipment is used to best portray or capture the personality of the individual being photographed. Specialized cameras in medium to large formats are commonly used by the portrait photographer. Use of equipment and exposure determination eventually become second nature to the experienced portrait photographer.

Wedding photography requires the use of flash equipment. Also, the photographer must anticipate when a photograph should be taken. Generally, the photographer will do this work on weekends.

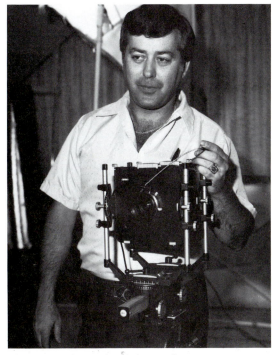

Fig. 48-1.   Portrait photographer.

Industrial or Commercial Photographer

The industrial or commercial photographer must be able to photograph a wide variety of subjects. This person must be thoroughly knowledgeable in 35 mm, medium format, and, in particular, view camera techniques. The industrial photographer is also called upon to make product advertising photographs, executive portraits, and close-up views of mechanical devices. Many industries have their own in-house studios and laboratories and employ their own full-time commercial photographers to operate them.

Photojournalist

Newspapers and magazines use hundreds of thousands of photographs each year. The photojournalist takes one or a series

Fig. 48-2. Sports events provide photojournalists with the opportunity to take exciting photographs. (Rockland Journal News)

of photographs to tell a story (Fig. 48-2). News photographers must work all hours, as they must be on the scene when news is breaking (happening). In many cases, the news photographer will write the accompanying story or caption as well as take the pictures. Local papers will usually accept unsolicited human interest photos from free-lance photographers. This is an excellent way for the novice (beginning) photographer to get photos published and build a reputation. The school yearbook and journalism classes are a good starting place if one's interest lies in this career field.

Photofinishing Technician

Most professional photographers who take still photographs or motion pictures do not do their own film or print processing. Instead, they send the film to commercial laboratories for processing. If one enjoys working in the darkroom, the field of photofinishing may be a good career choice. Many photofinishing laboratories employ and prepare high school graduates to work with specialized photofinishing equipment. Custom color print technicians with experience are in demand. This type of work requires a good eye for color balance and tonal contrast (Fig. 48-3).

Photo lab technicians may be called upon to mix chemicals, process prints as well as slides, and produce mural-size (large) photographs. In addition, they often mount and frame the finished product.

Photographic Retail Salesperson

A photographic background and a knowledge of photography equipment are usually required for a person to handle the sale of photographic equipment, whether in a specialty shop or in the camera area of a large department store (Fig. 48-4). A pleasant personality and a thorough understanding of photographic equipment

Fig. 48-3.   Photofinishing technician.

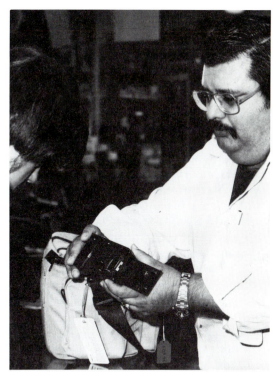

Fig. 48-4.   Photographic retail salesperson.

and print finishing are important qualities for a person interested in this type of career. Many photo store managers employ advanced amateur photographers for a summer job, during the Christmas holidays, or upon high school graduation. A store that specializes in photographic equipment and supplies may also have a studio or print-finishing capabilities. This type of store may also provide its employees with career preparation in these areas. Many leading professional photographers got their start in photo retail sales.

Camera Repairer

Today's modern cameras are sophisticated mechanical devices. Repair of photographic equipment requires experienced, highly trained technicians. This field has many openings for a person with mechanical ability. Preparation in this field can be gotten in a technical school or from an on-the-job training program. Many camera manufacturers offer their own training programs on their particular equipment. An understanding of optics and electronics is very helpful to the photographic repair technician.

Modern technology has produced a need for specialization in many career areas. This is particularly true for the photographic profession. Many broad classifications of photography have been discussed in which photographers earn their livelihood. In addition to these, photographers can choose to limit their work to medical, scientific, police and investigative, fashion, or underwater photography. These are just a few areas of photographic specialization. Whichever area of photography is chosen, you can be assured that photography will always be an exciting activity.

Camera Types and Formats

Many types of cameras are available to the amateur or beginning photographer. The most popular is the 35 mm single lens reflex camera (Fig. 49-1). Unit 51, "35 mm Camera Operation," will discuss this particular camera in detail. The professional photographer will probably have several cameras. Each camera has certain advantages and disadvantages. The professional will choose a camera on the basis of the photographic assignment. Select your camera in the same manner. Base the selection on the type of pictures most commonly taken. A brief discussion of the advantages and disadvantages and the common uses of different types of cameras will help you make a wise choice.

The Camera in General

All cameras have these characteristics:

1. They are light-tight boxes.
2. A shutter is used to control the amount of time that light is allowed to appear on the film.
3. A lens opening, or diaphragm, controls the quantity of light striking (appearing on) the film.
4. A transport mechanism usually advances a new frame after each exposure.

In addition to these characteristics, many modern cameras have the following features:

1. A built-in light meter to determine exposure
2. A film frame counter
3. A hot shoe for flash photography
4. Interchangeability of lenses

The Simple Camera

The simple camera has a fixed shutter speed. It also has a preset diaphragm, or aperture, setting. Typical examples of this type of camera are the Instamatic and the box camera. These cameras produce acceptable photographs during normal daylight conditions or indoors with the use of a flash unit. Film size is usually 110, 126, and occasionally 35 mm. This camera is inexpensive and very popular because it is not complicated to use.

Subminiature Format. This particular type of camera would be a spy camera or novelty camera. It produces a very small negative, usually 16 mm. It has few applications for the serious photographer.

Fig. 49-1. A typical 35 mm single lens reflex camera.

Fig. 49-2. Medium format camera: RB-67 SLR Mamiya-Sekor.

Miniature Format. The most popular camera for the amateur photographer is the 35 mm. The 35 mm camera can be divided into two main types: the rangefinder camera and the single lens reflex camera. Characteristics of the 35 mm camera include:

1. Inexpensive film
2. Built-in light meters
3. Light weight
4. Availability of many accessories (system approach)
5. Ease in focusing

Examples of the 35 mm format cameras include the following camera systems: Nikon, Pentax, Canon, Minolta, and Yashica.

Medium Format. The medium format camera is the most popular choice of the professional photographer. Its relatively larger size negative produces photographs that are sharper than those of the miniature camera. It also produces less grain than the miniature format. Medium format cameras can be twin lens reflex or single lens reflex. Typical examples of this style of camera are the Hasselblad SLR, the Mamiya-Sekor C330 TLR (Fig. 49-2), and the Bronica SLR.

Large Format. The large format camera is generally considered to be one that uses negatives 4 × 5 inches (102 × 127 millimeters) or larger. In each of the other camera formats, the size of the negative dictates the format of the camera. The 4 × 5 view camera is the "workhorse" of the commercial photographer. This large-size negative produces photographs that are practically without grain. The sharpness of the contact prints or enlargements made from a 4 × 5 negative cannot be easily achieved with smaller negatives. The 4 × 5 camera allows the professional photographer to control perspective by use of tilts, swings, rises, and shifts. Typical examples of the 4 × 5 camera are Linhoff, Calumet, and Omega. (See Fig. 49-3.)

Specialized Cameras. There are many cameras that perform a unique function or are designed for a particular application. The instant picture camera, such as the Polaroid Land camera or the Polaroid SX-70, provides photographs within minutes. Underwater cameras, such as the Nikonos, allow the photographer to photograph marine subjects. Scientific applications, such as for the space and medical industries, prompted the development of special photographic optical devices suitable for those particular needs. One of the earliest specialized cameras, the Polaroid Land camera, is shown in Fig. 49-4.

Fig. 49-3. A large format camera. (Calumet)

Fig. 49-4. An early model of the Polaroid Land camera.

Photographic Composition

Learning to see as the camera does takes practice. The camera records on film the scene as it actually appears, not as you, the photographer, "imagine" you see it.

Make photographs more appealing by deliberately controlling the camera angle, the objects to be photographed, and most important, the lighting. The organization and arrangement of all the elements that produce photographs are called *photographic composition*. There are some standard acceptable rules of composition that will produce better photographs. These guidelines are:

1. Avoid centering the horizon line when taking scenic shots. It is preferable to have one-third of the picture as sky and two-thirds as land, or just the opposite, two-thirds sky and one-third land. This is known as the *rule of thirds* (Fig. 50-1).

2. Keep photographs simple. One strong subject is preferable to (better than) many conflicting subjects (Fig. 50-2).

3. Place the subject in accordance with the rule of thirds. This will serve to attract and hold the attention of the viewer (Figs. 50-3 and 50-4).

4. Frame the subject so that the viewer's eye is drawn to the subject or to a strong center of interest (Fig. 50-5).

5. Create depth in photographs by using foreground objects, converging lines, and side lighting. Photographs are two-dimensional. The illusion of depth, or the third

Fig. 50-1.   A scenic shot with horizon line uncentered.

Fig. 50-2.   Example of strong center of interest.

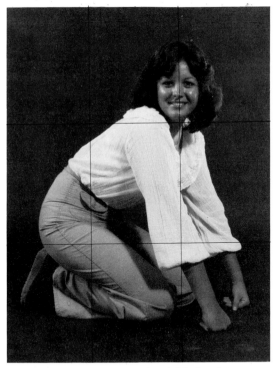

Fig. 50-3.   Diagram of rule of thirds placement.

Fig. 50-4.   Photographic examples of the rule of thirds, two photos.

dimension, is created by the photographer (Fig. 50-6).

6. Have the subject move into or look into the larger portion of the picture area. Having the subject moving or looking away from the center of interest will draw the

Fig. 50-5.   Example of framing the subject to direct the viewer's attention. (David Sykes)

Fig. 50-6.   The illusion of depth can be created by converging lines, as in this photograph. (Brian Stine)

viewer's attention out of the picture (Fig. 50-7).

7. Control of the background is a definite requirement in composition. A conscious effort must be made to look beyond the subject and view the background separately. Beginners are always amazed at what they find in the background when they print their first photographs. The *background* is as important to the overall composition of the photograph as the center of interest (Fig. 50-8).

Fig. 50-7. Example of subject moving across photograph. (Clark Godfrey)

A

Fig. 50-8. An uncluttered background will not detract from your subject. Learn to look behind the object to eliminate object mergers in the background.

B

Fig. 50-9. Examples of advanced darkroom enlarging techniques, used to improve composition, two photos.

An artist, when painting a picture, can change the actual scene to improve the overall composition by adding or subtracting certain elements. The photographer, while not having as much control, can do much the same thing in the darkroom. This manipulation is best done when the picture is taken. Certain darkroom techniques that can improve composition are:

1. Selective cropping (outer limit) of the negative
2. Dodging, or "burning in," areas of the print to make them appear lighter or darker
3. Vignetting (reducing), or eliminating unwanted background detail by darkroom enlarging methods (Fig. 50-9 a, b).

Good composition has basic guidelines that the beginning photographer should follow. Once mastered, they can be ignored if this is done for the specific purpose of improving the image or creating greater visual impact or communication.

Unit 51

35 mm Camera Operation

The 35 mm camera is probably the most popular camera for the advanced amateur photographer. The two types of 35 mm cameras are the rangefinder camera and the single lens reflex (SLR) camera. The 35 mm SLR, is the most popular because of through-the-lens (TTL) viewing and light meter reading. The rangefinder camera has the advantage of compactness and ease of operation (Fig. 51-1).

A beginner in photography will probably want to obtain a 35 mm SLR. This basic camera is a sophisticated optical tool that is part of a system.

After acquiring a good 35 mm camera, many students become confused about its operation. This unit should serve to eliminate some of that confusion.

Before loading the camera with film or attempting to take any pictures, be sure to read thoroughly the operator's instruction manual. This unit covers many of the basic principles which apply to most 35 mm SLR cameras and their operation.

Loading the Camera

Always load the camera in the shade or indoors. It if cannot be loaded in a subdued-light (darkened) area, shade the camera and film by turning your back to the sun.

Place the 35 mm film cassette in the camera so that the film leader, when engaged with the take-up reel, has its perforations engaged (caught) with both the top and bottom sprockets (Fig. 51-2). Many

Fig. 51-1. Examples of an SLR camera and a rangefinder camera.

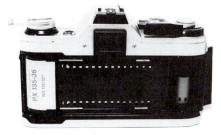

Fig. 51-2. Loading the 35 mm camera.

beginning photographers fail to advance the film far enough when loading the camera. Remember that the film counter will advance whether or not there is film in the camera. A simple way to check to determine whether the camera is loaded with film is to turn the film rewind knob gently. If the knob cannot be turned freely, there is film in the camera.

Aiming and Focusing the Camera

What is seen through the lens is what will be recorded on the film with an SLR camera. View the subject through the camera and compose the subject either vertically or horizontally (see Unit 50, "Photographic Composition"). Focusing the camera is accomplished by turning the focusing ring of the lens. When you take a picture of a person, focus on the eyes of the subject. For landscape or scenic photography, it is usually a good idea to focus at what is known as the hyperfocal distance. The *hyperfocal distance* is any distance that, when focused upon, will render objects close to the camera and objects at infinity in sharp focus. In this way, objects close to the camera as well as those at infinity (a great distance), will be in sharp focus. A small f-stop opening is necessary for a long depth of field.

The 35 mm camera body has been designed so that the left side of the camera is shorter than the right side. This design feature allows for a more comfortable grip of the camera (Fig. 51-3). When aiming and focusing the camera, use your arms and body as a tripod. Take a deep breath and gently press the shutter release button when taking a picture. Practice this before taking any pictures. There is a range scale on the camera lens. This scale gives the distance focused on in both meters and feet. It also has what is known as a depth-of-field scale. This will be discussed in more detail later.

Fig. 51-3. Front view of the 35 mm camera. Notice that the left side of the camera is wider than the right side. The right hand holds the camera; the left hand is used to focus.

Many photographers use the zone focusing or preset focusing technique when taking action photographs. Use the scale on the lens and preset the distance to approximately where the action will occur (happen). Then shoot the picture without focusing during the action. This will save time and allow you to concentrate on viewing and actually taking the picture.

Exposure Controls

Most adjustable 35 mm cameras allow the photographer to adjust the shutter speed and f-stop. The reason for purchasing a 35 mm camera is to obtain better optical quality and thus sharper pictures. But in addition to this primary reason, the ability to control depth of field and stop action are equally important. The shutter speed control allows you to stop action or purposely blur the subject. The f-stop, or aperture control, allows you to control depth of field. The combination of both these controls affects the exposure.

The sensitivity of film to light is an important factor. This is measured by the ASA, DIN, or ISO rating. Consult the instructions packaged with film that give the film's sensitivity. Program the camera's light meter after loading the camera and before taking pictures. (See Unit 52, "Photosensitive Materials.")

F-Stop

The f-stop on a camera controls the quantity (amount) of light that is allowed to pass through the lens and strike the film. The f-stop, or aperture opening, varies in size from a small to a large opening. Each f-stop will let in either twice as much, or half as much light as the adjacent setting. Opening or closing the aperture will allow you to control exposure and depth of field. The larger the f-stop opening (i.e., f-1.8), the shorter the depth of field. The smaller the aperture opening (i.e., f-16), the longer the depth of field. A lens is often referred to as either fast or slow. The speed of the lens is determined by its largest aperture. For example, f-1.8 would be the speed of many normal lenses on a typical 35 mm camera and is also faster than an f-2 lens (Fig. 51-4).

Shutter Speed

The shutter speed controls the amount of time the lens is held open. The length of exposure may vary from 1 second to 1/4000 of a second depending upon the camera manufacturer's instructions. This is a reciprocal, or fraction. However, on the camera, the numbers will not be fractions. It should be understood that what is seen on the camera body's shutter speed control is only the denominator of this fraction (Fig. 51-5).

Exposure is affected, as previously mentioned, by a combination of the shutter speed and the f-stop selected. As the shutter speed or the f-stop is changed, you will either double or half the amount of exposure. There are many possible combinations that will give the same exposure (Fig. 51-6).

Normal daylight exposure is determined by putting a 1 over the ASA or ISO of the film being used and setting the aperture to f-16. For example, Plus-X pan film, which has an ASA of 125, would have a normal daylight exposure of 1/125 of a second at f-16. This general rule holds true for most types of film. Tri-X film has an ASA of 400. Select a shutter speed of 1/500 of a second at f-16, since this is the closest shutter speed to this particular ASA.

Many cameras on the market today allow the photographer to preset the f-stop or shutter speed; the camera will then automatically select the best exposure. Many of these automatic cameras are shutter-speed priority, aperture priority, or a combination of the two. This is a feature to consider when selecting a 35 mm camera.

Fig. 51-4.   Examples of f-stop settings.

Fig. 51-5.   Shutter speed settings.

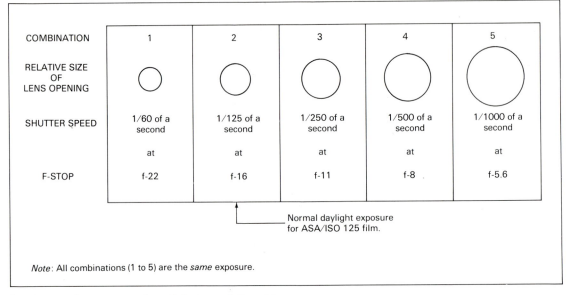

COMBINATION	1	2	3	4	5
RELATIVE SIZE OF LENS OPENING	○	○	○	◯	◯
SHUTTER SPEED	1/60 of a second	1/125 of a second	1/250 of a second	1/500 of a second	1/1000 of a second
	at	at	at	at	at
F-STOP	f-22	f-16	f-11	f-8	f-5.6

Normal daylight exposure for ASA/ISO 125 film.

Note: All combinations (1 to 5) are the *same* exposure.

Fig. 51-6. Shutter speed and f-stop relationships.

Unit 52

Photosensitive Materials

Film

Modern photographic roll film is produced under exacting standards of quality control. This enables the photographer to produce consistent results. Select a film on the basis of the specific photographic assignment. To make an intelligent selection, certain things should be considered.

Photographic film is rated according to its sensitivity to light. ASA, ISO, and DIN ratings designate the sensitivity of a particular film to light. ASA (American Standards Association); ISO (International Standards Organization); and DIN (the European designation for film sensitivity) values are listed in the packaging of most films. When selecting a film type, consider the following factors.

The beginning photographer should select a moderate-speed film, such as Plus-X

pan film. This film has a rating of 125 ASA and ISO. It has a built-in exposure latitude (variation) that allows the photographer some degree of error in exposure.

When taking pictures under existing light or low-light conditions, a high-speed film such as Tri-X pan (ASA 400) film should be used.

In portraiture (photos of people), or where the situation calls for extreme sharpness and clarity, a low-speed film is best to use. Panatomic-X, ASA 35, is a good choice in this situation.

The following terms relative to film should be understood.

1. Graininess is caused by the enlargement of the photographic image to a point where the clustering of metallic silver becomes apparent. Graininess increases as the sensitivity of the film increases.

■ **238**

2. *Sharpness* requires a slow-speed film. The silver halide particles in a slow-speed film, when developed, will produce a better-quality image because of less clustering. This produces enlargements that have better sharpness.

There are numerous manufacturers of film, such as Kodak, Ilford, and Fuji. Select one manufacturer and one film type and learn to become proficient with that type of film before experimenting with another (Fig. 52-1). Consult the manufacturer's recommendations for specific instructions.

Paper

Photographic enlarging papers are of two types: *resin-coated* (RC) and *fiber-based*. Each type has certain advantages and disadvantages. These are discussed later in this unit. There are a number of surfaces available to the photographer. A type F surface is a glossy surface, while a type N

Fig. 52-1.   Types of photographic film.

surface is a matte (softer) surface. The contrast in a photograph can be controlled by the photographer when selecting the enlarging paper. There are two ways to control contrast. The photographer can select a contrast-graded paper or can choose a multigrade paper. If the photographer chooses the latter, Polycontrast filters must be employed (used). Contrast-graded paper is numbered 0 through 5, with number 2 considered to be normal contrast.

Resin-coated paper is the accepted standard for amateur and professional photographers. This paper will not allow the photographic processing chemicals to penetrate (go through); thus, it is much easier to wash the print. This advantage significantly reduces the wash time, resulting in reduced processing time. Resin-coated paper can be air-dried or dried by forced hot air very quickly.

Fiber-based paper absorbs chemicals and thus requires substantially more wash time. Since fiber-based paper is usually heavier and is not resin coated, it requires ferrotyping. Ferrotyping requires the print to be placed in direct contact with a stainless-steel surface. The print should be thoroughly washed and placed in contact with the metal surface while it is wet to produce a glossy surface. If left to air-dry, it will curl and not be of a quality suitable for presentation. Fiber-based paper is recommended when toning of the print is called for, and many photographers consider it their first choice for portraits. The reason for this is that retouching is much easier with fiber-based paper.

Photographic enlarging paper for black-and-white printing has made many advances as modern technology has progressed. The creative photographer should carefully select the type, contrast grade, surface, and manufacturer with the utmost care. The final print can be enhanced (improved) or can suffer if the photographer is not knowledgeable about the printing medium (Fig. 52-2).

Fig. 52-2. Fiber-based and resin-coated enlarging paper.

Photographic Chemistry

As mentioned in other units, photographic chemistry is light-sensitive. Chemistry deteriorates for a number of reasons, including humidity, contact with air, and exposure to light. All these factors have detrimental effects on the shelf life of film-developing chemicals and print-processing chemistry.

Developers. Developers, either for paper or for film, are the most crucial (critical) stages in the development process. Developers have an extended shelf life in powder form or in stock solution. Once they are converted to a working solution, the shelf life is very limited. Extreme care should be taken to prolong their usefulness. Storage bottles or tanks should be kept full, sealed tightly, and stored in a subdued-light area. In addition, this area should be cool.

Stop Bath. Stop bath will retard, or as the name states, "stop" development. Stop bath is acetic acid. Indicator stop bath will turn purple when exhausted, hence the name "Indicator." This photosensitive material should be mixed from a stock solution to a working solution only as needed. Many photographers skip this chemical stage, preferring to use only water as a means of retarding film development or print processing.

Fixer. Fixer, which is also known as *hypo*, is a photosensitive chemical that is used to remove unexposed or undeveloped silver halide particles from film or paper. Use of this particular chemical represents a crucial stage in the film development process or print process. Extreme care should be taken to make sure that the fixer is at full strength. This can be accomplished by using a hypo checking solution. A simple method for determining its usefulness is to submerge an exposed but undeveloped piece of film into the fixer. If the film becomes clear within a few minutes, the fixer is suitable for the development process. Keep this chemical tightly sealed in a light-tight container and stored in a cool, dry area.

Light-sensitive, or photosensitive, materials require the user to be aware of the proper procedures and precautions for their proper handling. Nothing is more frustrating than taking good photographs, only to have them ruined by trying to develop the film or process the prints in exhausted (old) chemistry. It is equally disheartening to discover that the photographic film has been fogged by light or extreme temperature. Again, the pictures will become unprintable. Photographic enlarging paper is rather expensive. If proper care is not taken, an entire box can become exposed. Photosensitive materials are the media by which creative photographers express themselves. Protect your medium of expression (Fig. 52-3).

Fig. 52-3. Proper chemical storage. Keep bottles full and stored in a cool area.

Unit 53

Taking Photographs

Before one can take good photographs, one must understand some very important principles. Light, the medium with which the photographer works, should be studied. An understanding of the principles of light can make you a better photographer. Early morning light and late evening light are quite different from harsh bright afternoon light. There is an old saying that the best time to take pictures is early in the morning or late in the evening. Shadows are long and distinct when the sun has just risen or is about to set. After the sun has gone down, but before dark, is a good time to take outdoor portraits. The lighting is even, harsh shadows are eliminated, and contrast is reduced.

Light can be reflected. This principle of light is very important to the photographer. There are many times when the photographer will have to reflect light into shadowed areas or control lighting ratios by means of reflected light. Exposure is usually determined by a reflected light meter reading.

Light can be bent, or refracted. The camera lens design is based on these principles of light.

The more one understands about optics and physics, the better able one will be to communicate through the science of photography.

Camera and Subject Motion

Light travels in a straight line. This is another principle of light that applies to the choice of a shutter speed. If the subject one is taking a picture of is moving directly toward the lens, the reflected light from the subject is traveling in a straight line to the camera. A relatively slow shutter speed

Fig. 53-1. Blur caused by camera motion and subject motion. (Shutter speed too slow.)

can therefore be used without appreciable blurring of the picture. If, however, the subject is moving at right angles to the camera, a faster shutter speed should be used. Camera motion or subject motion will cause unsharp photographs if the choice of a shutter speed is not appropriate. Beginning photographers should use a shutter speed of 1/60 of a second or faster. Anything below this shutter speed requires the use of a tripod. This shutter speed will prevent blurring of the picture caused by the camera motion (Fig. 53-1).

Getting a sharp picture when the subject being photographed is in motion requires the photographer to have some experience. A photographer with experience in photographing moving objects can usually select an appropriate shutter speed for the subject being photographed. Shutter speeds of 1/250 of a second, 1/500 of a second, or 1/1000 of a second will stop action, thus producing a sharp image on the negative depending on the direction of motion by the subject and the speed of the subject.

The choice of a fast shutter speed requires opening up the lens. This will reduce the depth of field. A reduction in depth of field will blur foreground and background objects, but the fast shutter speed will keep a main subject in sharp focus. There are many times when a long depth of field and a fast shutter speed are called for. A faster-speed film may solve this problem.

Direction of Light

The direction of light should be selected by the photographer to create the desired effect. For example, if the photographer wishes to show texture, side lighting, which produces shadows in the texture, is appropriate. There are basically six directions of light: top, bottom, back, front, and either side. Front lighting gives pictures a very flat appearance. Back lighting tends to produce a halo effect and will separate the subject from the background, adding depth to a picture. Bottom lighting, which is used quite frequently in motion picture horror movies, produces unusual light patterns on the face. Control of the direction of light through the photographer's choice of camera angle or location can change significantly the message of the photograph (Fig. 53-2).

Subject Matter

Taking pictures of different subjects gives the beginning photographer valuable experience. The photographic techniques employed when taking portraits vary quite a bit from those used for close-up photography or sports and action photography.

Fig. 53-2. Example of side lighting.

Whatever the choice of subject matter, plan your photographs. Really good photographers do not "hope" for a good picture, or try to get a lucky shot. The following factors should be considered when you plan photographs:

1. Choose a film for the subject matter, time of day, and type of lighting.
2. Select the format of camera that is the most appropriate for what you will be taking pictures of.
3. Anticipate (imagine) how you will be composing the picture.
4. Be prepared with the necessary accessories, such as tripod, filters, and so on.
5. Determine the best time of day or type of lighting for the photograph you wish to take.
6. Control as much as possible the elements that make up the entire composition of the photograph.

The more prior planning you do, the better your pictures will come out.

Loading the Film Developing Tank

Fig. 54-1.   **Examples of developing tanks.**

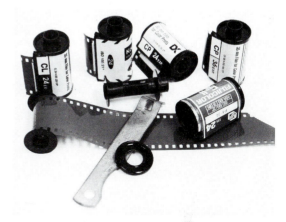

Fig. 54-2.   **Film cassettes.**

Exposed panchromatic film is sensitive to all light, and it must be loaded into the developing tank in complete darkness.

Many types of developing tanks are available for film processing. The stainless-steel and plastic types are very popular (Fig. 54-1). Select one type and practice loading it in daylight. After you feel secure in your ability to load the tank properly, try loading the tank with the eyes closed or in complete darkness. This initial (early) practice may save a very important photograph.

The film to be developed may be a 120 roll film, which can be easily separated from its paper backing. If, however, you are developing 35 mm film (which is sealed in a cassette), a can opener will be needed to remove the film from the cassette (Fig. 54-2). The developing tank, the top of the developing tank, the film cassette, and the can opener should be arranged so that they can be located in the dark while the photographer is loading the tank (Fig. 54-3).

Fig. 54-3.   **Arrangement of necessary items for film loading.**

After turning off the lights, let your eyes adjust to the darkness. Check for any possible light intrusions before opening the film cassette. Use a clean and dry developing tank and reel. Wet film is very difficult to load. Crimping the film while loading will cause half-moons on the developed film. Do not squeeze the film too much while loading. Make sure your hands are clean and dry before beginning, and avoid touching the film surface as much as possible. Oily fingers can ruin what might have been an excellent picture. If the film is loaded incorrectly, the chemical solution will not get to all parts of the film. This will result in partially developed film. The areas that touch stick together and are difficult, if not impossible, to separate after the film has been dried.

Loading the film-developing tank requires practice and patience. As has been mentioned, without a good negative, you cannot obtain a good print. The film-development stage is one of the most critical points in darkroom work.

Unit 55

Developing Black-and-White Film

Before it is developed, photographic film contains latent, or hidden, images. The process of development makes these images visible. This produces a negative that is the reverse of the actual scene. A print made from a negative produces a positive, which faithfully reproduces images as the human eye saw them. When the negative is made, any number of prints can be made from it. To develop black-and-white film successfully, the following chemicals are required (see Fig. 55-1):

1. Developer
2. Indicator stop bath
3. Fixer

4. Hypo clearing
5. Photo-Flo

The manufacturer of the film provides a data sheet in the film box. This data sheet mentions the recommended developers and gives a time and temperature chart (Fig. 55-2).

Developing Procedure

Developer. Developing is the most critical stage in the photographic process. The developing chemical causes the silver to darken, depending on the amount of light that exposed it when the picture was taken. Read the manufacturer's recommendations for the type of developers and the time required. Agitate the developing tank continuously for the first minute and for 15 seconds of each of the remaining minutes. Microdol-X and D-76 are popular developers for black-and-white film.

Indicator Stop Bath. This chemical is called Indicator stop bath because it "indicates" to the user that it has become exhausted when it turns purple. Thirty seconds to 1 minute will stop development.

Fixer. This chemical, also called hypo, removes any unexposed silver halide particles that were not exposed when the

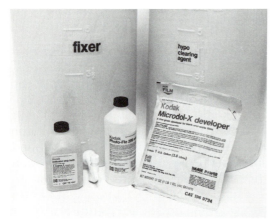

Fig. 55-1. Chemicals necessary for black-and-white film developing.

MINOR REVISION 6-83

KODAK PLUS-X PAN PROFESSIONAL FILM

PXP220 and PXP120

• Medium-speed panchromatic film with extremely fine grain and high resolving power • Medium contrast, wide exposure latitude
• Very high sharpness even at high degrees of enlargement
• Retouching surface on the emulsion side.

PXP220

Important: PXP220 film is for use in professional roll-holders and cameras designed to accommodate the longer length of film. Some cameras that take 120 film can be modified to accept the PXP220 film. The 220-size roll of film is about twice the length of the 120-size roll, and has no backing paper on the film itself. The film has a paper leader and trailer. Be sure that the processing equipment used can accommodate the extra length of 220-size film.

Caution: Since this film does not have the usual backing paper wound the length of the film, special care in loading and unloading the camera is necessary to avoid light fog on this sensitive emulsion. Load and unload the camera in subdued light.

PXP120

PXP120 film is a normal-length roll film, and it can be used in cameras accepting 120-size film.

To use this film, load and unload your camera in subdued light, never in direct sunlight or exceptionally strong artificial light.

Exposure

Speed:

ISO **125**
ASA **125**

These speed numbers are for use with meters and cameras marked for ISO and ASA Speeds or Exposure Indexes, in either daylight or artificial light. They will normally lead to approximately the minimum exposure required to produce negatives of highest quality.

Note: An ISO (International Standards Organization) film-speed number is given in anticipation of future worldwide use.

Outdoor Exposure Guide for Average Subjects: For shutter speed of 1/250.

Bright or Hazy Sun on Light Sand or Snow	Bright or Hazy Sun (Distinct Shadows)	Weak, Hazy Sun (Soft Shadows)	Cloudy Bright (No Shadows)	Open Shade† or Heavy Overcast
f/16	*f*/11*	*f*/8	*f*/5.6	*f*/4.0

*For backlighted, close-up subjects, increase exposure by 2 stops.
†Subjects shaded from the sun but lighted by a large area of clear, unobstructed sky.

Filter Factors: Multiply normal exposure by filter factor given below.

KODAK WRATTEN Filter	No. 6	No. 8	No. 11	No. 15	No. 25	No. 47	No. 58	Polarizing Screen
Daylight	1.5	2*	4	2.5	6	6	8	2.5
Tungsten	1.2	1.5	4*	1.5	4	12	8	2.5

*For a gray-tone rendering of colors approximating their visual brightness.

Flash Exposures: To determine the *f*-number for average subjects, divide the appropriate guide number by the distance (in feet) from flash to subject.

Guide Numbers for Blue Flashbulbs: Select the guide number on the flashbulb package for the film speed listed above, and for the type of reflector, shutter, and synchronization on the camera you are using.

Caution: Since bulbs may shatter when flashed, use a flashguard over the reflector. *Do not flash bulbs in an explosive atmosphere.*

Electronic Flash Guide Numbers:

Output of Unit (BCPS or ECPS)	350	500	700	1000	1400	2000	2800	4000	5600	8000
Guide Number for Trial	45	55	65	80	95	110	130	160	190	220

Adjustments for Long and Short Exposures:

If Indicated Exposure Time (seconds) is	Use: Either This Lens Aperture Adjustment	Use: Or This Adjusted Exposure Time (seconds)	And, in Either Case, Use This Development Time Adjustment From Normal
1/100,000	+1 stop	No Adjustment	+20%
1/10,000	+½ stop	No Adjustment	+15%
1/1,000	None	No Adjustment	+10%
1/100	None	No Adjustment	None
1/10	None	No Adjustment	None
1	+1 stop	2	−10%
10	+2 stops	50	−20%
100	+3 stops	1200	−30%

Processing

Handle in total darkness. However, a KODAK Safelight Filter No. 3 (dark green) in a suitable safelight lamp with a 15-watt bulb can be used for a few seconds after development is half completed, provided it is kept at least 4 feet from the film.

Develop for the approximate times given.

KODAK Packaged Developers	Developing Times (in Minutes)* — SMALL TANK—(Agitation at 30-Second Intervals)† 65°F 18.5°C	**68°F 20°C**	70°F 21°C	72°F 22°C	75°F 24°C	LARGE TANK—(Agitation at 1-Minute Intervals)† 65°F 18.5°C	**68°F 20°C**	70°F 21°C	72°F 22°C	75°F 24°C
HC-110 (Dilution B)	6	**5**	4½	4	3½	6½	**5½**	5	4¾	4
D-76	6½	**5½**	5	4½	3¾	7½	**6½**	6	5½	4½
D-76 (1:1)	8	**7**	6½	6	5	10	**9**	8	7½	7
MICRODOL-X	8	**7**	6½	6	5½	10	**9**	8	7½	7½
MICRODOL-X (1:3)	—	—	11	10	**9½**	—	—	14	13	**11**

*With properly exposed film, these suggested developing times should yield negative contrast suitable for printing with a diffusion enlarger or by contact. For printing with a condenser enlarger, a lower contrast (achieved by a reduction in development time) is usually desirable. For complete information about contrast control, see KODAK Publication No. F-5, *KODAK Professional Black-and-White Films.*
†Unsatisfactory uniformity may result with development times shorter than 5 minutes.
Note: Do not use developers containing silver halide solvents.

Rinse at 65 to 75°F (18.5 to 24°C) with agitation.
KODAK Indicator Stop Bath—30 seconds
or KODAK Stop Bath SB-5 —30 seconds

Fix at 65 to 75°F (18.5 to 24°C) with agitation.
KODAK Fixer —5 to 10 minutes
or KODAK Fixing Bath F-5—5 to 10 minutes
or KODAK Rapid Fixer —2 to 4 minutes
or KODAFIX Solution —2 to 4 minutes

Wash for 20 to 30 minutes in running water at 65 to 75°F (18.5 to 24°C). To minimize drying marks, treat in KODAK PHOTO-FLO Solution after washing. To save time and conserve water, use KODAK Hypo Clearing Agent.

Dry in a dust-free place.

Storage: Keep unexposed film at 75°F (24°C) or lower. Process film as soon as possible after exposure.

The Kodak materials described in this publication for use with KODAK PLUS-X Pan Professional Film PXP220 and PXP120 are available from those dealers normally supplying Kodak products for professional photography. Other materials may be used, but similar results may not be obtained.

Notice: This film will be replaced if defective in manufacture, labeling, or packaging. Except for such replacement, the sale or any subsequent handling of this film is without warranty or liability even though defect, damage, or loss is caused by negligence or other fault.

EASTMAN KODAK COMPANY, Rochester, N.Y. 14650
KP 65830f 6-83 Printed in the United States of America

Kodak, Plus-X, Wratten, HC-110, D-76, Microdol-X, Kodafix, and Photo-Flo are trademarks.

Fig. 55-2. Manufacturer's film data sheet.

STEP NUMBER	CHEMICAL NAME	TIME (BASED ON TEMPERATURE)
1	Mirodol	
2	Stop bath	30 seconds
3	Fixer	7 minutes
4	Rinse	1 minute
5	Hypo clearing	2 minutes
6	Wash	5 minutes
7	Dry	20 minutes

NOTES
1. Developing tank holds 16 ounces of chemical for all stages of developing.
2. Agitation during each stage is critical. Agitate for the first minute and thereafter for 15 seconds of each additional min if time permits.
3. Cleanliness is of extreme importance. Rinse each graduate, funnel, or tank after each stage of development.
4. Temperature range during developing must be within 68−75°F.

Fig. 55-3. Black-and-white film-developing chart.

picture was taken. This stage also serves to harden the negatives. Film may be exposed to white light after completing this stage. Typical types of fixers include Rapid Fix (a liquid) and powder fixer, which is slower than Rapid Fix. The normal fixing time is approximately 5 minutes.

Rinse. A short rinse in water of approximately 2 minutes is needed at this stage. Be sure that the temperature of the rinse is between 65° and 75°.

Hypo Clearing. This stage requires approximately 3 minutes. Hypo clearing reduces the normal amount of wash time needed.

Wash. Final washing takes 5 minutes. Keep the temperature within the recommended temperature range to avoid reticulation (the shattering of the emulsion layer as a result of a large deviation [change] in temperature between developing stages).

Photo-Flo. This chemical is known as a wetting agent. It prevents water spots from forming on the negatives. Two or three drops of this stock solution with 32 ounces of water will produce a working solution. One minute in this solution is sufficient.

Drying. The faster the negatives are dried, the better the grain structure. Dry negatives for at least 20 minutes (Fig. 55-3).

Developing film by time and temperature requires the utmost attention to detail. Without a good negative, there cannot be a good print. For the time spent at this critical stage, the reward is a superior-quality photograph.

Unit 56

Darkroom Equipment

The photographer's darkroom is often considered a mysterious and intriguing place by people who are not familiar with its contents. Subdued lighting, mechanical and optical devices, and strange chemical odors fascinate many photographic en-

Fig. 56-1. A typical darkroom.

Fig. 56-2. Print-processing area.

thusiasts. Many photographers find more personal enjoyment in darkroom manipulation of the print than in actually taking the pictures. In this unit we shall discuss the equipment that is needed in the school or home darkroom (Fig. 56-1).

Lighting

Many people believe that the darkroom walls should be black, but this is not correct. The darkroom should be a light color so that the reflected safe lights can provide enough illumination for proper activity. Reflected light will also help eliminate shadows at the enlarging station and the print-processing area. Safe lights are usually low wattage to avoid fogging the paper. The lighting may appear amber, orange, or red depending on the type of filtration used and the type of processing to be done. Black-and-white photographic paper is usually not sensitive to the entire light spectrum, as is panchromatic film. It can therefore be viewed safely under this special lighting. If photographic paper is left out for an extended period of time, it will become exposed. Replace the photographic paper in its light-tight box after selecting the amount needed for your job.

Processing Area

The print-processing area should have ample room for chemical trays and chemical storage as well as a source of hot and cold water for print washing. Cleanliness in this area is of the utmost importance. Never bring negatives or other dry materials to this area. Room should be provided for as many as five processing trays (8 × 10 or 11 × 14 inches [203 × 254 or 279 × 356 millimeters] in size) in this area (Fig. 56-2).

Enlarging Station

The enlarging station is a dry area. Therefore, wet material, such as test strips and partially processed prints, should never be taken to this area. To do so might ruin the negative or unprocessed enlarging paper.

The basic enlarging station consists of the enlarger, timer, printing easel, dodging tools, negative carriers, and an assortment of enlarging lenses (Fig. 56-3). There are many makes and models of enlargers.

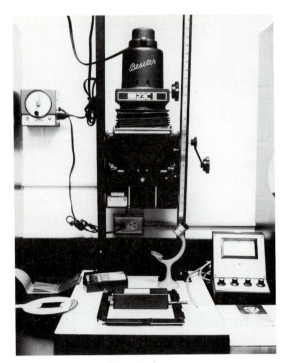

Fig. 56-3. Enlarging station.

include 8 × 10 inches (203 × 254 millimeters), 5 × 7 inches (127 × 179 millimeters), 4 × 5 inches (102 × 127 millimeters), 3½ × 5 inches (89 × 179 millimeters), and wallet formats.

Odds and Ends

In addition to the items already mentioned, the basic darkroom should also include the following:

1. Thermometers for accurate chemical temperature control, 65 to 75°F recommended

2. Print tongs for moving prints from one chemical stage to another without contaminating chemical stages or getting your fingers wet

3. Paper cutter to cut photographic enlarging paper accurately from standard sheet sizes into test strips or odd-size prints.

Safety in the Darkroom

Never run in the darkroom. Processing chemicals can make the floor extremely slippery.

Adjust your eyes to the low light level when you first enter the darkroom. When your eyes have adjusted, then and only then is it safe to proceed. This usually takes less than a minute.

Thoroughly familiarize yourself with the layout of the darkroom when the white lights are on. This will help you orientate (acquaint) yourself to the safe lights.

Read the manufacturer's operating instructions thoroughly before using a particular enlarger.

An enlarger is basically a camera in reverse. Exposure of the print is controlled by an f-stop opening of the enlarging lens, and the length of exposure is controlled by the setting of the timer in minutes and seconds. The easel has an image projected onto it by the enlarger. The size of the image is controlled by varying the height of the enlarger and the size of the lens. Many easels are adjusted to allow for odd-size prints. Generally, standard easel openings

Unit 57

Contact and Projection Printing

After a roll of film has been developed, prints can be made from negatives. A negative is the reverse of the tones of the actual scene. From negatives, positives are produced. They are called photographs. This process can be done two ways; by making contact prints or by making enlargements through projection of the image.

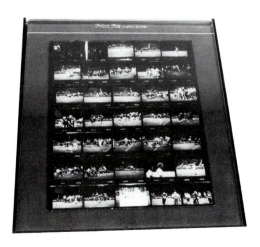

Fig. 57-1. Contact printing frame.

Contact Printing

A contact sheet, usually 8 × 10 inches (203 × 254 millimeters) in size, can be made with a clear piece of glass. A contact printing frame of either glass or Plexiglas makes the task much easier.

To make a contact print:

1. Place the enlarging paper, *emulsion side up*, on the enlarger's baseboard. The light source must cover the enlarging paper entirely.
2. Place the negatives, *shiny side up*, (dull side [emulsion side] down), directly on the enlarging paper. If the negatives have been stored in strips in a protective acetate sheet, the entire sheet can be placed directly on the enlarging paper.
3. Check the numbers on the negatives and be sure that they can be read.
4. Place a piece of glass on top of the negatives. The weight of the glass will hold them in direct contact with the photographic paper.
The contact printing frame (Fig. 57-1) is hinged to allow the sandwiching of both the negatives and the paper between the transparent top and its solid base.
5. Expose the paper. Project white light from the enlarger through the negatives.

Proper exposure requires trial and error. Exposure is controlled by the time.

The quantity of light is controlled by the f-stop setting of the enlarging lens. A recommended starting point would be f-8 at 5 seconds. See Unit 58 for development of contact and projection prints.

After you have processed the contact sheet, notice that you have produced a series of small prints exactly the same size as the negatives. A contact print is an extremely useful preview of the entire roll of film. It is an aid in selecting the negatives to be printed. Cropping can be done directly on the contact print. Storage of the negatives with the contact sheet aids in filing and later retrieval (Fig. 57-2).

To make a projection print:

1. Insert the negative (which has been placed in a negative carrier) between the light source and the lens of the enlarger (Fig. 57-3).
2. Check to see that the *shiny side is up.* (emulsion side down). Light from the lamp housing of the enlarger passes through the negative and is focused on the printing easel by adjusting the bellows of the enlarger.
3. Focus with the lens at its *widest opening.* The size of the projected image is determined by the size of the enlarging lens and the height of the enlarger above the printing easel.
4. Place the opening used for the size of print under the light beam. Compose the projected image within the opening of the easel (5 × 7 inches, 8 × 10 inches, etc.).
5. Focus the enlarger. Make sure that the entire opening is struck with light.
6. Determine the time of exposure. Exposure is a combination of the f-stop opening selected and the length of time the print is allowed to be exposed. Make a test strip first:
   a. Open the lens all the way.
   b. Close the lens three f-stops down from wide open.

Fig. 57-2. **Contact sheet.**

c. Set the timer for 5 seconds.

d. With the timer in the "off" position (so that no light comes from the enlarger), place a test strip in the opening of the enlarging easel.

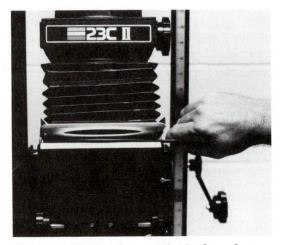

Fig. 57-3. **Placing the negative in the enlarger.**

e. Cover 75 percent of the test strip. Use an opaque material such as cardboard.

f. Expose the uncovered 25 percent for the 5-second duration by pressing the button on the timer.

g. Without moving the test strip, carefully move the piece of cardboard to reveal an additional 25 percent of the test strip.

h. Hit the timer again. The first 25 percent of the test strip has been exposed for 10 seconds; the second 25 percent for 5 seconds.

i. Repeat this procedure to obtain exposures of 20, 15, 10, and 5 seconds.

j. Examine the strip under white light to determine the correct exposure (Fig. 57-4).

7. After the test strip has been processed and the correct exposure has been calculated, set the timer accordingly and expose the paper. Making enlargements should be done only under safe light conditions. *Protect* unexposed photographic paper from any stray white light. *Before*

Fig. 57-4. Test strip. Four exposures at 5 seconds each.

placing the negative in the enlarger, make sure it is clean and free of dust.

In printing with 35 mm negatives, a 50 mm lens will probably be selected. A 120 negative requires a 75 mm lens. These are suggested lens sizes that will provide sufficient magnification. Most enlarging easels will allow standard-size prints to be made, either 5 × 7, 4 × 5, or wallet on one side of the easel, and an 8 × 10 print on the other side. With practice, speed and accuracy will be achieved in determining exposure and focusing.

Unit 58

Steps in Print Processing

Print processing uses many of the same chemicals that are used in the developing of film. Print processing is usually done with the chemicals in trays. Trays are usually 8 × 10 inches (203 × 254 millimeters), 11 × 14 inches (279 × 356 millimeters), and in some cases larger. The trays should be cleaned thoroughly before chemicals are put into them. The chemicals should rise high enough in the trays so that the prints will be completely submerged.

In processing prints, occasional agitation (stirring) of the chemicals is recommended for even development. Use print tongs when transferring the print from one chemical stage to another. Be sure to let the print drain before proceeding to the next stage. This will prevent contamination (spoiling) of the chemical liquids. Stock solutions are usually diluted (thinned) to make working solutions for many of the stages.

Developer

A typical developer for black-and-white print processing is Dektol. The development stage should be in the temperature

range of 65° to 75°. A stock liquid solution of Dektol should be mixed with two parts water to make a suitable working solution.

Prints should stay in this stage for 60 to 90 seconds. If the print is becoming too dark but has not stayed a minimum of 60 seconds in the developer, do not continue to process it. Instead, discard (throw away) it and reexpose another print for a shorter period of time or close down the enlarging lens. A minimum of 60 seconds is necessary for quality (good) black-and-white printing. It takes this long to develop proper contrast and tonal rendition as well as details in shadow areas. The developer, when exhausted (worn out), will turn dark and should be discarded.

Stop Bath

Indicator stop bath comes in a stock liquid solution. Two ounces of this solution will make 1 gallon of working solution when mixed with water. The working solution is used straight in the trays. Stop bath is yellow and will turn purplish-blue when exhausted. Prints should stay in this stage for approximately 30 seconds. Stop bath, as the name implies, stops the development abruptly.

Fixer

Fixer, also called hypo, can be purchased in powder or liquid form. The liquid form is called Rapid Fix. If the powder form is used, mix it with water below 80°F. The fixer stage removes unexposed or undeveloped silver from the print. It also hardens the print and makes it permanent. Prints should stay in the fixer stage for approximately 3 to 5 minutes. After this stage, the print can be safely viewed under white light.

Rinse

A 1-to-2 minute rinse stage is applied after the fixer stage. Be sure the rinse water is changed often. All chemicals for print processing should be in the same temperature range. This also includes the rinse stage.

Hypo Clearing

The hypo clearing stage reduces the amount of wash time necessary for the print. It reduces washing by two-thirds. Hypo clearing, which comes in powder form, or Orbit Bath, which is a liquid, can be used. Hypo clearing is used straight from the container. Discard the hypo clearing after each printing session. The print should stay in this stage for approximately 2 to 3 minutes.

Wash

The final wash stage takes about 5 minutes. Make sure the water is continuously being replaced with fresh water during the wash cycle.

Drying

Squeegee the prints after the wash stage to have them dry quicker. It is assumed that the printing is done on resin-coated (RC) black-and-white enlarging paper. The processing times and drying procedures recommended here are for RC paper. Fiber-based paper requires additional time for washing and drying. Fiber-based paper cannot be air-dried as resin-coated paper can. Use a hair drier or RC print drier to speed the drying process.

As the photographer becomes more familiar with the print-processing procedures, times, and chemicals, a system can be developed to have many prints going through the different stages at the same time. Check the strength of the chemicals periodically. One exhausted stage can ruin the print and waste time.

Unit 59

Mounting and Print Finishing

Presenting and displaying photographic work gives you, the photographer, a feeling of pride and accomplishment. People appreciate photographic work much more if it is printed well, mounted properly, and displayed correctly (Fig. 59-1). Nothing detracts more from a good photograph than improper presentation of the finished product.

Print Finishing

After a good negative is selected to produce an enlargement for possible display,

Fig. 59-1.   Example of properly displayed print.

make every effort to produce (make) a technically perfect print. This perfect enlargement is usually 8 × 10 inches (203 × 254 millimeters) or larger as necessary for display. It should be free of dust spots, have proper contrast and tonal rendition, and adhere to the basic rules of composition.

If the final print has white spots caused by dust on the negative during printing, a process known as spotting should be used to eliminate them. Spotting dyes come in liquid form as well as powder form. In liquid form, they are packaged in groups of three; blue, black, and brown are the colors provided. The No. 3 black Spottone is used most often and can be purchased separately. Use a very fine pointed brush, such as a No. 000, when applying the dye to the print. A technique known as *stippling* is used to fill in the spot by touching the spot with the point of the brush. Practice is necessary to develop the right technique. Beginners usually make the spot too dark. Start off with a diluted amount of dye. It is much easier to add more dye than to have to etch the dye out if it becomes too dark (Fig. 59-2).

Mounting

Mounting the finished print can be done by any of the following methods:

1. Dry mounting with a hot press. This requires dry mounting tissue and a dry mounting machine. The temperature varies according to the manufacturer's specifications set forth for each particular type of dry mounting tissue. In most cases, the temperature will be approximately 200°F.

2. Mounting by means of a spray adhesive. The spray adhesive should be applied to both the photograph and the mounting surface. Let the adhesive dry until tacky; usually 1 minute is sufficient. This produces a permanent bond between the photograph and the mounting material. Spray mounting adhesive is expensive, and precautions (care) should be taken

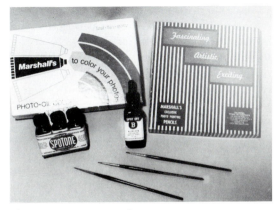

Fig. 59-2.   Spottone dyes.

since this material is very flammable (burns quickly).

3. Cold mounting with pressure-sensitive materials. This is a relatively easy technique for mounting photographs. It is, however, more expensive than any other method.

4. Gluing with wallpaper paste. This method is time-consuming and messy and requires an additional mounting material on the opposite side of the mounting board to prevent the mounted photograph from curling.

Whichever method is used, be sure to select a mount board that will complement and enhance the photograph. Never center a photo when mounting. Leaving more area at the base of the photo gives stability to the print and tends to attract attention.

Once the photo is securely applied to the mount board, it is advisable to place either a single or a double matte over the print. Matting photos gives them a professional touch and significantly improves their visual impact (appearance).

Unit 60

Tools for the Photographer

Introduction

A wide assortment of accessories is available for cameras (Fig. 60-1).

What to buy, when to buy it, and why it should be purchased are questions the beginning photographer must be concerned with. This unit will help you make wise decisions that can save time, money, and frustration.

Accessories to the basic camera can afford the photographer the opportunity to become more creative.

Gadgets and accessories such as filters, tripods, and lenses will stimulate the imagination and broaden one's photographic capabilities. Keep one thing in mind as a general rule: Purchase only what is needed, when it is needed. Do not purchase a motor drive merely for the sake of ownership. It is, however, useful for sports or action photography.

Before purchasing any major piece of photographic equipment, know the brands available, the relative cost of the item, and the advantages and disadvantages of the various models.

Fig. 60-1. Camera accessories.

Start your research by reading the advertisements in photographic magazines. Browsing through speciality camera stores and asking questions of the salespeople are also excellent methods of learning about current camera accessories. Avoid impulse buying if possible.

The wide variety of camera accessories makes photography a rewarding and exciting hobby, but be careful when buying camera accessories. The old saying, "You get what you pay for" is very true with photographic equipment.

In the following discussion of camera accessories, it is assumed that a modern 35 mm SLR camera is used with built-in light meter and the capability of interchangeability of lenses.

Filters

One of the following filters should be kept on the lens at all times: skylight, ultraviolet (UV), or haze. Any of these filters will protect what is probably the most important and expensive parts of the camera: the lens. The camera body is essentially little more than a transport mechanism used to align the film behind the lens opening. Good pictures cannot be obtained without a good unscratched, undamaged, clean lens. Protect your investment by using one of these filters at all times. These filters require little if any exposure compensation (adjustment); with through-the-lens metering, the light meter will automatically adjust for this compensation as necessary (Fig. 60-2). The modest investment made for one of these filters is well worth the money.

Probably the most important filter a photographer can have for color photography is the polarizer. The polarizer acts much the same way as a pair of sunglasses. It makes the blue sky darker and makes white, billowy clouds stand out. It reduces reflections in water or glass. It is a highly recommended filter for landscape and scenic photography.

These are the filters used for black-and-white photography:

1. Yellow. This filter darkens a blue sky and makes clouds stand out. One f-stop is

Fig. 60-2. Assorted filters.

needed to compensate (adjust) for light loss when using a yellow filter.

2. Red. The red filter acts much the same way as the yellow, but with a much more dramatic increase in contrast. Three f-stops are needed to compensate for light loss when using a red filter.

3. Green. This improves skin tones in outdoor portraiture. It also lightens green foliage. Two f-stops compensate for loss of light when using a green filter.

Through-the-lens light metering adjusts for the filter. Filters are useful and very necessary for the creative photographer.

Lens Hood

An inexpensive but often overlooked addition to the lens is a flexible lens hood. Metal or rigid lens hoods dent easily and do not have the advantage of collapsibility (folding) found in the flexible lens hood. A lens hood will prevent lens flare and thus improve pictures. Better light meter reading and improved color saturation can be obtained by using a lens hood.

Camera Cases

Having acquired basic additions to the camera, one must find a means (method) for their storage, protection, and transport. The obvious solution is the purchase of a camera case or gadget bag. The color, style, and size of the camera case is a personal decision. In general, it is recommended that an aluminum attache-style camera case be used by a photographer who does quite a bit of traveling. They are quite expensive, but they provide an added measure of protection for the camera system. Lightweight and flexible gadget bags made of nylon or other synthetic cloth material are becoming more popular. They are usually padded and compartmentalized, with areas set aside specifically for film, lenses, and the camera itself (Fig. 60-3).

Lenses

Normal Lenses. For the beginning photographer, getting to know the limitations of the normal lens that came with the camera should be the first objective. For the 35 mm format, a normal lens would be approximately 45 to 55 millimeters in focal length. The view through this lens is roughly what the human eye sees. A normal lens is a general-purpose lens. It is used for full-length pictures of people as well as scenic or action photography.

Wide-angle Lenses. When one is taking pictures in confined spaces, a wide-angle lens is the solution to the problem of including all of the subject matter. A wide-angle lens, such as the 28 mm or 35 mm, provides an increased field of view and an increased depth of field. For the photojournalist, the wide-angle lens is almost a necessity. For architectural and scenic photography, a 24 mm or 28 mm wide-angle lens has become the standard.

Telephoto Lenses. The telephoto lens is the reverse of the wide-angle. This lens allows the photographer to increase the

Fig. 60-3.   Assorted camera cases.

camera-to-subject distance and still fill the frame with the subject. The telephoto range starts at approximately 100 mm. The standard 135 mm lens has become very popular as a moderate telephoto lens and is used extensively for head-and-shoulders portraiture. It affords the photographer a good working distance from the subject, and it eliminates the facial distortion that is evident when a normal lens is used. For sports photography, the telephoto lens is a *must*. A lens in the 200 mm to 250 mm range would be a good choice for the sports or wildlife photographer. Telephotos are inherently very slow. This drawback can be overcome through the use of faster film or "push" processing. In many cases, sports photographers do both. Tri-X pan film, black-and-white, ASA 400, pushed to 1600 ASA and developed in Acufine developer, is almost a standard for a night sports photographer who uses the long telephoto lens.

Zoom Lenses. A zoom lens has the capability of covering many different focal lengths. With this capability, you can compose a shot without having to change the camera position. In addition, the zoom lens eliminates the need for many lenses with different focal lengths. The zoom lens is more expensive than a fixed-focal-length lens. Typical zoom lenses can cover the wide-angle through medium-telephoto range (that is, 40 mm to 80 mm) or medium-telephoto through long-telephoto range (70 mm to 210 mm).

Macro Lenses. When buying a new camera, consider the purchase of a macro (close-focusing) lens instead of a normal lens. The price is a bit higher, but the advantage of having this capability of close focusing is well worth the added expense. Other, less expensive camera accessories can afford the photographer the opportunity to do close-up work. Bellows, supplementary lenses, or extension tubes give the normal lens the capability of close focusing. Each of these may be purchased for much less than a macro lens. They do have disadvantages, however. The supplementary lenses, for example, are not as optically sharp as a macro lens. The bellows requires the photographer to calculate exposure loss. Extension tubes allow only a certain degree of magnification and also require exposure compensation, which the light meter will not do.

Teleconverters. Doublers or triplers allow the photographer to increase the focal length by two (2×) to three (3×) times. These converters are relatively inexpensive compared with a telephoto lens. A normal lens can be converted from a 50 mm to a 150 mm by the insertion of a tripler between the camera body and the lens. Select a converter that is compatible with the lens mount. It should be automatic, and it is a quality optical device. There is some light loss from the addition of the teleconverter, but the light meter will automatically compensate for it.

Lens Tips

1. Buy brand-name equipment if possible.
2. Buy lenses that are automatic.
3. Make sure the lens is coated.
4. Shop around for the fastest lens possible for the money you have to spend.
5. Use a shutter speed fast enough to eliminate camera motion blur. Generally go one shutter speed higher than the focal length of the lens you are using. That is, a 135 mm lens requires a shutter speed of 1/250 of a second or faster.
6. Use a tripod for anything longer than a 135 mm lens.
7. Protect lenses. Put filters on the lenses, use front and rear lens caps, and keep the lenses in their cases when not in use.
8. Be careful not to cross-thread the lens when installing it on the camera body. Never force the lens.
9. Clean lenses only when necessary.

Unit 61

Discussion Topics on Photographic Technology

1. List six career opportunities available in the photographic industry.

2. What are the four formats of cameras? Give three advantages of each.

3. List the steps necessary for film developing.

4. List the steps necessary for print processing.

5. Name three classifications of lenses and give the approximate focal length of each.

6. Name five controls on the 35 mm camera and explain their functions.

7. Name two common types of black-and-white film. Give their ISO and ASA rating.

8. List five pieces of equipment used in the darkroom.

9. What is the purpose of fixer?

10. Why is agitation of chemicals important in film developing and print processing?

11. Exposure is determined by which three factors?

12. What affects depth of field?

13. Name three ways to mount photographs. Which is best? Why?

14. List five accessories used by the photographer.

15. Describe the major use of any three filters.

16. Name three early pioneers in photography. Give one contribution that each made.

17. Which format of camera is the most popular and why?

18. What is the slowest shutter speed you can use when holding the camera?

19. Why is the speed of a lens important to photographers?

20. Explain why photographers should be concerned with the direction of light.

21. Name three recent features used on modern cameras.

22. What four things do all cameras have in common?

23. What specialized camera can give you instant photographs?

24. List five guidelines for improving photographic composition.

25. What three darkroom printing techniques could be used to improve print quality and composition?

26. Explain the zone or prefocusing technique.

27. What six factors should be taken into account when planning your photographs?

28. Explain how a contact sheet is made.

29. How do you determine if the camera is loaded with film?

PRODUCTION TECHNOLOGY

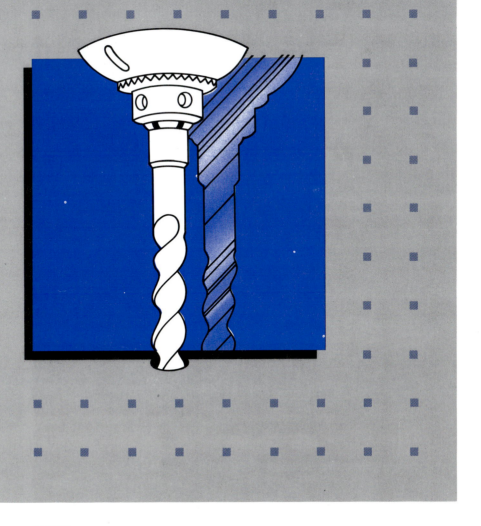

Section 6
Woods
Technology

Unit 62

Introduction to Woods Technology

People who work with wood find it an enjoyable and meaningful activity. Wood is the most plentiful and renewable natural resource. It is used extensively in industries, school laboratories, and home workshops. Skill in woodworking can lead to a rewarding hobby or a professional career.

When working with wood, it is necessary to know how to read simple working drawings. The elements (parts) of a drawing are made up of many different types of lines that give instructions and describe the use of project views. The working drawing (sketch) is a technical language. It shows and explains the various parts of a project activity or product.

As one studies the *technology* (concept) of woodworking, interest and curiosity will be broadened. One will want to know more about jobs and careers and personal possibilities in the wood and lumber industries.

Unit 63

Woods Technology-related Careers

There are many exciting career opportunities in the areas of forests, woods, and wood-products industries. Some involve conservation occupations, such as *forestry aides* and *range managers*. Others pertain to the lumbering industry, which includes *loggers, sawyers,* and people connected with the transportation of the *raw material* (logs). There are *craftspersons, technicians, semiskilled workers,* and *laborers* of all types. *Architects, contractors, carpenters, painters, paperhangers, brick masons, plumbers, electricians,* and *floor-covering* employees are active in the construction industry.

There are more than 1 million workers who earn a living directly from forest products. Many more millions of persons depend on forest products for jobs in transportation, construction, utilities, selling, and business interests. Tens of thousands are employed in professional and technical occupations as teachers, research technicians, and scientists. Approximately 12 million people, either directly or indirectly, provide us with the products that come from the great natural resource of timber. Researchers continue to develop new wood products.

In factory production of materials for construction there are *supervisors, managers, craftspersons, assemblers, inspectors,* and *laborers* (Fig. 63-1). Most of these occupations are associated with labor unions, and therefore there are *specific* (definite) requirements and certain legislation that affect them.

Building Trades

The building-trades *craftspersons* (journeymen) make up the largest group of *skilled* workers in the labor force. Approximately one-third of all skilled workers are in the building trades. The more than two dozen skilled major building trades include *carpenters, painters, plumbers, pipe fitters, bricklayers, operating engineers* (construction machinery operators), and *construction electricians.*

Each of these categories has more than 100 000 workers. There are approximately 1 million carpenters (Fig. 63-2). Other workers in the building trades are marble

Fig. 63-1. A supervisor explains the operating functions of the new hot-press control panel used in the plywood industry to a group of technicians. (American Plywood Association)

Fig. 63-2. Carpenters and building-trades people can see day-to-day progress in what they build. (Southern Pine Association)

cutters, terrazzo workers, glaziers (glass installation workers), and stone masons.

Building-trades craftspersons work in the areas of maintenance, repair, alteration of homes and other types of buildings, highways, airports, and missile and space programs (Fig. 63-3).

Membership in building trades consists primarily of *journeymen* (skilled craftspersons). They must have a high degree of skill and a good knowledge of assembly

Fig. 63-3. Building-trades people are involved in the missile and space programs, building mock-ups out of wood, steel, and aluminum. (North American-Rockwell)

Fig. 63-4. Woodworking offers many careers in millwork production.

and construction. Often they are assisted by apprentices and laborers.

Thousands of building-trades workers are employed in factories, mines, stores, hotels, and other business establishments (Fig. 63-4). Most work for contractors; however, some are self-employed. Many acquire (learn) the skills of the trade by working as laborers and helpers, often through an apprenticeship program. Others gain skills attending high school trades classes, vocational classes, or technical institutes.

The usual entry into the building trades is through *apprenticeship* (Fig. 63-5). The formal registered apprenticeship agreement generally sets up a training period of from two to five years of continuous employment and training. In addition, the apprentice must attend related-subjects classes for a minimum of 144 clock hours a year. Such apprentices learn the trade by working with journeymen. Classroom instruction varies with the trade. It usually includes information about the history of the trade and the characteristics of the materials of the trade. The apprentice studies related mathematics, basic principles of engineering, sketching, drafting, safety, and other specialty courses.

The apprentice is usually paid 50 percent of the journeyman's rate of pay upon beginning the apprenticeship. The rate in-

creases at 6- or 12-month intervals until it reaches about 90 percent of the journeyman's rate. Frequently, local unions will recognize advanced standing. This is based on trade skills that may have been

Fig. 63-5. The usual entry into building trades is through apprenticeship. This requires a period of several years of employment and training. (Southern Pine Association)

acquired through formal education or in the armed forces.

The occupational outlook for the building trades will improve in the next *decade* (10-year period). This is due to population increase, a higher standard of living, and newer techniques of building construction (Fig. 63-6). Technological advances in construction tools and equipment will increase the efficiency of building-trades workers. The building trades offer especially good opportunities for those who do not wish to go to college but are willing to spend several years learning a skilled occupation (Fig. 63-7). Since building-trades-workers are usually paid on a hourly basis, their total wages tend to *fluctuate* (go up and down). Changes in general business

Fig. 63-7. The building trades offer good opportuities for those who take wood technology courses in school. (Georgia-Pacific Company)

Fig. 63-6. Modular-home construction is a new development that has improved the occupational outlook for the building trades. (American Plywood Association)

conditions and the seasonal nature of construction work affect their income.

Conservation Careers

People interested in forestry and conservation careers should pursue a bachelor's degree program in a college or university. These occupations are usually professions that require an academic background or preparation (Fig. 63-8). Careers in forestry

Fig. 63-8. Careers in forestry and conservation offer great professional opportunities. This researcher is measuring the growth of a tree.

Fig. 63-9. Management of company-owned and privately owned tree farms offers an excellent occupational livelihood. (California Redwood Association)

Fig. 63-10. There will always be ample tree growth because of research in scientific reseeding and planned harvesting, as indicated by the block-cutting program shown in this photograph. (U.S. Forest Service)

and conservation are generally associated with state and national government employment.

A few people with a background in forestry and conservation are employed by large, privately owned lumbering industries that produce their own *raw materials* (trees). In the northwestern part of the United States, there are many company and privately owned tree farms (Fig. 63-9) where forestry-trained people readily find employment. Conservation of woodlands has become so scientific that it is predicted there will always be ample tree growth as a result of scientific *reseeding* and *planned harvesting* (cutting) (Fig. 63-10).

Wood-products Industries

Wood-products industries were one of the first industrial and business enterprises in the United States. Careers in this broad area have increased as the scientific know-how in research has developed better and wider uses for the products of trees (Fig. 63-11). Billions of dollars are invested in land, equipment, and salaries of people

Fig. 63-11. Developments in the woods-products industries depend on expanded scientific research programs. (American Plywood Association)

who work in the production and manufacture of products made of wood.

Wood is the product of sun, soil, and water. The supply can be maintained through scientific forest management. Forest-products industries rank fourth in the number of full-time employees in industry in the United States. Millions work in the areas of *growing*, *harvesting*, *transporting*, *processing*, *distributing*, and *selling* wood products. The approximate total number of persons who make a *livelihood* (living) directly or indirectly from wood products is estimated at 15 million (Figs. 63-12 and 63-13). More than one-third of the *lumber* supply for the world, one-half of the *plywood*, two-fifths of the *wood pulp*, and half of the *paper* (Fig. 63-14) and *paperboard* are produced in the United States.

High school graduates can find employment in the woods industries. They must take advantage, however, of all educational opportunities. This includes on-the-job training. There are excellent careers in the areas of *general forestry*, *wood technology*, *wood-products engineering*, *distribution* and *merchandising*, *construction*, *furniture manufacture*, *pulp* and *paper technol-*

Fig. 63-13. These new seven-story lignin evaporation towers produce by-products with many uses after the initial production of alcohol from wood pulp. This involves woods-chemistry research. (Georgia-Pacific Corporation)

Fig. 63-14. Chemists, scientists, engineers, technicians, and production people work together to produce paper and the vast assortment of all kinds of paper products that form the basic product of cellulose. (Weyerhauser Company)

Fig. 63-12. Many thousands of craftspersons are employed in furniture-manufacturing technology. (National Association of Furniture Manufacturers)

ogy, packaging, and *tree-farm management.* Students should prepare for a career in the woods-related areas by taking basic English, science, mathematics, drafting, and all the woods courses available in junior and senior high school.

Forests of the United States

The forests of the United States give people a very valuable resource. There are over 1000 species of trees in our country. Only about 100 of these species are used for lumber or woods products.

Forests

About one-third of the land in the United States is forest land. There are over 755 million acres (3.14 million square kilometers) in 10 forest regions (Fig. 64-1).

The *west coast forest* has an area of giant redwoods on the northwest coast of California (Fig. 64-2). This forest produces about one-third of our lumber, one-fifth of our pulpwood, and nearly all our fir plywood. Douglas fir is the most common fir species. Other woods are lodgepole pine, incense cedar, western hemlock,

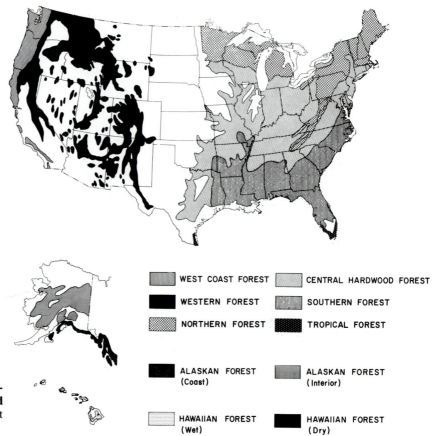

WEST COAST FOREST   CENTRAL HARDWOOD FOREST

WESTERN FOREST   SOUTHERN FOREST

NORTHERN FOREST   TROPICAL FOREST

ALASKAN FOREST (Coast)   ALASKAN FOREST (Interior)

HAWAIIAN FOREST (Wet)   HAWAIIAN FOREST (Dry)

Fig. 64-1. Forest regions of the United States. (American Forest Institute)

Fig. 64-2. Woods technology depends on well-managed redwood forests like these for a profitable economy. (Georgia-Pacific Corporation)

western red cedar, red alder, and Sitka spruce.

The *western forest* produces nearly one-fourth of our lumber supply. Three pine species are ponderosa, Idaho white, and sugar. Other timber trees are Douglas fir, spruce, western larch, and western red cedar. Aspen, a hardwood, grows here.

The *northern forest* produces about one-tenth of our lumber and almost one-fifth of the pulpwood. Softwoods include red and jack pine, white and red spruce, eastern hemlock, white cedar, white pine, and balsam fir. The hardwoods are black, yellow, and paper birch; maple; oak; black cherry; black gum; and aspen.

The *central hardwood forest* is the largest forest area. About one-twentieth of the lumber and one-tenth of the pulpwood come from there. Hardwoods are red and black gum, yellow poplar, oak, beech, red maple, hickory, elm, ash, and sycamore. Softwoods are the shortleaf and Virginia pine, and red cedar.

The *southern forest* produces about 60 percent of the pulpwood and nearly one-third of the lumber. The most common softwoods are shortleaf, longleaf, loblolly, and slash pines and bald cypress. Hardwoods are red and black gum; red, white, water, live, and pin oak; white cedar; willow; cottonwood; ash; and pecan.

The *tropical forest* is the smallest. Hardwoods are bay, mangrove, eucalyptus, and mahogany.

The *Alaskan coast forest* gives us pulpwood and lumber. Most of the production is western hemlock, Sitka spruce, western red cedar, and Alaska yellow cedar.

The *Alaskan interior forest* is still too far away to be useful. White and black spruce are softwoods. White birch, aspen, and several poplars make up the hardwoods.

The *Hawaiian wet forest* produces wood for lumber, furniture, and souvenirs. These are made from koa, tree fern, kukui, tropical ash, and eucalyptus.

The *Hawaiian dry forest* is somewhat noncommercial. Algarroba, koa, haole, wili wili, and monkeypod trees grow here.

Wood Species

People employed in the wood industry must know about the make-up and uses of woods. The species of wood you will read about in this unit are among the most common ones found in the United States. The descriptions have both common names and Latin names. The general shape of each tree, with the left side in summer foliage and the right side in winter outline, is pictured in Figs. 64-3 through 64-15. The leaf or needle of each species and its fruit or nut are also shown.

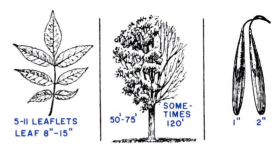

Fig. 64-3. Ash, white (*Fraxinus americana*). Bark: dark gray or gray-brown, deeply furrowed. Wood: heavy, hard, elastic, tough, brown. Habitat: rich, moist, cool woods; fields and riverbanks; Nova Scotia to Minnesota; Florida to Texas. Uses: agricultural implements, furniture, and oars.

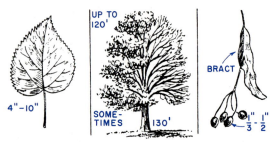

Fig. 64-4. Basswood, or linden (*Titilia americana*). Bark: deep brownish-gray. Wood: soft, straight-grain, light-brown, easily worked. Habitat: rich woods or fertile soils; Maine to Georgia, west to Texas, and north to Lake Superior. Uses: general woodenware, furniture, wood pulp, and mat fiber from inner bark.

Fig. 64-5. Poplar, yellow, or tulip tree (*Liriodendron tulipifera*). Bark: brownish-gray, round-ridged. Wood: pale buff, close, straight-grain, light, soft, easily worked. Wood that does not readily split, warp, or shrink. Habitat: rich, moist, soil; from Rhode Island to Michigan, south to Georgia and Arkansas. Uses: interior cabinet work and an excellent core for veneers.

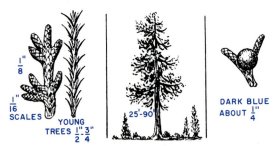

Fig. 64-6. Cedar, red (*Juniperus virginiana*). Bark: light ruddy brown. Wood: light, soft, brittle, close-grained, fragrant, durable. Habitat: all soils, swamps to rock ridges; Nova Scotia to South Dakota, south to Florida and Texas. Uses: cedar chests, wood closets, fence posts, and lead pencils.

Fig. 64-7. Birch, yellow (*Betula lutea*). Bark: on young trees, thin, papery scales of silver or yellow; on old trees, large, thin, dull plates of bark, grayish in color. Wood: heavy, strong, hard, close-grained. Habitat: rich uplands, swamps, stream banks; from Minnesota to Newfoundland, as far south as North Carolina and Tennessee. Uses: distinctive, pleasing grain figures for woodworking, furniture, flooring, and interior finishes.

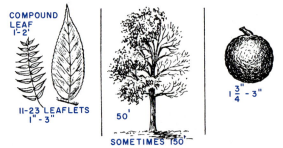

Fig. 64-8. Walnut, black (*Jugians nigra*). Bark: thick, dark brown, deeply divided, broad, round-ridged. Wood: deep brown, hard, heavy, slightly brittle, nonwarping, even-textured. Polishes well. Edible, tasty nutritive nut. Habitat: rich woodlands of Missouri, Kansas, Illinois, Indiana, Ohio, Kentucky, and Tennessee. Uses: fine furniture, woodworking, boat building, gunstocks, and veneers.

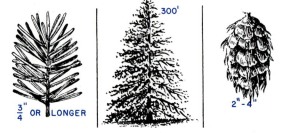

Fig. 64-9. Fir, Douglas (*Pseudotsuga taxifolia*). Bark: dark, gray-brown, rough. Wood: light, ruddy, or tan-yellow. Habitat: Rocky Mountains to Pacific coast, and central British Columbia to Northern Mexico. Uses: construction purposes and railroad ties and piles.

■ 268

Fig. 64-10. Gum, sweet, or red (*Liquidambar styraciflua*). Bark: gray-brown, deeply furrowed. Wood: hard, heavy, close-grained, reddish-brown. Sap made into chewing gum and medicine. Habitat: rich, wet lowlands; Connecticut to Florida, west to Kansas. Uses: interior paneling and furniture, stained to resemble mahogany.

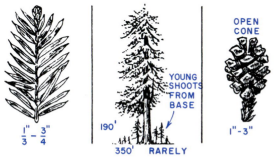

Fig. 64-11. Redwood (*Sequoia sempervirens*). Bark: deep cinnamon-brown, gray-tinted. Wood: crimson-brown, soft, brittle, straight-grained, easily worked. Habitat: Pacific coast, within 20-mile-wide fog belt from southern Oregon to Monterey County, California. Tallest tree in the world, reaching 200 to 350 feet. Sometimes lives to be 1200 to 1400 years old. Uses: interior finish, woodworking, and outdoor structures such as silos, barns, tanks, bridges, pipelines, flumes, mill roofs, and cooling towers.

Fig. 64-12. Maple, sugar (*Acer saccharum*). Bark: light, brown-gray, deeply furrowed. Wood: heavy, hard, strong, close-grained, easily polished. Sap makes maple sugar. Habitat: rich woods, rocky hillsides; every state east of the Mississippi River but rare in the south. Uses: interior finish, floors, turnery, shipbuilding, shoe lasts, and fuel.

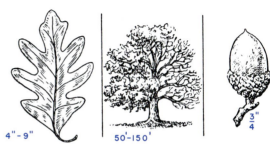

Fig. 64-13. Oak, white (*Quercus alba*). Bark: whitish-gray, firm, deeply furrowed. Wood: strong, heavy, tough, hard, pale brown. Habitat: dry uplands, sandy plains, gravelly ridges; eastern half of United States. Uses: building, furniture, floors, beams, and shipbuilding.

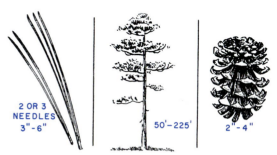

Fig. 64-14. Pine, western yellow (*Pinus ponderosa*). Bark: light russet-red, scaly surfaced. Wood: hard, strong, light-colored. Habitat: open, parklike forests, dry and moist soils; from southern British Columbia south through the western Rocky Mountain region to northern Mexico. Valuable lumber tree, sometimes living 500 years. Uses: sashes, doors, frames, siding, knotty paneling, exterior and interior finish, crates, boxes, wood novelties, toys, and caskets.

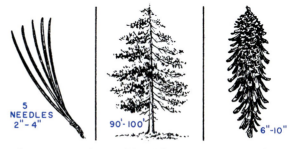

Fig. 64-15. Pine, white (*Pinus strobus*). Bark: rough, gray-brown, small-segmented. Wood: pale, buff-yellow, soft, durable, easily worked. Habitat: light, sandy soil; throughout northwestern United States from Iowa to Minnesota, east to southeastern Canada; Appalachian Mountains to northern Georgia. Uses: building.

Forest Products

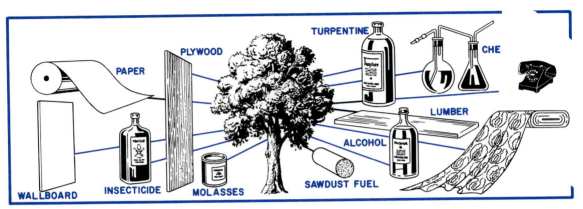

Fig. 65-1. A few products made from trees. (American Forest Institute)

More than 10 000 products are developed from the natural resource of wood. Figure 65-1 shows only a small number, which grows larger as research finds new uses for wood. Scientists develop many different products from forests (Fig. 65-2). The raw materials of trees are used to make furniture, books, magazines, sporting equipment, turpentine for paint, and resins for soap. Some additional items are plastics, rayon, and photographic film.

Veneer and Plywood

Wood veneers are used to put wood-grain designs on plywood. Plywood has great strength and does not warp or bend easily. It is being used more and more for furniture and for building construction (Fig. 65-3).

Fig. 65-2. Scientific development of forest product resources is essential. (American Plywood Association)

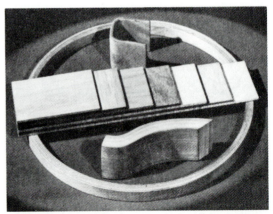

Fig. 65-3. Plywood can be press-formed into many different shapes. (American Forest Institute)

Veneer. Veneer is a very thin sheet of wood. It is cut from a log with either a straight blade or a rotary cutter. The straight blade *shears* (cuts) thin sheets of wood. Veneer cut this way may be matched into patterns on plywood (Fig. 65-4). In the rotary way of cutting veneer, the log is turned against a long, sharp steel blade. This is like unwinding a roll of wrapping paper.

Cabinets, table tops, and panels in furniture are often made from plywood with interesting veneer designs. (See Fig. 65-4.)

Plywood. Plywood is made by gluing and pressing together three or more sheets of thin wood. The grain in each sheet of wood runs crosswise in every other layer. Plywood has great strength with less weight than many other building materials. Plywood sheets are sold in different sizes and thicknesses (Fig. 65-5).

Building engineers sometimes use special kinds of plywood for industrial building and for outside covering. The special materials are *bonded* (glued) with synthetic resin glues that stand up to sun or water.

Laminated Arches and Beams

Building construction has grown a great deal because of the many new kinds of *laminated* (layer-on-layer) arches and beams. Laminating allows different building shapes. At the same time, the shapes stay strong, span large spaces, and look quite beautiful (Fig. 65-6).

Paper Products

About 6 percent of the timber harvest is used for making pulp. Paper is made from pulp (Fig. 65-7). This book would not be here if there were no forests. Technicians in scientific laboratories have made many types of *laminated* paper products. These can be made strong by *impregnating* (soak-

Fig. 65-4. Exotic figures in veneer paneling can beautify the interior of any office or home. (U.S. Plywood Corporation)

Fig. 65-5. Plywood with alternating grain direction; this ensures dimensional stability. (Fine Hardwoods Association)

Fig. 65-6. This spectacular home overhangs high ground; it is supported by a series of sturdy wooden beams. (Douglas Fir Plywood Association)

Fig. 65-7. The paper industry is dependent on the forests.

Fig. 65-8. A variety of beech parts used in high-voltage electrical operations. These are threaded rod and nuts, cable clamps, tap-changer base, oil circuit-breaker arcing chamber, insulating collars for a motor, and a mounting panel. (Permali, Incorporated)

ing) sheets of paper with resin. The sheets are then heated and placed under pressure. The results can be materials with the strength of metal. They often can take the place of metal because they can be shaped.

Insulation and Fuel

All parts of the tree have some useful purpose. In many up-to-date mills, the waste sawdust and shavings are made into insulating materials, wallboard, and pressed-wood products.

Rayon and Woollike Fabrics

Scientists have learned that the *cellulose* fiber of the tree can be treated to make one of the ingredients of rayon. This material is useful for making cords that form the body of rubber tires. It can also be woven into cloth or fabric. Other products that can be made from wood cellulose are blankets and coats.

Scientific Treatment of Wood

Wood is made up of two classes of substances known as *cellulose* and *lignin*.

About two-thirds of all wood is cellulose. The other third is lignin.

Paper and rayon are only two of the articles made from cellulose. Alcohol, felt fabric, food protein, glycerine, gunpowder, imitation leather, lacquer, plastics, sugar, and yeast come from chemical and mechanical treatments of wood. Figures 65-8 and 65-9 show products made from wood on which some chemicals have been used.

Scientists have found that lignin is a good tanning agent for leather. It is also used to make a water softener, as a base for fertilizer, and in a compound called *vanillin*.

Products from Trees

A few products from trees are drying agents, gum (Fig. 65-10), dyes, solvents,

Fig. 65-9. Stabilite, a pressure-impregnated wood for model and pattern carving, is used to make an intricate forging pattern. (Georgia-Pacific Corporation)

Fig. 65-10. Researchers measure gum pressure and yields from a pine tree. (U. S. Forest Service)

sugar, and spirits. All of these are needed in many manufacturing processes. For example, oleoresin, used by the paint industry, comes from southern pine trees.

Turpentine and resins come from an *extract* of trees. This extract is a molasseslike substance that drips from the tree after the bark has been chipped. Wood alcohol and acetone are also by-products of trees.

Unit 66

Metrics in Woodworking and Other Soft Materials

The full change to the metric system will probably be much slower in woodworking and plastics than in metalworking. Most wood products, particularly furniture and homes, are not the most important part of worldwide trade. There is less need to build or make these products to metric standards. You should, however, be able to understand metric measurement when you study and experiment with wood. Begin by studying the metric tables for woodworking in Figs. 66-1 and 66-2.

Tools and Equipment

Tools used for measuring, such as the bench rule, marking gage, try square, combination square, folding rule, and tape, need to be replaced with tools that have both U.S. customary and metric measurements. Try to use measuring tools with the numbered lines that stand for millimeters (Fig. 66-3). For most activities in woods and other soft materials, work to the nearest millimeter. That is usually close enough for most project activities (Fig. 66-4).

CONVERSION TABLE FOR WOODWORKING

CUSTOMARY (ENGLISH)	METRIC				
	ACTUAL	ACCURATE WOODWORKERS LANGUAGE	TOOL SIZES	LUMBER SIZES	
				THICKNESS	WIDTH
$1/32$ in	0.8 mm	1 mm bare			
$1/16$ in	1.6 mm	1.5 mm			
$1/8$ in	3.2 mm	3 mm full	3 mm		
$3/16$ in	4.8 mm	5 mm bare	5 mm		
$1/4$ in	6.4 mm	6.5 mm	6 mm	6 mm	
$5/16$ in	7.9 mm	8 mm bare	8 mm		
$3/8$ in	9.5 mm	9.5 mm	10 mm		
$7/16$ in	11.1 mm	11 mm full	11 mm		
$1/2$ in	12.7 mm	12.5 mm full	13 mm	12 mm	
$9/16$ in	14.3 mm	14.5 mm bare	14 mm		
$5/8$ in	15.9 mm	16 mm bare	16 mm	16 mm	
$11/16$ in	17.5 mm	17.5 mm	17 mm		
$3/4$ in	19.1 mm	19 mm full	19 mm	19 mm	
$13/16$ in	20.6 mm	20.5 mm	21 mm		
$1/8$ in	22.2 mm	22 mm full	22 mm	22 mm	
$15/16$ in	23.8 mm	24 mm bare	24 mm		
1 in	25.4 mm	25.5 mm	25 mm	25 mm	
1 $1/4$ in	31.8 mm	32 mm bare	32 mm	32 mm	
1 $3/8$ in	34.9 mm	35 mm bare	36 mm	36 mm	
1 $1/2$ in	38.1 mm	38 mm full	38 mm	38 mm	(or 40 mm)
1 $3/4$ in	44.5 mm	44.5 mm	44 mm	44 mm	
2 in	50.8 mm	51 mm bare	50 mm	50 mm	
2 $1/2$ in	63.5 mm	63.5 mm	64 mm	64 mm	
3 in	76.2 mm	76 mm full		75 mm	75 mm
4 in	101.6 mm	101.5 mm		100 mm	100 mm
5 in	127.0 mm	127 mm			125 mm
6 in	152.4 mm	152.5 mm			150 mm
7 in	177.8 mm	178 mm bare			
8 in	203.2 mm	203 mm full			200 mm
9 in	228.6 mm	228.5 mm			
10 in	254.0 mm	254 mm			250 mm
11 in	279.4 mm	279.5 mm			
12 in	304.8 mm	305 mm bare			300 mm
18 in	457.2 mm	457 mm full	460 mm		
24 in	609.6 mm	609.5 mm			
36 in	914.4 mm	914.5 mm		Panel stock sizes	
48 in or 4 ft	1219.2 mm	1220 mm or 1.22 m		1220 mm or 1.22 m	
96 in or 8 ft	2438.4 mm	2440 mm or 2.44 m		2440 mm or 2.44 m	

Fig. 66-1. A conversion table for woodworking. Most metric sizes will round to a full millimeter.

WOODWORKING	

Units of Measurement	millimeter (about $1/25$ in) for all dimensions on drawings
	    for thickness and width of lumber
	    for size of panel stock
	meter (about 10 percent longer than a yard)
	    for lengths of lumber
	liter (about 6% more) to replace the quart
	kilogram (about 2.2 times) to replace the pound

Tools Replacement	All measuring tools—no change in other tools
Machine Changes	Add a metric *scale* (rule) next to the customary scale.
Lumber Thickness	$1/4$ inch becomes 6 millimeters $1/2$ inch becomes 12 millimeters 1 inch becomes 25 millimeters
Lumber Lengths	6 feet becomes 1.8 meters 8 feet becomes 2.4 meters
Panel Sizes	4 × 8 feet becomes 1220 × 2440 millimeters

Fasteners	No actual change in size
	    1 inch nail becomes 25 millimeters
	    2 inch screw becomes 50 millimeters

Drills	Metric sizes are available. However, customary sizes are close enough for metric use.

Fig. 66-2. These are the major changes that must be made in woodworking to use the metric system.

Fig. 66-3. Always select a 300-mm rule that has the numbered lines graduated in millimeters.

Other tools, such as saws, planes, chisels, and hammers, will not need to be replaced. A 2-inch plane will become a 50-millimeter plane. A 6-inch coping saw will be known as a 150-millimeter coping saw.

Machines can be adjusted to depth and width of cut. To change from U.S. customary to metric measurement, add a metric tape next to the customary tape. In this way the settings can be made in metric or in U.S. customary (Fig. 66-5).

Material Sizes

Lumber and other wood materials will be given in millimeter size instead of inches for thickness and width. Lumber lengths will be given in meters rather than feet. The size of a two-by-four board will be listed as 50 × 100 millimeters in the metric system. A dry two-by-four that measures $1^{1}/_{2}$ inches by $3^{1}/_{2}$ inches in the customary system will be listed as 38 by 89 millime-

A MILLIMETER IS ABOUT HALF WAY BETWEEN THE $\frac{1}{32}$" & THE $\frac{1}{16}$"

$\frac{1}{2}$ MILLIMETER IS ABOUT HALF WAY BETWEEN THE $\frac{1}{32}$" & THE $\frac{1}{64}$"

(ENLARGED 5 TIMES ACTUAL SIZE.)

$\frac{1}{2}$ mm   1 mm

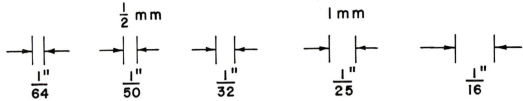

$\frac{1}{64}$" $\frac{1}{50}$" $\frac{1}{32}$" $\frac{1}{25}$" $\frac{1}{16}$"

Fig. 66-4.   The millimeter is a very small measurement. (The L. S. Starrett Company)

MACHINE WOODWORK
GAGE AND SCALE CONVERSION

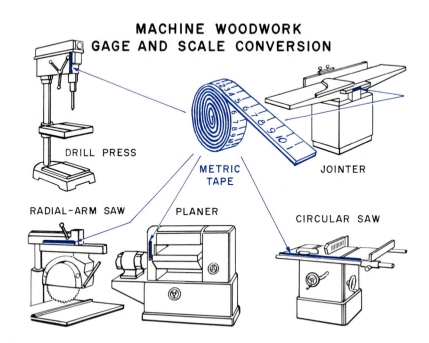

DRILL PRESS

METRIC TAPE

JOINTER

RADIAL-ARM SAW   PLANER   CIRCULAR SAW

Fig. 66-5.   Add a piece of metric tape or a short metal rule to all machines.

ters. When metric standards are ready, this size will probably be 35 by 90 millimeters.

Standard plywood sheets now measure 4 × 8 feet. They will become 1200 × 2440 millimeters in the metric system. When

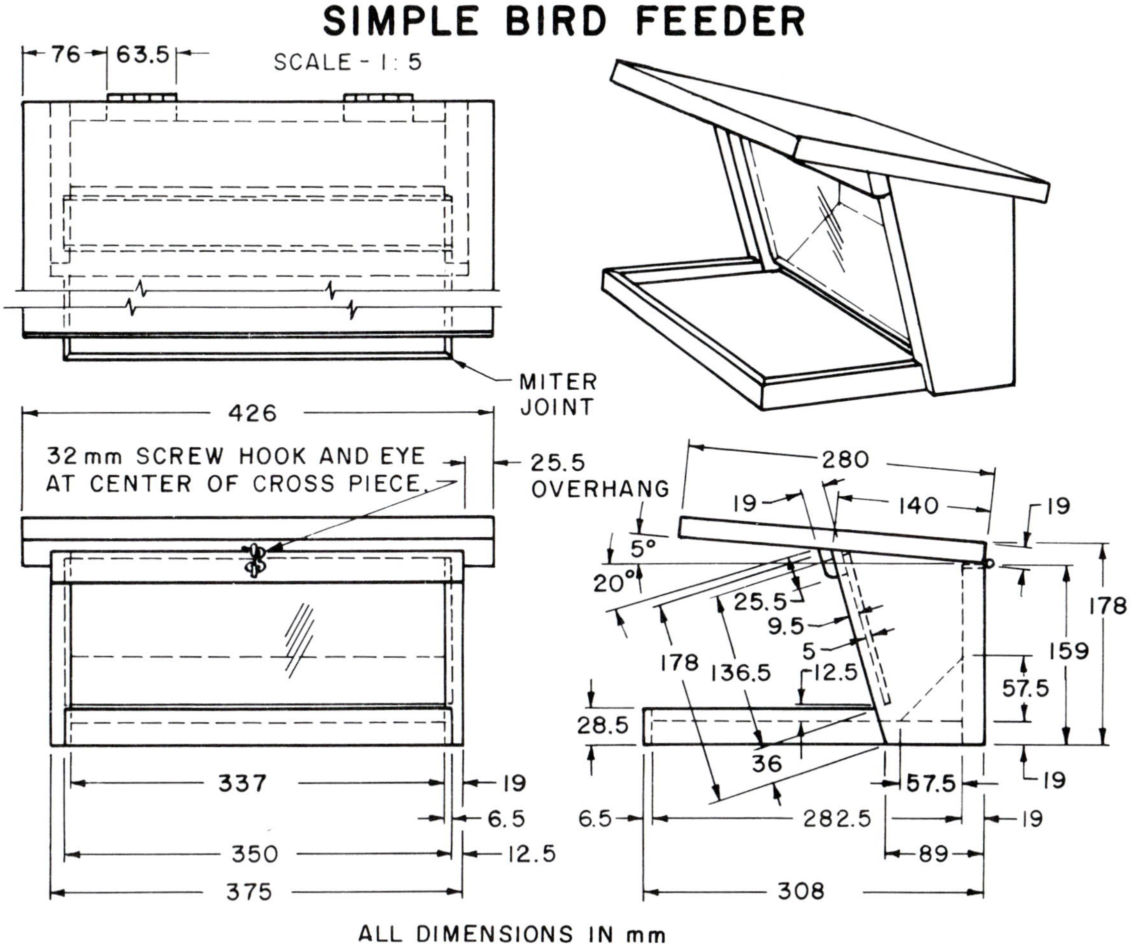

SIMPLE BIRD FEEDER

Fig. 66-6. This drawing for a bird feeder has only metric dimensions. When the measurement calls for a half millimeter (0.5 mm), mark to the middle between two full millimeter gradations.

the change to the metric system is complete, all building materials will be based on a 100-millimeter unit, or *module*. Panel sizes will become 1200 × 2400 millimeters. Most other materials, such as nails, screws, and other types of fasteners, will not really change in size. These will have a metric measurement added. For example, a 1-inch nail will become a 25-millimeter nail, and a 3-inch wood screw will become a 75-millimeter wood screw.

Working Drawings

There are three ways of making woodworking drawings metric. The best way is to use an *all-metric* drawing. Here the dimensions are given in millimeters (Fig. 66-6). A full millimeter is $1/25$ inch. It is more accurate than the $1/16$-inch unit now used for most woodworking projects. A half millimeter is just about $1/50$ inch.

The second way shows the metric dimensions on the drawing itself. It also has a *readout chart*. The chart shows the customary *equivalent*. The third way of making woodworking drawings metric is to use dual dimensioning. This way shows both the customary (inch) and metric (millimeter) units for all dimensions.

Unit 67

Purchasing Lumber

You should know how to *estimate* (figure) the amount and the cost of wood and other materials needed for making your project activity. Design and plan your project activity before you fill out a plan sheet. Be sure to make a correct list of materials.

Lumber Information

Using U.S. customary measurements, most lumber is sold by the *board foot* (bd ft). A board foot is a piece of wood 1 inch thick, 12 inches wide, and 12 inches long. One board foot represents 144 cubic inches. When you buy a board foot of lumber, however, it usually measures $3/4$ inch thick, $11\,1/2$ inches wide, and 12 inches long. The thickness and the width are smaller because of cutting, sawing, and planing. A board that is figured as 2 inches thick is really only $1\,1/2$ inches thick. A board a little less than 1 inch thick is usually figured as 1 inch thick. Plywood, however, is sold by the square foot. The numbers used to describe its thickness and width are actual measurements. Thus, $1/2$-inch plywood is actually $1/2$ inch thick.

In a bill of materials, the thickness and the width of a board are always figured in inches. Lengths for short pieces may also be figured in inches. Lengths for long pieces are usually figured in feet.

Lumber prices are usually given by the lumber industry in amounts per 1000 board feet. A listing of 980/M means that 1000 board feet (M) costs $980. At this price, 1 board foot costs 98 cents. Your teacher will most likely give you the price per board foot.

In the metric system, most lumber will probably be sold by the cubic meter (m^3). A cubic meter is much larger than a board foot. It measures 1 meter long $\times$ 1 meter wide $\times$ 1 meter deep. It takes a little less than 424 board feet to make a cubic meter. To change board feet to cubic meters, multiply by 0.002 359. Plywood will be sold by the square meter rather than the square foot.

Types of Lumber

Softwoods are used in building and construction. *Hardwoods* are used for cabinetmaking, furniture, and architectural woodworking.

Grading. The National Hardwood Association has set the *grading* (rating) rules for hardwood lumber. The grade of lumber is the amount of lumber in a piece that can be used. One side must be "clear." (It must be free of knots and blemishes.) The reverse side must be "sound." (It must be free of rot and other defects that might weaken the board.)

The term *Firsts* is used to describe the highest grade of hardwood lumber. *Seconds* is used to describe the next highest grade. *Selects* is the third grade. Other grades are *No. 1 Common, No. 2 Common, Sound Wormy, No. 3A Common,* and *No.*

3B Common. These are general grades. There are exceptions and special rules for different species of trees.

Softwoods have more grades than hardwoods. Because they are graded by several lumber associations and organizations, softwoods are sometimes not easy to *classify* (put into order). The Western Pine Association classifies western white pine as *Select, Shop,* and *Common.* Each of these grades has other grades within it.

Surfaces. Terms that point to the treatment given to lumber are *Rough, S2S,* and *S4S.* Rough means that lumber is in the rough as it came sawed from the mill. S2S indicates that it has been planed on both sides. S4S tells that it has been planed on four sides (both surfaces and both edges).

Methods of Drying. *Air-dried* (AD) means that lumber has been dried naturally in the air (Fig. 67-1). Drying time may take from weeks to months, depending on the type of lumber and the dryness needed. *Kiln-dried* (KD) refers to lumber that has been dried in a *kiln* (a kind of furnace or oven). Kiln drying is the quicker method (Fig. 67-2).

Methods of Cutting. *Plain sawing* and *quarter sawing* are two ways of cutting lumber into boards (Fig. 67-3). Plain sawing is the more common way. Quarter sawing costs more than plain sawing because of the way the log is handled and cut.

Plywood

Panels of plywood are most often made up of three, five, or seven *plies* (thicknesses). Plywood sells by the square foot. The price depends on the thickness, the glue, and the kind of veneer used on the surfaces. The panels vary in thickness from 1/8 to 3/4 inch (3 to 19 millimeters). A 3/8-inch (9-millimeter) thickness of plywood usually has three plies. Thicker panels have five or sometimes seven plies.

Fig. 67-1. Lumber stacked in a mill yard for air drying. Note the space left between the vertical stacks for air to circulate. (Georgia-Pacific Corporation)

Fig. 67-2. Stacks of lumber being rolled out from a heated kiln (furnace or oven). (Weyerhaeuser Company)

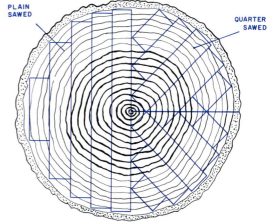

PLAIN SAWED

QUARTER SAWED

Fig. 67-3. Two methods of sawing lumber: left, plain sawing; right, quarter sawing.

Grades of Plywood. The two most common grades of fir plywood are called *Sound 1 Side* and *Sound 2 Sides.* Fir-face veneers have four grades: A, B, C, and D. The A grade is best. The quality of fir plywood panels is based on the grades of the surface veneer. For example, A-D means that the face veneer (A) is the best grade and that the back veneer (D) is the poorest grade. A-A means that both faces are the best grade of fir.

Fir plywood for outdoor use is marked EXT (for *exterior*) on one side. White pine plywood is manufactured under grading rules that are not quite the same.

Hardwood plywood is usually graded as *Good 1 Side* (G1S) and *Good 2 Sides* (G2S). G1S means that one of the faces is good. G2S means that both faces are good.

Board-foot Measure

Estimate (figure out) how much lumber you will need. This estimate should show the rough size of the lumber. You will also need to figure the number of board feet in each piece and the total number of board feet. Then you can figure the cost of the project. Do so by multiplying the total number of board feet by the cost per board foot.

When the length is given in *linear* (running) *feet,* use the following formula:

$$\frac{\#Pcs \times T'' \times W'' \times L'}{12} = bd\ ft$$

where #Pcs = number of pieces
T″ = thickness in inches
W″ = width in inches
L′ = length in feet

Example: To find the board feet in four pieces, 2 inches × 6 inches × 4 feet:

$$\frac{4 \times 2 \times \overset{2}{\cancel{6}} \times 4}{\underset{3}{\cancel{12}}} = 16\ bd\ ft$$

When the length is given in linear *inches,* use this formula:

$$\frac{\#Pcs \times T'' \times W'' \times L''}{12 \times 12} = bd\ ft$$

Example: To find the board feet in four pieces, 1 × 10 × 18 inches:

$$\frac{4 \times 1 \times \overset{5}{\cancel{10}} \times \overset{3}{\cancel{18}}}{\underset{3}{\cancel{12}} \times \underset{2}{\cancel{12}}} = 5\ bd\ ft$$

Cubic-meter Measure

You can estimate the amount of lumber needed in metric dimensions. *Multiply the dimensions, as decimals of a meter, by the number of pieces.* Use this formula:

Pcs. × thickness × width × length = m³
(No.)   (m)   (m)   (m)

or #Pcs. × Tm × Wm × Lm = m³

Example: To find the cubic meters in three pieces of lumber, 20 mm × 300 mm × 1500 mm:

$$3 \times 0.02 \times 0.3 \times 1.5 = 0.027\ m^3$$

Estimating Finishes

Finishes must be included in a cost estimate. The price of the finishing materials for a project activity is very often about 20 percent of the cost of the lumber. This cost is added to the cost of the lumber and other expenses.

Other Costs

There are more items to figure in the total cost of a project activity. You may need nails, screws, and other fastenings. You may need sandpaper, steel wool, glue, dowel rods, and special hardware. The cost will depend on purchase prices and how much is used.

Safety in the Woods Laboratory

Working with wood is fun if you have respect for *safety*. This means you should learn to care for yourself, the tools, and the materials. You should follow safe-conduct rules. More accidents happen around ten o'clock in the morning than at any other time of the day. Studies show that more accidents happen on Wednesdays, except on days right before and after vacations.

The wood chisel is the hand tool that causes most injuries in woodworking. The jointer is the power tool that causes the greatest number of accidents. Unskilled people or beginners have more injuries than people who have skill. Carelessness and the wrong use of tools are the causes.

The general safety rules that follow apply to both hand and machine tools. Special rules for using machine tools are talked about in the units dealing with these machines.

Physical Safety

1. Use your leg and arm muscles to lift heavy objects. Never depend on your back muscles. Ask someone to help you.
2. Test the sharpness of edge-cutting tools on wood or paper, not your hand.
3. Be careful when you are using your thumb as a guide when cutting with a saw.
4. Always cut *away* from your body when using a knife.

Clothing Safety

1. Wear a shop or laboratory apron or some other protective clothing when you work.
2. Tuck in or remove your tie and roll up your sleeves.

3. Wear safety goggles or glasses when operating power tools or where airborne debris may be present.
4. Tuck in shirttails or any loose clothing.
5. Wear a hat or tie up long hair.

Tool Safety

1. Place tools in the proper order on a bench top. The cutting edges should be pointed away from you. Sharp tools should not rub against each other or stick out over the edges of the bench.
2. Keep screwdrivers properly pointed to prevent injury to hands and to the wood fiber.
3. Fasten handles firmly on planes, hammers, mallets, files, and chisels.

Materials Safety

1. Always fasten or hold wood properly. You may use a vise with clamps or sawhorses.
2. Put waste pieces of lumber in a storage rack or a scrap box.
3. Keep oily or finishing rags in closed metal containers.

Laboratory Courtesy

1. Report an accident immediately. There should be no time wasted in giving first aid.
2. Warn others to stay out of the way when you handle long pieces of lumber.
3. Walk carefully—*do not run*—in the industrial laboratory or shop or home workshop. Running can be dangerous.
4. Carry only a few tools at a time, especially if they are sharp.

Unit 69

Measuring and Laying Out Lumber

Correct (accurate) measuring is a necessity when working with wood. It is a skilled craft. The inch, foot, millimeter, and meter are the measurements used in school laboratories and in most industries in the United States.

Read Units 17 and 66 for information on the metric system. Metric measurement is expanding in use; Fig. 69-1 shows markings in both *customary* and *metric* measurements. Many measuring tools used in woodworking are divided into the customary sixteenths, eighths, quarters, and halves of an inch as well as being marked in millimeters.

Tools

The tools most often used for measuring and laying out are the wooden or steel *bench rule* (Fig. 69-2), the *steel square* (Fig. 69-3), the *try square* (Fig. 69-4), the *combination square* (Fig. 69-5), the *zigzag rule*

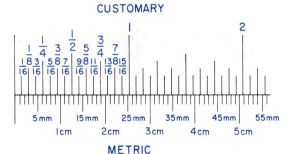

Fig. 69-1. **Customary and metric equivalent measures of length.**

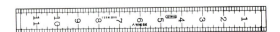

Fig. 69-2. **A wooden bench rule.**

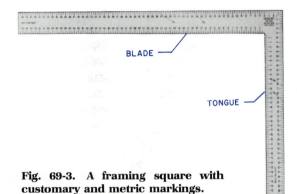

Fig. 69-3. **A framing square with customary and metric markings.**

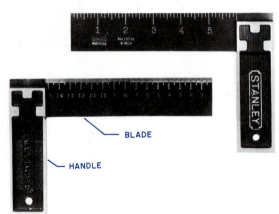

Fig. 69-4. **Customary- and metric-marked try squares.**

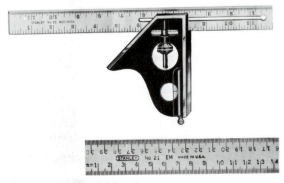

Fig. 69-5. **A combination square.**

Fig. 69-6. A zigzag folding rule with customary and metric measuring systems.

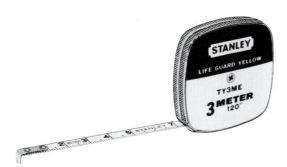

Fig. 69-7. A flexible steel rule.

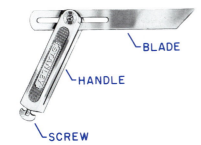

BLADE

HANDLE

SCREW

Fig. 69.8 A bevel.

THUMBSCREW

PIN

BEAM

HEAD

Fig. 69-9. A marking gage.

Fig. 69-10. A marking knife.

(Fig. 69-6), the *flexible steel tape* (Fig. 69-7), the *bevel* (often called the *T bevel*) (Fig. 69-8), the *marking gage* (Fig. 69-9), and the *marking knife* (Fig. 69-10).

Laying Out Length

1. Select a board with few cracks or checks (Fig. 69-11).

2. Square a line across the end of the board to avoid checks or cracks (Figs. 69-11 and 69-12). Hold the blade of the square against the edge of the board. Mark the line with a sharp knife or a pencil by pressing against the tongue of the square.

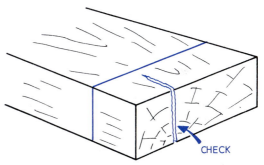

CHECK

Fig. 69-11. The end of a board marked to avoid checks or cracks.

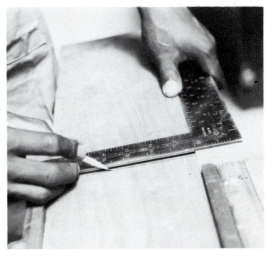

Fig. 69-12. Squaring a line across a board.

Fig. 69-13. Measuring and laying out with a rule.

3. Lay out the length you want (Figs. 69-13 and 69-14). Place the rule on its edge to get a more *accurate* (correct) measurement. Mark the length with a sharp pencil.
4. Square the line you just marked. Follow the instructions in step 2.

Fig. 69-14. Measuring with a flexible steel rule.

Laying Out Width

1. Measure and mark the desired width with a measuring tool (Fig. 69-15). You can divide a board into any number of equal widths. Lay the rule on edge across the board in a diagonal position (Fig. 69-16).

Gaging Width and Thickness

1. Adjust the marking gage to the correct distance (Fig. 69-17). Check the setting against a rule to make sure it is accurate.
2. Push the marking gage on the wood to make the mark (Fig. 69-18). Figure 69-19 shows another way to mark the board to width. Hold the head of the gage firmly against the edge of the wood while you *scribe* (mark) a light line.

Marking an Edge

Mark a line around the edge of the board, and continue the face line (Fig. 69-20).

Laying Out an Angle

1. *Fix* (adjust) the bevel to the angle you need (Fig. 69-21). Then fasten the screw on the handle. This tool is useful for marking any angle that is not 90° (a right angle).
2. Hold the handle firmly against the face or the edge of the board. Mark the wood along the edge of the blade.

Fig. 69-15. Marking width.

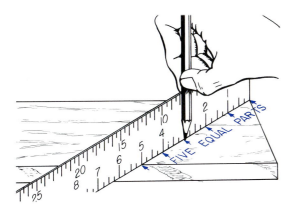

Fig. 69-16.   Dividing a board into equal widths.

Fig. 69-19.   Marking the width on a board by using a rule and pencil.

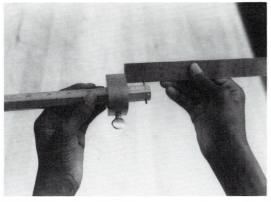

Fig. 69-17.   Checking the accuracy of a marking gage with a rule.

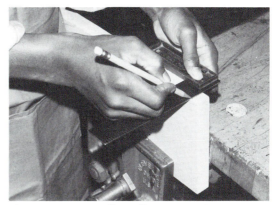

Fig. 69-20.   Extending a line across the edge of a board.

Fig. 69-18.   Scribing a line on a board using a marking gage.

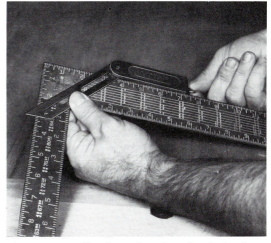

Fig. 69-21.   Adjusting a bevel to the desired angle against the square.

Unit 70

Sawing across or with the Grain of Wood

The first saw used for cutting was most likely a stone with sharp, jagged edges. It was one of the earliest tools used by people. Today the steel saw is used extensively for construction, cabinetmaking, and furniture building.

Fig. 70-2. Points per inch on a crosscut handsaw.

Tools

Woodworkers' handsaws include the crosscut saw, the ripsaw, and the backsaw, or cabinet saw. Figure 70-1 shows a handsaw with the parts labeled. How rough or how fine a saw is depends on the number of points per inch (Fig. 70-2).

Crosscut Saw. This type of saw is used to cut *across* the grain of wood (Fig. 70-3). The teeth are set and filed to make a saw *kerf* (cut). The saw kerf must be wider than the thickness of the blade. This lets the blade move freely in the cut without sticking. A crosscut saw with 8 to 10 points per inch cuts wood easily.

Ripsaw. The ripsaw is used for cutting *with* the grain of the wood. Figure 70-4 shows how the teeth are filed and set and how they make a kerf. Five to seven points per inch cut easily.

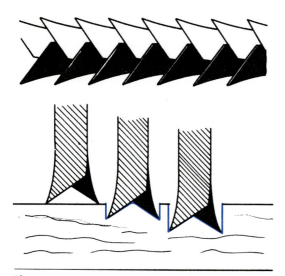

Fig. 70-3. Top and side views of crosscut teeth and the *kerf* (cut) made by them.

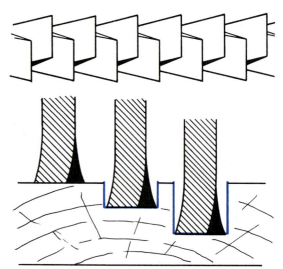

Fig. 70-4. Top and side views of ripsaw teeth and the *kerf* (cut) made by them.

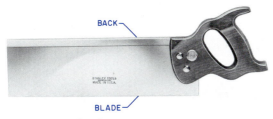

Fig. 70-5. A backsaw.

Backsaw. Figure 70-5 shows this fine-cutting crosscut saw. The blade is stiffened with a thick back. The backsaw is used in fine cabinet work and for making joints.

Crosscutting

1. Lay out and mark the board.
2. Fasten the board in a bench vise if it can be held in this way. Very wide or long boards may be laid across sawhorses.
3. Place the heel of the crosscut saw near the cutting line on the *waste* side of the wood (Figs. 70-6 and 70-7). Pull the saw while you guide it with the thumb of your other hand.
4. Make several short cuts. Check with a try square to see that the saw blade is cutting at right angles to the face of the board (Fig. 70-8).
5. Continue cutting. Use long strokes. Cut at a 45° angle to the board.
6. Finish sawing with short, easy strokes. This helps keep the waste side from split-

Fig. 70-6. A board held in a vise for crosscutting.

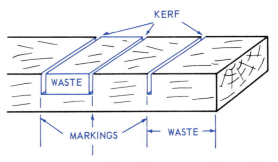

Fig. 70-7. Examples of waste portions of a board.

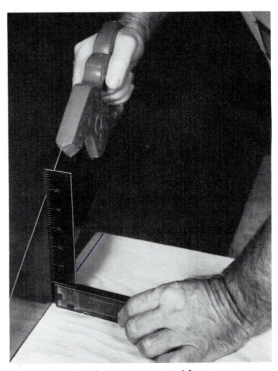

Fig. 70-8. Testing a saw cut with a try square.

ting or breaking. If possible, hold the waste part of the board safely with your other hand (Fig. 70-9).

Ripping

1. Mark the board to be cut.
2. Fasten the board in a bench vise or hold it on a sawhorse (Figs. 70-10 and 70-11).

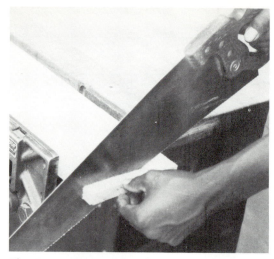

Fig. 70-9.   Crosscutting the end of a board.

Fig. 70-10.   Gripping a board in a bench vise.

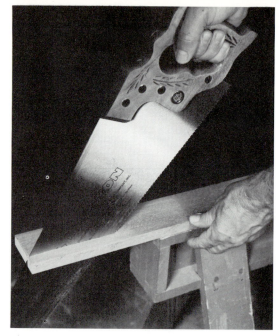

Fig. 70-11.   Ripping a board on a sawhorse.

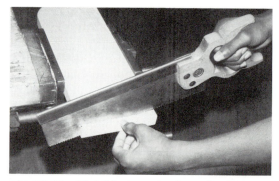

Fig. 70-12.   Finishing a cut using a backsaw.

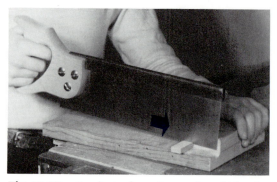

Fig. 70-13.   Sawing a board on a bench hook.

3. Start the cut by pulling the ripsaw toward you. Hold the saw so that the cutting edge is about 60° to the face of the board. Cut the board on the waste side of the mark.

4. Continue cutting the board. Use short, easy strokes to prevent splitting. If possible, hold the waste portion of the board safely with your other hand.

Cabinet Sawing

1. Lay out and mark the board.

2. Fasten the board in a vise (Fig. 70-12). It may also be held firmly against a bench hook (Fig. 70-13).

3. Start cutting with the backsaw in the same way you started crosscutting.

Sharpening, Assembling, and Adjusting Planes

The plane is one of the best tools for smoothing wood. There are many types of planes. All are *assembled* (put together), adjusted, and used in just about the same way. You will probably use jack, block, and rabbet planes. Figure 71-1 shows the parts of a plane.

Planes

Jack Plane. This plane (Fig. 71-2) is popular because of its size and usefulness. It is about 14 inches (356 millimeters) long.

Junior Jack Plane. This plane is narrower and shorter than the jack plane. It is light in weight and is easily handled.

Block Plane. This small plane is built somewhat differently from the jack plane but is adjusted similarly (Fig. 71-3). It is best for planing end grain and for doing many small jobs. The length is about 7 inches (178 millimeters).

Rabbet Plane. The rabbet plane (Fig. 71-4) is also called a *bullnose plane* because of the way it looks. The rabbet plane is often

Fig. 71-2.   A jack plane.

Fig. 71-3.   A block plane.

Fig. 71-4.   A rabbet plane.

used for planing in close places and for fitting joints.

Sharpening, Assembling, and Adjusting

Knowing how to sharpen, assemble, and adjust a plane is important. Follow these instructions to learn how.

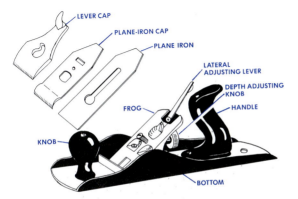
Fig. 71-1.   Exploded view of plane parts.

LEVER CAP
PLANE-IRON CAP
PLANE IRON
LATERAL ADJUSTING LEVER
DEPTH ADJUSTING KNOB
FROG
HANDLE
KNOB
BOTTOM

Sharpening

1. Test the plane iron for sharpness (Fig. 71-5). It should cut paper easily with a *shearing* (sideways) motion. If it is dull, it needs to be sharpened.

2. Grind the edges of the plane iron on a grinder (Fig. 71-6). Do this until all nicks are removed. Use a lubricant (oil or water, depending on the type of grinder) to carry the metal particles away. This lubricant will also keep the tool cool.

3. Test the cutting edge for the correct grinding angle (Fig. 71-7).

4. Test the cutting edge for squareness with a try square (Fig. 71-8).

5. *Whet*, or *hone*, the plane iron. Do this by placing the bevel side down on an oilstone. Hold the plane iron so that both the toe and heel ride on the oilstone. Move it in a *circular* (round) motion. Keep oil on the stone to float away any steel particles (Fig. 71-9).

6. Turn the plane-iron blade on the back (flat) side. Move it back and forth gently on the oilstone (Fig. 71-10).

7. Remove the *burr* (wiry edge). Pull the

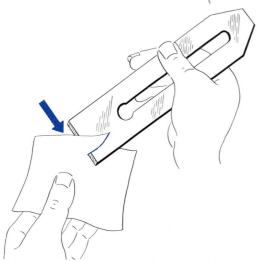

Fig. 71-5. Test the sharpness of the plane-iron cutting edge on a piece of paper.

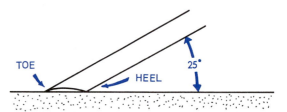

Fig. 71-7. The angle for grinding the plane iron.

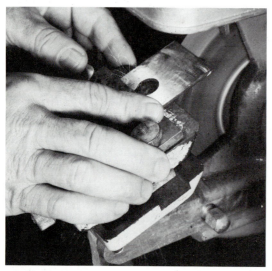

Fig. 71-6. Sharpening a plane iron by grinding the edge.

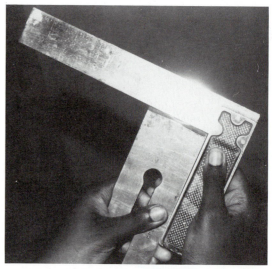

Fig. 71-8. Testing the cutting edge of a plane iron for squareness with a try square.

Fig. 71-9. Whetting the cutting edge of the plane iron on an oilstone.

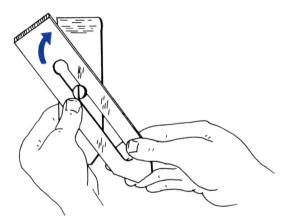

Fig. 71-11. Assembling the plane-iron cap and the plane iron.

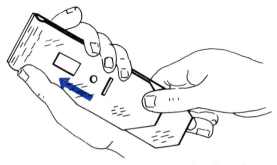

Fig. 71-12. Aligning the cap on the plane iron.

sharpened edge carefully across a piece of wood. Pulling the blade on an oily leather strap will also make a sharp edge and help remove the wiry burr. This is sometimes called *stropping*.

Assembling

1. Place the plane-iron cap on the flat side of the plane iron. Put the screw in the slot (Fig. 71-11).

2. Pull the plane-iron cap back and turn it straight with the plane (Fig. 71-12).

3. Slide the cap toward the cutting edge of the plane iron. Do not push it over the edge of the blade.

4. Adjust and tighten the plane-iron cap with a screwdriver. The cap should be placed about $1/16$ inch (1.5 millimeters) from the cutting edge of the blade.

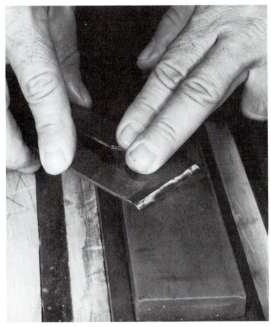

Fig. 71-10. Whetting the back side of a plane iron on an oil stone.

5. Place the assembled blade and plane-iron cap in the plane. Put the plane iron with its bevel side down on the "frog." Be sure that the plane iron is properly placed on the *lateral adjusting lever.*

6. Lay the lever cap over the plane-iron assembly so that the screw slides in the slot.

7. Tighten the lever cap to hold the entire assembly.

Adjusting

1. Move the plane iron with the lateral adjusting lever. The cutting edge should be *parallel* (in the same line) with the slot in the bottom of the plane (Fig. 71-13).

2. Turn the adjustment nut right or left to set the cutting edge for the correct depth.

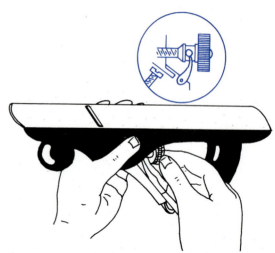

Fig. 71-13. **Adjusting the cutting depth so that the cutting edge of the plane iron is parallel with the slot in the bottom of the plane.**

Unit 72

Planing Lumber

There are steps to be followed for planing the surfaces, edges, and ends of wood pieces. A board has been *squared* when all surfaces, edges, and ends are at 90° angles to one another. A squared board is true and smooth. Figure 72-1 shows by number the six basic steps for planing the faces, edges, and ends of a board.

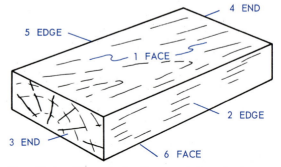

Fig. 72-1. **The six basic steps in planing a board.**

Tools

The tools used for planing lumber are the jack plane, the try square, the framing square, the marking gage, the rule, the crosscut saw, the ripsaw, and the backsaw.

Planing the First Surface (Step 1)

1. Select the best surface (face) of the board.

2. Place the board on the workbench and fasten it between a vise dog and a bench

stop. Place the board so that you can plane in the direction of the grain (Fig. 72-2).

3. Adjust the cutting edge of the plane iron to cut evenly but not too deeply.

4. Plane the surface until it is smooth (Fig. 72-3).

5. Test the surface to see that it is flat. Use the blade of a try square (Fig. 72-4) or the tongue of a framing square.

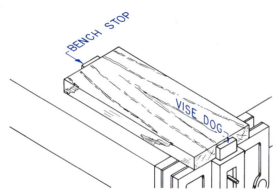

Fig. 72-2. A board fastened between a vise dog and a bench stop.

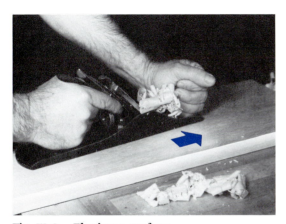

Fig. 72-3. Planing a surface.

Fig. 72-4. Testing for flatness.

Fig. 72-5. Testing diagonally for wind (twist).

6. Test the surface across the corners. See if there is a *wind* (twist) (Fig. 72-5).

Planing the First Edge (Step 2)

1. Select the best edge of the board. This will probably be the one that needs the least planing.

2. Fasten the board in a vise with the best side up. The direction of the grain should be in the direction that you will plane, that is, away from you.

3. Plane the edge until it is square with the planed surface (Figs. 72-6 and 72-7).

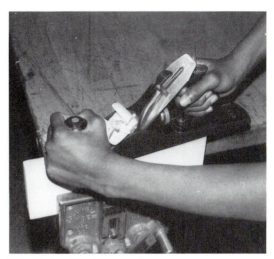

Fig. 72-6. Planing an edge.

Fig. 72-7.   Pressure points for planing an edge.

Notice the places to put pressure on the plane for starting and finishing the stroke.
4. Test the edge with the face for squareness (Fig. 72-8).

Planing the First End (Step 3)

1. Select the best end of the board.
2. Fasten the board in a vise with the best end up.
3. Choose the method you will use to plane this end. Follow one of these three steps:
  a. Cut the end to make a *chamfer*, as shown in Fig. 72-9. The chamfer should be cut from the unfinished edge. You can then plane in the direction of the arrow (Fig. 72-9) without splitting the edge.
  b. Clamp a narrow piece of scrap wood against the unfinished edge (Fig. 72-10). Plane in the direction of the arrow. This will keep the outer edge from splitting off.
  c. Plane two-thirds of the distance across the end from one side. Then reverse the direction of the stroke (Fig. 72-11). The edges will not split off if the plane is lifted before planing all the way across. Use a block plane on narrow boards.
4. Test the end for squareness to the planed face (Fig. 72-12) and planed edge (Fig. 72-13).

Planing the Opposite End (Step 4)

1. Measure the board for the length you want. Mark it. Allow an extra $1/16$ inch (1.5

Fig. 72-8.   Testing an edge for squareness.

Fig. 72-9.   Corner chamfered for end planing.

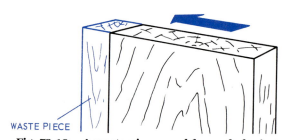

Fig. 72-10.   A waste piece used for end planing.

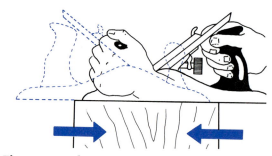

Fig. 72-11.   Planing end grain from both directions.

Fig. 72-12. Testing an end for squareness to the face.

Fig. 72-13. Testing an end for squareness against the edge.

millimeters) for sawing and planing to the line.

2. Square a line across the board for the length.

3. Cut off the extra lumber with either a crosscut saw or a backsaw.

4. Plane the cut end to the line so that it will be square with the planed face and edge. Test for squareness.

Planing the Opposite Edge (Step 5)

1. Measure and mark the board for width. Do this with a rule or a marking gage.

2. Cut off the extra lumber with a ripsaw if necessary. Allow about $1/16$ inch (1.5 millimeters) extra for planing to the line.

3. Plane this edge to the line so that it is square with the working face and with both ends.

Planing the Last Surface (Step 6)

1. Mark this surface board for thickness with a marking gage. Mark the gage line on both edges and both ends.

2. Plane this last surface to the gage line.

3. Test this surface for smoothness and squareness to the other edges and ends.

Unit 73

Shaping a Chamfer and a Bevel

A chamfer and a bevel look somewhat alike. A *chamfer* is made to decorate an edge. A *bevel* may be either an edge deco-

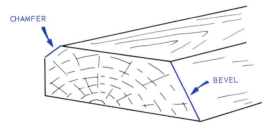

Fig. 73-1. A chamfer and a bevel.

ration or a way to fit two boards together at an angle. Figure 73-1 shows the difference. The chamfer is usually planed to a 45° angle. The bevel may be made to any angle.

Tools

A marking gage, a sharp pencil, a sliding bevel, and a jack or a block plane are the tools needed for making a chamfer or a bevel.

Fig. 73-2. Gaging a line with a pencil without damaging the grain of the wood.

Laying Out a Chamfer or a Bevel

1. *Gage* (mark) the line or lines lightly with a marking gage or with a sharp pencil. This outlines the chamfer or bevel. Figure 73-2 shows how to draw a gage line with a pencil. This is better than using a marking gage. The spur point of a gage may harm the grain of the wood.

2. Set the bevel tool to the angle you want.

Planing and Testing

1. Fasten the board in a vise.

2. Plane the chamfer or bevel to the marked line (Fig. 73-3). If the board or block is small, fasten it in a hand-screw clamp, as shown in Fig. 73-4. It may then be planed with a small block plane.

3. Test the angle of the chamfer or bevel with a sliding bevel tool (Figs. 73-5 and 73-6).

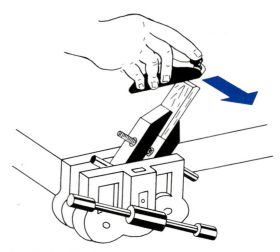

Fig. 73-4. Planing a chamfer on a small piece of wood with a block plane.

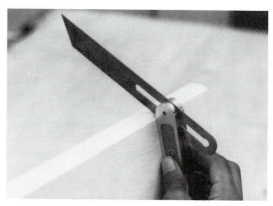

Fig. 73-5. Testing the angle of the chamfer with a bevel tool.

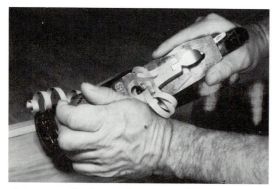

Fig. 73-3. Planing a chamfer to the marked line with a jack plane.

Fig. 73-6. Testing the angle of the bevel with a bevel tool.

Shaping and Forming with a Spokeshave

Fig. 74-1. Spokeshave.

The spokeshave (Fig. 74-1) shapes and forms. It was used many years ago to make spokes for wheels. Its main use now is for forming *concave* (inside-curved) and *convex* (outside-curved) edges on boards. It may be used for making the bows and hulls of model boats, for example.

The cutting edge is sharpened like the plane iron. The blade of the spokeshave may be adjusted to different depths. This will let you control the cutting of a curved surface. You may push or pull this tool, whichever is easier.

Assembling and Adjusting

1. Test the blade for sharpness. If necessary, sharpen it as you did for the plane-iron blade.
2. Put the blade in the frame. Fit the slots on the adjusting nut.
3. Place the lever cap over the blade and slide it under the lever-cap screw.
4. Tighten the blade with the thumb-screw.
5. Adjust for the proper cutting depth with the adjusting nuts.
6. Test the cutting depth on a piece of scrap wood.

Shaping and Forming

1. Fasten the board securely in a bench vise.
2. Use the spokeshave to trim or smooth the curved edge to the exact pattern line (Figs. 74-2 and 74-3). Push or pull this tool. Make a shearing cut.

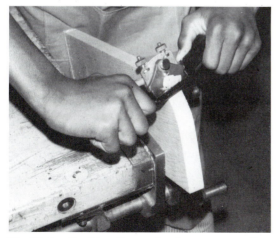

Fig. 74-2. Pushing the spokeshave to smooth a convex curve.

Fig. 74-3. Pulling the spokeshave to smooth a concave curve.

Unit 75

Laying Out and Forming Irregular Pieces

Wood projects sometimes have irregular or curved pieces. Often an irregular pattern may be drawn right on the piece of wood. It is best to make a full-sized drawing on paper or cardboard. Then cut it out to use as a *template* (pattern). Project drawings in books are drawn to a scale to fit the page. Dimensions are given in full size.

Fig. 75-2.   Trammel points.

Tools

The following tools are needed for laying out and forming irregular pieces:

Compass, Dividers, and Trammel Points. The compass or dividers (Fig. 75-1) are used to lay out small circles, to divide spaces equally, to scribe arcs, and to transfer measurements. Larger arcs often require *trammel points* (Fig. 75-2).

Coping Saw. The coping saw (Fig. 75-3) cuts thin boards or plywood.

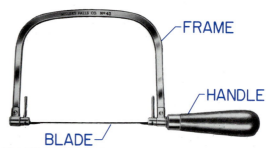

Fig. 75-3.   Coping saw.

Fig. 75-4.   Compass saw.

Compass Saw. The compass saw (Fig. 75-4) cuts inside curves where a coping saw cannot be used. The cut is usually started by *boring* (drilling) a hole near the line to be sawed.

Wood Files, or Cabinet Files. Common wood files are available in many shapes and sizes (Fig. 75-5). Files are used for smoothing edges and making small curves. The cutting surfaces have rows of teeth in patterns (Fig. 75-6). Make sure that

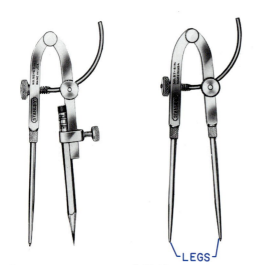

Fig. 75-1.   Compass and dividers.

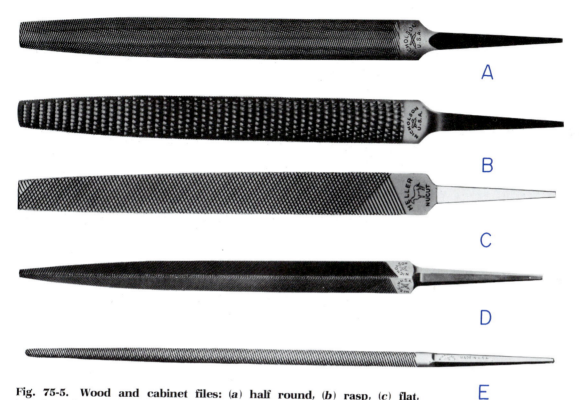

A

B

C

D

E

Fig. 75-5. Wood and cabinet files: (*a*) half round, (*b*) rasp, (*c*) flat, (*d*) triangular, and (*e*) round.

there is a handle on every file. Figure 75-7 shows two kinds of surface-forming tools that have action similar to that of files.

File Card. The file card (Fig. 75-8) has steel bristles. It is used to clean the teeth on a file.

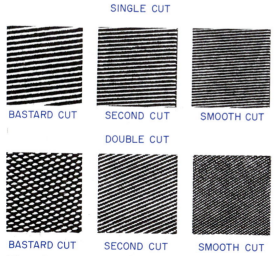

SINGLE CUT

BASTARD CUT   SECOND CUT   SMOOTH CUT

DOUBLE CUT

BASTARD CUT   SECOND CUT   SMOOTH CUT

Fig. 75-6. Patterns of teeth on single-cut and double-cut files.

A

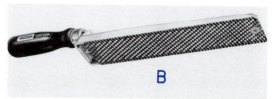

B

Fig. 75-7. Surface-forming tools: (*a*) block plane, and (*b*) surface.

Fig. 75-8. Cleaning the file with a file card.

Marking Curves, Arcs, and Circles

1. Adjust the dividers, compass, or trammel points to the desired radius of the arc, curve, or circle (Fig. 75-9).

2. Scribe the arc, curve, or circle as shown in Figs. 75-10 and 75-11. Place a piece of cardboard under the leg of the dividers that does not move. This will protect the wood surface.

Marking Equal Distances

1. Set the dividers for the distance that is to be *duplicated* (made the same), or *stepped off.*

2. Lay out, or step off, these equal distances.

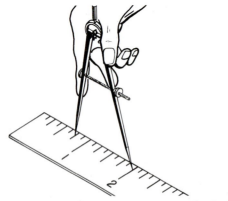

Fig. 75-9. Setting dividers for a desired distance.

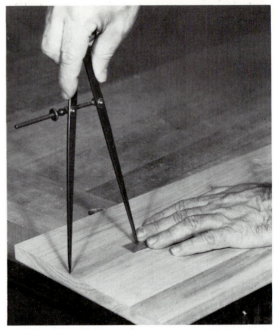

Fig. 75-10. Scribing an arc with dividers.

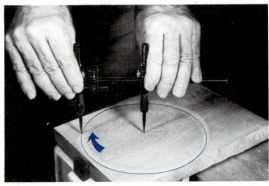

Fig. 75-11. Scribing an arc with trammel points.

Laying Out a Pattern

1. Lay out the curved pattern right on the board or on paper or cardboard.

2. If the pattern has been drawn on paper or cardboard, cut out the template with scissors.

3. If a template is used, trace around it (Fig. 75-12).

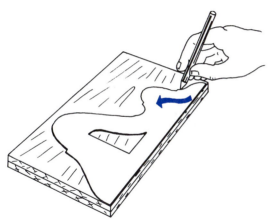

Fig. 75-12. Tracing around a template on wood.

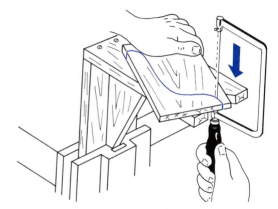

Fig. 75-13. Sawing on a V block.

Cutting with a Coping Saw or Compass Saw

1. Lay out the irregular design or pattern. Make it right on the wood or mark around a template.

2. Fasten the board in a vise or on a V block. When using the V block, hold the board *securely* (tightly) with one hand or with a hand clamp.

3. Check the coping saw to make sure the blade has been put in with the teeth pointing toward the handle.

4. Saw the board with firm strokes (Fig. 75-13) just outside the pattern line.

5. To cut a pierced design:

   a. Bore or drill a small hole on the waste part of the board near the design line.

   b. Remove the blade from the frame.

   c. Insert it through the hole.

   d. Fasten it to the frame again.

6. Saw out the pierced design (Fig. 75-14). Figure 75-15 shows how to cut with a compass saw.

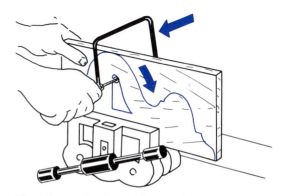

Fig. 75-14. Cutting with a coping saw.

Forming with the File and the Surface-forming Tool

1. Select a medium-coarse wood file or a surface-forming tool of the desired shape for the first smoothing.

Fig. 75-15. Making an internal cut with a compass saw.

2. Fasten the wood in a bench vise.

3. Push the file or the surface-forming tool over the edge of the wood. Use a forward-

and-side motion (Figs. 75-16 through 75-18). This makes a shearing cut that keeps the edges from *splintering* (breaking off in pieces).

4. Continue filing the irregular edge. Use a medium-coarse file until you get a semi-smooth finish.

5. Finish smoothing the edges with a smooth-cut file.

6. Test the irregular edges often. Use a try square to see that they are square with the face of the board.

7. Clean the files often with a file card. This will keep the cutting edges or teeth of the file in good condition. (See Fig. 75-8.)

Fig. 75-17. Dressing a curve with a file surface-forming tool.

Fig. 75-16. Filing a curved edge.

Fig. 75-18. Dressing a curve with a block plane-forming tool.

Unit 76

Cutting and Trimming with a Wood Chisel

A sharp wood chisel is used to cut, fit, and shape. This tool must have a sharp cutting edge and a correct bevel (slant). A *dull* chisel causes *more injuries* than any other tool, and it should be used with care and with great concern for safety.

Tools

Chisels. The two wood chisels most often used are the *socket*, or *firmer*, and the *tang*. These names tell the way that the handle is fastened to the blade (Fig. 76-1 a

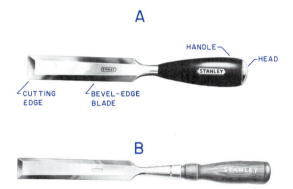

A

HANDLE — ┐ ┌— HEAD

STANLEY   STANLEY

CUTTING
EDGE

BEVEL-EDGE
BLADE

B

STANLEY   STANLEY

Fig. 76-1. (*a*) Wood chisel with reinforced tang handle. (*b*) Socket chisel.

and b). The width of the blade indicates the size of the chisel. The *ranges* (sizes) are from $1/8$ to 1 inch (3.17 to 25.4 millimeters) by eighths of an inch and from 1 to 2 inches (25.4 to 50.8 millimeters) by quarters of an inch.

> **CAUTION:** Never put your hand in front of a chisel while you are using it.

Mallet. A wood or fiber mallet (Fig. 76-2) is used for adding more pressure in chiseling.

Horizontal Chiseling

1. Fasten the board firmly in a bench vise or on a bench top.
2. Push the chisel with your right hand. Guide the blade with your left hand (Fig. 76-3). Use the forefinger and thumb of the guide hand as a brake. Be sure the bevel of

Fig. 76-2. Mallet.

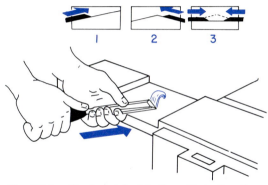

Fig. 76-3. Horizontal chiseling. Note the three chiseling steps.

the chisel is turned *up* when it is used in this way. Always make the cut away from yourself.
3. Continue to make thin strokes (cuts). When you cut across a board, follow the steps shown in the insets of Fig. 76-3.

Vertical Chiseling

1. Fasten the board in a bench vise.
2. Hold the flat side of the chisel against the wood in a *vertical* (up-and-down) position (Fig. 76-4). If you are chiseling or

Fig. 76-4. To do vertical chiseling, push the chisel with one hand and guide the blade with the other.

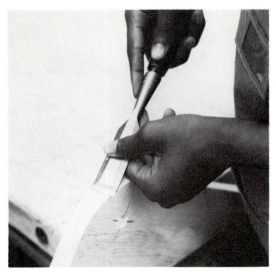

Fig. 76-5. **Trimming an outside curve with a chisel. The bevel edge must be up.**

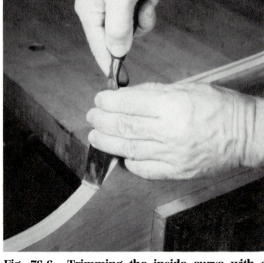

Fig. 76-6. **Trimming the inside curve with a chisel. The bevel edge should be against the wood.**

cleaning out a mortise joint, first bore holes a little smaller than the space to be cut out.

3. Hold the chisel with your right hand. Guide the blade with your left hand (Fig. 76-4). The guide hand will serve as a brake.

4. Push the chisel. Make a shearing cut, as shown in Fig. 76-4. Use a mallet to drive the chisel only when necessary.

Curved Chiseling

1. Fasten the board securely in a bench vise.

2. Push the chisel with a shearing motion when you cut a round corner (Fig. 76-5). Make several short strokes. Be sure the beveled edge of the chisel is turned *up*.

3. On a concave edge, trim by holding the beveled side of the chisel against the wood (Fig. 76-6). Use your left hand to hold the chisel against the board. Push the chisel with your right hand.

CAUTION: Always cut in the same direction as the wood grain.

Unit 77

Smoothing a Board by Scraping

The scraper removes bumps and dents left by the plane. The surfaces and edges of a board are *scraped* to smooth the wood. Uneven wood grains, burls, and knots are smoothed with this tool. The scraper dif-

fers from a chisel or plane in that this tool has a filed or burred edge. Learn to sharpen the edge of a scraper blade. A blade that has been correctly sharpened will make thin shavings.

Fig. 77-1. Scraper blades are available in different shapes.

Fig. 77-2. A burnisher.

Tools

There are several types of scraping tools. Only the scraper blade will be described in this unit. Other types are cabinet and pull scrapers.

Scraper Blade. The most frequently used scraper blade is rectangular (Fig. 77-1). Hand scrapers are thin pieces of high-grade steel that can bend. They are either pulled or pushed. The blade is held with both hands. A burnisher (Fig. 77-2) is used to smooth filed scraper-blade edges (see Fig. 77-6).

Sharpening a Scraper-blade Edge

1. Fasten the hand-scraper blade in a vise.
2. Drawfile the edge by pulling at a 90° angle with a single-cut mill file (Fig. 77-3).

Fig. 77-3. Drawfiling the edge of a scraper blade.

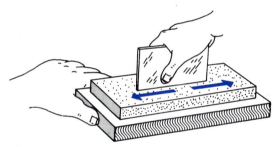

Fig. 77-4. Whetting a scraper blade on an oilstone.

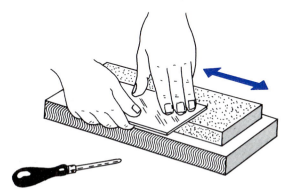

Fig. 77-5. Removing the filed burr on an oilstone.

3. Whet (sharpen) the filed edge. Move the scraper blade back and forth on an oilstone (Fig. 77-4). Guide the blade at *right angles* to the stone.
4. Place the scraper blade flat on the oilstone. Move it back and forth to remove the burr (Fig. 77-5). Then turn the blade over and remove the burr from the other side.
5. Place the scraper blade in a vise. Turn both edges slightly with a burnisher, as shown in Fig. 77-6.

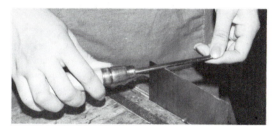

Fig. 77-6. Burnishing a scraper-blade edge.

Fig. 77-7. Scraping a wood surface with a scraper blade.

Fig. 77-8. Pulling a scraper blade.

Hand-scraping

1. Hold the scraper blade firmly between the thumb and fingers (Fig. 77-7). Spring the blade to a slight curve. Hold it at an angle of about 45°.

2. Push or pull (Figs. 77-7 and 77-8) the scraper blade over the wood.

Unit 78

Boring and Drilling Holes

Holes are bored or drilled in wood for screws, bolts, dowels, inside sawing, and decoration. Bits and drills are discussed in this unit.

Tools

Brace. The brace is shown in Fig. 78-1. It holds any of the bits that have a square *tang* (shank).

Hand Drill. The hand drill is used for drilling holes ¼ inch (6.35 millimeters) or less in diameter (Fig. 78-2). A straight-shank drill is used with this tool.

Automatic Drill. This drill (Fig. 78-3) is sometimes used in place of the hand drill.

Auger Bit. Standard auger bits (Fig. 78-4) vary in length from 7 to 10 inches (178 to 254 millimeters). Auger bits are sized by sixteenths of an inch. The common range is ¼ to 1 inch (6.35 to 25.4 millimeters).

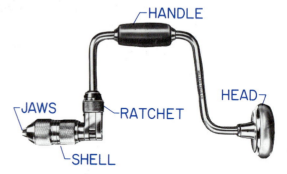

Fig. 78-1. A brace.

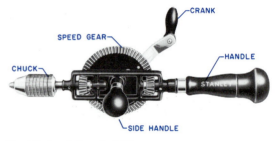

Fig. 78-2. A hand drill.

Fig. 78-3. Automatic, or push, drill.

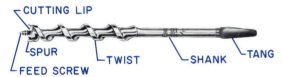

Fig. 78-4. Auger bit.

Fig. 78-5. Iron drill bit.

Fig. 78-6. Expansive bit.

Fig. 78-7. Straight-shank drill bit.

Fig. 78-8. Automatic drill bit.

Fig. 78-9. Drill and countersink bit.

Fig. 78-10. Awl for starting a hole.

Fig. 78-11. Adjustable metal depth gage.

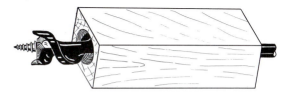

Fig. 78-12. Wooden depth gage.

The number stamped on the square tang shows the bit size in sixteenths of an inch. For example, a bit with 9 stamped on it will cut a hole $9/16$ inch (14.28 millimeters) in diameter. One marked 4 will cut a $1/4$-inch (6.35-millimeters) hole.

Iron Drill Bit. The iron drill (Fig. 78-5) is used to drill holes in both metal and wood.

Expansive Bit. The expansive bit (Fig. 78-6) has a scale on the cutter bit. It is adjustable to cut holes of various diameters in excess of 1 inch (25.4 millimeters).

Straight-shank Drill Bit. Figure 78-7 shows the straight-shank drill. This drill is used with the hand drill. Fractional-size drills are marked by sixty-fourths of an inch. The smallest is $1/16$ inch (1.58 millimeters). Woodworkers usually have an assortment up to $1/2$ inch (12.7 millimeters).

Automatic Drill Bit. This type of bit (Fig. 78-8) fits into the automatic drill (Fig. 78-3). It is used for drilling small holes.

Drill and Countersink. This bit drills and countersinks in one operation (Fig. 78-9).

Awl. This tool, also known as a scratch awl, is helpful for starting a hole (Fig. 78-10).

Depth Gage. The depth gage shown in Fig. 78-11 is one of the commercial types. A very simple gage can be made by boring lengthwise through a piece of wood (Fig. 78-12).

Boring a Hole

1. Mark the place to bore the hole. Start the hole with the point of an awl. This

Fig. 78-13. Starting a hole with an awl.

gives the point of the bit a better hold (Fig. 78-13).

2. Select the correct size of auger bit for boring the hole.

3. Fasten the bit in the chuck of the brace (Fig. 78-14).

4. Place the point of the bit on the spot marked for the center of the hole. Make a few turns in a clockwise direction with the brace to start the hole (Fig. 78-15). Do the same for drilling holes vertically and horizontally.

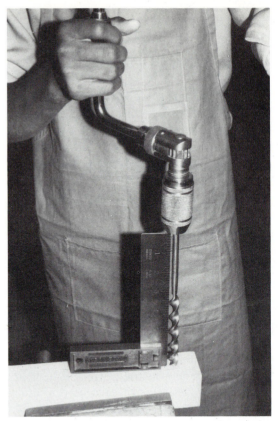

Fig. 78-15. Boring a hole vertically. Note the use of a try square as a guide.

5. Check the angle of boring by testing it with a square against the bit (Fig. 78-15).

6. Bore carefully until the feed screw begins to come through on the back of the board (Fig. 78-16, step 1).

7. Remove the bit from the hole. Turn the brace in a counterclockwise direction.

8. Bore through from the back of the board to make a clean-cut hole (Fig. 78-16, step 2). Figure 78-17 shows what will happen if the bit goes all the way through the wood without being reversed.

Holes may be bored right through a board if you place a piece of scrap wood behind the board (Fig. 78-18). Also do this when you bore with an expansive bit.

Use a shortened dowel auger bit with the doweling jig to bore holes for dowel joints.

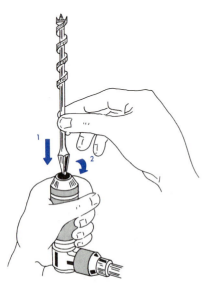

Fig. 78-14. Fastening an auger bit into the brace chuck.

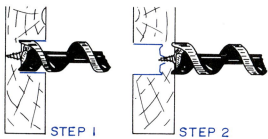

STEP 1   STEP 2

Fig. 78-16. The correct way to bore a hole.

Fig. 78-17. The incorrect method of boring a hole.

Fig. 78-18. Boring a hole horizontally. Note the wood scrap behind the board for protection.

Boring to a Specified Depth

1. Fasten a square-tang bit of the desired diameter in a brace.

2. Fasten the adjustable metal depth gage on the bit. Put it at the depth you wish to bore the hole. (See Fig. 78-11.)

3. Check the depth against the rule to make sure it is correct.

4. Bore the hole until the depth gage just touches the surface of the board.

5. Remove the bit. Clear loose pieces from the hole.

Drilling a Hole

1. Locate and mark the hole.

2. Pick a straight-shank drill bit of the correct diameter. An automatic drill with automatic drill bits may be used instead of the hand drill.

3. Fasten the drill bit into the hand drill or automatic drill.

4. Place the bit on the mark. Hold the drill steady. Then turn the crank in a clockwise direction at a constant speed (Fig. 78-19). A hole drilled with an automatic drill is shown in Fig. 78-20.

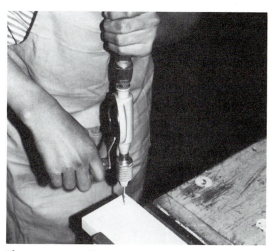

Fig. 78-19. Drilling a hole with a hand drill.

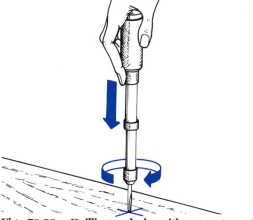

Fig. 78-20. Drilling a hole with an automatic push drill.

Unit 79

Fastening with Wood Screws

Wood screws are one way to fasten boards. When screws are used to assemble (put together) a project activity, the product can be taken apart easily (if necessary) and then reassembled.

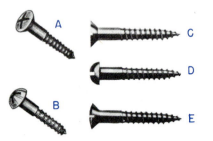

Fig. 79-1. Types of wood screws: (*a*) oval Phillips head, (*b*) round Phillips head, (*c*) flat slotted head, (*d*) round slotted head, and (*e*) oval slotted head.

The three most common types of screws for joining wood are shown in Fig. 79-1. These are *round-head, flat-head,* and *oval-head* screws. All three kinds can be bought with a straight slot or Phillips head, which has cross slots.

Screws vary in length from 1/4 inch to 6 inches (6.35 to 152.4 millimeters). They are graded from 0 to 24 in gage sizes. These numbers refer to the diameter of the shank. Most screws are made of steel and then plated. Brass screws are also manufactured. They are used where *moisture* (dampness) is a problem, as in boats.

Screws are usually packaged in boxes of 100 or 1 gross (144). The boxes are labeled to show length, type, material, number of screws, and gage. Figure 79-2 gives infor-

SIZES OF BITS OR DRILLS TO BORE HOLES FOR WOOD SCREWS

NUMBER (GAGE) OF SCREW	APPROXIMATE DIAMETER OF SCREW SHANK		FIRST HOLE (SHANK) TWIST-DRILL SIZE		FIRST HOLE (SHANK) AUGER-BIT SIZE		SECOND HOLE (PILOT) TWIST-DRILL SIZE		SECOND HOLE (PILOT) AUGER-BIT SIZE	
	in	mm	in	mm	in	mm	in	mm	in	mm
1	5/64	1.98	5/64	1.98	—	—	—	—	—	—
2	3/32	2.38	3/32	2.38	—	—	1/16	1.59	—	—
3	3/32	2.38	7/64	2.78	—	—	1/16	1.59	—	—
4	7/64	2.78	7/64	2.78	—	—	5/64	1.98	—	—
5	1/8	3.18	1/8	3.18	—	—	5/64	1.98	—	—
6	9/64	3.57	9/64	3.57	—	—	3/32	2.38	—	—
7	5/32	3.97	5/32	3.97	—	—	7/64	2.78	—	—
8	11/64	4.37	11/64	4.37	—	—	7/64	2.78	—	—
9	11/64	4.37	3/16	4.76	—	—	1/8	3.18	—	—
10	3/16	4.76	3/16	4.76	—	—	1/8	3.18	—	—
12	7/32	5.56	7/32	5.56	1/4	6.35	9/64	3.57	—	—
14	15/64	5.95	1/4	6.35	1/4	6.35	5/32	3.97	—	—
16	17/64	6.75	17/64	6.75	5/16	7.94	3/16	4.76	—	—
18	19/64	7.54	19/64	7.54	5/16	7.94	13/64	5.16	1/4	6.35

Fig. 79-2. Sizes of bits or drills to bore holes for wood screws.

mation for choosing screws, drill and auger bits, and shank and pilot holes.

Tools

The tools needed to fasten with screws are the same as those described in Unit 78, "Boring and Drilling Holes." Other tools needed are a screwdriver, a countersink bit, and a combination wood drill and countersink.

Screwdriver. Screwdrivers (Figs. 79-3 and 79-4) and screwdriver bits (Fig. 79-5) come in many sizes. The tip should fit the slot of the screw. The screwdriver for the Phillips-head screw (Fig. 79-4) has a special top to fit the cross slots.

Countersink Bit. The angle of the cutting edges of this bit fits the shape of the head of a flat-head screw. This tool (Fig. 79-6) *en-*

Fig. 79-3. A flat-blade screwdriver.

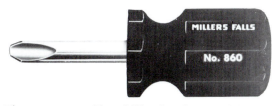

Fig. 79-4. A stubby Phillips-head screwdriver.

Fig. 79-5. A flat-blade screwdriver bit for use with a brace. Phillips-head bits can also be used.

Fig. 79-6. A countersink bit may be used with a brace or fitted with a file handle.

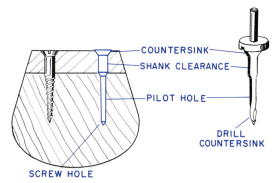

Fig. 79-7. A combination wood drill and countersink.

larges (makes bigger) the part of the hole near the surface. In that way, a flat-head or an oval-head screw will fit *flush* (level) with the surface of the board.

Combination Wood Drill and Countersink. This combination tool (Fig. 79-7) does several things at once. It makes the screw pilot hole, the hole for the screw shank, and the countersink for the screw head. It fits the chuck of a hand drill, an electric drill, or a drill press. Drill countersinks come in 24 sizes that match the diameter of standard wood screws.

Fastening Boards with Screws

1. Mark the place for the screw hole. A mark made with an awl guides the bit or drill (Fig. 79-8).

If the combination drill and countersink is used, steps 2 through 6 may be left out.

2. Select the correct size of bit for drilling or boring the shank hole. The bit size should be large enough to clear the shank of the screw.

3. Fasten the bit in the brace or the drill in the hand drill. Make the shank hole in the first board (Fig. 79-9).

4. Place the boards for the joint in position. Mark the pilot hole with an awl.

5. Bore or drill the pilot hole.

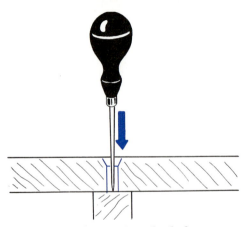

Fig. 79-8.   Marking for the pilot hole.

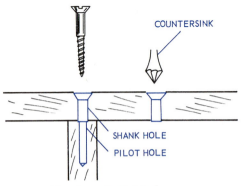

COUNTERSINK

SHANK HOLE

PILOT HOLE

Fig. 79-9.   Shank and pilot holes.

6. Countersink the shank hole slightly if a flat-head or oval-head screw is used (Fig. 79-10).

7. Select a screwdriver that fits the screw-head slot.

Fig. 79-10. Countersinking for a flat-head screw.

8. Fasten the screw with the screwdriver. Hold the screwdriver firmly and in line with the screw (Fig. 79-11). A screw will turn more easily if you first coat it with soap or paste wax. Figure 79-12 illustrates driving a screw with the screwdriver bit and brace.

Fig. 79-11.   Driving a screw with a flat-blade screwdriver.

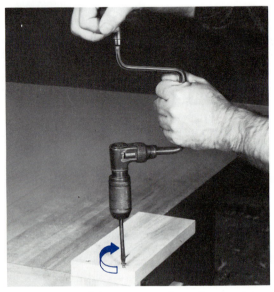

Fig. 79-12.   Driving a screw by using a screw-driver bit and brace.

Fastening with Nails

Nails are driven, set, or pulled. Skill in using the hammer is needed. Knowing how to choose the correct nail for a specific (definite) use is essential.

The kinds of nails used most often in woodworking are shown in Fig. 80-1. *Finishing nails* have small heads. They may be *set* (driven in) with a nail set and covered with putty or a wood plastic. *Casing nails* have cone-shaped heads. They are used for *interior* (inside) trim and for cabinetwork. *Brads* are small finishing nails. They vary in length from ¼ inch to 1¼ inches (6.35 to 31.75 millimeters) and are used to nail thin boards together. *Box nails* are thin and have flat heads. They were first used for nailing together wooden boxes. *Common nails* have flat heads and are a little larger in diameter than box nails.

Nail sizes are usually given by the term *penny*. The abbreviation of penny is d. It is believed that this term came from the weight of a thousand nails. For example, 1000 6-penny, or 6d, nails weigh 6 pounds, which is equal to 2.72 kilograms (kg).

Figure 80-2 is a nail chart showing sizes and lengths. A 2-penny, or 2d, nail is 1 inch (25.4 millimeters) long. For each additional penny, add ¼ inch (6.35 millimeters) in length up to 3½ inches (88.9 millimeters). A 10-penny nail is 3 inches (76.2 millimeters) long.

Remember these rules when you nail:

1. The length of the nail should be about three times the thickness of the first board the nail goes through.

2. The nail should not be too large, or it will split the wood.

3. When driving nails through hardwood, such as birch, maple, and oak, drill a very small pilot hole through the first board.

Tools

The tools used for driving and pulling nails are the claw hammer and the nail set.

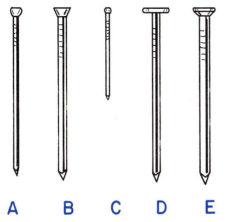

Fig. 80-1. Kinds of nails: (*a*) finishing, (*b*) casing, (*c*) brad, (*d*) box, and (*e*) common.

SIZE (PENNY)	LENGTH	
	in	mm
2	1	25.40
3	1¼	31.75
4	1½	38.10
5	1¾	44.45
6	2	50.80
7	2¼	57.15
8	2½	63.50
9	2¾	69.85
10	3	76.20
12	3¼	82.55
16	3½	88.90
20	4	101.60
30	4½	114.30
40	5	127.00

Fig. 80-2. Nail sizes.

Claw Hammer. The most frequently used sizes of claw hammers are from 12 to 16 ounces (397 to 454 grams) (Fig. 80-3). The size of the hammer you should use depends on the size of the nail.

Nail Set. This tool (Fig. 80-4) sets the head of a finishing nail, casing nail, or brad. The tip is slightly concave to keep it on the nail head.

Driving Nails

1. Choose the proper type and size of nail.
2. Hold the nail in place with one hand. Hold the hammer handle firmly near the end. Make the first light blow (Fig. 80-5).

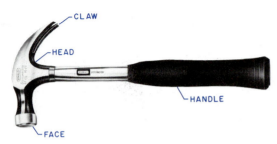

Fig. 80-3.   Claw hammer.

Fig. 80-4.   Nail set.

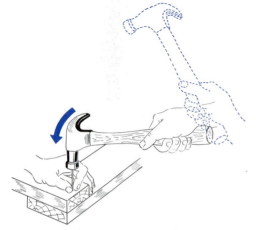

Fig. 80-5.   Starting to drive a nail.

3. Take your hand away from the nail. Continue to hit the nail directly on the head until it is driven flush with the wood (Fig. 80-6). If necessary, you can hold two boards together more securely if you drive nails in at an angle (Fig. 80-7).
4. Set the head of finishing and casing nails about 1/16 inch (1.5 millimeters) below the surface of the wood (Fig. 80-8).
5. Clinch nails on rough construction jobs, as shown in Fig. 80-9.
6. Fill the hole with putty or wood plastic if the nail has been set.

Pulling Nails

1. Slip the claw of the hammer under the head of the nail (Fig. 80-10).

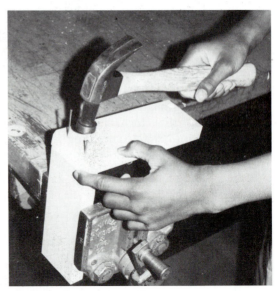

Fig. 80-6.   Driving a nail.

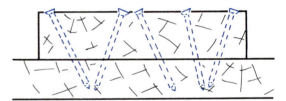

Fig. 80-7.   Holding two boards by driving nails at an angle.

Fig. 80-8. Setting a nail.

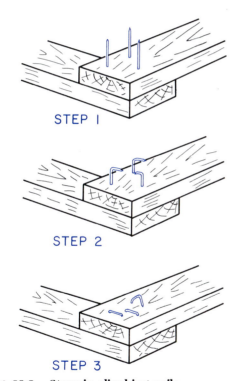

STEP 1

STEP 2

STEP 3

Fig. 80-9. Steps in clinching nails.

Fig. 80-10. Pulling a nail out with a claw hammer.

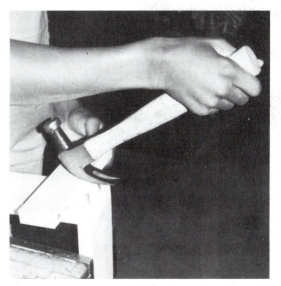

Fig. 80-11. An extra block of wood under the hammerhead gives increased leverage when pulling.

2. Pull the hammer handle until it is nearly vertical, or 90° with the board.

3. Sometimes the nail is too long to come out. To pull it free, slip a block of wood under the head of the hammer (Fig. 80-11). This will add to the leverage and lessen the strain on the hammer handle. To avoid harming the surface, use a softwood block.

Joints and Joining

There are many basic joints and variations of them. A piece of furniture or cabinet-work almost always has one or more types of joints. Each joint has a use of its own. Each requires laying out, cutting, fitting, and putting together. Study the several joints pictured in Fig. 81-1. Use them to fit your own needs.

Tools

The miter box with stiff-backed saw, the doweling jig, and the dowel pointer are special tools used in making joints.

Miter Box. The commercial miter box has a stiff-backed saw (Fig. 81-2). The saw can be adjusted to cut the desired angle.

Doweling Guide. This is used for boring dowel-rod holes (Fig. 81-3 and 81-4). These are commercial types.

Dowel Pointer. The dowel pointer rounds the ends of dowels (Fig. 81-5). A dowel with a shaped end fits into a hole easily.

Fig. 81-2.   A miter box with stiff-backed saw.

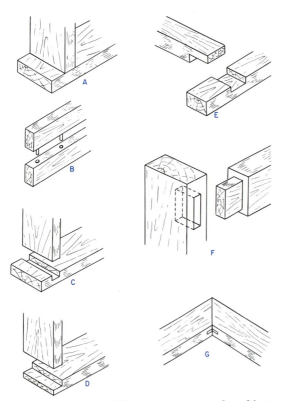

Fig. 81-1.   A few of the common woodworking joints: (*a*) butt, (*b*) dowel, (*c*) dado, (*d*) rabbett, (*e*) lap, (*f*) mortise and tenon, and (*g*) miter.

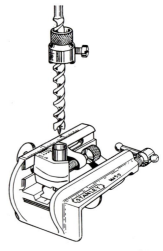

Fig. 81-3.   Doweling guide with a bit stop.

Fig. 81-4. Self-centering doweling guide.

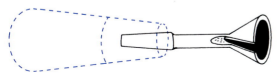

Fig. 81-5. Dowel pointer fitted into a file handle.

Joints

Butt Joint. This joint is simple to make. It can be held together with nails, screws, or dowel pins.

To make a butt joint:

1. Square the end of the board that is to be butted against another board.

2. Mark the exact location of the joint on the surface of the second board.

3. Select the type of fasteners you wish to use: nails, screws, dowels, or other fasteners.

4. If you use nails for the joint, pick the best kind. Drive them so that the points barely go through the first board.

5. Put the pieces to be joined in the correct place. Fasten at least one of the pieces in a vise, and finish driving the nails into the second board.

6. Use a try square to see that the joint is at right angles. Check the joint with a sliding bevel if it is more or less than 90°.

7. If the joint is fastened with screws, refer to Unit 79, "Fastening with Wood Screws."

8. Butt and miter joints also may be held together with corrugated or special metal fasteners (Figs. 81-6 and 81-7).

Fig. 81-6. Fastening a butt joint with a corrugated fastener.

Fig. 81-7. Fastening a miter joint with a special metal fastener.

Edge and Dowel Joints. Dowels are often used in joining furniture parts. Three common types of dowel joints are shown in Fig. 81-8.

■ **317**

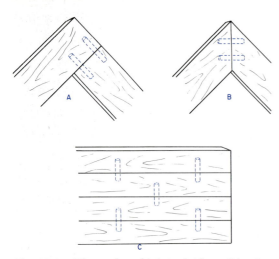

Fig. 81-8. Three dowel joints: (*a*) butt, (*b*) miter, and (*c*) edge.

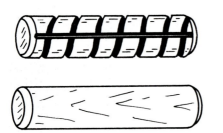

Fig. 81-10. Testing the edge for straightness.

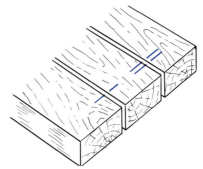

Fig. 81-11. Boards marked for edge gluing or dowel joints. Note how the end grain direction alternates.

Fig. 81-9. Two types of dowel pins.

Dowel rods are usually made from birch or maple. Many diameters and lengths may be obtained. A special grooved dowel pin spreads glue better than a smooth one (Fig. 81-9).

To make an edge joint *without* dowel pins:

1. Plane the edges of the boards to be fastened together. They must be straight, true, and square with the surface of the boards (Fig. 81-10).

2. Arrange the boards so that the surface grain runs in the same direction on both pieces. The end grain of the boards should run *alternately* (in an every-other way), as shown in Fig. 81-11. This lessens warping when the boards are glued.

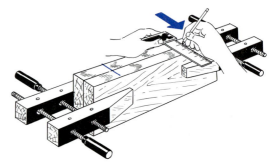

Fig. 81-12. Marking edges for the location of dowel pins.

3. Mark the boards as shown in Fig. 81-11. This will keep the pieces in order.

An edge joint *with* dowels needs the following steps also:

4. Clamp the boards that are to be glued (Fig. 81-12).

5. Mark lines across the edges every 12 to 18 inches (305 to 457 millimeters) with a pencil. Use a try square. These lines locate the places for dowels. Dowels should be at least 2 to 4 inches (50 to 100 millimeters) in from the ends of the boards (Fig. 81-12).

6. Set the marking gage to half the thickness of one of the boards.

7. Mark across the lines made in step 6 (Fig. 81-13). Keep the head of the marking gage against the matched faces of the boards.

8. Select an auger bit the same size as the dowel rod. For most edge joints, the diameter will be ³/₈ inch (9.5 millimeters).

9. Fasten the bit in the brace. Bore holes to the depth you want for the dowel rods. You may do this with or without the use of a doweling guide, but it is easier when one is used. Dowels for edge joints are usually

2 to 3 inches (50 to 75 millimeters) long. Control this depth with a depth gage (Fig. 81-14).

10. Fasten the doweling guide on the edge of the board. It should line up with the marks put on in steps 5 to 7.

11. Bore all holes. Usually, 1¹/₂ to 1³/₄ inches (38 to 45 millimeters) is deep enough.

12. Test the dowel rod in the hole. If it fits too tightly, sand it some, as shown in Fig. 81-15.

13. Place the dowel on a bench top. Make a slight flat place with one or two cuts of a plane (Fig. 81-16). This will let extra glue escape from the bottom of the hole.

14. Saw the dowels to proper length. This should be ¹/₄ inch (6 millimeters) shorter than the overall depth of the two matched holes.

15. Taper both ends of the dowels with a dowel pointer (Fig. 81-17).

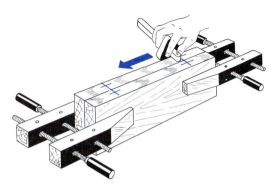

Fig. 81-13.   **Marking centers for dowels with a marking gage.**

Fig. 81-14.   **Boring a dowel hole using a doweling guide and depth gage.**

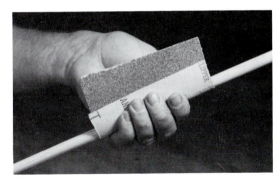

Fig. 81-15.   **Sanding a dowel rod to fit a hole.**

Fig. 81-16.   **Planing a slight flat place on a dowel for glue to escape.**

16. Place the dowels in the holes. Make a trial assembly of the joint to see if it fits (Fig. 81-18). If the boards pull up snugly, the joint is ready to be glued.

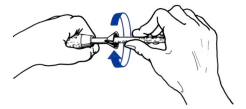

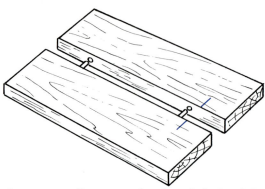

Fig. 81-17. **Tapering the end of a dowel pin with a dowel pointer.**

Fig. 81-18. **Alignment of a doweled-edge joint so that the dowel pins fit.**

Unit 82

Gluing and Clamping

Boards can be glued edge to edge to make larger surfaces. They may also be glued face to face to add thickness. If glued joints are made properly, they can be as strong as or stronger than the wood.

Kinds of Glues

General glues include *animal, casein, plastic resin, polyvinyl resin, resorcinol resin,* and *fish* glues. Others are *blood albumen, starch, all-purpose cements,* and *adhesives.* Some bind wood to metals, plastics, and glass. For example, contact cement is widely used for gluing plastic laminates, such as Formica, to wood.

Glues, adhesives, and cements have many trade names. Manufacturers give instructions for their own brands. Read these instructions carefully. Follow the suggested methods in using any product.

Kinds of Clamps

The clamps most often used in cabinet-work and furniture making are the *cabinet,*

or *bar, clamp* (Fig. 82-1), the *hand-screw clamp* (Fig. 82-2), and the *C clamp* (Fig. 82-3).

Each type works by a hand-screw adjustment. Each has its own use. You should have enough clamps ready when you glue a project activity together. This is most important when using quick-setting glues. You should also have someone to help you.

Fig. 82-1. **Cabinet, or bar, clamp.**

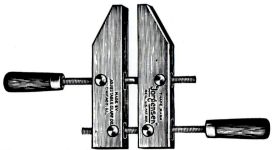

Fig. 82-2. **Hand-screw clamp.**

Gluing and Clamping

1. Mix or prepare the glue or adhesive according to the directions.

2. Adjust all clamps to fit the pieces.

3. Make small blocks for use with bar clamps. These will protect the boards from clamp-pressure marks.

4. Fit all pieces together first *without* glue. Make any fitting adjustments that are needed.

5. Use a brush to spread the glue quickly and evenly on the parts.

6. Put together the parts of the joint properly. Fasten the clamps. Place the bar clamps 12 to 15 inches (305 to 380 millimeters) apart (Fig. 82-4). Use hand-screw and C clamps on the ends of the boards. This will keep them from buckling. Use a piece of paper between the clamps and the board. The paper will keep them from sticking together. With C clamps, use a scrap board so as not to mark the boards being glued. Use several hand-screw and C clamps when clamping boards to build up thickness (Figs. 82-5 and 82-6).

7. Remove the extra glue before it hardens. This may be done with a scraper blade, a wood chisel, a used plane iron, or a scrap piece of wood.

8. Wipe the joint clean with a damp cloth. Remove all traces of glue on the parts you can see. All glue must be removed before the finish is put on.

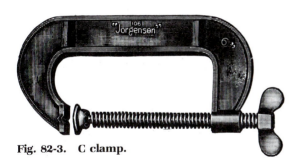

Fig. 82-3.   C clamp.

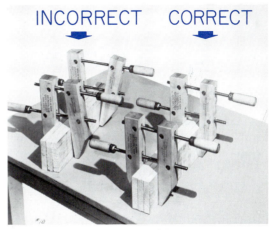

Fig. 82-5.   The correct and incorrect arrangements of hand-screw clamps for maximum pressure.

Fig. 82-4.   Arrangement of bar clamps for edge gluing.

Fig. 82-6.   Holding glued boards with C clamps.

Making Drawers

Cabinets and furniture often have one or more drawers. The instructions that follow are for building a simple drawer. The *rabbet joint* is the one used most often in building a simple drawer. Figure 83-1 gives the dimensions used to cut the rabbet joint.

Making a Drawer

1. Study the working drawing. Learn the exact height, width, and length of the drawer.

2. Select wood for the front. It should be the same as that used for the cabinet or piece of furniture.

3. Select wood for the sides and back. Poplar is a good kind of wood for this purpose.

4. Select material for the bottom. It may be a three-ply panel $1/4$ inch (6.35 millimeters) thick or hardboard $1/8$ inch (3.17 millimeters) thick.

5. Cut and square the pieces for the front, sides, and back. Follow the dimensions in your drawing. A good thickness is about $5/8$ inch (16 millimeters) for the sides and back and $3/4$ inch (19 millimeters) for the front.

6. Lay out and cut the rabbet joints for the ends of the front piece. The combination dado, tongue, and rabbet joint shown in Fig. 83-2 is a firm, sturdy type but is more difficult to make.

7. Cut *plows* (horizontal grooves) in the side pieces for the drawer bottom. Make them to a depth of one-third to one-half the thickness of the side stock. You may cut them with a handsaw and chisel. You may also cut them with a circular saw.

8. Cut the same kind of plow in the back of the drawer front to hold the bottom panel.

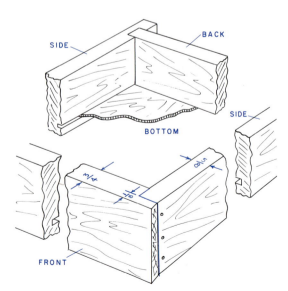

Fig. 83-1. Details of simple drawer construction.

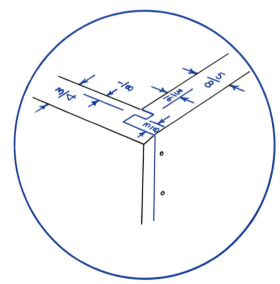

Fig. 83-2.   A combination dado, tongue, and rabbet drawer corner joint.

9. Lay out and cut the vertical dado joints on the inside of each drawer. These hold the back piece. The joint should be about ½ to 1 inch (12.5 to 25 millimeters) in from the back edges. Cut to a depth of about one-half the thickness of the side.

10. Lay out and cut the panel for the drawer bottom. Measure it from the insides of the grooves.

11. See whether the drawer parts fit together. The joints should fit snugly.

12. Take the pieces apart.

13. Fasten the front of the drawer upright in a bench vise.

14. Start two or three brads into the drawer sides for the joints. Resin-coated brads hold best.

15. Put glue on the joint. Place the drawer side on the front piece. Drive the brads to hold the joint firmly.

16. Fasten the opposite side to the drawer front in the same way.

17. Apply glue to the dado joints for the back piece. Nail the back piece in place with brads.

18. Remove extra glue from all joints. Wipe these areas carefully with a damp cloth.

19. Slide the drawer bottom in place from the back of the drawer through the grooves. Do not glue the bottom in place.

20. Check to see that the drawer is square by using a try square.

21. Fasten the bottom panel to the back piece. Use one or two brads. Drive them up from the bottom.

22. Try the drawer to see that it fits the opening and works easily.

23. Dress all joints and drawer parts with sandpaper.

Unit 84

Fastening Hardware

The choice of hardware as part of the final trim on a project activity is a matter of use and personal preference. Hinges and other pieces of cabinet hardware must be placed or inserted accurately. Most manufacturers of these items include special instructions for installing them. Typical pieces of hardware used in woodworking are drawer pulls, hinges, cabinet catches, furniture glides, modern table leg brackets, and ferrules (metal reinforcing rings often used on the feet of legs that have been turned on a lathe).

Drawer Pulls and Knobs

Drawer pulls come in many forms, materials, patterns, and sizes. They may be made from wood, plastic, composition materials, or metals (Fig. 84-1).

Fig. 84-1. Fastening one of several types of drawer pulls.

Pulls and knobs come with screws for fastening. The *single-post*, or *screw*, *knob* is easy to put on. Mark the exact place and drill a hole of the proper size. Pulls with two screws are harder to fasten. Most packets have complete instructions with measurements for drilling the holes.

Hinges

The hinges pictured in Fig. 84-2 are only a few of many types in use today. The *butt hinge* (A) is a popular one. It requires fitting and *gaining* (chiseling out) in the cabinet or frame. Some are *swaged* (shaped to lay flat); others are not.

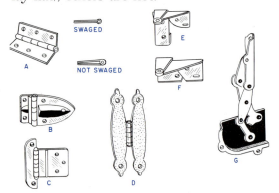

Fig. 84-2. A few commonly-used hinges: (*a*) butt, (*b*) half-surface, (*c*) and (*d*) surface, (*e*) and (*f*) concealed, and (*g*) combination chest hinge and lid support.

Fig. 84-3. Fastening a half-surface hinge.

The *half-surface hinge* (B) is put in so that only half of it is seen. The other half is placed underneath or behind the door.

Surface hinges (C and D) are often used. They are easy to put in. They are placed on the surface and can be seen.

Concealed hinges (E and F) are used on better kinds of furniture. They are easy to mark and install.

The *combination hinge* (G) is used on chests. This type is both a hinge and a lid support. Figure 84-3 shows how a half-surface hinge is put on.

Cabinet Catches

Some cabinet catches are *friction* (A), *magnetic* (B), *roller* (C), *spring-wedged* (D), and *spring-button* (E) (Fig. 84-4). Fastening a tutch latch is shown in Fig. 84-5.

Modern Table-leg Assembly

Figure 84-6 shows the metal parts of tapered-round legs for end tables and coffee tables. The following instructions will help you put them together:

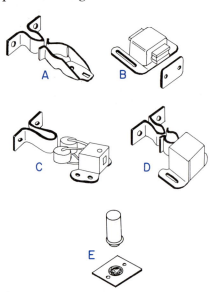

Fig. 84-4. Five types of cabinet hinges: (*a*) friction, (*b*) magnetic, (*c*) roller, (*d*) spring-wedged, and (*e*) spring-button.

Buy or turn a straight, tapered leg according to the dimensions shown on the drawing.

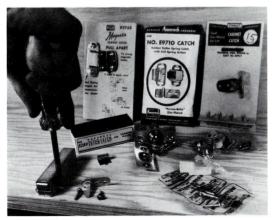

Fig. 84-5. Fastening a tutch latch, and a display of other cabinet catches.

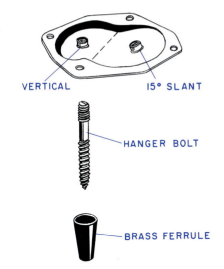

VERTICAL   15° SLANT

HANGER BOLT

BRASS FERRULE

Fig. 84-6. A table leg hanger bolt and a brass ferrule for the end of the leg.

2. Fit the *ferrule* (metal cap) on the foot of the leg (Fig. 84-7). This will give a finish to the foot of the leg, but it is not really needed.

3. Fasten the table leg bracket with screws to the underside of the table top. You can get leg brackets for fastening legs at an angle or vertically. The bracket shown in Figs. 84-6 and 84-7 can be fastened either vertically or at an angle.

4. Drill a pilot hole into the top of the leg for the hanger bolt.

5. Screw the hanger bolt into the leg. Let about 1/4 inch (6 millimeters) stick out.

6. Screw the legs into the table leg bracket.

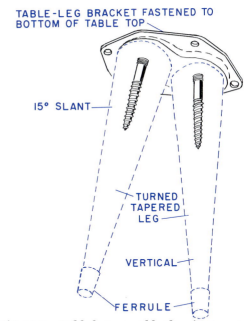

TABLE-LEG BRACKET FASTENED TO BOTTOM OF TABLE TOP

15° SLANT

TURNED TAPERED LEG

VERTICAL

FERRULE

Fig. 84-7. Table leg assembly showing two positions for the bracket, a leg, and the ferrule.

Sawing on the Circular, or Table, Saw

The electric circular saw in Fig. 85-1 shows the important parts found on nearly all such machines.

Safety

1. Use the circular saw *only* after you have your teacher's permission to do so.

2. Make sure the correct blade is on the saw. Use a *crosscut blade* for cutting *across* the grain of wood, a *ripsaw blade* for cutting *with* the grain, or a *combination blade* for both ripping and crosscutting.

3. Keep the safety guard in place and adjusted.

4. Set the saw blade to extend $1/8$ to $1/4$ inch (3 to 6 millimeters) above the board to be cut.

5. Stand to one side of the saw so that you will be out of the way if the board kicks back.

6. Use a push stick for ripping whenever you can.

7. Do not make adjustments while the blade is turning.

8. Do not reach behind the saw blade to pull a board through.

9. Never try to saw without using the *ripping fence* (support for ripping) or the *cut-off guide* (support for crosscutting).

10. Roll up your sleeves and tuck in any loose clothing.

11. Wear clear goggles or a face shield.

Saw Blades and Accessories

Three of the circular saw blades used most often are pictured in Fig. 85-2. The *ripsaw*

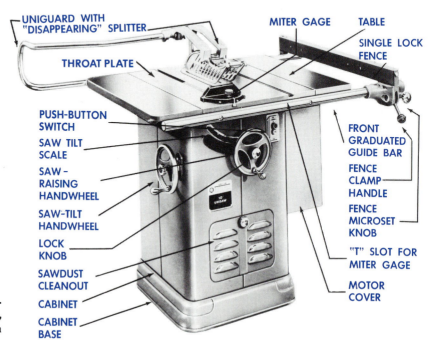

Fig. 85-1. A 10-in (254.0-mm) table, or circular, saw equipped with a safety guard.

UNIGUARD WITH "DISAPPEARING" SPLITTER

MITER GAGE

TABLE

SINGLE LOCK FENCE

THROAT PLATE

PUSH-BUTTON SWITCH

SAW TILT SCALE

SAW-RAISING HANDWHEEL

SAW-TILT HANDWHEEL

LOCK KNOB

SAWDUST CLEANOUT

CABINET

CABINET BASE

FRONT GRADUATED GUIDE BAR

FENCE CLAMP HANDLE

FENCE MICROSET KNOB

"T" SLOT FOR MITER GAGE

MOTOR COVER

blade is used to *rip* (cut lengthwise, with the grain). The *crosscut*, or *cutoff, blade* cuts across the grain. The *combination blade* may be used to crosscut, rip, and miter. The set teeth are made up of crosscut and rip teeth.

Figure 85-3 shows a dado-head set. There are two $\frac{1}{8}$-inch (3.17-millimeter) thick outside blades, two $\frac{1}{8}$-inch (3.17-millimeter) thick chipper cutters, one $\frac{1}{16}$-inch (1.58-millimeter) thick chipper cutter, and one $\frac{1}{4}$-inch (6.35-millimeter) thick chipper cutter. When fully assembled with all *chipper blades* (cutters), this dado head cuts a *dado* (groove) $\frac{13}{16}$ inch (20.63 milli-

meters) wide (Fig. 85-4). The throat plate on the saw table must be replaced with one that will take the dado head.

Ripping

1. Adjust the saw blade to cut about $\frac{1}{8}$ to $\frac{1}{4}$ inch (3 to 6 millimeters) higher than the thickness of the board.
2. Move and fasten the ripping fence to the distance you need from the saw blade (Fig. 85-5). Check the distance with a rule.
3. Turn on the switch and let the blade come to full speed.
4. Make a trial cut on a piece of scrap wood. Check it for accuracy. Allow at least $\frac{1}{16}$ inch (1.5 millimeters) for dressing with a hand plane or a jointer.
5. Lay the board on the saw table. Place it against the ripping fence. Push it with a steady pressure into the blade (Fig. 85-6). Make sure that the saw guard is in place. Stand to one side of the board.
6. Use a push stick when you rip boards less than 4 inches (100 millimeters) wide. This is a safety precaution (Fig. 85-7).
7. Turn off the switch. Wait until the saw blade stops and then remove the board.

Crosscutting

1. Place the cutoff guide (miter gage) in the groove on the top of the saw table.
2. Check the guide to see that it is set at a 90° angle.

Fig. 85-2. **Three types of circular saw blades: left, combination; center, rip; right, crosscut or cutoff.**

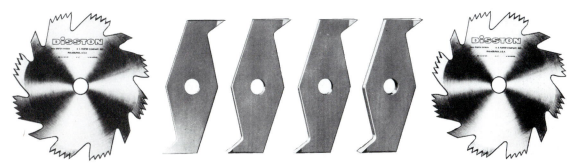

Fig. 85-3. **A dado-head set.**

Fig. 85-4. An assembled dado head.

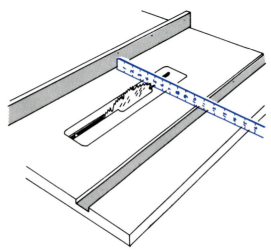

Fig. 85-5. Ripping fence set for the width of the board to be ripped.

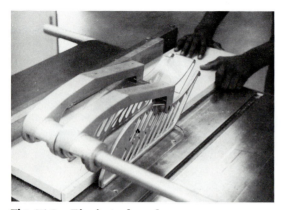

Fig. 85-6. Ripping a board.

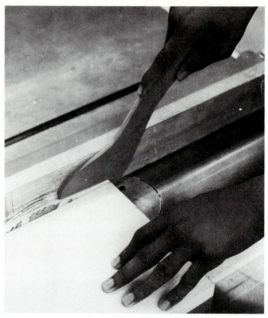

Fig. 85-7. Using a push stick when ripping narrow pieces of wood. The guard has been removed for clarity.

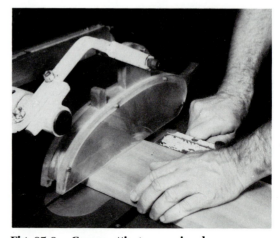

Fig. 85-8. Crosscutting on a circular saw.

3. Mark the board for the cut.

4. Hold the board firmly against the cutoff guide.

5. Turn on the switch and let the blade come to full speed.

6. Push the cutoff guide and the board into the saw (Figs 85-8 and 85-9). Saw on the waste side of the marked line. Move

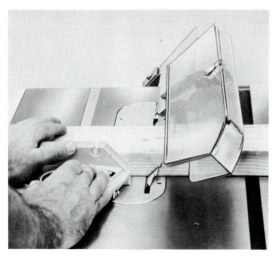

Fig. 85-9. Crosscutting a bevel on a circular saw.

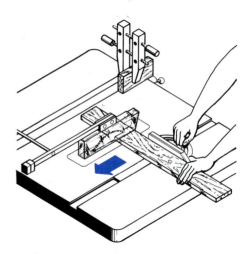

Fig. 85-10. Crosscutting several pieces to the same length. Note the block clamped on the ripping fence.

the board into the blade slowly and evenly until the cut is completed.

7. Fasten a block on the ripping fence if several pieces are being cut to the same length (Fig. 85-10). This will give *clearance* (extra space) for the cut pieces between the saw blade and the ripping fence.

8. Turn off the switch. Wait until the saw blade stops and then remove the boards.

Miters, bevels, chamfers, rabbets, dados (Fig. 85-11), and other cuts can be made on a table saw. If you need to make these cuts, ask your teacher to show you how.

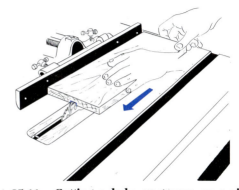

Fig. 85-11. Cutting a dado, or groove, on a circular table saw.

Unit 86

Sawing on the Band Saw

Straight or curved wood pieces can be cut on the band saw. The important parts of this machine are shown in Fig. 86-1. The size of the band saw is known by the diameter of the wheels. The 14- to 24-inch (355.6- to 609.6-millimeter) band saw is the one usually used in schools because it is easy to work.

Safety

1. Get permission to use the band saw.
2. Always use a sharp blade that is in good working order.
3. Keep the safety guards in place.
4. *Feed* (push) the work into the band-saw blade firmly and slowly.

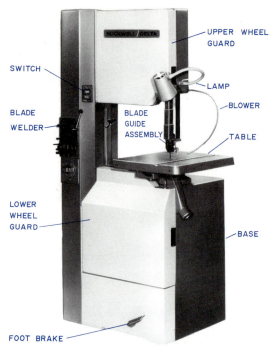

Fig. 86-1. A 24-in (610-mm) band saw. Note the welding attachment on the left column for brazing band-saw blades.

4. Push the board slowly and firmly against the saw blade. Make the cut on the waste side of the line. (Fig. 86-2). Leave about 1/16 inch (1.5 millimeters) for final smoothing.

5. To cut sharp curves, make a few *relief cuts* first to prevent the saw blade from binding on the curve (Fig. 86-3).

Fig. 86-2. Ripping on the band saw.

5. Make cuts on curves slowly. A short, sudden twist may break the blade.

6. Check to see that the upper guide clears the board to be cut by about 1/4 inch (6 millimeters).

7. Tuck in loose clothing and roll up your sleeves.

8. Wear clear goggles or a face shield.

Sawing

1. Mark the board.

2. Adjust the upper guide on the band saw. Clear the thickness of the board by about 1/4 inch (6 millimeters).

3. Turn on the switch. Let the blade come to full speed before you start to cut.

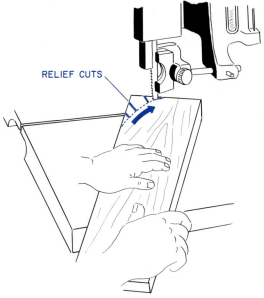

Fig. 86-3. Sawing a curve on the band saw. Note the use of relief cuts.

Sawing with Scroll and Saber Saws

The table scroll saw (Fig. 87-1) and the portable saber saw (Fig. 87-2) are used for cutting inside or outside curves. The size of the table model is determined by the distance from the saw blade to the back of the frame.

The saber saw is easy to use. It can be taken to the place where you are working. The blade moves up and down in the portable saber saw just as it does on the table model. Blades for both kinds cut such materials as wood, metal, leather, rubber, plastic, insulating material, and composition board.

The makers of these saws have special methods for putting in or taking out blades. Study their instructions carefully.

Safety

1. Always get permission before you use the scroll or saber saw.

2. Make certain the electric connection is grounded before using the portable saber saw.

3. Set all adjustments as directed by the maker.

4. Fasten the blade in the saw. The blade in the table scroll saw should be fastened with the teeth pointing down. With the portable saw, the teeth should point up.

5. Hold the board firmly when you cut it.

6. Protect your clothing. Roll up your sleeves and tuck in any loose clothing.

7. Wear clear goggles or a face shield.

Sawing on the Scroll Saw

1. Mark, lay out, or transfer the design to the board or boards.

2. Fasten the blade, with the teeth pointing down, in the lower chuck.

3. Release the tension for the top chuck. Fasten the upper end of the blade in it.

4. Adjust the tension of the blade according to instructions.

5. Lay the board or boards on the scroll saw table against the saw blade. Lower and adjust the top guide so that it just about clears the top of the board.

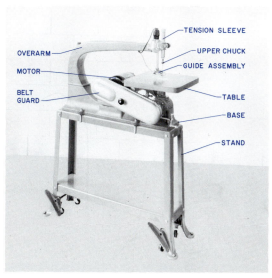

Fig. 87-1.   A 24-in (610-mm) scroll saw.

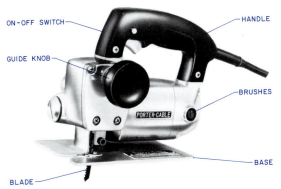

Fig. 87-2.   Portable electric saber saw.

6. Turn on the switch. Gently move your work against the moving blade. Start the cut outside of the pattern marking to allow for edge dressing (Fig. 87-3).

7. Drill a hole in the waste portion for inside-design cuts (Fig. 87-4). Put the blade through this hole. See steps 2 and 3.

8. Cut the inside design as you read in step 6.

9. Fasten *duplicate* (alike) parts of thin boards together with brads in the waste portion. Saw them at the same time to make the cut pieces alike (Fig. 87-5).

Sawing with the Portable Saber Saw

1. Lay out and mark the board.

2. Insert and fasten a blade that is right for your job into the saber saw chuck.

3. Connect the plug in an electric outlet. Make sure the connection is grounded.

4. Place the forward (front) edge of the portable-saw base on the board when you cut from the edge. Drill a small hole. Place the saw blade through the hole when you start the cut from the center of the board.

5. Flip the switch to start the motor. Hold the saber saw firmly. Start cutting with a forward and downward pressure (Fig. 87-6). You can cut freehand on curved or straight lines. Many portable saber saws have circle and straight cutting guides. Use them only after proper instruction.

Fig. 87-4. **Making an inside cut with a scroll saw.**

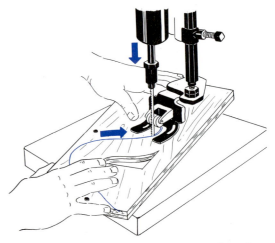

Fig. 87-5. **Cutting duplicate pieces at the same time on a scroll saw.**

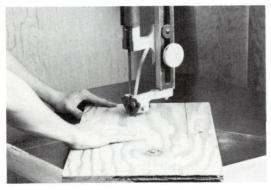

Fig. 87-3. **Sawing on a scroll saw.**

Fig. 87-6. **Cutting with a portable electric saber saw.**

Unit 88

Planing on the Jointer

The jointer is an electric machine that does the work of a hand plane. The skilled worker will often cut rabbets, tapers, bevels, chamfers, and molding on it. You will learn how to plane the faces and edges of a board with the jointer.

The important parts of the jointer are shown in Fig. 88-1. Its size is determined by the widest cut it can make. The front table closest to the person working the jointer can be moved up or down to change the depth of the cut. The rear table is adjusted to the height of the cutter head.

Safety

1. Get permission from your teacher to use the jointer.
2. Always keep the safety guard in place.
3. Use the jointer only to plane boards longer than 12 inches (304 millimeters). Plane shorter ones by hand.
4. Use the push block when you plane the faces of a board.
5. Hold the board firmly against the fence when you plane edges.

6. Make certain the jointer blades are sharp.
7. Take a firm, balanced position to the left of the machine. Never stand at the end of the table. A board may kick back.
8. Do not plane the end grain of narrow boards on a jointer.
9. Always try to plane with the grain.

Planing

1. Adjust the front table for a cut of about $1/32$ inch (1 millimeter) for surface planing. Cut $1/16$ inch (1.5 millimeters) for edge planning.
2. Test the position of the fence for squareness to the table surface with a try square (Fig. 88-2). Adjust it if needed.
3. Place the safety guard in position over the cutter blades. Make sure that it works right.
4. Turn on the switch. Make a trial cut on a piece of clean scrap wood. Make any needed adjustments for depth.

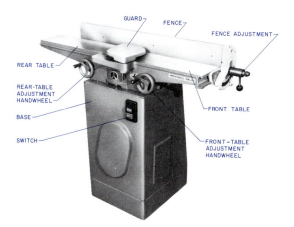

Fig. 88-1.   A 8-in (203-mm) jointer.

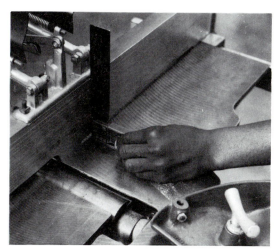

Fig. 88-2.   Testing the squareness of the fence with the table.

5. Place the board flat on top of the front table for planing the surface.

6. Feed the board slowly and firmly over the cutter blades. Push it through until all of the surface has been planed (Fig. 88-3).

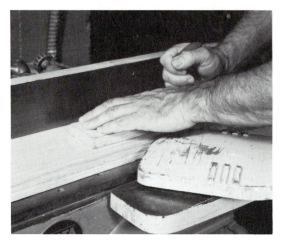

Fig. 88-3.   Planing the surface of a board on the jointer. Note the use of the wooden push block for safety.

Use a push stick to protect your hands from the cutter blades.

7. To plane an edge, place the board on its edge over the front table. Hold it firmly against the fence. Feed it slowly over the cutter blades (Fig. 88-4).

8. To cut chamfers and bevels on the jointer, tilt the fence to the angle you want. Use the bevel tool to test the angle. Then do the same as in step 7.

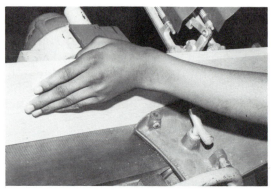

Fig. 88-4.   Planing the edge of a board.

Unit 89

Boring and Drilling Holes with the Electric Drill

The drill press (Fig. 89-1) and the portable electric drill (Fig. 89-2) are used for boring and drilling holes. Both machines can be used on wood and other materials. The important parts of both are shown. The lower speeds of the drill press are used for drilling and boring metals and other hard materials. The faster speeds are used for wood.

The drill bit may be used in the portable electric drill or the drill press. Both machines have chucks for holding bits. Most portable drill chucks will take drill bits up to ¼ inch (6.35 millimeters) in diameter. Some others, however, take bits up to ½ inch (12.7 millimeters). Bits are usually tightened in the chuck with a chuck key.

Boring holes larger than ½ inch (12.7 millimeters) in wood on the drill press can be done easily with a reduced-shank machine bit (Fig. 89-3) or a spade-type wood auger bit (Fig. 89-4).

Safety

1. Get permission from your teacher to use either machine.

2. Always remove the key from the chuck before you start either machine.

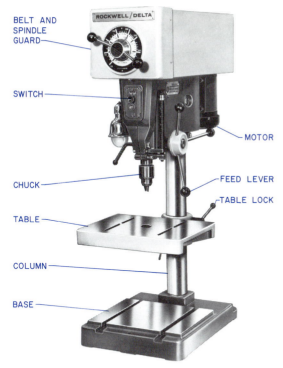

Fig. 89-1. A bench-model drill press with the basic parts labeled.

BELT AND SPINDLE GUARD

SWITCH

MOTOR

CHUCK

FEED LEVER

TABLE LOCK

TABLE

COLUMN

BASE

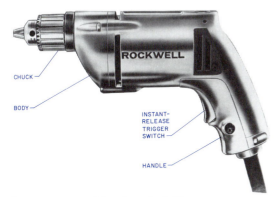

Fig. 89-2. Portable electric drill.

CHUCK

BODY

INSTANT-RELEASE TRIGGER SWITCH

HANDLE

Fig. 89-3. A machine bit with a reduced shank.

Fig. 89-4. A spade-type wood auger bit.

3. Ground the portable electric drill connection.

4. Do not use larger bits than the makers of the machine suggest.

5. Fasten the bit firmly in the chuck before using either machine.

6. Use goggles or a face shield when you work at the drill press.

7. Hold the work firmly when you drill or bore holes.

8. Do not push the drill into the work too fast or it may burn the drill bit and your project activity.

Boring and Drilling on the Drill Press

1. Lay out and mark the center for drilling or boring a hole.

2. Select a bit of the correct size. Fasten it in the chuck.

3. Place the board on the table of the drill press. Adjust the table to the correct height. If you plan to bore or drill a hole through the board, place a piece of scrap wood underneath it (Figs. 89-5 and 89-6).

Fig. 89-5. Drilling a hole on the drill press.

Fig. 89-6. Drilling a hole at an angle on a drill press. This requires either tilting the head mechanism or tilting the table.

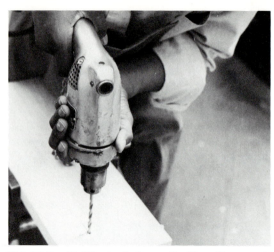

Fig. 89-7. Drilling a hole vertically with a portable electric drill.

4. Check to see that the bit is firmly fastened in the chuck.

5. Turn on the switch.

6. Hold the board securely. Use slow, even pressure when you feed the bit into the wood. If the wood smokes, use less pressure.

Drilling with a Portable Electric Drill

1. Follow steps 1 and 2 for drilling on the drill press.

2. Put the point of the bit in the starting hole before you turn on the motor.

3. Turn on the trigger switch and drill the hole (Fig. 89-7). Use a block of wood to back up the board if you plan to drill all the way through it.

4. Pull the bit from the hole with the motor still on. Then turn off the switch.

Unit 90

Wood Turning

Wood turning is a skill that will reward you with interesting and attractive pieces for a project activity. This unit gives instructions for turning a straight-tapered leg and a bowl. When advanced turning is planned, ask your teacher for specific (definite) procedures. Also study a book on wood turning. Beginners learn to turn wood between *live* (rotating) and *dead* (stationary, or nonmoving) centers. They also use a metal faceplate.

Figure 90-1 shows the important parts of a wood lathe. Find out from your teacher how to control the speed or make adjustments on it. It is advisable also to read a book about working on a lathe. A general

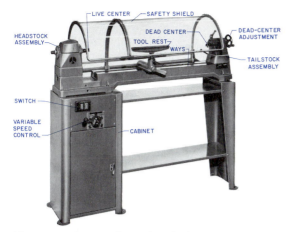

Fig. 90-1. A wood-turning lathe with labeled parts.

Fig. 90-2. A set of wood-turning chisels: (*a*) 0.25-in (6.35-mm) square-nose, (*b*) 0.75-in (1905-mm) skew, (*c*) 0.50-in (12.7-mm) roundnose, (*d*) 0.50-in (12.7-mm) gouge, (*e*) 0.50-in (12.7-mm) spear, (*f*) 0.50-in (12.7-mm) skew, (*g*) 0.12-in (3.18-mm) parting, and (*h*) 0.75-in (19.05-mm) gouge.

rule to keep in mind is: The *larger* the stock (wood piece) to be turned, the *slower the speed of the lathe.*

Wood-turning Tools

Some of the most commonly used wood-turning chisels are shown in Fig. 90-2.

The *gouge* is used for rough turning. It is mostly used for *reducing* (making smaller) stock between centers.

The *skew chisel* is used to make a shearing cut after the stock has been reduced with the gouge.

The *parting tool* cuts grooves with straight sides and a flat bottom. It cuts by scraping action.

The *roundnose chisel* is a scraping tool used mostly in rough turning. It also makes grooves and is best for faceplate turning.

The *diamond-point chisel* is a scraping tool. It is used wherever a cut must be made for a sharp-pointed groove.

The *outside caliper* (Fig. 90-3) is a measuring device used for checking the diameter.

The *slipstone* (Fig. 90-4) is used to *whet* or *hone* (sharpen) the turning tools.

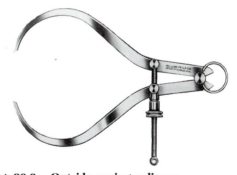

Fig. 90-3. Outside spring calipers.

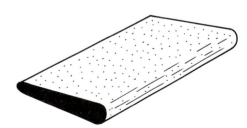

Fig. 90-4. Slipstone.

1. Get permission from your teacher to use the wood lathe.

2. Tuck in loose clothing. Roll up your sleeves. Do not wear a necktie.

3. Wear clear goggles or a face shield.

4. Keep the turning chisels sharp.

5. Stand firmly and well balanced on both feet.

6. Adjust the lathe to its slowest speed for the start of rough turning. Turn up the speed as the work smooths.

7. Lubricate the dead-center end with lubricating oil or beeswax. This will help prevent burning the wood as it turns between centers.

8. Always hold the tool firmly on the tool rest with both hands.

9. Turn the stock by hand before you turn on the electric power. This will keep the stock from *jamming* (getting stuck) against the tool rest.

10. Always stop the lathe when you measure or use the caliper.

Straight, or Spindle, Turning

Stock (wood) for turning between centers must be well centered for balance.

Preparing Stock for Turning

1. Pick stock about 1 inch (25.4 millimeters) longer than the size it will be when it is finished. The piece of wood should be nearly square. Leave at least 1/4 inch (6 millimeters) for turning down to the finished diameter.

2. Draw diagonal lines on both ends of the stock (Fig. 90-5).

3. Make a saw cut 1/8 inch (3 millimeters) deep on the marked lines on both ends (Fig. 90-6).

4. Place the stock to be turned on a solid surface or hold it in a bench vise.

5. Remove the live center from the headstock of the lathe.

6. Place the live center on the saw-cut

Fig. 90-5. Marking diagonal lines to locate the center.

Fig. 90-6. Sawing on the diagonal lines to complete the center marking.

grooves on one end of the stock. Tap it a few times with a mallet (Fig. 90-7). This will seat the prongs of the live center in the saw cuts.

7. Remove the live center. Put it back in the headstock of the lathe.

8. Make a small hole with an awl on the end of the stock opposite from where the line center was placed.

9. Put two or three drops of lubricating oil in this hole. Soap or beeswax also works well.

10. Place the grooved end of the stock against the live center. Hold it in place with your left hand.

Fig. 90-7. Driving the spur point, or live center, in place.

11. Move the tailstock up to about 1 inch (25 millimeters) of the end of the wood. Clamp it to the bed of the lathe frame.
12. Turn the handwheel on the tailstock until the point of the dead center fits into the hole made with the awl.
13. Tighten the handwheel on the tailstock until the piece to be turned is well fastened (Fig. 90-8).

Turning between Centers

1. Adjust the tool rest a little above the lathe's live and dead centers. (See Fig. 90-8.)

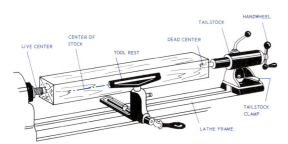

Fig. 90-8. Stock fastened between centers for spindle turning.

Move the wood by hand. Make sure that there is at least $\frac{1}{8}$ inch (3 millimeters) of *clearance* (extra space) between the wood and the tool rest.

Adjust the lathe to run at a slow speed. Start the motor.

Set the caliper to the diameter you need.

Place the parting tool on the tool rest. Cut into the wood (Fig. 90-9) until the proper diameter is reached (Fig. 90-10). Always stop the motor when you check dimensions with the caliper.

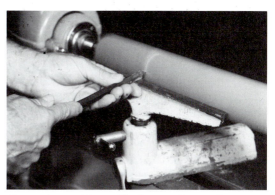

Fig. 90-9. Cutting with the parting tool.

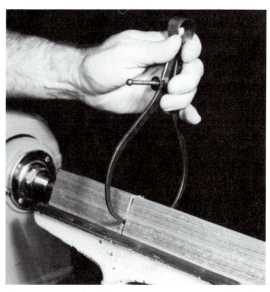

Fig. 90-10. Checking measurements with the calipers.

Fig. 90-11.   Rough-cut turning with the gouge.

Fig. 90-13.   Marking lines on the turned spindle stock.

6. Place the gouge on the tool rest. Start making the rough cut. Move the gouge from left to right or from right to left (Fig. 90-11). Note the angle of the gouge.

7. Continue rough cutting until the piece of wood is round.

8. Set the caliper against a rule to check the diameter of the work (Fig. 90-12).

9. *Shape* (cut) the stock with a gouge, parting tool, or roundnose chisel until the caliper just slips over it. *Turn* (cut) the whole length to the diameter as set in step 8.

10. Mark the places for the shoulders or ends and for any other cuts you plan to make. Use a pencil and a rule (Fig. 90-13).

The stock should be turning slowly while you do this.

11. Place the parting tool firmly on the tool rest. Push it into the wood until the set caliper barely slips over it (Fig. 90-14).

12. Cut the turned leg to the tapered shape with a skew chisel or with a scraping tool (Fig. 90-15).

Sanding

1. Use a quarter of a sheet of medium (80) or fine (120) *abrasive paper* (sandpaper) or emery cloth.

2. Move the tool rest away from the work.

3. Start the lathe at a slow speed.

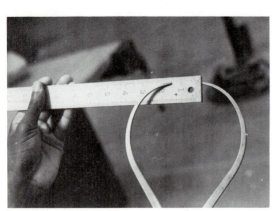

Fig. 90-12.   Setting the calipers to the desired diameter.

Fig. 90-14.   Cutting with the parting tool and checking the measurement with the calipers.

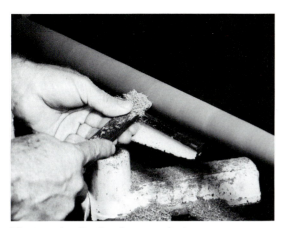
Fig. 90-15. Cutting by scraping.

4. Sand the turned parts. Fold the abrasive paper or cloth and hold it against the work, as shown in Fig. 90-16.

5. Use extra-fine (180) sandpaper for final sanding.

6. Stop the lathe and sand the turning, with the grain, by hand. This will remove the cross-grain scratches that were made while the work was revolving.

Faceplate Turning

Turning wood on the faceplate requires careful positioning of the stock on the faceplate.

Fig. 90-16. Sanding a spindle turning on the lathe.

Preparing Stock for Turning

1. Find a piece of wood that is the right thickness, width, and length. When the stock is marked for cutting on the band saw, it will look like Fig. 90-17. Allow about ½ inch (12.7 millimeters) extra in width and length and ⅛ inch (3 millimeters) in thickness.

2. Plane one face smooth.

3. Draw diagonals on the block to locate the center (Fig. 90-17).

4. Use a compass or dividers to lay out a circle that represents the diameter of the bowl or faceplate turning. The center of the circle is the point where the diagonals cross.

5. Cut off the waste stock with the band saw. (See Fig. 90-18.)

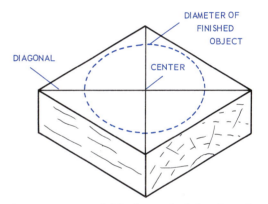
Fig. 90-17. Wood block marked for faceplate turning.

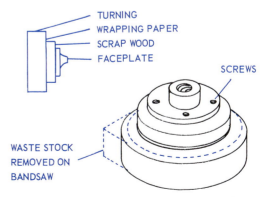
Fig. 90-18. Faceplate fastened to wood block.

6. Select a faceplate to fit the block.

7. Glue a ³/₄-inch (19-millimeter) thick piece of scrap wood to the smooth surface of the block of wood. Use a piece of thick wrapping paper in the joint, as shown in the inset of Fig. 90-18. When the turning is finished, the project can be easily taken apart from the scrap piece with a wood chisel.

8. Fasten the faceplate to the scrap block with screws. The screws should not go beyond the piece of scrap wood.

Turning On the Faceplate

1. Remove the live center and fasten the faceplate and wood-block assembly on the live-center spindle of the lathe head.

2. Adjust the tool rest so that it is parallel to the face of the wood block. It should be about ¹/₄ inch (6 millimeters) away from the block and about ¹/₈ inch (3 millimeters) down from the center.

3. Turn the lathe on to a slow speed.

4. Smooth the face of the wood with a gouge. You may also use a roundnose scraping tool (Fig. 90-19).

5. Turn off the motor. Reset the tool rest parallel with the *ways* (bed) of the lathe. Set it about ¹/₄ inch (6 millimeters) from the outer edge and ¹/₈ inch (3 millimeters) down from the center.

6. Turn the wood block by hand to see that it does not hit the tool rest.

7. Turn on the motor. Smooth the edge with a gouge (Fig. 90-20) or a roundnose tool.

8. Finish turning according to the design or drawing. Use the correct turning chisels for each cut (Figs. 90-21 and 90-22).

9. Sand the turning on the downward side of the faceplate.

10. Stop the lathe and remove the faceplate with your turned project activity. Separate the project activity from the scrap wood.

Fig. 90-19.   Smoothing the face with a gouge.

Fig. 90-21.   Turning on the face with a round-nose chisel.

Fig. 90-20.   Turning the edge with a gouge.

Fig. 90-22.   Turning on the edge with a round-nose chisel.

Sanding with Portable Belt and Orbital Sanders

The two most commonly used types of sanders are the portable belt sander (Fig. 91-1) and the orbital sander (Fig. 91-2). They are used to sand wood smooth.

The sanding belt on the belt sander runs over pulleys at both ends. The orbital sander works in a circular (orbital) pattern on the motion of the base pad. The pad is covered with abrasive paper, and it moves back and forth and in a slightly circular motion at the same time. It sands in any direction. The manufacturers of these types of sanders furnish handbooks that give instructions on how to adjust and use them correctly.

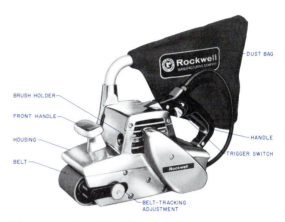

Fig. 91-1.   Portable electric belt sander.

Safety

1. Get permission from your teacher to use the electric sander.
2. Be sure the electric connection is grounded before you use a portable sander.
3. Always disconnect the electric power source before you change belts or abrasive paper.
4. Hold the sander away from the wood when you turn on the switch. You may then lower it to the work. Also, lift it away from the work before you turn off the power.

Sanding with the Belt Sander

1. Fasten the board or project activity firmly.
2. Select a sanding belt of the proper *grit* (degree of coarseness). The grit grades are most often listed as *coarse* (Nos. 30 to 50), *medium* (Nos. 100 to 120), and *fine* (Nos. 140 to 180). Start with either a coarse- or

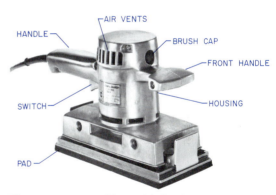

Fig. 91-2.   Portable orbital electric finishing sander.

medium-grit grade. See the *abrasive* (sanding) chart in Fig. 92-1.
3. Place the sander on its left side. Release the tension on the idler (front) pulley with the lever or screw adjustment on the sander.
4. Slip the abrasive belt over the rear pulley and the idler pulley. Make sure that the arrow on the inside of the belt is pointing the way the belt is to turn.

5. Release the idler pulley so that it puts tension on the belt.

6. Connect the plug to the electric power outlet. Start the motor. The sander should lie flat on its left side. Fix the alignment screw so that the belt runs evenly on the pulleys. Turn off the motor.

7. Lift the sander with both hands. Turn on the trigger switch.

8. Lower the sander to the wood surface. Guide the machine over the surface with both hands. Its own weight creates enough pressure on the surface.

9. Sand only in the direction of the grain (Fig. 91-3).

10. Work the sander back and forth over a fairly wide area. To avoid making dents, do not pause or stop in any one spot.

11. Change sanding belts. Continue sanding with finer grits until the surface is smooth.

Sanding with the Orbital Sander

1. Fasten the board or project activity firmly.

2. Select abrasive paper or cloth of the right grit. Start out with coarse- or medium-grit grade.

3. Fasten the abrasive paper or cloth on the base pad with the holding clamps.

4. Connect the plug to the electric power outlet.

5. Lift the sander up off the bench. Start the motor.

6. Set the sander down evenly on the

Fig. 91-3. Sanding with the portable belt sander.

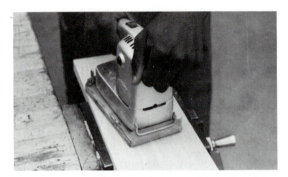

Fig. 91-4. Sanding with an orbital finishing sander.

project activity or board and move it back and forth (Fig. 91-4).

7. Guide the sander with the handle. Use both hands until you get used to working with it. The weight of the machine is enough pressure for most sanding. Do not bear down on the machine while sanding.

8. Change the abrasive grit or cloth. Sand with finer grit until the surface is smooth.

Unit 92

Preparing Wood for Finishing

A smooth, final finish depends on full sanding of all the parts you can see. The project activity should be sandpapered fully with fine grades of abrasive paper even though the parts were sanded before they were put together.

Abrasives

Sandpaper is the abrasive material most often used on wood. The grit on the paper looks like sand, but it is really crushed flint or quartz of gray-tan color. *Garnet paper*, which is reddish in color, lasts longer than flint paper. *Emery cloth* is tough and black in color.

Abrasive papers and cloth are graded from *fine* to *coarse*. Figure 92-1 lists the grit and number classifications of abrasives.

Preparing the Surface

1. Carefully check all board surfaces, edges, and ends to see that planer marks have been removed with the plane or scraper plus fine abrasive papers.
2. Remove all traces of glue, especially around the joints.
3. Moisten any dents in the wood if the fiber has not been broken. Let the wood dry naturally.
4. Fill small knots, holes, and cracks by pressing in colored wood plastic or wood

---

ABRASIVE GRADING CHART (Comparison of mesh and grit numbers)

CLASSIFICATION AND USE	ARTIFICIAL SILICON CARBIDE, ALUMINUM OXIDE	NATURAL GARNET	FLINT (QUARTZ)	EMERY
Extra-coarse (sanding coarse wood texture)	12 16 20	16(4) 20($3\frac{1}{2}$)		
Very coarse (second stage in sanding wood texture)	24 30 36	24(3) 30($2\frac{1}{2}$) 36(2)	Extra-coarse	Very coarse
Coarse (third stage in sanding wood texture)	40 50	40($1\frac{1}{2}$) 50(1)	Coarse	
Medium (removing rough sanding texture)	60 80 100	60($\frac{1}{2}$) 80(0) 100($2/0$)	Medium	Coarse Medium
Fine (first stage in sanding before applying finish)	120 150 180	120($3/0$) 150($4/0$) 180($5/0$)	Fine	Fine
Very fine (second stage in sanding before applying finish)	220 240 280	220($6/0$) 240($7/0$) 280($8/0$)	Extra-fine	
Extra fine (rubbing between finish coats)	320 360 400 500 600	320($9/0$) 400($10/0$)		

Fig. 92-1. Abrasive grading chart (comparison of mesh and grit numbers).

dough. Pick the color that will match the wood most closely when a finish is put on the project activity.

5. Smooth the hardened wood plastic to the wood surface. Use abrasive paper. Be careful to not sand a low spot on the wood.

Preparing Abrasive Paper

1. Tear a piece of abrasive paper into four equal parts (Fig. 92-2).

2. Make a block of wood for holding the sandpaper. The block should be about ³/₄ inch × 2¹/₂ inches (19 × 64 millimeters). You may also use a factory-made hand sander.

3. Fold the quarter sheet of sandpaper around the block so that you can hold it with your hand. See Figs. 92-3 and 92-4. Do not tack sandpaper to the block because the paper will have to be changed often.

Sanding

1. If possible, fasten the board or project activity securely on the bench.

2. Sand all flat surfaces (Fig. 92-3) and edges (Fig. 92-4) *with the grain.* Use an even pressure. Do not sand across the grain or in a circular motion. This can harm the wood fiber. Sand first with medium-grade paper. Finish sanding with fine and extra-fine grit.

3. Sand the edges and the ends in the same way.

4. Sand irregular or shaped edges by holding the abrasive paper in your hand (Fig. 92-5).

5. Rub lightly all exposed parts with a moist sponge to raise any loose fibers. When it is dry, sand it with very fine sandpaper.

6. Check all surfaces, edges, and ends that you can see. Make sure that they have been completely sanded before you put on the finish.

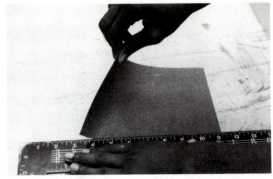

Fig. 92-2. Tearing abrasive paper against a metal edge.

Fig. 92-3. Sanding a wood surface with the grain.

Fig. 92-4. Sanding a flat edge.

Fig. 92-5. Sanding a rounded edge.

Selecting Materials for Finishing Wood

Finishes protect and beautify wood. Study the different kinds of finishing materials. Learn how to select and use them. Make sure that all exposed surfaces, edges, and ends of the wood project activity have been fully sandpapered and are smooth.

Materials

The materials used in putting on finishes include brushes, abrasive paper, linseed oil, turpentine, or commercial paint thinner, alcohol, steel wool, rubbing compound, lacquer thinner, wax and wet-dry abrasive paper. Other finishing materials are stains, wood filler, shellac, lacquer-base finishes, varnishes, urethane or polyurethane sealer, paint, and enamel.

Brushes. Use a good brush to get a high-quality finish. A brush with bristles set in rubber will put on finishes better than a cheap brush. Clean the brush when you are through with it. The solvent should be the thinner for the finish you have put on. Alcohol is a solvent for shellac; turpentine or paint thinner, for enamel, varnish, urethanes, or paint; and lacquer thinner, for lacquer-base finishes.

Linseed Oil. This is a product of flaxseed. It is often used to bring out the rich color of walnut, mahogany, and cedar. It often takes the place of stain on these woods. Linseed oil is also used to put a lasting finish on gunstocks.

Turpentine. Turpentine is refined to a liquid from the sap of the longleaf pine tree. It is often used as a thinner for paints, enamels, urethanes, and varnishes. It is also used for cleaning the brushes used to apply these finishes.

Alcohol. This thinning solvent is made from ethyl and wood alcohol. It is used to thin shellac and for cleaning the brushes used to apply shellac.

Steel Wool. Steel wool is available in rolls or pads. Grades vary from No. 0000 (very fine) to No. 3 (coarse). It is sometimes used instead of abrasive paper to rub down the finish between coats.

Pumice Stone. This is a light-colored powder made from lava. The best grades for rubbing finishes are No. FF and No. FFF. It is mixed with rubbing oil, paraffin oil, or water and is used for rubbing final finishes.

Rottenstone. Rottenstone is a dark gray powder made from shale. It cuts much finer than pumice stone. Rottenstone is mixed with rubbing or paraffin oil to smooth the final finish after pumice stone has been used.

Rubbing Oil. Rubbing oil is usually made of paraffin oil that is to be mixed with either pumice or rottenstone.

Wax. Wax comes in liquid or paste form. Paste wax makes a heavier final coating than liquid wax. Both types are made from a base of beeswax, carnauba wax, paraffin, and turpentine.

Woods

Select a wood that will take the kind of finish you want. The wood will be different for each project activity. It should be picked for its make-up and on the basis of whether its grain is open or closed.

Open-grained Wood. This type of wood includes ash, oak, walnut, and mahogany. These woods need a paste wood filler rubbed into the pores to make a smooth finished surface.

Walnut is one of the most beautiful cabinet woods. Filler should be stained to the color of the wood before it is rubbed in.

Mahogany gives a lovely natural finish or may be stained to a brown or red tone. The filler should be colored to match the tone.

Close-grained Wood. Alder, birch, cherry, fir, gum, maple, pine, and poplar are types of close-grained woods. They do not need a paste wood filler before final finishing.

Alder, gum, and poplar can be easily stained to look like mahogany or walnut. They take a paint or enamel finish well.

Birch and maple are good woods for natural light or blond finishes. Either type can be stained.

Cherry is a hardwood that takes a beautiful finish. It is considered one of the finest cabinet woods.

Fir may be finished in a natural tone or stained to almost any color. It takes paint and enamel very well.

Pine makes a very fine base for paint and enamel finishes.

Unit 94

Staining Wood

Stain applied to wood gives color (tone). On less-expensive furniture, the wood is often stained to look like oak, walnut, or mahogany. Real woods, such as walnut, oak, mahogany, cedar, birch, and maple, have a beautiful natural color and do not require stain. Their rich tone can be enhanced (made more beautiful) by applying a coat of linseed oil or Danish (tung) oil. There are nationally sold brands of both types.

Oil, *water*, and *spirit* (alcohol-based) stains are widely used. Oil stain comes ready-mixed, is easy to apply, and produces an attractively colored base for further finishing. It does not raise the grain of the wood; because of this, it is the only staining method mentioned here.

The end grain of wood, which is open and porous, must first be given a coat of linseed or Danish (tung) oil. This prevents the end grain from turning many shades darker than the surface will be. The oil stain is then applied.

Applying Oil Stain

1. Select the desired color of stain.
2. Shake the stain container well before opening it.
3. Brush some stain on a piece of scrap wood from the project activity to test the color.
4. Use a medium-size brush to put the stain on all of the parts you can see. Brush with long, even strokes (Fig. 94-1). You can get the same results by using a cloth. Always follow closely the instructions on the container.
5. Wipe off extra stain quickly with a cloth (Fig. 94-2).
6. Let the stain dry overnight.

Fig. 94-1. Applying oil stain with a brush.

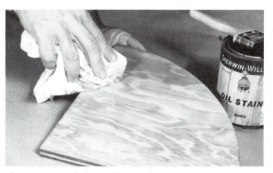

Fig. 94-2. Wiping off surplus stain with a cloth.

Unit 95

Applying Wood Filler

Wood filler comes in liquid or paste form. It fills and seals the pores of open-grained woods such as walnut and mahogany. Wood filler also forms a base for a smooth surface. Wood filler is not needed for close-grained woods such as alder, gum, and poplar.

Paste wood filler is made from *silex* (a ground silicon) mixed with linseed oil, paint drier, and turpentine. Filler should be colored to the tone of the final finish. Use colors in oil to tint it. To apply wood filler:

1. If the filler needs thinning, mix it with turpentine. It should be like thick cream.
2. Add the desired color in oil, a little at a time. Stir it until the tint is correct.
3. Apply the filler on the exposed surfaces (those you can see) with a stiff-bristle brush (Fig. 95-1). Rub the filler into the wood by brushing or wiping *across* the grain.
4. When the filler looks dull, wipe off the *surplus* (extra) across the grain. Use a piece of burlap about 10 to 12 inches (254 to 305 millimeters) square (Fig. 95-2).

Fig. 95-1. Applying wood filler with a stubby brush.

Fig. 95-2. Wiping off surplus filler with burlap.

5. Next, wipe the grain with a clean cotton cloth. Remove the cross-grain strokes.
6. Clean the filler from corners and grooves with a short, pointed stick.

7. Let the filled project activity dry at least 24 hours before going on with more finishing.

Unit 96

Applying Shellac and Lacquer-base Finishes

Shellac is a finishing material that has been used for centuries. It is easy to apply, dries quickly, and rubs out smoothly, but it is not waterproof. There are two colors of shellac, white and orange. White shellac has been bleached and is more expensive. It is used as a sealer and also as a natural finish. First coats of shellac may be thinned with alcohol. Alcohol is also used to clean the brush.

Newer kinds of lacquer-base finishes, such as Deft, are applied like shellac. They serve as bar top finishes and do not water stain. They seal (close the wood pores), prime (the first coat), and also make the final finish. They may be brushed on or sprayed. Follow the instructions given in this unit for applying shellac and lacquer-base finishes. Water will not mar lacquer-base finishes.

To brush on a finish:

1. Clean the surface. If stain, oil, and filler have been put on, make sure they are dry.
2. Pour a little white shellac or lacquer-base finish into an open container, such as a cup.
3. Thin the shellac with alcohol. Thin the lacquer-base finish with lacquer thinner. Use equal amounts of thinner and finishing material. Stir the mixture.
4. Brush the thinned mixture on the entire project (Fig. 96-1). Put it on evenly and quickly. It dries rapidly.
5. Let this coat dry at least 1 to 2 hours. It will act as a sealer coat.

Fig. 96-1.   Brushing on a thin, even finish with a long-bristle brush.

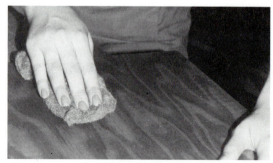

Fig. 96-2.   Smoothing the finish with fine steel wool.

6. Rub smooth with very fine sandpaper or with No. 0000 fine steel wool (Fig. 96-2).
7. Wipe the surface clean with a dry cotton cloth.
8. Apply another coat right from the container if you want a shellac finish or one of the lacquer-base finishes.
9. Let it dry at least 24 hours. Then put wax on the project activity.

Applying Varnish, Urethane, and Penetrating Wood Finishes

Varnish finishes have been applied on wood for a long time. A newer type of wood finish is *urethane* or *polyurethane*. Both varnish and urethane finishes are final coatings. These are tough and generally heat-resistant. They dry to a hard, lasting surface.

Transparent *spar* (water resistant) finishes are the most commonly used. They may be rubbed with a paste mixture of pumice or rottenstone in oil or water to produce a smooth finish. Control of dust, temperature, and humidity (moisture) is important for any satisfactory finish. Most of these final finishing coatings are available in glossy (bright) or satin (less bright) finishes.

Another popular finish is *tung oil.* It is actually a balanced blend of tung oil and other fine oils mixed with high-quality varnish or urethane. This type is hand rubbed with No. 0000 steel wool to produce an attractive surface.

A *penetrating* wood finish differs in that it goes right into the pores of the wood. Penetrating wood finishes give the wood a more natural appearance, and the surface wears well. Always follow the directions printed on the container because each company's product differs slightly in its basic formula and method of application.

Applying Varnish or Urethane Coating with a Brush

1. Carefully clean the surfaces to be varnished. Any finishing done before this time should be dry now.
2. Pour a small amount of varnish or urethane into a cup or other open container.

Fig. 97-1. Rubbing a finish with very fine (No. 400) abrasive paper.

3. Thin the first coat. Follow the instructions on the container.
4. Put the first coat with a good-grade, fine, long-bristle brush. (See Fig. 96-1.) Put it on evenly with long strokes.
5. Allow the coat to dry at least 24 hours.
6. Rub the surface lightly. Use a very fine wet-dry sandpaper or garnet paper and water (Fig. 97-1). This sanding will smooth the surface for the next coat.
7. Put on two more coats without thinning. Do not sand the last coat. Let it dry for 8 hours between coats.

Rubbing a Varnish or a Urethane Surface

1. Make a thin paste of No. FFF pumice stone and rubbing or paraffin oil.
2. Rub this paste back and forth with the grain on the varnished surface. Use a soft cloth pad. Rub until all traces of brush marks or other marks are gone.
3. Wipe the surface clean to remove the pumice paste.

4. Mix a thin paste of rottenstone and either rubbing or paraffin oil

5. Rub this on the coated surface. A fine *luster* (glow) will come if you keep on rubbing. You can use a factory-made rubbing compound instead of these mixtures.

6. Apply a coat of high-grade paste furniture wax. Allow it to dry for 20 minutes. Then polish it with a clean cotton cloth (Fig. 97-2).

Wiping on a Penetrating-wood Finish

1. Clean all the surfaces that are to be finished. Be sure to remove any sanding dust from corners.

2. Pour a small amount of penetrating finish into a cup or other container.

3. Put plenty of the finish on the surface with a cloth or a brush. Follow the instructions on the container.

4. Let it dry about 20 to 30 minutes. Then wipe off the excess with a clean, dry cloth.

5. Let this finish dry overnight.

6. Rub the finish lightly with No. 0000 fine steel wool. (See Fig. 96-2.)

7. Put on a second coat of the penetrating finish. If the wood is open-grained, rub

Fig. 97-2.   Polishing a waxed surface.

in paste wood filler before the second coat is put on. Allow the second coat to dry for 4 hours.

8. Put three more coats on your project activity. Wait at least 4 hours between coats. Rub each coat smooth with fine steel wool. Let the last coat dry overnight.

9. Rub the last coat smooth with fine steel wool.

10. Put on a coat of paste furniture wax. Let it dry for 20 minutes. Then polish the surface with a clean cotton cloth.

Unit 98

Painting and Enameling

Paint and enamel protect and decorate. They are often used on less expensive woods, when a *transparent finish* (one you can see through) may not be wanted. Paint is often put on exterior surfaces or on project activities that are used outdoors. Enamel is better for interior trim and project activities that will be used indoors. You can buy either gloss or semigloss enamel.

Enamel produces a harder finish than

paint because it has some varnish in it. Both paint and enamel come in many colors. You can also tint them yourself to get additional colors.

Manufacturers of paints and enamels suggest certain thinners for their products. Always read the instructions on the container. You will be told how many coats to apply. The instructions will tell you how much time is needed for drying between coats. Many manufacturers also give directions for putting on the paint or enamel.

> **CAUTION:** Be sure the painting and finishing area is well-ventilated. Do not discard oily rags in a confined area.

Mixing and Applying Paint and Enamel

1. Plane, scrape, or sand the surface to be painted or enameled.

2. Read the directions on the container before you open it. You will learn the correct mixture, how to paint or use the enamel, and the drying time.

3. Put on a primer coat if the directions call for it. Shellac makes a good primer coat.

4. Shake the paint or enamel can well. Pry off the lid. Stir it until it is well mixed and blended.

5. Add the thinner recommended on the can if needed.

6. Choose a good brush.

7. Dip the brush into the paint so that about three-fourths of the bristle length is in the paint. Press the brush against the edge of the can to remove some of the extra paint or enamel.

8. Brush the paint or enamel on the wood with long, even strokes (Fig. 98-1). It should cover smoothly and evenly. Do not let it run.

9. Let the coat dry well. Allow the time suggested in the instructions for complete drying.

10. Sand the coat smooth with fine sandpaper. Then wipe the surface free of dust with a clean cloth.

11. Put on a second and a third coat if needed. Do not sand the final coat.

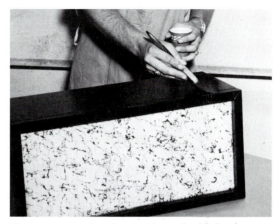

Fig. 98-1.   Apply paint or enamel with long, even brush strokes.

Unit 99

Upholstering a Slip Seat

A *removable* slip seat (one that can be taken off) is easy to upholster (Fig. 99-1). It is often used on dining room chairs and stools. The upholstering part of your project activity could be the seat of a wood or metal stool or chair.

Materials

Upholstering materials for the slip seat include webbing, burlap, rubberized hair padding or a polyurethane foam sheet, cotton, muslin, tacks, and cover upholstering fabric (cloth). These are sold in upholstering shops.

Tools

Special upholstery tools are an upholsterer's tack hammer (Fig. 99-2), a webbing stretcher (Fig. 99-3), and a pair of strong heavy-duty scissors (Fig. 99-4).

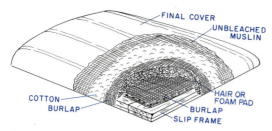

Fig. 99-1. Cross section of an upholstered slip seat.

Fig. 99-2. A magnetic-head upholstery hammer.

Fig. 99-3. A webbing stretcher.

Fig. 99-4. Heavy-duty fabric shears.

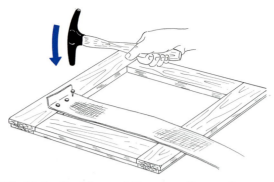

Fig. 99-5. Tacking webbing to a slip-seat frame.

Fig. 99-6. Stretching webbing on a frame.

Making a Slip Seat

1. Make a wooden frame to fit the stool or chair. Use stock ³/₄ inch × 2¹/₄ inches (19 × 57 millimeters). A good joint for this frame is the glued and doweled butt joint. (See Fig. 81-4.)

2. Use a plane to shape or round the outer top-side edges of the frame.

3. Tack one end of the webbing to the frame. Use No. 12 upholstery tacks (Fig. 99-5). Space the webbing strips from ¹/₂ inch to 2 inches (13 to 50 millimeters) apart. Drive three tacks through the first layer. Fold back ³/₄ inch (19 millimeters) of webbing. Drive in two more tacks.

4. Stretch the webbing tight with the webbing stretcher (Fig. 99-6).

5. Drive three tacks in the other end of the stretched webbing. Cut the webbing with the scissors about ³/₄ inch (19 millimeters) beyond the tacks. Fold it back and drive in two more tacks.

6. Tack the rest of the strips of webbing. The strips should make an *interwoven* (crisscross) pattern (Fig. 99-7).

7. Cover the webbed section of the frame with close-grained burlap. Fasten it in place with No. 8 tacks (Fig. 99-8).

Fig. 99-7.   Webbing completed on a frame.

Fig. 99-9.   Rubberized-hair pad of polyurethane foam sheet cut to fit the frame.

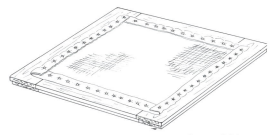

Fig. 99-8.   Burlap fastened over the webbing.

Fig. 99-10.   Underneath view showing a finished corner.

8. Cut a piece of rubberized hair or polyurethane foam 1 inch (25 millimeters) thick to form a padding over the frame (Fig. 99-9).

9. Cover the pad with burlap. Tack it under the frame (Fig. 99-1). Cut away the burlap on the corners to reduce the *bulk* (thickness).

10. Cut a layer of cotton padding and place it over the tacked burlap.

11. Cover the cotton padding with a piece of unbleached muslin. Tack it tightly under the frame. (See Fig. 99-1.)

12. Put the decorative cover fabric in place. Tack it under the frame. Fold the corners, as shown in Fig. 99-10.

13. Tack a piece of dark, glazed cambric to the underside of the seat.

14. Fasten the slip seat in place.

Unit 100

Discussion Topics on Woods Technology

Hand Woodworking

1. List the differences between veneer and plywood.

2. List 12 career opportunities in the areas of forests, woods, and wood-products industries.

3. What type of workers make up the largest group of skilled workers in the labor force?

4. On what day of the week do most accidents happen in industrial arts laboratories or shops? Why?

5. What hand tool causes the most injuries in woodworking? Why?

6. Name 12 safety rules that apply to hand woodworking.

7. How many sixteenths are there in $^1/_4$, $^3/_8$, $^7/_{16}$, $^3/_4$, $^7/_8$, 1, $1^1/_8$, $1^3/_8$, $1^1/_2$, and $1^5/_8$? How many millimeters in 2.7 centimeters?

8. Illustrate in a drawing how ripsaw teeth are different from crosscut-saw teeth. Tell about the cutting action of each.

9. Are there more sawtooth points or saw teeth per inch? Why?

10. Where did the backsaw get its name?

11. What is meant by the term *wind?*

12. List the six general steps for planing a board.

13. Name six commercial forests in the United States and give at least three species of wood in each location.

14. Give three uses for lignin.

15. Define a board foot.

16. Explain the meaning of each of the following abbreviations: (a) S2S, (b) S4S, (c) A-D, (d) K-D, (e) bd ft, (f) M, and (g) FAS.

17. How many degrees are in a right angle?

18. What is a saw kerf?

19. What does the number 8 stamped on the heel of a saw mean?

20. Make sketches to illustrate the difference between a chamfer and a bevel.

21. What tool is used to test the angle of a chamfer or a bevel?

22. Name three tools that can be used for forming or shaping an edge.

23. What is the difference between an auger bit and a drill bit?

24. What kind of a bit is often used for boring holes larger than 1-inch (25.4 mm) in diameter?

25. What information must you know before buying wood screws?

26. Why do you drill shank and pilot holes? What is the purpose of each?

27. List five common types of nails and brads. Explain the uses of each type.

28. List two places where you have seen each of the following joints used: (a) butt, (b) dowel, (c) rabbet, (d) dado, (e) cross lap, (f) mortise and tenon, and (g) miter.

29. List two reasons for using dowels when you make an edge joint.

30. What kind of wood is used for making most dowels? Why?

31. List four types of hinges.

32. Name three natural abrasive materials. Which is the most commonly used?

33. What grait of abrasive paper would you use after applying the first coat of brushed varnish or urethane?

34. List four kinds of cabinet catches.

35. Give two reasons for applying finishes.

36. Name three widely-used wood stains. Which is the most common? Why?

37. What solvents can you use to thin (a) shellac, (b) varnish, (c) polyurethane or urethane, and (d) lacquer-base finishes?

38. Why do shellac and lacquer finishes dry so rapidly?

39. List two kinds of wood that require filler to produce a smooth finish and three kinds of woods that do not require filler.

40. Why are rubbing compounds used?

41. What are the differences between a varnish finish, a urethane finish, and a penetrating finish?

Machine Woodworking

42. List 10 general safety rules for using woodworking machines.

43. Name the parts of the circular saw.

44. Tell why you should use the safety guard on the circular saw.

45. Name the parts of the band saw.

46. What is the purpose of making relief cuts when sawing on the band saw?

47. Name the important parts of the table scroll saw.

48. What are the parts of a jointer?

49. Name the important parts of both the drill press and the portable electric drill.

50. List the parts of a wood-turning lathe.

51. What kind of stone is used to sharpen, or hone, wood-turning tools?

52. What are two general kinds of turnings that can be done on the wood lathe?

Section 7

Plastics
Technology

Unit 101

Introduction to Plastics Technology

Plastics are not a natural material like wood. They are not found in the environment, but they are as important as wood and metal. Plastics are manufactured by people. The many types of plastic materials discussed in Unit 105 (see Fig. 105-10) have been developed since 1868, the year cellulose nitrate was discovered. Figure 101-1 illustrates an important twentieth-century discovery.

The plastics industry is a billion-dollar business. Production has increased over 200 percent during the past 10 years. This industry has become increasingly important in the development of automobiles, airplanes, missiles, and communications. The pioneering flights of Lindbergh and Byrd were made possible through the use of plastics, and the numerous outer-space flights, including the lunar (moon) landings and more recent space shuttle flights, were successful because of the use of plastics as part of the essential materials.

Fig. 101-1. Discovery of fluorocarbon polymers in 1938 was made by Dr. Roy Plunkett (right), who holds the original patent. Technician Jack Rebok (left) helped. Chemist Robert McHarness did early fluorocarbon research. In this photograph Plunkett and Rebok reenact the discovery. (E. I. du Pont de Nemours and Company)

Plastics are used in insulation and are the materials used for parts in such items as telephones, telegraphs, radios, radar, sonar, and communications satellites. They are also used to make foam cushions, furniture, easy-to-clean upholstery, tile, translucent lighting, and many types of floor coverings. These are only a few of the thousands of uses of plastics.

There are two types of plastics. *Thermoplastics* can be poured and molded. *Thermosetting plastics* must be bent or machined to change its shape. Both types will be discussed in this section.

Unit 102

Plastics Technology-related Careers

Plastics technologies continue to expand through research, product development, and production (Fig. 102-1). It is estimated that over 6000 companies in the United States deal with plastics. This huge industry sells billions of dollars worth of finished plastic products yearly. The production for one year of only three plastic materials—*ethylene*, *vinyl*, and *styrene*—amounts to more than 2 billion pounds (900 million kilograms). The manufacture of all synthetic (artificial) plastic and resin materials for one year is well over 16 billion pounds (7 billion kilograms).

Celluloid was the first type of plastic manufactured for commercial use. Photographic film was manufactured from this plastic, as were window curtains for early automobiles. Since 1868, when cellulose nitrate was invented, plastics have expanded in use in nearly all industries, especially after World War II.

Careers

The plastics industry often has fewer properly trained technical people than it needs. Plastics manufacturers are always looking for *mold makers* and *diemakers.* These are two of the more highly skilled careers in the industry. These workers must have the skills of machinists and toolmakers.

Fig. 102-1. Chemists analyze a plastic formula on polymer action during heat treatment. (E. I. du Pont de Nemours and Company)

Plastics engineering is an occupation that is often in need of people. Very few colleges or universities prepare engineers specifically for the plastics industry. The plastics engineer should have experience with and information about *molding and product design, process engineering, mold and process automation, material analysis and selection*, and *machinery design*. The graduate *chemist* and *chemical engineer* can offer ideas and skills to the industry because they know about many kinds of chemicals needed to make plastics.

Other occupational categories in the plastics industry are *supervisors, mold and product designers and drafters, mold-setup technicians, production engineers,* and *inspection and quality-control technicians* (Fig. 102-2). There are also *processing-accessory and finishing-machinery specialists, color- and material-mixing personnel,* and *fabricators.* Occupational opportunities in the plastics industry should improve greatly during the next 10 years. This is because of the stress on the development of better synthetic materials. There are about 280 000 people employed in the plastics industries.

Fig. 102-2. Plexiglas windshield for military helicopters being fabricated.

Unit 103

Industrial Production of Plastics

Thousands of companies in the United States manufacture, fabricate, or do research on plastics. There are three general categories of plastics industries. One is the plastic-materials *manufacturer,* who produces the basic resin or compound. Another is the *processor,* who converts plastic into a solid shape. The third is the *fabricator and finisher,* who further fashions and decorates the plastic.

Manufacturers of Plastic Materials

The basic job of the manufacturers of materials is to make plastic from chemical compounds. This material takes the form of powder (Fig. 103-1), granules, pellets (Fig. 103-2), flakes, and liquid resins. It is processed into finished products. Some materials companies also form the resin into sheets, film (Fig. 103-3), rods and tubes.

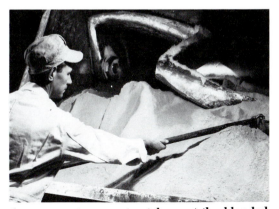

Fig. 103-1. An operator takes out the blended plastic powder from a mixer. (Tennessee Eastman Company)

Processors

The classifications of plastics processors are molders, extruders, film and sheeting processors, high-pressure laminators, reinforced-plastic manufacturers, and coaters.

Fig. 103-2. Plastic pellets are sorted by a screen to obtain a standard size. They will be loaded into drums for shipment to a plastic manufacturer. (Tennessee Eastman Company)

Fig. 103-4. This worker is taking the finished pieces off an injection-molding press. The product being removed is a group of transparent polystyrene box halves. (Creative Packaging Company)

Fig. 103-3. A big bubble of vinyl plastic is blown by air through a special die. It will then be flattened and rolled into packaging film. (Goodyear Tire and Rubber Company)

Molders produce finished plastic products by forming the plastic in molds of the desired shapes (Fig. 103-4). There are about 2000 companies in this group, making it the largest.

Extruders have two groupings: (1) those who produce sheets, rods, film, tubes, special shapes, pipe, and wire covering, and (2) those who produce threadlike plastic filaments. These are woven into cloth for such items as automobile seat covers and plastic upholstering materials.

Film and *sheeting processors* make film and vinyl sheeting by *calendering* (pressing between rollers), casting, or extruding. High-pressure laminators develop sheets, rods, and tubes from paper, cloth, and wood that are *impregnated* (saturated, or filled) with resin chemicals.

Reinforced-plastics manufacturers deal with *rigid* (stiff) structural plastics and with molded and formed plastic products. The liquid resins (polyesters, epoxies, phenolics, and silicones) are combined and then made stronger by adding glass fibers, asbestos, synthetic fibers, or sisal. There are about 500 producing companies in this group.

Coaters cover fabric, metal, and paper with plastics. This is done by calendering, spread-coating, preimpregnating, dipping, and vacuum deposition.

Fabricators and Finishers

Fabricators and finishers use plastic sheets, rods, tubes, and special forms to make finished products. These include industrial parts, signs, and jewelry. Vacuum forming is one of the more important methods of fabricating sheet materials. Sheet stock is formed into television lenses, airplane canopies, scientific aids, and special-purpose items (Fig. 103-5).

One of the newer areas of the plastics industry is made up of companies that produce plastic sheeting and film to make

Fig. 103-5. Acrylic sheet material being formed and fabricated to make this scientific plasma exhaust vessel. (Rohm and Haas Company)

shower curtains, rainwear, inflatable furniture, upholstery, and luggage. There are about 3000 plastics producers in the fabricator-and-finisher group.

Unit 104

Methods of Processing Plastics

Industrial methods of processing plastics include (1) blow molding, (2) calendering, (3) casting, (4) coating, (5) compression molding, (6) extrusion, (7) high-pressure laminating, (8) injection molding, (9) pulp molding, (10) reinforcing, (11) rotational molding, (12) solvent molding, (13) thermoforming, and (14) transfer molding.

Blow Molding

In *blow molding*, a thermoplastic material is stretched and then hardened against a mold. The two general methods are *direct* and *indirect*. In the direct method, a mass of *molten* (hot-liquid) thermoplastic material is formed into the rough shape of the

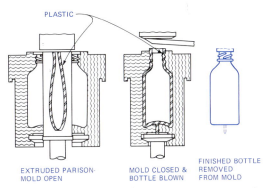

PLASTIC

EXTRUDED PARISON-MOLD OPEN

MOLD CLOSED & BOTTLE BLOWN

FINISHED BOTTLE REMOVED FROM MOLD

Fig. 104-1. Blow molding. (Society of the Plastics Industry, Incorporated)

finished product. This shape is then put into a mold, and air is blown into the plastic (Fig. 104-1). It must be cooled before

Fig. 104-2. This plastic bottle has been blow molded. (Allied Chemical Company)

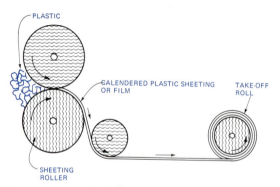

Fig. 104-3. Calendering. (Society of the Plastics Industry, Incorporated)

Fig. 104-4. Vinyl sheeting being removed from the calender. This sheeting will be sent to a fabricator to make racing stripes for auto decoration. (Goodyear Tire and Rubber Company)

being removed from the mold (Fig. 104-2). The indirect method uses a thermoplastic sheet that is first heated and then clamped between a die and a cover. Air pressure forces the plastic into contact with the properly designed die.

Calendering

In *calendering* processes, thermoplastics are rolled into film or sheeting. *Film* is any thickness up to and including 10 mils, or 0.254 millimeters. (A *mil* is 0.0254 millimeter, or $1/1000$ inch.) *Sheeting* refers to thicknesses over 10 mils. Calendering is also used to apply a coating to materials (Fig. 104-3).

In the manufacture of film or sheeting, the plastic compound passes between a series of several large heated revolving rollers. These rollers squeeze it into the desired thickness (Fig. 104-4). Plastic coating is applied to fabric or other material. The coating compound passes through two horizontal rollers on a calender. The uncoated material is passed through two

bottom rollers. The product that comes out is a smooth film fastened to the fabric or other material.

Casting

Casting is different from molding because no pressure is used. The plastic material is heated to a fluid mass. Then it is poured into molds (Fig. 104-5). The product is then *cured* (hardened) and removed from the mold.

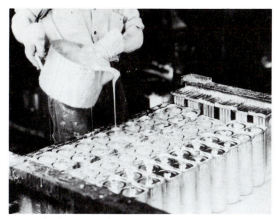

Fig. 104-5. Hot fluid plastic is cast into molds where it is cured. (Catalin Corporation)

Coating

Coating is the process of applying thermoplastic or thermosetting materials onto fabric, leather, wood, paper, metal, glass, and ceramics. The different materials are coated by using a knife, spray, or roller or by dipping or brushing.

Material can be *spread-coated* as it is passed over a roller and under a long blade. The coating compound is placed on the material in front of the knife and is then spread out (Fig. 104-6).

When material is coated with a roller, one horizontal roller picks up the plastic coating solution and puts it on a second roller. The second roller then deposits it on the material (Fig. 104-7). Coatings are also sprayed or brushed onto materials.

Compression Molding

Compression molding is the most frequently used method of forming thermosetting plastic. The material is squeezed into the mold shape with heat and pressure (Fig. 104-8). Plastic molding powder, mixed with such fillers as wood flour, cellulose, and asbestos, is compressed into a mold that is then closed. The pressure on the plastic makes it flow throughout the heated mold. After cooling,

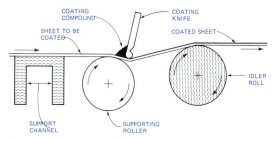

Fig. 104-6. Coating. (Society of the Plastics Industry, Incorporated)

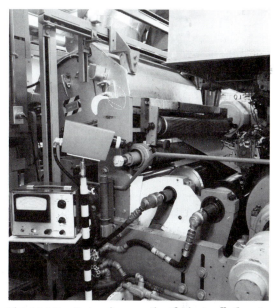

Fig. 104-7. Extrusion coating being rolled onto sheeting. (E. I. du Pont de Nemours and Company)

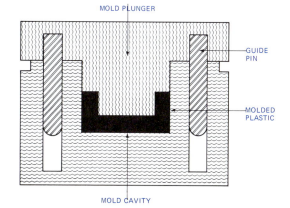

Fig. 104-8. Compression molding. (Society of the Plastics Industry, Incorporated)

the mold is opened, and the plastic object is removed.

Extrusion

In the *extrusion* process, thermoplastic materials are formed into sheeting, film, tubes, rods, and filaments. Dry plastic is loaded into a hopper. It is then fed into a heated chamber, where it is moved by a continuously revolving screw. At the end of the heating chamber, the molten plastic is forced out through a small opening or die that has the desired shape (Fig. 104-9). The plastic extrusion is cooled as it is fed onto a conveyor belt. Other uses of extrusion molding include applying the plastic coating onto the object to be covered or onto plastic extrudes in the form of a tube or sheet (Fig. 104-10).

High-pressure Laminating

In *high-pressure laminating*, heat and pressure are used to form thermosetting plastic materials. Reinforcing materials, such as paper, cloth, wood, and glass fibers, are held together by thermosetting plastics. The first step in the process is *impregnating* (saturating, or filling) the reinforcing materials with plastic.

The impregnated sheets are stacked between highly polished steel plates. These are put under heat and pressure to produce a flat surface (Fig. 104-11).

Resin-treated reinforcing sheets are wrapped around heated rods to produce plastic tubing. This assembly is then cured in an oven.

Injection Molding

Injection molding is the best method for forming thermoplastic material. The plastic is fed into a hopper and then goes into a heating chamber. A plunger then shoves the plastic through a heating chamber as the material softens to a fluid. At the end of

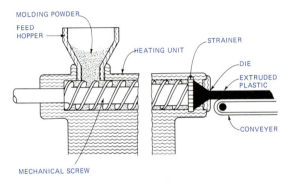

Fig. 104-9. Extrusion molding. (Society of the Plastics Industry, Incorporated)

Fig. 104-10. Polyethylene being extruded in a blown-film application. (Tennessee Eastman Company)

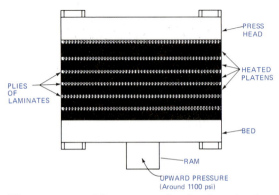

Fig. 104-11. High-pressure laminating. (Society of the Plastics Industry, Incorporated)

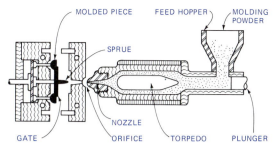

Fig. 104-12. **Injection molding.** (Society of the Plastics Industry, Incorporated)

MOLDED PIECE   FEED HOPPER   MOLDING POWDER
SPRUE
GATE   ORIFICE   NOZZLE   TORPEDO   PLUNGER

Fig. 104-13. **Plastic parts being removed from an injection mold.** (E. I. du Pont de Nemours and Company)

the chamber, a nozzle feeds the fluid into a cool, closed mold (Fig. 104-12). As the plastic cools to a solid, the mold opens and the finished plastic object is *ejected* (released) (Figs. 104-12 through 104-14).

Pulp Molding

In *pulp molding;* a porous form about the size of the finished article is lowered into a container of pulp, plastic resins, and water. The water is drawn through the porous form by a vacuum. This causes the

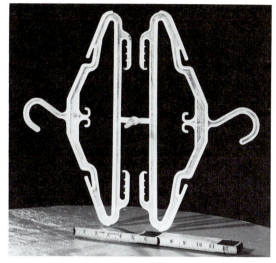

Fig. 104-14. **Injection molding can make many objects that are the same all at one time.** (Society of the Plastics Industry, Incorporated)

pulp and resin mixture to be drawn to the form and *adhere* (stick) to it. When enough pulp has been drawn into the form, it is removed and molded into the final shape.

Reinforcing

Reinforcing uses thermosetting plastics in producing *reinforced* (strengthened) plastic materials. Plastic is used to bind together glass fibers, paper, or cloth for reinforcing.

Reinforced plastics are light weight but very strong. They are easy to work with because they do not need much pressure and heat for curing. Molds are usually the basic form around which reinforced plastics are built (Figs. 104-15 and 104-16). Units 117 through 123 present useful ways of working with reinforced plastics.

Rotational Molding

In *rotational molding*, a measured amount of resin is *charged* (forced) into a warm mold that turns in an oven (Fig. 104-17). The plastic is evenly spread throughout

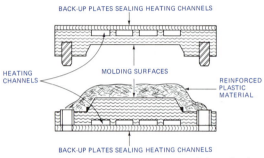

Fig. 104-15. **Reinforcing.** (Society of the Plastics Industry, Incorporated)

Fig. 104-16. **Molded fiber glass reinforced plastics (FRP) escorter (people mover), telephone booths, and benches will hold their shape. They will not blister, rot, or mildew.** (Owens-Corning Fiberglas Corporation)

Fig. 104-17. **A technician pours powdered polyethylene compound into a rotational mold.** (U.S.I. Chemicals)

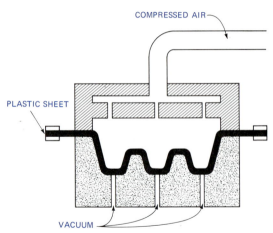

Fig. 104-18. **Sheet thermoforming.** (Society of the Plastics Industry, Incorporated)

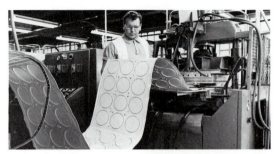

Fig. 104-19. **Thermoforming plastic packaging components.** (Creative Packaging Company)

the mold by *centrifugal force* (the force of circular motion). The heat melts and fuses the plastic *charge* (material) to the shape of the cavity or mold to make the objects.

Thermoforming

In *thermoforming*, a thermoplastic sheet is heated to a *pliable* (workable) state. Air, vacuum, or mechanical forming shapes it to the contour of the mold (Figs. 104-18 and 104-19). These are very common and useful processes for making plastic products.

Solvent Molding

In *solvent molding*, a layer of plastic film sticks to the sides of a mold when the mold is put into a solution and then is taken out. This also happens when the mold is filled with a liquid plastic and then emptied. One product that is made with this type of molding process is the bathing cap. Thermoplastic materials are best for this process.

Transfer Molding

Transfer molding is somewhat like compression molding. However, the plastic is heated to a point of *pliability* (workability) before it reaches the mold. It is then forced into the closed mold by a plunger (Fig. 104-20). Thermosetting plastics are often used in this process. This method was developed to make intricate parts with small, deep indentations.

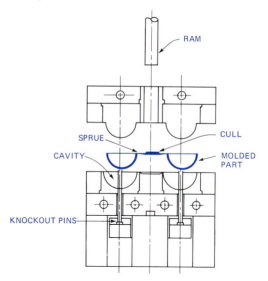

Fig. 104-20. **Transfer molding.** (Society of the Plastics Industry, Incorporated)

Unit 105

Types of Plastics

Plastics are either *thermoplastic* or *thermosetting*. There are widespread and growing uses for each. Both types of plastics are lightweight, have a range of color and physical properties, and can be used in mass-production methods (Figs. 105-1 through 105-9).

Thermoplastic materials and products *become soft* when placed in enough heat and *harden* when cooled, no matter how often the process is repeated. In this group are ABS (acrylonitrile-butadiene-styrene), acrylic, the cellulosics, ethylene-vinyl acetate, fluorocarbon, ionomer, nylon, parylene, phenoxy, polyallomer, polycarbonate, polyethylene, polyphenyl oxide, polyamide, polypropylene, polystyrene, urethane, and vinyl.

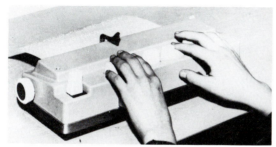

Fig. 105-1. **A one-piece multipart frame molded of high-strength phenolic has replaced a heavy metal frame for this Braille typewriter.**

Fig. 105-2. **Face-wide sunshade is made up of colorfully printed cellulose acetate sheet.**

Fig. 105-3.   Moon rock displayed at the Smithsonian Institution is protected by clear cast acrylic tube. (Cadillac Plastic and Chemical Company)

Fig. 105-6.   Polyester-reinforced fiber glass canoe. A layer of polyurethane foam provides extra flotation for the canoe.

Fig. 105-4.   A polystyrene foam cover over a water treatment plant to reduce odors. (Dow Chemical)

Fig. 105-7.   Acetal resin keeps the shoe cleat from locking into turf.

Fig. 105-5.   Molten saran polymer being extruded from circular dies, cooled quickly, and then transformed into large bubbles. Saran Wrap is made from polyvinylidene chloride. (Dow Chemical)

Thermosetting plastic materials are set into *permanent* (unchanging) shape when heat and pressure are applied to them during forming. Reheating *does not soften* these materials. They include alkyd, amino (melamine and urea), cold-molded, epoxy, phenolic, polyester, and silicone.

Figure 105-10 lists the many plastic materials in their order of development.

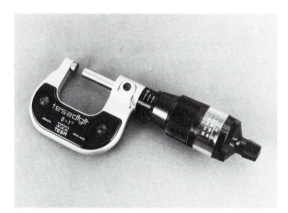

Fig. 105-8. This micrometer is made from du Pont delrin. (E. I. du Pont de Nemours and Company)

Fig. 105-9. Polycarbonate visor on the astronaut's space suit has good optical properties.

PLASTIC MATERIALS GENERALLY IN ORDER OF THEIR DEVELOPMENT

MATERIAL	TYPICAL USE	MATERIAL	TYPICAL USE
Cellulose nitrate	Eyeglass frames		
●*Phenol-formaldehyde*	Telephone handsets	●*Acrylonitrile-*	
Casein	Knitting needles	*butadiene-styrene*	Luggage
Alkyd	Electrical bases	Allylic	Electrical connectors
Cellulose acetate	Toothbrushes	●*Polyurethane*	Foam cushions
●*Polyvinyl chloride*	Wall coverings	Acetal	Automotive parts
●*Urea-formaldehyde*	Lighting fixtures	High-density	
Ethyl cellulose	Flashlight cases	polyethylene	Milk bottles
●*Acrylic*	Brush backs	●*Polypropylene*	Safety helmets
Polyvinyl acetate	Flashbulb linings	Polycarbonate	Appliance parts
Cellulose acetate		Chlorinated poly-	
butyrate	Packaging	ether	Valves and fittings
●*Polystyrene*	Housewares	Polyallomer	Typewriter cases
Nylon	Gears	Phenoxy	Bottles
Polyvinylidene		Ionomer	Skin packaging
chloride	Packaging film	Polyphenylene oxide	Battery cases
●*Melamine-*		Polyimide	Bearings
formaldehyde	Tableware	Ethylene-vinyl	Adhesives and
●*Unsaturated polyes-*		acetate	coatings
ter	Boat hulls	Parylene	Insulating coatings
●*Low-density*		Polysulfone	Electrical and
polyethylene	Packaging		electronic parts
Fluorocarbons	Industrial gaskets	Thermoplastic	Electrical and
Silicone	Motor insulation	polyester	electronic parts
Cellulose propionate	Pens and pencils	Polybutylene	Piping
●*Epoxy*	Tools and jigs	Nitrile barrier resins	Bottles

●Twelve most widely used plastic material categories according to the Society of Plastics Industry, Inc.

Fig. 105-10. Plastic materials in the order of their development.
(Society of the Plastics Industry, Incorporated)

Illumination in Plastics

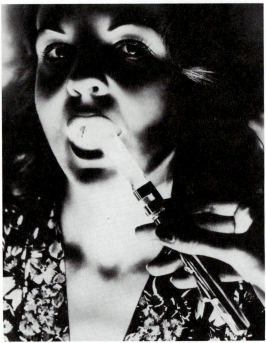

Fig. 106-1.   A light-piping acrylic plastic medical instrument. (Rohm & Haas Company)

Fig. 106-2.   An end table with illuminated legs. The light comes from a small electric light located underneath the top.

The properties of acrylic (Plexiglas and Lucite) plastic permit the passage and control of light through this synthetic material. Light can be *refracted* (deflected), *intensified*, or *directed* through long lengths and around curves. Figure 106-1 shows an excellent example of these characteristics. The acrylic end table in Fig. 106-2 shows where light illuminates the curvature (bend) of the plastic legs. These features are used in many commercial applications, such as medical instruments, signs, and displays.

Theoretically, acrylic plastic can pipe light indefinitely. However, dust, dirt, and the amount of light that can be carried through a given thickness and distance limit the possibilities. Plastic signs 6 feet (19.68 meters) long have been illuminated with excellent results. Molded dials 12 to 14 inches (305 to 356 millimeters) in diameter carry light satisfactorily.

The several illustrations presented in this unit describe conditions that must be considered when taking advantage of light-piping qualities. Figure 106-3 shows limitations that must be observed when bending acrylic plastic in order to maintain (hold) efficient *transmission* (passage) of light.

It is possible to permit the escape of light at selected, planned locations on the surface of plastic by *sanding, scribing,*

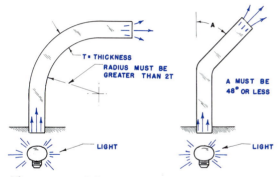

Fig. 106-3. Light-carrying limitations of curves and angles.

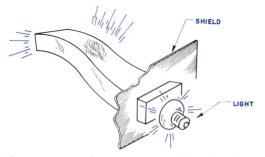

Fig. 106-4. Surface treatment allows the escape of light.

Fig. 106-5. Sanded edges on the column of this lamp release light.

carving, or *engraving.* Figure 106-4 illustrates this feature. Figure 106-5 shows the escape of light on the sanded edges of the lamp column support.

If intensive (bright) light is to be released in an area, the plastic should be tapered and sanded, as shown in Fig. 106-6.

It is sometimes desirable to increase the thickness of acrylic plastic at the edge or end in order to *funnel* (direct) light into the thin section, as indicated in Fig. 106-7. The grooves in this illustration merely indicate how light can escape when there is a break in the surface.

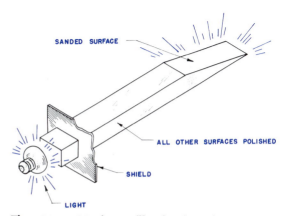

Fig. 106-6. Maximum illumination of an area.

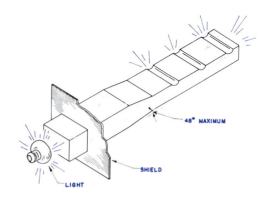

Fig. 106-7. Increasing illumination for thin plastic.

Laying Out and Transferring Patterns

The steps in marking and transferring a pattern onto a plastic sheet are important. If the protective covering is on the plastic, this process is easy. If the covering has been removed, you may use rubber cement to put on another covering. This paper sheet serves as a base on which to make (mark) and transfer the pattern.

If there is no covering on the plastic sheet, a scriber, awl, or dividers can be used to lay out the pattern directly on the plastic. You should understand and be able to use both the U.S. customary and the metric systems of measurement. See Units 17 and 66.

Tools

The few tools needed for laying out a pattern on plastics are used often in the industrial laboratory. They are the try square, rule, pencil compass, pencil, and scriber.

Laying Out and Transferring

1. Check to see that the plastic material has a paper coating on one side or both

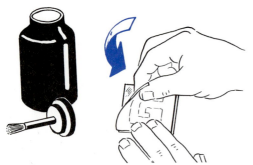

Fig. 107-1. Rubber cementing a pattern design to plastic.

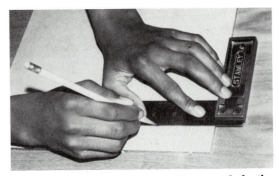

Fig. 107-2. Squaring the line on covered plastic.

Fig. 107-3. Laying out a straight line with an awl or scriber directly on plastic.

sides. If it is covered, start to mark the layout. If the paper covering has been removed, apply a coating of rubber cement on the plastic and press a piece of paper over it (Fig. 107-1).

2. Mark the lines on the paper-covered plastic (Fig. 107-2). This process is like laying out a pattern on wood. You can also scribe lines right on the plastic (Fig. 107-3).

3. Irregular or curved patterns should first be drawn on a piece of cardboard. A design or pattern may be made larger from this *template* (pattern).

Place the template flat on the paper-covered plastic surface. Mark around it with a pencil (Fig. 107-4). If there is no protective paper covering, use a china pencil, awl, or scriber.

Do not remove the paper covering until marking, cutting, drilling, and edge dressing have been completed. The paper protects the surface of the plastic from minor scratches.

Fig. 107-4. Marking around a template on plastic.

Unit 108

Cutting and Sawing Plastics

Just about all types of plastic materials can be cut. You can use both wood- and metal-cutting tools and machines. However, plastics dull tools, especially the tools used in woodworking. Because of this, the tools need to be sharpened often.

Fig. 108-1. A special acrylic cutting tool for plastic.

Tools

A special tool for cutting acrylics is shown in Fig. 108-1. The coping saw and the backsaw are also used to cut plastics. A useful holding device is the wood V block (Fig. 108-2), which can be made in the laboratory. The V block may be held in a bench vise or fastened to the bench top.

Cutting to a Straight Line with the Special Acrylic-plastics Cutting Tool

1. Lay out and mark the straight line to be cut. (See Unit 107, "Laying Out and Transferring Patterns.")

2. Using a straight edge as a guide, place the point of the special acrylic-plastics cutting tool at the beginning of the cut. Use firm pressure and *pull* the cutting-tool point the full width or length of the marking (Fig. 108-3). This requires 5 to 10 re-

Fig. 108-2. A wood V block for holding plastic while cutting with a coping saw.

peated cuts, depending on the thickness of the plastic sheet: 5 times for 0.125 or $\frac{1}{8}$ inch (3.17 millimeters) and up to 10 times for 0.25 or $\frac{1}{4}$ inch (6 millimeters).

Fig. 108-3. Scribing and cutting a sheet of acrylic plastic. It may take 7 to 10 times of scribing for a sheet ¼ in (6.35 mm) thick.

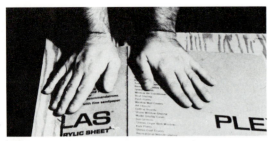

Fig. 108-4. Breaking a sheet of acrylic plastic over ¾ in (19 mm) diameter wood dowel.

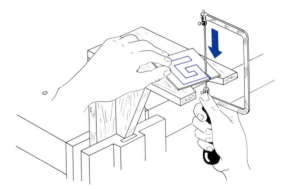

Fig. 108-5. Sawing plastic on the V block.

Fig. 108-6. Cutting plastic rod with a backsaw.

3. Place the plastic sheet over a ¾-inch (19-millimeter) dowel. Keep the scribed line *up*. Hold the widest part of the plastic sheet down with one hand. Apply a *downward pressure* on the short side of the cut line with the other hand (Fig. 108-4).

Sawing with Hand Tools

1. Use a pencil to lay out the pattern or design on the paper-covered plastic. (See Unit 107, "Laying Out and Transferring Patterns.")

2. Hold the plastic firmly on a V block. Cut with vertical strokes of the coping saw (Fig. 108-5). The teeth of the blade should point toward the handle.

3. Hold flat stock, rod, or cylindrical material in a woodworker's bench vise when cutting with a backsaw (Fig. 108-6).

Sawing with Power Tools

Study the several units in the woods and metals technology sections before trying to use machine power tools.

1. Lay out the line, pattern, or design on the paper-covered plastic with a pencil. See Unit 107.

2. Select the power or machine tool most suited for making the planned cut. Check with your teacher before doing this.

Plastic materials may be cut on the circular saw (Fig. 108-7), scroll saw (Fig. 108-8), band saw (Figs. 108-9 and 108-10), or portable electric saber saw (Figs. 108-11 and 108-12).

When sawing on the scroll saw, it may be necessary to use a soap-and-water solution on the cut to prevent overheating. Apply this solution to the cut with an oil can.

Figure 108-13 shows the use of a metal cutting lathe on plastic material.

Fig. 108-7. Sawing acrylic plastic sheet on a circular saw. Use a crosscut blade recommended for plywood, veneers, and laminates.

Fig. 108-11. Making a straight cut on a plastic sheet with a saber saw. A straight-edge board clamped to the plastic makes a good guide.

Fig. 108-8. Sawing acrylic sheet on a scroll saw.

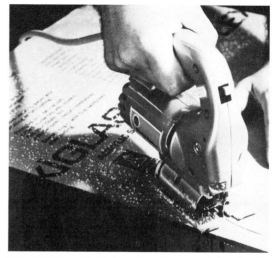

Fig. 108-12. Making a free-form cut on plastic sheet with a saber saw. The saw blade should have at least 14 teeth per 1 in (25.4 mm).

Fig. 108-9. Sawing plastic sheet on a bandsaw.

Fig. 108-10. Cutting a pattern freeform from a plastic sheet on a bandsaw.

Fig. 108-13. Cutting acrylic plastic on the metal-cutting lathe.

Dressing Plastic Edges

The edges and ends of plastic sheets, rods, and cylinders must be dressed smoothly. Dressing makes close-fitting joints and improves the appearance of the project activity.

Tools

Cut and sawed edges can be smoothed with a block plane, a file, or abrasive paper. Garnet and silicon carbide are good abrasive papers to use.

Dressing Edges

1. Smooth the edges of the piece of flat plastic stock. Use either a block plane (Fig. 109-1) or a flat mill file (Fig. 109-2). Protect the plastic with paper or cloth while it is being held in the vise. Dress the ends of rods and cylinders with a file or with abrasive paper.

2. Dress the edges further with fine No. 120 to No. 220 abrasive or wet-dry paper (Figs. 109-3 and 109-4). If a polish is wanted, buff the piece on a power-driven cloth buffer (Fig. 109-5) or with a buffing attachment on an electric hand drill. Use rouge or tripoli compound for a polishing agent.

3. Test the edges and the ends for straightness with a square (Fig. 109-6).

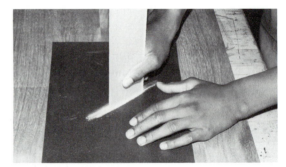

Fig. 109-3.   Dressing an edge on abrasive paper.

Fig. 109-1.   Dressing a plastic edge with a block plane.

Fig. 109-2.   Dressing a plastic edge with a flat mill file.

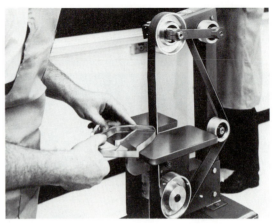

Fig. 109-4.   Dressing a curved plastic edge on a belt sander.

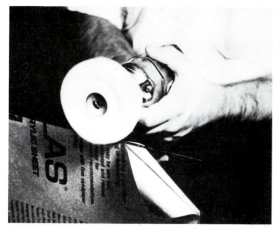

Fig. 109-5. Polishing a plastic edge with a buffer.

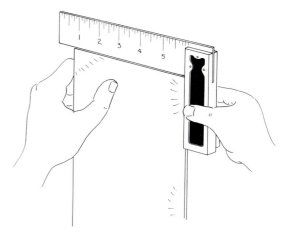

Fig. 109-6. Testing squareness.

Unit 110

Drilling Holes in Plastics

Holes may be drilled in plastic materials with the same drill bits and tools that are used on metal. (See Unit 132.) Specially ground bits can also be used for plastics. A soap-and-water solution makes drilling easier, especially when you are using electric power tools. This solution keeps the plastic from overheating. Holes in very thin sheet-plastic material can often be punched with a hollow punch or a leatherworker's punch.

Drilling a Hole

1. Mark the location for the hole.
2. Use a center punch to make a dent in the plastic for starting the hole (Fig. 110-1).
3. Choose and fasten the correct size drill bit in the chuck of the hand drill, portable electric drill, or the drill press.
4. Drill the hole (Figs. 110-2 through 110-4).

Fig. 110-1. Center punching for drilling a hole.

Fig. 110-2. Drilling a hole with a hand drill.

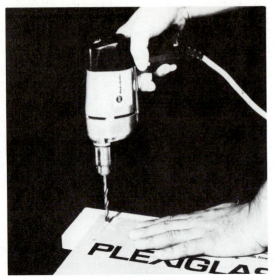

Fig. 110-3. Drilling a hole with a portable electric hand drill.

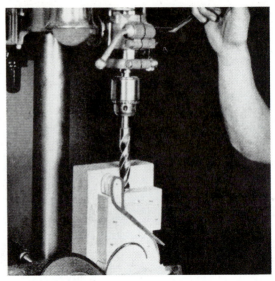

Fig. 110-4. Drilling a hole with the help of a specially made wooden guide.

Unit 111

Tapping and Threading Plastics

Plastic parts may be fastened together with machine screws. It is necessary to drill and tap holes for this type of assembly. Use machine screws with a National Coarse thread. A description of taps and dies, with the procedures for using each, is given in Unit 140, "Cutting Threads." The tap-drill sizes given in the table (Fig. 111-1)

TAP DRILL SIZES MOST USED FOR PLASTICS

| GAGE SIZE OR DIAMETER OF SCREW | NC THREADS PER INCH | TAP DRILL— ANCHOR HOLE | | PILOT HOLE DRILL SIZE |
		APPROX. 75% FULL THREAD	NEAREST FRACTIONAL SIZE	
6	32	35	$7/64$	25
8	32	29	$9/64$	16
1/4	20	7	$13/64$	$17/64$
1/8 pipe	NPT*	$11/32$	$11/32$	$27/64$

*National Taper Pipe series.

Fig. 111-1. Tap drill sizes most used for plastics.

are the ones you will use most often. See a metalworking tap drill chart for other sizes.

Tools

The tools used for tapping are the tap and the tap wrench (Fig. 111-2). Threading is done with the die and the diestock (Fig. 111-3). Also review Unit 140. A soap-and-water solution is a good lubricant to use when tapping and threading. Turn the tap or die once or twice. Then reverse it to break the chip. Continue until the depth of thread is reached.

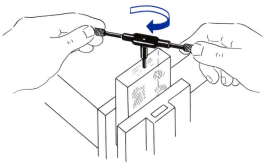

Fig. 111-2.   Tapping the hole in a piece of plastic.

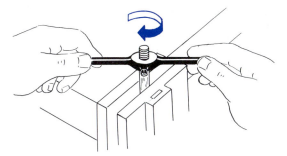

Fig. 112-3.   Threading a plastic rod using a die and diestock.

Unit 112

Surface Decorating

Plastic surfaces may be decorated in several ways. You can *scribe* lines or *cut* shallow saw kerfs. Scribing and sawing should be done before the parts are fastened together.

Tools

The tools for scribing lines are the awl, the dividers, and the straightedge. *Saw-kerf* (cut) lines can be made easily with a backsaw or a circular saw.

Decorating Surfaces

1. Scribe straight lines on the surface of the plastic with an awl. Hold the straightedge firmly with one hand and make the line with the awl (Fig. 112-1). Scribe curved lines with the dividers.

2. Cut a shallow saw kerf with a backsaw (Fig. 112-2). This will produce a wider line

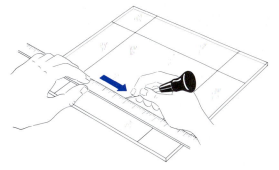

Fig. 112-1.   Scribing a straight line for surface decoration.

than a scribed one. Clamp a straightedge or a board even with the marked line on the piece of plastic. This edge will serve as a guide for sawing.

A saw cut or groove can also be made on the circular saw (Fig. 112-3).

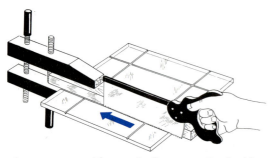

Fig. 112-2. Making a shallow saw kerf with a backsaw for surface decoration.

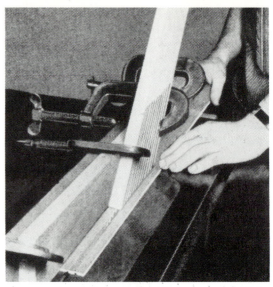

Fig. 112-3. Cutting a decorative groove on a circular saw.

Forming Plastics

Acrylics (Lucite and Plexiglas brands) are easy to heat and form. Dress and polish all edges before the forming or bending operation. Figures 113-1 through 113-3 show several objects that have been heated and free-formed.

Thermoplastic Material

Acrylics are thermoplastic. They can be formed and re-formed. These materials always return to their original sizes and shapes when they are heated unless they are held or put in a form until cool. Thermoplastic material must be heated to temperatures of 220° to 300°F (105° to 150°C) in an oven or over a heating device. This action is referred to as *plastic memory*.

CAUTION: Do not overheat the plastic, or it will blister.

Fig. 113-1. A salad serving set preformed from acrylic-sheet stock.

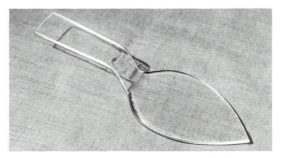

Fig. 113-2. An acrylic plastic server.

Fig. 113-3. An acrylic serving tray.

Tools

The tools needed for forming acrylic plastics are a heating device and forms built to make the desired shapes. Heating may be done with an oven, an electric hot plate, a reflector spot lamp, or a strip heater element (Fig. 113-4). This can be bought or made.

Forming Acrylics

1. Make a wooden form in the desired shape and sand it smooth.

2. Heat the oven to between 200° and 300°F (93° and 149°C). If there is no temperature control, use an oven thermometer to get the desired temperature.

3. Remove masking tape from the plastic before heating it.

4. Put a piece of asbestos sheet on the oven shelf to support the softened plastic.

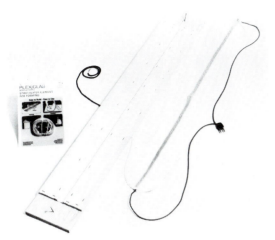

Fig. 113-4. A strip heater element for forming plastic.

Place the acrylic plastic in the oven after the proper temperature has been reached. Let it stay in the oven for several minutes until it is *pliable* (softened).

5. Take the soft plastic from the oven being careful of the hot surfaces.

6. Form the soft plastic by twisting to any shape (Figs. 113-5 and 113-6). Cool the plastic with water. Plastic may be shaped in a wooden form (Figs 113-7 through 113-9). It may also be shaped freehand (Figs. 113-10 through 113-14).

7. To make angular bends, heat the plastic over an electric element or under a heat lamp until it becomes soft (Fig. 113-14).

Fig. 113-5. Removing a soft acrylic plastic piece from an oven.

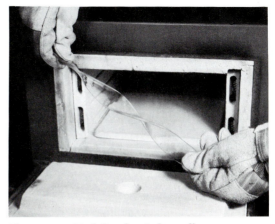

Fig. 113-6. Twisting heated acrylic.

Fig. 113-7. Shaping heated plastic in a wooden form.

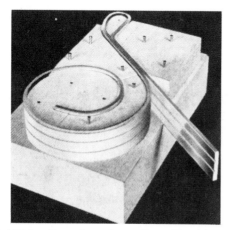

Fig. 113-8. Forming a heated acrylic piece in a special wooden form.

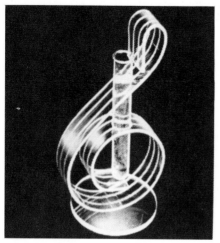

Fig. 113-9. A plastic music-motif flower holder. This was formed in Fig. 113-8.

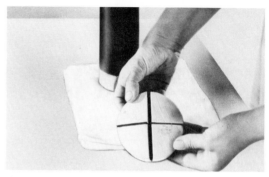

Fig. 113-10. A plywood split disk makes a simple mold for freehand forming. The disk parts are held together with masking tape.

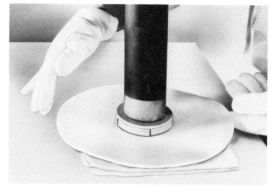

Fig. 113-11. A heated plastic disk is placed on cloth or flannel to protect the bench top. A wooden disk is placed in the center with a heavy weight on it.

Fig. 113-12. Heated plastic is lifted in four sections and squeezed in from the base to make a bottom for the bowl. As the plastic begins to cool and harden, the curves are formed. The split plywood disk can be removed easily. Wear gloves for this process.

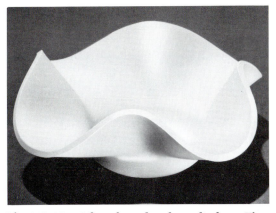

Fig. 113-13. A free-form bowl results from Figs. 113-11 and 113-12.

Fig. 113-14. Heating a plastic piece over an electric element for forming.

Unit 114

Carving Plastics

Transparent (clear) acrylic plastics, such as Lucite and Plexiglas, can be carved to produce interesting effects. A flower or other design can be carved on the inside of a block or sheet of this material. *Internal* (inside) carving gives an appearance of depth, or a three-dimensional quality, to plastic. The plastic is hollowed in a design, colored, and then filled with plaster of paris (Figs. 114-1 through 114-3).

Tools

You will need an electrically driven high-speed rotary cutting tool (Fig. 114-4). A

Fig. 114-1. A plastic box with an internally carved lid, and a carved paper weight.

Fig. 114-2. A plastic bookend with an internally carved decoration.

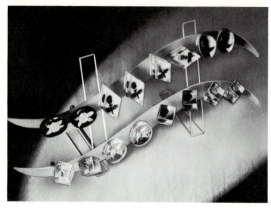

Fig. 114-3. An assortment of internally carved earrings.

Fig. 114-5. Drills and burrs for internal carving.

Fig. 114-4. An electric high-speed cutting tool.

Fig. 114-6. Three of the more common shapes of drills and burrs for carving.

Fig. 114-7. Holding a block while carving internally from the bottom side.

motor-driven flexible-shaft machine can also be used. The cutting is done with long, tapered, internal-carving drill bits (Fig. 114-5). Three of the more common shapes of drills and burrs are shown in Fig. 114-6.

Internal Carving

1. Select the design to be carved.

2. Choose the right tapered internal-carving drill. Fasten it in the chuck of the electric drill.

3. Turn on the motor and make some practice cuts in scrap pieces of clear plastic. Hold a thick piece of plastic in one hand and the electric drill in the other (Fig. 114-7). You may also drill down from the top, as shown in Fig. 114-8. For internal carving, follow the steps outlined in Fig. 114-9 to form a flower pattern.

4. Tint the carved decoration with an aniline or special acrylic dye. Use a medicine dropper (Fig. 114-10). Be sure that all loose chips have been shaken out of the carved space. Allow the dye to remain about 1 minute and then pour it out. Follow the manufacturer's instructions given with the plastic dye you use.

5. Fill the carved area with wet plaster of paris (Fig. 114-11).

Fig. 114-8. Internal carving from the top.

Fig. 114-10. Applying dye to the carved portion with a medicine dropper.

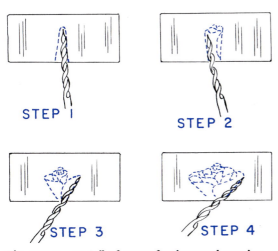

STEP 1

STEP 2

STEP 3

STEP 4

Fig. 114-9. Detail of steps for internal carving.

Fig. 114-11. Filling and packing the colored carved area with wet plaster of paris.

Unit 115

Fastening Plastics

Bonding (cementing) is a quick and easy way to fasten plastic parts. Cementing works especially well with acrylic plastics (Lucite and Plexiglas). Bonding agents include ethylene dichloride and methylene dichloride. Both form solid, firm joints.

Plastics companies also recommend bonding and cementing solutions (Fig. 115-1). Using a colored liquid produces special effects.

Machine screws are used if the project activity is to be taken apart for some rea-

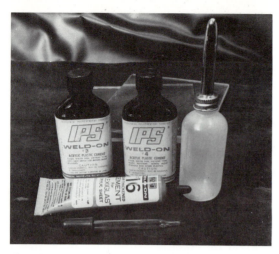

Fig. 115-1. Commercial joint-fastening cements with a needle squeeze bottle and a medicine dropper.

Fig. 115-2. A needle squeeze bottle.

Fig. 115-3. Flowing cementing liquid into an acrylic plastic joint with a brush.

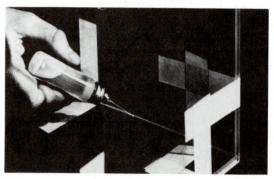

Fig. 115-4. Applying liquid cement to a joint with a needle squeeze bottle. Masking tape, as shown, helps hold the parts together.

son. *Self-tapping* screws are useful for putting together plastic parts. It is not necessary to tap holes for the self-tapping type. There are *drive* screws that fasten metal findings (pins and clasps) firmly to plastics.

Tools

The tools needed for fastening depend on the method used. You will need a needle squeeze bottle (Fig. 115-2), an eye dropper (Fig. 115-1), or a small brush (Fig. 115-3) for cementing joints. If the soak-cement method is followed, you will need a container made of metal, glass, or ceramic. It should be larger than the plastic edge or surface

Cementing and Bonding a Joint

Three of the easiest methods of bonding acrylic parts are (1) applying the liquid to the joint, (2) applying a cement bead to the edge, and (3) soak-cementing.

Applying Liquid

1. File or sand the plastic edges to be cemented. This will make a clean surface for a firm joint.
2. Place the plastic parts in the correct position (Fig. 115-3). Masking tape will help position the parts. See Fig. 115-4.

3. Apply cementing liquid with a small pointed brush (Fig. 115-3) or use a needle squeeze bottle (Fig. 115-4).

4. Allow the joint to *set* (become firm) for at least 5 minutes. Continue with the other joints of the project, following the same procedure as for the first.

Applying a Cement Bead

1. Follow steps 1 and 2 for applying liquid.

2. Attach masking tape to the back of the vertical piece and also to the top of the horizontal piece. See Fig. 115-5. This will form a hinge.

3. Tip back the vertical hinged piece. Squeeze the thickened cement into the length of the joint under the vertical piece (Fig. 115-5).

Another procedure for applying thickened cement is to put it along the edge of the vertical piece before it is positioned on the base (Fig. 115-6).

Soak-cement Method

1. Place the joint or edge to be bonded in a tray. Rest the joint edge on brads. Pour just enough ethylene dichloride into the tray to touch the edge of the piece (Fig. 115-7). Allow the edge to rest in the solution about $2^1/_2$ minutes.

Fig. 115-5. Applying a thickened clear-cement along a joint.

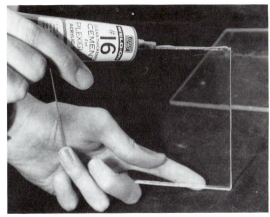

Fig. 115-6. Applying a thickened clear-cement bead along a plastic edge.

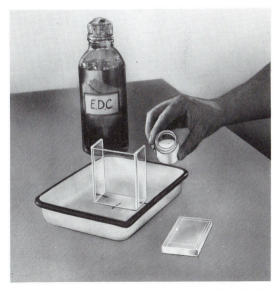

Fig. 115-7. Preparing an edge of acrylic plastic for a soak-cement joint.

2. Place the softened plastic edge in the proper position to make the joint (Fig. 115-8). Put a weight on the piece to force bubbles out of the cemented joint (Fig. 115-9). Let it set for about 10 minutes. You can then handle it with care.

3. If you are building a box, soak the other joints and follow the procedure shown in Figs. 115-10 through 115-13. Figure 115-13 shows an assembled box.

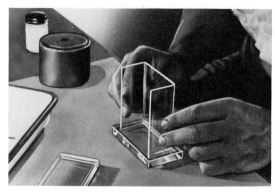

Fig. 115-8. Arranging the softened plastic edge to make the joint.

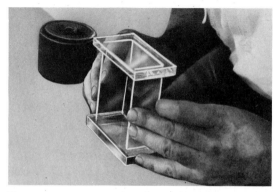

Fig. 115-11. Arranging the softened edge into position.

Fig. 115-9. Placing a weight on the cemented piece to force bubbles out of the joint.

Fig. 115-12. Placing a weight to make a firm joint.

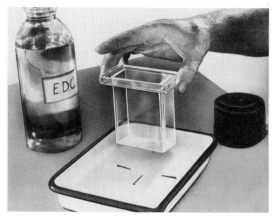

Fig. 115-10. Preparing the opposite end of the box for cementing.

Fig. 115-13. The completed plastic box.

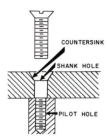

Fig. 115-14. The machine-screw assembly for a plastic joint.

Fastening with Machine Screws

1. Locate, center-punch, and drill holes for fastening with machine screws (Fig. 115-14). When you use a flat-head screw, countersink the hole for the head.
2. Tap the anchor hole with a tap the same size as the machine screw. Finish tapping the bottom of the hole with the bottoming tap.
3. Assemble the plastic pieces in their proper positions and fasten with machine screws.

Fastening with Self-tapping Screws

1. Select the correct screw size. Refer to Fig. 115-15, which shows the diameters of self-tapping screws.
2. Mark, center-punch, and drill the shank and pilot holes.

RECOMMENDED HOLE SIZES FOR SELF-TAPPING SCREWS

SCREW DIAMETER	HOLE DIAMETER, IN	DRILL-SIZE NUMBER
4	0.093	42
6	0.120	31
8	0.144	27
10	0.169	18

Fig. 115-15. Recommended hole sizes for self-tapping screws.

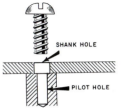

Fig. 115-16. The assembly for a self-tapping screw.

3. Place the plastic parts in their proper positions. Drive in the self-tapping screw with a screwdriver (Fig. 115-16).

Fastening with Drive Screws

1. Select the proper size drill for the screw. See Fig. 115-17.
2. Drill the hole deep enough to allow the screw to go to its full depth.
3. Place the metal *finding* (part) in position. Drive the screw into the hole with a light-weight hammer (Fig. 115-18).

RECOMMENDED DRILL SIZES FOR DRIVE SCREWS

SCREW DIAMETER	HOLE DIAMETER, IN	DRILL-SIZE NUMBER
00	0.052	55
0	0.067	51
2	0.086	44
4	0.104	37

Fig. 115-17. Recommended drill sizes for drive screws.

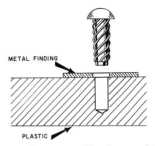

Fig. 115-18. The assembly for a drive-screw fastening.

Coloring Plastics

Clear acrylic plastic is easy to color. The tints of the colors can be controlled from light to dark tones. Colors will make your project activity more attractive.

Types of Dyes

Dyes come in many tints and colors. Multicolored effects can also be obtained. Colors can be bought in liquid or powder form. Some must be heated. Others can be put on right from the container or as they are mixed. Cold dyes come mostly as liquids. The manufacturers furnish directions for the proper use of their products. Follow their instructions or those given here.

The instructions given in this unit are for coloring with cold liquid dye. The only tool that is needed is a porcelain or earthenware vat.

Cold Method of Coloring

1. Pour the coloring solution into a porcelain or earthenware vat. The container should be large enough so that the entire project activity can be colored at once.
2. Test the tint. Dip a scrap piece of clear plastic into the solution. Note the number of minutes needed to get the desired tint.

3. Clean the project activity with a clean cloth to remove dirt and grease.
4. Dip the plastic project activity in the cold dye solution. Leave it there until you get the tint you want (Fig. 116-1).
5. Remove the project activity from the liquid and rinse it with tap water. This will stop the coloring action.
6. Allow the project activity to dry a few minutes. Blot any extra moisture with a soft, dry cloth. This will prevent water stains.
7. Put on a coat of hard-finish paste wax. Polish the project activity with a soft cloth.

Fig. 116-1.   Coloring clear acrylic plastic in the dye solution.

Fiber Glass Activity

Your activities in plastics thus far have probably been limited to acrylic material and how to work and experiment with it. It is one of the basic types of plastics. This and the next six units will be concerned with fiber glass material. Fiber glass is classified as a reinforced type of polyester plastic.

Fiber glass is thought to have been discovered thousands of years ago. It was probably first noticed when a sailor poked under a fire on the beach. When the sailor withdrew the stick he or she probably noticed a thin strand extending from the end of the stick to a molten puddle of sand on the ground. This thin strand was the first fiber glass.

How Fiber Glass Products Are Made

Fibers of glass are made by forcing molten glass through tiny holes in a die. The small glass strands are then woven into threads, and the threads are made into cloth. The cloth fabric is shaped and then coated with polyester (plastic) resin to form a fiber glass and plastic laminate. Fiber glass laminates can now be made into many shapes to form useful and decorative products.

Products Made of Fiber Glass

The automobile in Fig. 117-1 gives an example of how fiber-reinforced laminates

Fig. 117-1. Shaded areas mark present and future uses of molded fiber-reinforced plastics in automobiles. A variety of polyester fibers and resins make tailor-made materials for specific functions. (Contours Unlimited)

are used. Laminates can also be formed into fishing rods, serving trays, bowls, automobile bodies, and many other products. Objects made of fiber glass are very light in weight, are good insulators, and are very strong. They do not rust or corrode, and they are highly resistant to shock.

Unit 118

Materials Needed for Fiber Glass (Polyester) Activity

Polyester plastics are thermosetting materials because they harden into permanent shapes. Plastic resin is mixed with a catalyst that produces the hardening, or curing, process when applied to a shape.

Forms of Fiber Glass

Fiber glass is made into two basic forms—*cloth* and *mat* (Fig. 118-1). The *cloth* form is made by weaving many fiber glass threads together. Fiber glass *mat* results when the threads are chopped and then fall freely onto a flat surface. The thick layer of threads is then *bonded* (held) together to form a mat.

Fiber glass mat is better for some project activities because it can be pulled apart

easily. This allows the material to form into shapes that are not possible with the cloth.

Fiber glass cloth is stronger and may be better for objects that must have strength. It is also easier to handle

Fig. 118-1. Fiber glass cloth and mat.

Plastic Resin and Catalyst

Plastic resin is used to bind the glass fiber together to form fiber glass and plastic laminate (Fig. 118-2). Liquid polyester resin is the most commonly used form. This type of resin will stick to wood, metal, paper, fabric, and many other materials.

A catalyst is needed to harden, or cure, plastic resin. When a small amount is added, a chemical action produces heat, resulting in hardened resin.

Coloring, Cleaning Materials, and Waxes

Oil colors are often used to add color to plastic resin (Fig. 118-3). The resin is clear, and the added color often makes an object more attractive.

Acetone or lacquer thinner is also needed when working with fiber glass. Brushes used for applying plastic resin should be thoroughly cleaned with either of these products.

If a fiber glass project activity is made by using a mold, a good paste wax is needed. The mold must be waxed completely before the fiber glass and plastic resin are applied. The wax allows the object to be removed easily.

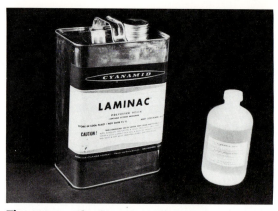

Fig. 118-2.   Plastic resin and catalyst.

Fig. 118-3.   Wax, lacquer thinner, oil colors in tubes, and oil color in a can.

Unit 119

Molds for Fiber Glass

One-piece or two-piece molds are required to hold fiber glass to the shape of the object while plastic resin is applied. These types are basic in using laminate, which is fiber glass in the form of either cloth or mat.

One-piece Molds

A one-piece mold is the easiest to make. An open one-piece mold (Fig. 119-1) is

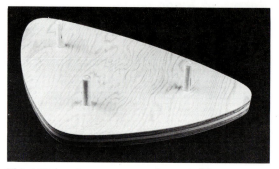

Fig. 119-1.   An open one-piece mold.

used to make the draped tray. The laminate touches this mold only on the dowel pins and around the edges of the base. This mold will not give a finish as smooth as the solid one-piece mold will produce (Fig. 119-2), but it allows for free-form creativity in project activities.

A solid one-piece mold will produce one smooth surface on the object where the laminate is in contact with the mold. The smoothness of the laminate surface depends on the smoothness of the mold. It must be made with great care and should be sanded smooth, finished well, and waxed before use.

Two-piece Molds

The two-piece mold (Fig. 119-3) is the hardest to construct, but it makes the project smooth on both surfaces. It is difficult to construct because an *allowance* (extra amount) is needed for the thickness of the fiber glass and the plastic resin used. Both surfaces of the mold must be smooth and waxed before use. This will prevent the fiber glass from sticking.

Fig. 119-2. Two solid one-piece molds.

Fig. 119-3. A two-piece mold.

Unit 120

Making an Open One-piece Mold for a Draped Tray

Unit 119, "Molds for Fiber Glass," described two basic molds, or forms, used for shaping fiber glass (polyester) objects. The procedure outlined here will help in building an open one-piece mold on which you can fabricate an interesting tray.

The Mold Base

1. Sketch the full size of the shape of your tray on paper (Fig. 120-1).
2. Cut out the shape to use as a pattern.
3. Transfer the pattern to a piece of plywood ¾ inch (19 millimeters) thick (Fig. 120-2).

4. Cut out the base shape of the mold on a jigsaw or band saw.
5. Sand the edges and surfaces of the mold base smooth.

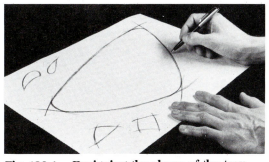

Fig. 120-1. Designing the shape of the tray.

Fig. 120-2. Drawing around the pattern on wood.

The Mold Supports

1. Cut three pieces of ³/₈-inch (9.5-millimeter) dowel rod. They should measure about ¹/₂ inch (13 millimeters) longer than the depth of the tray.

2. Locate and mark three points on the plywood base for tray feet. Do not make them too close to the edge. Keep them far enough from each other so that the tray will stand firm and not tip.

3. Drill or bore ¹/₂-inch (13-millimeter) deep holes in the three marked locations on the plywood base. Use a ³/₈-inch (9.5-millimeter) drill or bit.

4. Sand one end of each dowel round. This way, the dowels will not push through the patterned fabric and fiber glass.

5. Tap the three dowels into the holes with a wooden mallet (Fig. 120-3).

6. Finish the mold with shellac, lacquer, or Deft.

7. Wax the mold with a good-quality paste wax. Figure 120-4 shows a mold, or form, for making a draped tray.

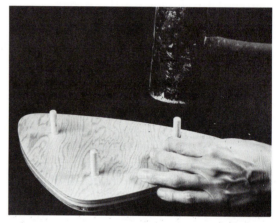

Fig. 120-3. Tapping the dowels into the wood base.

Fig. 120-4. The completed open mold.

Unit 121

Attaching the Material to the One-piece Mold

The following three units show how to make a simple tray by using the basic one-piece mold described in Unit 120, "Making an Open One-piece Mold for a Draped Tray." The general procedure for making a fiber glass (polyester) object with a solid or a two-piece mold would be somewhat the same. The activity in this unit is a free-form experiment. It lets you freely create what you want to make. After some experience, you will find that you can make *variations* (changes) for each process.

1. Select a patterned fabric. Any cotton print material will work.

2. Cut the fabric so that it will drape over the mold with a surplus of about 2 or 3 inches (50 or 75 millimeters), as shown in Fig. 121-1.

3. Staple the patterned fabric to one edge of the mold (Fig. 121-2).

4. Stretch the patterned fabric over the dowel rods. Put one staple in the opposite edge. Now gently stretch the fabric to remove all wrinkles.

5. Completely staple the fabric on all edges in this position. The fabric-covered mold should be wrinkle-free when completed (Fig. 121-3).

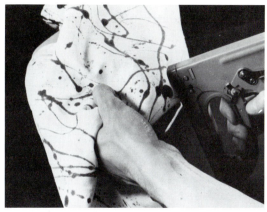

Fig. 121-2.   Stapling the pattern fabric to the mold.

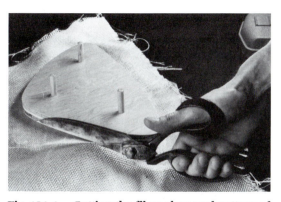

Fig. 121-1.   Cutting the fiber glass and patterned fabric.

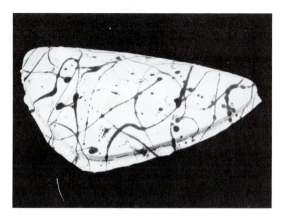

Fig. 121-3.   The mold covered with the fabric.

Unit 122

Mixing and Applying Plastic Resin

When mixing the ingredients for fiber glass, make certain that you have read *thoroughly* and followed the instructions given on the containers. These may change a little with different manufacturers. You must use exact amounts of both the resin and the catalyst.

1. Pour 2 ounces (56.8 milliliters) of plastic resin into a container (Fig. 122-1).

2. Add colors to the resin at the time, if desired. Do not add too much coloring agent. If you do, it will take the resin longer to *cure* (harden). Add just enough color to make the resin *translucent* (allowing light to pass through).

Adding Catalyst to the Resin

3. Using an eyedropper or a container with a small opening, add from 8 to 16

Fig. 122-1. Pouring the resin.

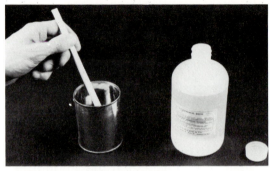

Fig. 122-3. Stirring the resin with the catalyst added.

Fig. 122-2. Adding the catalyst.

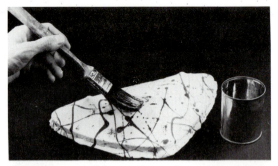

Fig. 122-4. Applying resin to the pattern fabric.

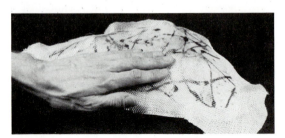

Fig. 122-5. Laying fiber glass cloth on top of the pattern fabric.

drops of catalyst to the 2 ounces (59 milliliters) of resin (Fig. 122-2). *Read the directions* on the catalyst container for the exact amount of catalyst to add. If too much is added, the resin will cure before it is applied to the project.

4. Stir the mixture right after you add the catalyst (Fig. 122-3). Make sure the contents are well mixed.

Applying the Resin

5. Apply the resin to the patterned fabric with a soft brush (Fig. 122-4). Paint an even coating of the resin on all the exposed fabric.

6. Lay the fiber glass cloth on top of the patterned fabric. Press it down gently so that it sticks to the resin or the fabric (Fig. 122-5).

7. Brush on an even coat of resin onto the fiber glass (Fig. 122-6).

8. As you apply the resin, check the *laminate* (layers) for air bubbles. These form where there are gaps between the patterned fabric and fiber glass cloth. Remove them by pressing on the bubble. This will force the air out through the fiber glass.

9. Let it cure (Fig. 122-7).

10. Clean the brush right away with acetone or lacquer thinner. If the resin hardens on it, it cannot be removed.

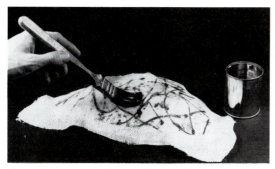

Fig. 122-6. Applying resin to the fiber glass cloth.

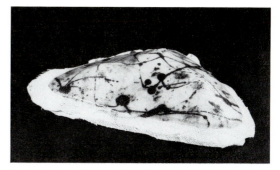

Fig. 122-7. The mold with pattern fabric, fiber glass cloth, and resin applied.

Trimming and Finishing the Tray

Basically, trimming and finishing depend on the use of proper grits of abrasives. The steps given here serve as a guide.

1. Carefully remove the staples that hold the laminate to the mold (Fig. 123-1). Use pliers and a screwdriver.
2. Remove the tray from the mold. Make sure the tray does not stick to the mold anywhere.

Trimming the Edges

3. Trim off the *excess* (extra) fiber glass and patterned fabric with a pair of tin snips (Fig. 123-2). If the laminate is too thick for tin snips, use a backsaw or other fine-toothed saw.
4. File the edges with a cabinet file or a smooth mill file.
5. Sand the edges, first with fine-grit abrasive paper (120), next with very fine (220), and finally with extra-fine (360) (Fig. 123-3).
6. Finish sanding the entire tray with fine wet or dry abrasive paper. Use water when sanding.

Finishing with Plastic Resin

1. Mix enough plastic resin to cover the entire tray.

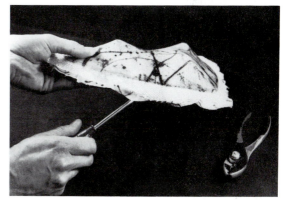

Fig. 123-1. Removing the tray from the mold.

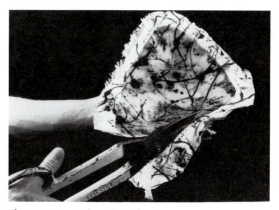

Fig. 123-2. Trimming the tray.

Fig. 123-3.   Sanding the tray edges.

Fig. 123-4.   Applying the final cost of resin.

2. Apply the clear plastic resin with a soft brush. Put it on the bottom first.

3. Turn over the tray. Apply a coat to the inside (Fig. 123-4). Make certain you cover the edges of the tray also.

4. Allow the resin-coated tray to cure (Fig. 123-5). This is the basic procedure used to finish all fiber glass objects.

Fig. 123-5.   The completed tray.

Unit 124

Discussion Topics on Plastics Technology

1. Describe a way to color acrylics.

2. What type of plastic is fiber glass?

3. Describe the qualifications and duties of a plastics career that interests you.

4. Name the three general categories of plastics industries. Give their functions.

5. List 10 industrial methods of processing plastics. Briefly describe each.

6. Name 12 basic types of plastics and give an example of each.

7. What are the two general classifications of plastics? How do they differ?

8. What is meant by illumination in plastics? List three applications where illumination in plastics would be desirable.

9. What is the advantage in using the soak-cement process when fastening acrylic plastic parts?

10. Describe the differences in the effects produced by scribing and sawing lines.

11. Why is a cooling agent needed when you cut plastic material with a scroll saw?

12. Why is it necessary to dress the edges of plastic after cutting it with a saw?

13. Can an auger bit be used satisfactorily for boring a hole in plastic stock? Explain your answer.

14. What lubricant is suitable when plastic material is tapped and threaded? How does a cooling agent affect the cutting action?

15. How is acrylic plastic material heated for forming?

16. Name three methods of fastening plastic pieces.

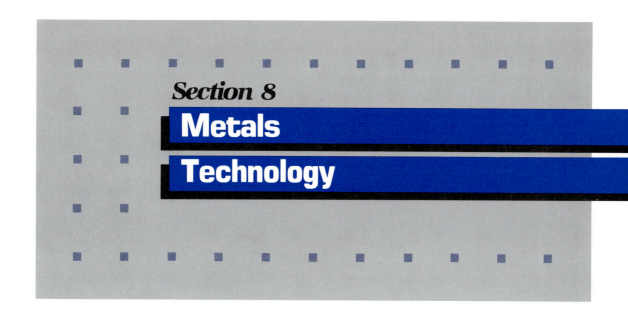

Section 8

Metals

Technology

Unit 125

Introduction to Metals Technology

We would not be able to live as well as we do without the metalworking industries. Well over 8 million people work in some way with metals and metal products. You too may someday earn a living as an engineer, as a skilled metalworker, or as a person who sells metal products (Fig. 125-1).

Products made of metal, such as kitchenware, bicycles, appliances, automobiles, and farm equipment, are used every day. You could not get along without planes, trains, bridges, and many other metal items. There are over 20 tons (18 000 kilograms) of metal in use for every person in the United States. Most of the products used each day have some metal parts. And even if they do not, they were most likely made by machines with metal parts.

There are many ways that metal can be shaped or formed. The four most impor-

Fig. 125-1.   This "man of steel" sculpture is a good example of the use of stainless steel as an art form. (Inland Steel Company)

tant ways of shaping metal are the following (Fig. 125-2 a):

1. *Cold forming* is changing the shape and form of metal when it is cold (at room temperature). Some of the most commonly used cold-forming processes are cutting, bending, extruding, stamping, pressing, and drawing metals. Stamping, pressing, and drawing are done by placing sheet metal in a *die* (a device used for cutting, stamping, or forming metal). The metal is then pressed into shape. (See Fig. 125-2 b.)

2. *Hot forming* is rolling, hammering, or squeezing heated metal into a shape. Forging is a hot-forming process used to shape metal parts that must be able to take great stress. Axles and crankshafts for engines are two such metal parts. (See Fig. 125-2 c.)

Fig. 125-2 b.   **Cold forming.**

COLD FORMING

DRAWING

HOT FORMING

FORGING

CASTING

POURING

MACHINING

LATHE

Fig. 125-2 a.   **Four methods of shaping metal.**

Fig. 125-2 c.   **Hot forming.**

3. *Casting* is forming metal by pouring *molten* (liquid) metal into a *cavity* (a hole or hollow) in sand, ceramics, or metal. Castings are made in *foundries*. Most machine parts are made this way (Fig. 125-2 d).

4. *Machining* is shaping metal to a certain size (Fig. 125-2 e). The most common machining processes are drilling, grinding, turning (on lathes), milling, and shaping (planing).

As you work with metalworking machines, try to find out how their parts were made. What parts were made by forging? What parts were machined? Were the sheet-metal parts made by stamping?

In metalworking, you will learn about many of these ways of making metal products. You will learn to use hand tools like the ones used by metalworkers. Some of your work will be like that done by sheet-metal workers. Or it may be like that done by artisans who make and repair art metal and jewelry (Fig. 125-3). If there is enough time, you may make a casting in the foundry. You may forge a piece of metal into shape by heating and forming it to the kind of object you want. You may also machine a part on the lathe or put together parts by welding. If you take advanced metalworking, you will study many areas more carefully. In a machine-shop class, you will learn to work the drill press, the lathe, the metal shaper, the milling machine, and the grinder. You will do both oxyacetylene and arc welding, which are ways of joining metals by heating and melting the edges.

Products you build in metalwork follow the same basic steps industry uses in manufacturing any commercial product (Fig. 125-4).

Fig. 125-2 d.   Casting.

Fig. 125-2 e.   Machining.

Fig. 125-3.   Artistic metal objects beautifully designed and well made are treasured possessions. (Knob Creak)

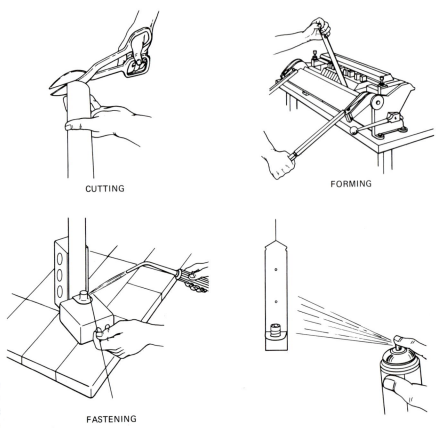

CUTTING

FORMING

FASTENING

FINISHING

Fig. 125-4. The four basic steps in producing any metal product: cutting, forming, fastening, and finishing.

Unit 126

Metrics in Metalworking

All the largest manufacturing industries in the United States are changing over to the metric system. Most of these companies have plants around the world. Since all other major industrial countries use the metric system, it is easier for any company in the United States to use this one measuring system in all of its plants. Also, products made to metric measurement can be sold and used in other metric countries. It is important to learn about the metric system as you study metalworking. Figure 126-1 shows what metric changes are needed in metalwork.

Measuring Tools

All measuring tools, such as rules, micrometers, and other precision measuring instruments, come in metric units. The rules most often used are the 300-millimeter rule (a little shorter than the 12-inch rule) and the 150-millimeter rule (a little shorter than the 6-inch rule). For most metalwork, rules should be divided into full millimeters (about 1/25 inch). The numbered lines are marked in millimeters, such as 10, 20, 30, 40, and so on. Do not use a metal rule marked in centimeters. It is

METRICS IN METALWORK

Units	millimeter (mm) (about 1/25 inch)	For all dimensions on drawings
		For all thickness and width of materials
		For tool sizes
	meter (m) (about 10% longer than a yard)	For lengths of materials
	degree Celsius (°C) (about half of the Fahrenheit readings for high heats, such as heat-treating, 600°F is about 300°C)	For temperature
	kilogram (kg) (about 2.2 pounds equals 1 kilogram)	For mass (weight)
Hand Tools	Measuring tools including rules, micrometers, vernier calipers in millimeters and hundredths of a millimeter.	
Machine Changes	Change gears for cutting metric threads. Add dual-reading dials to machines.	
Fasteners	Use ISO metric fasteners, such as M6 × 1 instead of 1/4-20 NC.	
Drills	Metric drills are available. However, customary drills are close enough in size for most work.	
Taps and Dies	Set of metric taps and dies for cutting metric threads.	
Material Sizes	All thickness, widths, and diameters in millimeters. No gages will be used. All lengths in meters.	

Fig. 126-1. Examples of the change to metric in metal work.

more difficult to read. For more *precise* (exact) work, a rule marked in half millimeters can also be used (Fig. 126-2).

The best way to read a standard metric micrometer is as follows: The pitch of the spindle screw is one-half millimeter (0.5 mm), so one *revolution* (turn) of the thimble advances the spindle toward or away from the anvil the same 0.5-millimeter distance (Fig. 126-3 a). The reading line on the sleeve is marked in millimeters (1 mm). Every fifth millimeter, from 0 to 25, is num-

bered. Each millimeter is also divided in half (0.5 mm). It requires two revolutions of the thimble to advance the spindle a full millimeter (1.0 mm).

The beveled edge of the thimble is marked in 50 divisions. Every fifth line, from 0 to 50, is numbered. Since one revolution of the thimble advances or withdraws the spindle 0.5 millimeter, each thimble graduation equals $1/50$ of 0.5 millimeter, or 0.01 millimeter. Thus, two thimble graduations equal 0.02 millimeter,

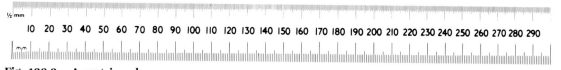

Fig. 126-2. A metric rule.

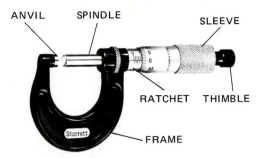

Fig. 126-3 a. A metric micrometer with parts named. (L. S. Starrett Co.)

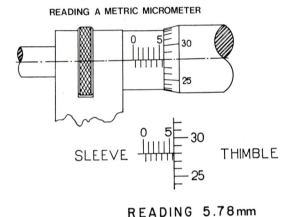

READING A METRIC MICROMETER

SLEEVE   THIMBLE

READING 5.78mm

Fig. 126-3 b. Reading a metric micrometer.

three graduations equal 0.03 millimeter, and so on. You can read the micrometer. Just add the number of millimeters and half millimeters you see on the sleeve to the number of hundredths of a millimeter shown by the thimble graduation that lines up exactly with the reading line on the sleeve (Fig. 126-3 b).

Machine Changes

All machine tools must be changed so that machining can be done to metric measures. Dual-reading dials can be added to lathes and other machine tools. The settings can be made either in inches (U.S. customary) or millimeters (metric) (Fig. 126-4). To cut threads on a lathe, it is necessary to have different gears in the gear box. Once a standard lathe is converted to cut metric threads, it cannot cut customary threads. Some lathes now operate in both systems.

Material Sizes

In metric units, material thickness and width are shown in millimeters. The length of stock is shown in meters (Fig. 126-5). All kinds of metals will be described using the same terms. Flat pieces will be described in terms of thickness and width. Round, square, and *hexagonal* (six-sided) pieces will be described in terms of their diameter or distance across flat sides (Fig. 126-6). Sheet stock will have thickness given in millimeters. This will take the place of gage numbers. This will be an important change because now there are about 35 different gages used in metalworking. For example, a 16-gage sheet of copper (nonferrous) is not the same thickness as a 16-gage sheet of steel (ferrous). With metric conversion, all materials will be measured with the same standard.

Threads, Fasteners, and Wrenches

The United States has taken up the 25 most commonly used ISO thread sizes

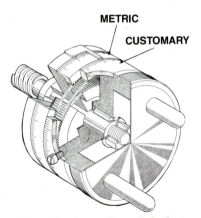

METRIC

CUSTOMARY

Fig. 126-4. A dual reading dial that can be added to lathes, milling machines, or grinders. One dial reads in thousandths of an inch and the other dial in hundredths of a millimeter.

THICKNESS		WIDTH	
in	mm*	in	mm*
1/16	1.6	3/8	10
1/8	3	1/2	12
3/16	5	5/8	16
1/4	6	3/4	20
5/16	8	1	25
3/8	10	1 1/4	30
7/16	11	1 1/2	40
1/2	12	1 3/4	45
9/16	14	2	50
5/8	16		
11/16	18		
3/4	20		
7/8	22		
1	25		
1 1/8	28		
1 1/4	30		
1 5/16	32		
1 3/8	35		
1 1/2	40		
1 3/4	45		
2	50		

*Replacement sizes based upon ANSI B32.3-1974.

Fig. 126-5. Metric material sizes. The term *flat* **will replace such terms as** *band* **and** *bar*.

ROUND, SQUARE, AND HEXAGONAL BAR SIZES

in	mm*
1/8	3
1/4	6
3/8	10
1/2	12
5/8	16
3/4	20
1	25
1 1/4	30
1 1/2	40
1 3/4	45
2	50

*Replacement sizes based upon ANSI B32.4-1974.

Fig. 126-6. Inch and metric sizes used for round, square, and hexagonal bar stock.

shown in Fig. 126-7. Terms for metric threads are different from those for customary threads (Fig. 126-8). For example, M6 × 1 stands for the following: M is the symbol for an ISO thread; 6 is the diameter in millimeters (not quite 1/4 inch); and 1 is the pitch in millimeters. (*Pitch* is the distance from the top of one thread to the top of the next; this one is about 25 threads per inch.)

The pitch for metric threads is found by measuring the distance between the crests of the threads. It is not done by counting the number of threads per inch, as you do for customary fasteners. (See Unit 140, "Cutting Threads."). Metric bolts are made with either hexagonal or 12-spline heads. Fixed metric wrenches have openings in millimeters (equal to the distance across the head of the fasteners).

Metric Drills

Metric drills are available. Many customary drill sizes are close enough so that you can choose one of these instead of a metric size (Fig. 126-9).

Taps and Dies

To cut metric threads, a full set of metric taps and dies is needed. The sizes used most often will be those in the range from 2 millimeters to 24 millimeters (M2 × 0.4 to M24 × 3).

Drawings

To build products to metric measurement, you must have drawings that give dimensions in millimeters. The three kinds of drawings that will be used most often are (1) a *dual-dimensioned* drawing, which shows both the metric and customary dimensions on the drawing itself, (2) a *metric drawing with a customary readout chart* (Fig. 126-10), and (3) an *all-metric dimensioned drawing* (Fig. 126-11).

NOMINAL SIZE	INTERNAL THREAD MINOR DIA		TAP DRILL DIA	NOMINAL SIZE	INTERNAL THREAD MINOR DIA		TAP DRILL DIA
	MAX	MIN			MAX	MIN	
M1.6 × 0.35	1.321	1.221	1.25	M20 × 2.5	17.744	17.294	17.5
M2 × 0.4	1.679	1.567	1.6	M24 × 3	21.252	20.752	21.0
M2.5 × 0.45	2.138	2.013	2.05	M30 × 3.5	26.771	26.211	26.5
M3 × 0.5	2.599	2.459	2.5	M36 × 4	32.270	31.670	32.0
M3.5 × 0.6	3.010	2.850	2.9	M42 × 4.5	37.799	37.129	37.5
M4 × 0.7	3.422	3.242	3.3	M48 × 5	43.297	42.587	43.0
M5 × 0.8	4.334	4.134	4.2	M56 × 5.5	50.796	50.046	50.5
M6 × 1	5.350	4.917	5.0	M64 × 6	58.305	57.505	58.0
M8 × 1.25	6.912	6.647	6.8	M72 × 6	66.305	65.505	66.0
M10 × 1.5	8.676	8.376	8.6	M80 × 6	74.305	73.505	74.0
M12 × 1.75	10.441	10.106	10.2	M90 × 6	84.305	83.505	84.0
M14 × 2	12.210	11.835	12.0	M100 × 6	94.305	93.505	94.0
M16 × 2	14.210	13.825	14.0				

Note: All dimensions are in millimeters.

Fig. 126-7. These 25 sizes will be used for fastening all metric products in the United States. To obtain the correct size tap drill, merely subtract the pitch from the diameter. The M6 × 1 will be used to replace the 1/4-20NC.

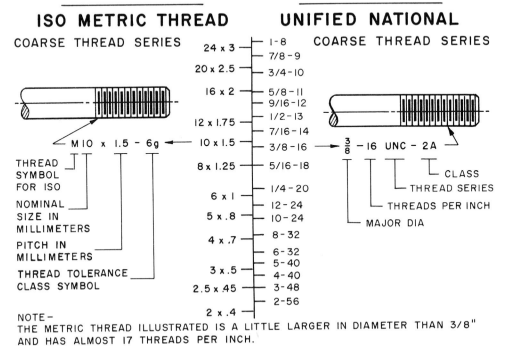

COMPARISON OF COMMON THREAD SIZES

ISO METRIC THREAD

COARSE THREAD SERIES

UNIFIED NATIONAL

COARSE THREAD SERIES

24 x 3 — 1-8
— 7/8-9
20 x 2.5 — 3/4-10
16 x 2 — 5/8-11
— 9/16-12
— 1/2-13
12 x 1.75 — 7/16-14
10 x 1.5 — 3/8-16
8 x 1.25 — 5/16-18
6 x 1 — 1/4-20
— 12-24
5 x .8 — 10-24
4 x .7 — 8-32
— 6-32
— 5-40
3 x .5 — 4-40
2.5 x .45 — 3-48
— 2-56
2 x .4

M 10 x 1.5 - 6g

THREAD SYMBOL FOR ISO

NOMINAL SIZE IN MILLIMETERS

PITCH IN MILLIMETERS

THREAD TOLERANCE CLASS SYMBOL

3/8 - 16 UNC - 2A

CLASS
THREAD SERIES
THREADS PER INCH
MAJOR DIA

NOTE—
THE METRIC THREAD ILLUSTRATED IS A LITTLE LARGER IN DIAMETER THAN 3/8" AND HAS ALMOST 17 THREADS PER INCH.

Fig. 126-8. The comparison of the common ISO threads and the Unified threads.

METRIC	CUSTOMARY	METRIC	CUSTOMARY	METRIC	CUSTOMARY
1.00	60	5.00	9	9.00	(T)
1.20	3/64	5.20	6 or 13/64	9.20	(23/64)
1.40	54	5.40	3	9.40	U
1.60	1/16	5.60	2 or (7/32)	9.60	V
1.80	50	5.80	1	9.80	W
2.00	47	6.00	B or (15/64)	10.00	(X) or (25/64)
2.20	44	6.20	D	10.20	Y
2.40	3/32	6.40	1/4 or E	10.50	Z
2.60	38	6.60	G	10.80	(27/64)
2.80	34 or 35	6.80	H	11.00	(7/16)
3.00	31	7.00	J	11.20	—
3.20	1/8	7.20	9/32	11.50	29/64
3.40	29	7.40	L	11.80	(15/32)
3.60	9/64	7.60	(N)	12.00	—
3.80	25	7.80	(5/16)	12.20	(31/64)
4.00	22	8.00	O	12.50	—
4.20	19	8.20	P	12.80	(1/2)
4.40	17	8.40	Q or (21/64)	13.00	(33/64)
4.60	14	8.60	R	14.00	(9/16)
4.80	12 or (3/16)	8.80	S	15.00	(19/32)

Fig. 126-9. While you can purchase metric drills, customary drills will work for most jobs.

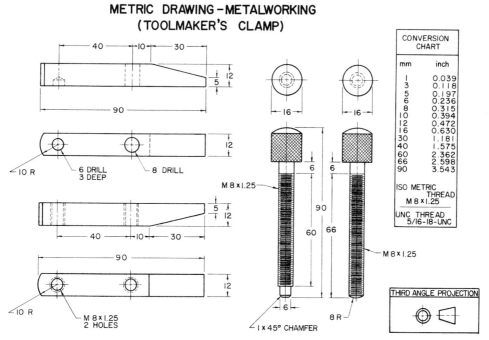

METRIC DRAWING – METALWORKING
(TOOLMAKER'S CLAMP)

Fig. 126-10. A drawing of a toolmaker's clamp with metric measurements and a customary readout or conversion chart.

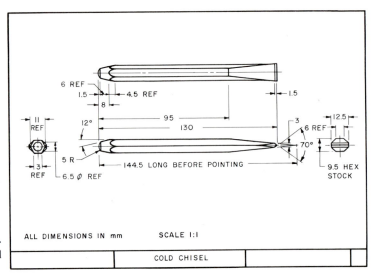

Fig. 126-11. An all-metric drawing of a cold chisel.

6 REF
1.5
8
4.5 REF
1.5
11 REF
12°
95
130
3
6 REF
12.5
70°
3 REF
5 R
6.5 ⌀ REF
144.5 LONG BEFORE POINTING
9.5 HEX STOCK

ALL DIMENSIONS IN mm   SCALE 1:1

COLD CHISEL

Unit 127

Metals Technology-related Careers

Careers and jobs in the metals technology industries are as different (Fig. 127-1) as the many kinds of metals and their *alloys* (mixtures of metals). Technology plays a very important part in finding new metal alloys for uses in new products. Research is a very important part of this huge worldwide industry (Fig. 127-2).

Steel is the backbone of an industrial country. There is hardly a product in use that has not been made from steel or made by steel machines.

There are about 700 000 workers in the iron and steel industry. Jobs range from unskilled to highly skilled and professional. The industry uses blast and steelmaking furnaces. There are also rolling mills that finish steel products into many shapes, such as sheet, strips, bars, and rods.

Steel plants are usually large. About 70 percent of all steel employees work in plants that have more than 2500 workers.

Fig. 127-1. Metalworkers check the design of a car with an all-aluminum body. (Reynolds Metal Company)

Fig. 127-2. **Finding out the right amount and kind of metals that are best for a product takes research and an analysis laboratory.** (Western Electric Company)

Fig. 127-3. **This student is learning how metal products are manufactured.** (Inland Steel Company)

There are over 1000 kinds of jobs related to the iron-and-steel industry. Some workers make iron and steel that is then turned into finished or semifinished products (Fig. 127-3). Others care for machines and equipment.

There are people in *clerical* (office-work), *sales, professional, technical, administrative, supervisory, management,* and *research* positions. Some other names of jobs are *stock-house workers, skip persons, stove tenders, blowers, keepers, cinder persons, melters, charging-machine operators, hot-metal crane operators, ladle crane operators, steel pourers, ingot strippers, heaters, soaking-pit crane operators, blooming-mill rollers, manipulator operators, shear persons, assorters, equipment operators, inspectors,* and all types of *machine operators.* It would be impossible to list all of the jobs in this unit. A full listing is given in the *Dictionary of Occupational Titles* (DOT), published by the U.S. Office of Education, Washington, D.C.

Aluminum Industry

There are about 100 000 workers in the aluminum industry. Aluminum was at one time thought to be an *exotic* (rare) metal with only limited use. Today, however, it is mass-produced in amounts second only to iron and steel.

Aluminum is used in products ranging from appliances and cooking utensils to automobiles, aircraft, and aerospace vehicles (Fig. 127-4). More recent uses are in building siding, structural members, con-

Fig. 127-4. **Aluminum windows are widely used in new and old homes.** (Reynolds Metal Company)

tainers, and electric cables. Aluminum use is growing greatly because of research and development activities.

Some aluminum companies have their own *bauxite* (aluminum-ore) mines. They run huge fleets of vehicles to carry the bauxite from the mine to the mill. At the mill, the ore is refined into *alumina* and then reduced to aluminum. The aluminum is then formed into semifinished and finished products. Other companies buy alumina from outside sources.

Employment in the aluminum industry falls into several groups. There are many jobs that deal with smelting and transforming aluminum into products. The maintenance and service of machinery and equipment account for another important group of jobs. There are also many *clerical, sales, technical, supervisory, professional*, and *administrative* positions.

The largest number of employees in the aluminum industry work in plants that process the metal. Some of the job classifications are *anode persons, potliners, potpersons, tappers, hot-metal crane operators, scale persons, remelt operators, casting operators, scalper operators, soaking-pit operators, rolling-mill operators,* *annealers, stretcher-leveler operators, radiographers, wire-draw operators, extrusion-press operators, electronics mechanics, millwrights, maintenance machinists,* and *maintenance welders*. This is only part of the list, but it gives an idea of job titles. Getting into these jobs is usually done on the unskilled level for *in-service* (on-the-job) training.

Professional, technical, and related operations use the abilities of *chemists, metallurgists, mechanical engineers, electrical* and *industrial engineers, technicians* of all types, and *business-oriented personnel* (Fig. 127-5). These professional and technical occupations most often require a college degree.

Machining Occupations

Machinery works make up the largest single occupational *cluster* (group) in the metal-working trades. There are over 1 million people employed as *machinists, toolmakers* and *diemakers, instrument makers, machine-tool operators, setup persons,* and *layout persons*.

The most common machining job involves operating machine tools (Fig. 127-6).

Fig. 127-5. This computer data center needs business-oriented people to run it. (Exxon Office Systems Co.)

Fig. 127-6. Machinists must know how to operate many kinds of machines and work to close tolerances. (Republic Steel Corporation)

The most common types of machine tools are lathes, grinding machines, drilling and boring machines, milling machines, shapers, broachers, and planers. The working *tolerances* (ranges of variation) in using these machines are very small. The work must be highly accurate, sometimes to one ten-millionth of an inch. Skilled toolmakers and diemakers, instrument makers, machinists, and layout people spend much time doing precision handwork (Fig. 127-7).

The usual way of entering skilled machining occupations is through apprenticeship. This is a period of on-the-job training. There may also be instruction, through which the new worker can learn all about the chosen trade. Most employers prefer apprentices who have at least a high school diploma and maybe some trade or vocational school experience.

The future for workers such as machinists, machine-tool operators, toolmakers and diemakers (Fig. 127-8), instrument makers, and setup and layout persons will change as new uses and ways of working with metal are found. It is not thought that there will be many new jobs in this category. But there is always need for well-prepared and well-educated workers.

Welders and Oxygen-and-arc Cutters

Many of the metal parts used in automobiles, airplanes, spacecraft, household appliances, and thousands of other products are made by welding. Structural metal used in the aerospace industry and in bridges, buildings, storage tanks, and hundreds of other objects is often welded (Fig. 127-9). Broken metal parts are repaired by welds. Somewhat like welding is oxygen and arc cutting. These are often called *flame cutting*. This is a basic way to cut heavy metal.

The skilled welder is able to plan and lay out work from drawings, blueprints, or other *specifications* (plans). A welder must have a working knowledge of the welding properties of any type of metal. There are over 40 different welding processes. The three basic categories are *arc*, *gas*, and *resistance* welding.

Fig. 127-7. Machine operators are skilled workers. (Caterpillar Tractor Co.)

Fig. 127-8. There is a constant need for toolmakers and diemakers. (Inland Steel Corporation)

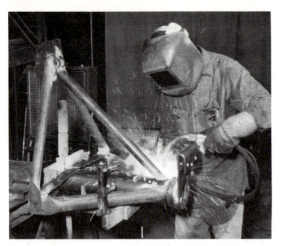

Fig. 127-9. Welding a steel bracket requires a skilled worker.

There are about 500 000 welders and oxygen-and-arc cutters in the United States. About three-fourths work in manufacturing industries.

It usually takes several years of training to become a skilled welder. People going into the welding trade often start with simple hand-welding production jobs. When they get enough experience, they move into the semiskilled classification. With further training, they can pass examinations to become qualified welders. It is suggested that a person attend a trade or vocational school or a technical institute to learn the basics of welding.

The employment outlook for welders is good. Newer welding processes are always being developed. These require better-trained welders and open up thousands of jobs every year. This is particularly true in the construction industry, where the building of welded metal structures continues to grow.

Sheet-metal Workers

Sheet metal is a very important building material (Fig. 127-10). The use of sheet metal of all kinds is growing because air conditioning is now used widely. People are needed to work with these products. Workers are needed to make a great many air ducts that must be cut and installed.

Sheet-metal workers usually work for firms that make equipment from sheet metal. Or workers may have jobs with construction contractors. Apprenticeship is usually the first step in this trade. There is often a very good opportunity for a sheet-metal worker to become a boss or own a small business.

The job outlook for a sheet-metal worker is good because of new kinds of building and new uses for sheet metals.

Jewelers and Watchmakers

Jewelers and watchmakers are skilled craftspersons. The jeweler makes or repairs rings, pins, necklaces, bracelets, and other precious and semiprecious jewelry. Many make original designs and pieces of jewelry from gold, silver (Fig. 127-11), platinum, and precious and semiprecious stones.

The watchmaker cleans, repairs, and adjusts watches, clocks, chronometers, and electrochemical and other types of timepieces. About 18 000 jewelers and 16 000 watchmakers make a living in the United States from these skilled crafts.

Fig. 127-10. These sheet-metal workers are cutting and forming parts from aluminum.

One may either take courses in vocational and trade classes or serve an apprenticeship to an *artisan* (highly skilled person) to enter these trades. *Horology* is the science of making instruments to tell time. Many schools have courses and programs to prepare a person for this craft. Employment in the categories of jeweler and watchmaker promises to be steady. There are hundreds of job openings every year because of the need to replace experienced workers who retire or leave the field for other reasons.

Fig. 127-11. A silversmith designing jewelry.

Unit 128

Metalworking Safety

The metalworking area has many tools and machines that can be dangerous. Before starting to work in the shop, read the following safety rules.

1. *Dress correctly.* Always roll up your sleeves, tuck in or remove loose clothing, and wear an apron or a shop coat. Keep your hair cut short or cover it with a tight cap. Wear goggles or other protective clothing for welding, grinding, and foundry work (Fig. 128-1).

2. *Follow instructions.* The safe way to do a job is to do it the correct way. Watch the way your instructor does it first. Then follow the instructor's advice. Be safe instead of sorry. Never use a power tool without first getting permission.

3. *Do not roughhouse.* Accidents can happen when you are off guard.

4. *Take care of small injuries.* Small cuts, burns, scratches, and splinters are very common in the metal shop. Have them looked at and cared for right away.

5. *Protect your eyes.* In metalwork, there is great danger of injury from flying chips and abrasives. Always wear goggles or eye shields when grinding, buffing, chipping, or pouring hot metal.

There are many other safety rules for each tool and machine. *Follow these rules:*

1. Concentrate on your job.
2. Stay alert.
3. Look things over carefully.
4. Stand clear of the work place.
5. Keep the work area clean.
6. Do not wear loose and flowing clothes.

Fig. 128-1. Working safely with tools and equipment is important. Most metal industries require workers to wear hard hats and safety glasses as required by OSHA standards. (Republic Steel Corporation)

The Story of Steel, Copper, and Aluminum

The metals that are most important to our industrial life and growth are iron and steel, copper, and aluminum. They are also put together with other metals to form *alloys*.

Iron and Steel

Billions of tons of steel are made in this country. Steel is used for everything from railroads to razor blades. The biggest users of steel are motor-vehicle manufacturers, construction companies, railroads, and metal-container makers. Everyone uses steel in some form, but few know how it is made (Fig. 129-1).

The main ingredient in steel is iron. The first step in making steel is the *smelting* of (melting the metal out of) iron ore. Smelting is done in blast furnaces to get iron from ore.

Iron ore is only one of the four raw materials needed to make iron. The other three are fuel (usually coal in the form of coke), a fluxing material (usually limestone), and air. About 2 tons of iron ore, 1 ton of coke, ½ ton of limestone, and 4 tons of air are needed to make 1 ton of *pig* (crude) iron.

The blast furnace that makes this pig iron is a large steel shell. Nearly 100 feet (30 meters) high, it is lined with a kind of brick that can stand a great amount of heat. It is *charged* (loaded) through the top with layers of coke, ore, and limestone. Each blast furnace has three or more stoves to heat air for the hot blast. Air is forced through the highly heated *flues* (chimneys) of these stoves. It is brought to the blast furnace at high temperatures. This heated air enters the furnace near the base. Along with the burning coke, it makes a very hot fire near the bottom of the furnace. This heat melts the ore. The limestone acts as a flux to take out the *impurities* (things not wanted) from the melted ore.

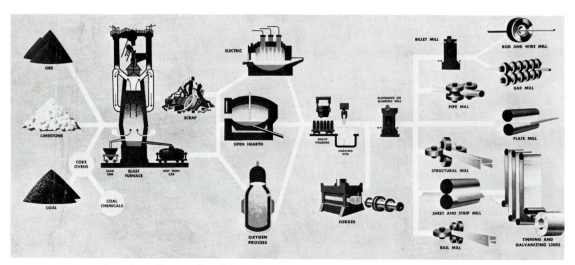

Fig. 129-1.   The flowchart for steel making.

Limestone joins with the impurities of the ore to form *slag*. Since the slag is lighter than the molten iron, it floats on top. This slag is drawn off through the *cinder notch*, leaving the clean iron. Clean iron is then drawn off through the *iron notch* at the bottom. This is called *tapping*. During these operations, the iron picks up carbon.

A blast furnace operates all the time. The raw materials are fed in at the top as quickly as needed to provide enough hot metal for tapping, usually every 4 to 6 hours. From 100 to 125 tons of liquid iron are taken from the furnace at each tapping. The liquid iron flows through troughs into huge ladles set on railroad cars. This product of the blast furnace is usually called *hot metal*. After it is poured into shapes and hardened, it is called *pig iron*.

Steel is made from either hot metal or pig iron. There are three ways of making steel: the basic-oxygen process, the open-hearth process, and the electric-furnace process.

Basic-oxygen Process (BOP). In the basic oxygen furnace, a *lance* (metal tube) is lowered after the furnace is charged. The water-cooled tip of the lance is usually about 6 inches (152 millimeters) above the charge. The lance blows oxygen down from the top. The oxygen mixes wth the carbon and other unwanted elements and starts a high-temperature churning action. This quickly burns out the impurities and changes the metal into high-quality steel. By this method, steel of good quality can be made at a very rapid rate, one charge per hour.

Open-hearth Process. In the open-hearth process, heat is directed over the metal instead of through it. The hearth looks like a large rectangular basin holding 50 to 175 tons of metal. A charge of limestone and scrap is placed in the furnace. The fuel is lit and burns and is then di-rected downward, over the metal. After the scrap metal is nearly melted down, hot metal from the blast furnace is added, and the refinement of the metal continues. Additional hot metal, manganese, or other alloying elements are added to make different kinds of steel.

Electric-furnace Process. Steel is made in an electric furnace the same way as it is made in the open hearth (Fig. 129-2). But it is usually made in smaller amounts, and electricity is used for heating. The kinds of steel made in the open hearth can be made in the electric furnace, but electric furnaces are also used to make special kinds of steel.

Continuous Casting. This is the fastest and most efficient method of producing steel (Fig. 129-3).

Finished Product. After the steel is made, it is cast into molds called *ingots*. The next steps in using steel depend on what is to be made. For example, if the steel is to be rolled into sheets, the ingot molds are allowed to cool enough to stand alone. The molds are then taken off, and the ingots are sent to pits, where they are soaked. Here they are held until they are

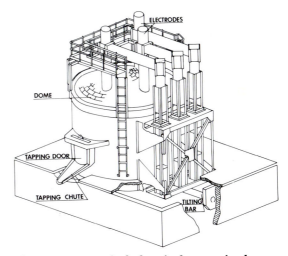

Fig. 129-2. A typical electric furnace is shown in a cutaway diagram.

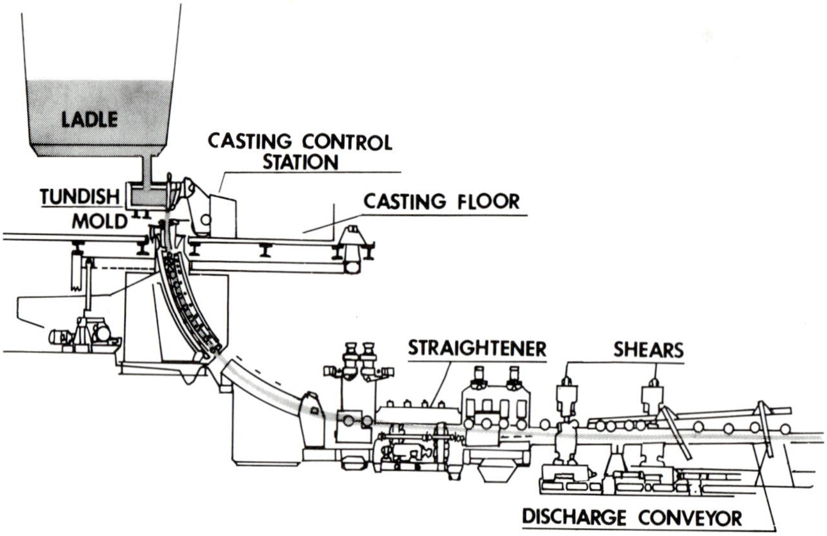

Fig. 129-3. Continuous casting works as follows: Molten steel flows from tundish into copper mold, where light "skin" forms to contain the gradually cooling steel. Further cooling takes place when strand goes through the guide rolls and straightener.

heated throughout. Big rollers called *blooming mills* then roll the steel into long strips about 4 inches (100 millimeters) thick and 3 feet (915 millimeters) wide. The ends are cut off, and the rest is cut into pieces about 7 feet (2130 millimeters) long. These pieces, called *slabs*, are sent to reheating furnaces. Here they are heated to a certain temperature. The slabs then slide down a slope to *roll tables* that carry them through *roll stands.* When a slab passes through these stands, it is stretched to a thin strip of steel about 500 to 600 feet (150 to 180 meters) long.

These strips are rolled into coils. They are then ready for other operations, such as annealing, pickling, and cold rolling. Other shapes of steel are also made in ways like this.

Sheets to be used for roofing, pails, and other products that may rust are cleaned in an acid bath. They then are coated with a thin film of pure zinc to make *galvanized sheet.* This process is known as *hot dipping.* Sheets that are coated with an alloy of tin and lead are called *terneplate.* Sheets to be used for such purposes as cans and kitchen utensils are covered with a thin coat of pure tin and are called *tinplate.*

Copper

Copper is one of the oldest and most useful metals. It has been used since the beginning of history for utensils, tools, and weapons. Today, its biggest use is for conducting electricity.

The copper in the United States is obtained by both underground and open-pit mining. The mined copper has different grades of purity. All grades of copper ore must be purified. High-grade ore is sent to smelters. The copper comes from the smelters 99 percent pure. It has very small amounts of gold, silver, and other substances in it. These can prevent the copper from being a good conductor of electricity. It must therefore go through another process, which is called *refining.* The refined copper is poured into bars or ingots. All forms of copper products are made from these ingots. At this point, zinc may be added to make brass, or tin may be added to make bronze. The copper or alloy is then rolled into sheets and strips or made into tubing, rod, or wire.

Because it is a very good conductor of electricity, copper is very important in the electrical industry. Electricity is carried by

Fig. 129-4. **Copper is widely used for electrical wiring.** (Western Electric Company)

copper wires to give us the comforts and conveniences we need (Fig. 129-4).

Copper is also a very good conductor of heat. Therefore, it is also used in cooking utensils, heating elements, and furnace systems. Copper has many uses in transportation and in military life, too.

Aluminum

Aluminum is widely used today. Aluminum is in the wrappers on candy bars, in the bridges you cross, in toothpaste tubes, in boats, and in jet planes. This metal is often taken for granted. Yet less than 100 years ago, aluminum cost over $500 a pound. It was worth more than gold. Now there is more aluminum produced than any other metal.

Bauxite ore, from which aluminum is made, is found in large amounts in almost every country. The problem has always been to refine the metal from the ore. Most other metals can be refined directly. Aluminum is gotten only after a long, difficult process of purifying and refining.

Fig. 129-5. **Aluminum is a good light material for making canoes.**

Aluminum is in great demand for the following reasons:

1. It is light, only about a third as heavy as iron or steel (Fig. 129-5).
2. It conducts heat quickly and well. It is used in electrical work and for cooking utensils.
3. It does not rust.
4. It reflects heat and light, and therefore is a good insulating material.
5. It is *nonmagnetic* and can be worked easily. Aluminum is not strong enough by itself for some uses. Other materials, such as copper, silicon, manganese, magnesium, and chromium, are added to make many aluminum alloys. These alloys have thousands of uses in industry.

Metals and Alloys

A *metal* is one of the basic elements in nature. An *alloy* is a combination (mixture) of two or more metals. For example, brass is a mixture of copper and zinc. In the shop, we call all of these materials *metals* even though some are actually alloys (Fig. 129-6 a and b).

There are two basic kinds of metals and alloys. Those made from iron are called *ferrous.* Those without iron are called *nonferrous.*

Sheet stock and wire are measured in the customary system by the gage sizes shown in Fig. 129-7. In the metric system,

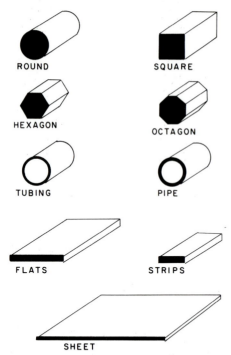

ROUND

SQUARE

HEXAGON

OCTAGON

TUBING

PIPE

FLATS

STRIPS

SHEET

Fig. 129-6 a. Common shapes of most metals.

Fig. 129-6 b. Selecting the correct kind of metal to use for a project activity is very important.

all thicknesses are given in millimeters. All iron and steel sheet and plate stock is measured with the United States Standard (U.S.S.) gage. The Brown & Sharpe, or American, gage is used for all nonferrous metals except zinc. Sheet and wire are usually measured with a metal disk that has gage openings cut around its edge.

Make sure that the burr is filed off the sheet before measuring. Also, be sure that you are using the correct gage for the material. The gages look alike, but you can tell them apart by the names stamped on the sides.

Ferrous Metals

Wrought Iron. This is almost pure iron, with a little slag in it. It is used for ornamental ironwork and for many business purposes. It contains no carbon, and so it is almost rustproof.

Low-carbon Steel. This is sometimes called *mild steel.* It contains 20 to 30 points of carbon (100 points equals 1 percent), or less than one-third of 1 percent. Low-carbon steel is manufactured in the following forms:

1. *Black iron sheet.* The sizes most often used are 26 to 22 gage for light projects and 20 to 18 gage for heavier projects.
2. *Band, or strap, iron.* This is rectangular in shape with round edges. The sizes used most often are $1/8 \times 1/2$ inch (3.18 $\times$ 12.7 millimeters), $1/8 \times 3/4$ inch (3.18 $\times$ 19.0 millimeters), and $1/8 \times 1$ inch (3.18 $\times$ 25.4 millimeters).
3. *Rounds and squares.* Common sizes are from $1/4$ inch to $1 1/2$ inches (6.35 to 38.1 millimeters).
4. *Galvanized steel (iron).* This is sheet stock with a zinc coating that keeps it from rusting. For thin project activity parts, 28 to 26 gage is needed; for heavier projects activities, 20 to 18 gage.
5. *Tinplate.* Kitchen utensils are often made from tinplate, which is mild-steel sheet covered with tin. Common thicknesses are IC (about 30 gage) and IX (about 28 gage).

High-carbon Steel. This steel has about 60 to 130 points of carbon. It is made into small tools and parts. These must be hardened and tempered before they are useful.

SHEET METAL THICKNESS

	NONFERROUS		FERROUS	
GAGE	in	mm*	in	mm*
16	0.050	1.20	0.0598	1.60
18	0.040†	1.00	0.0478	1.20
20	0.032‡	0.80	0.0359	0.91
22	0.025§	0.65	0.0299	0.80
24	0.020¶	0.50	0.0239	0.65
26	0.015	0.40	0.0179	0.45
28	0.012	0.30	0.0149	0.40
30	0.010	0.25	0.0120	0.35
32	0.007	0.18	0.0097	0.22

*Replacement sizes based upon ANSI B32.3-1974.
†32-ounce
‡24-ounce
§20-ounce
¶16-ounce

Fig. 129-7.   Sheet metal thickness in gage sizes, inches, and millimeters.

You should choose this steel if you are going to make a cold chisel, a center punch, a knife, or another such project activity. *Octagonal* (eight-sided) tool steel is used for punches and chisels.

Nonferrous Metals

Copper. This is a beautiful reddish-brown metal often used for metal project activities. It forms easily into such objects as bowls and trays. The thicknesses used most often are 16-ounce (which means that the copper weights 1 pound per square foot) and 24-ounce (1½ pounds per square foot). Sheets or rolls of copper are usually bought soft or half hard.

Brass. This alloy of copper and zinc is bright gold in color. It saws easily but is more brittle than copper. It is often combined with copper or used alone for objects that need not be formed too deeply (Fig. 129-8).

Nickel Silver. Sometimes called *German silver*, this metal looks almost like silver. It is an alloy of copper, nickel, and zinc. The base of some tableware (knives, forks, and spoons) is made of nickel silver that is then covered with a coating of pure silver.

Aluminum. A bright, bluish metal, aluminum is a good art metal. Choose 1100,

Fig. 129-8.   Brass is used for making this candleholder.

Fig. 129-9. These trays and containers are made of aluminum. (McDonal Products Company)

which consists of pure aluminum, or 3003, which has some manganese in it. Many other aluminum alloys are not soft. Aluminum is easily formed, but is hard to solder (Fig. 129-9).

Perforated and Expanded Metal

This stock comes in several different metals (usually steel or aluminum) and in many patterns. Perforated metal has a design stamped in it (Fig. 129-10). Expanded metal is made by cutting slits in the metal and pulling it open to expand it. The common thicknesses are 20, 22, and 24 gage.

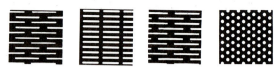

Fig. 129-10.   Perforated metal.

Unit 130

Measurement and Layout

The first step in building a metal product is to measure and mark out the correct size and shape of the material. You should also understand and be able to use the metric system of measurement as explained in Unit 17. There are many tools for measuring and marking (Fig. 130-1). Some of these tools you learned about in woodworking. Many measuring tools have both U.S. customary and metric markings.

Customary Rules

One-piece Metal Rule. This is either a 6- or 12-inch *steel rule* (Fig. 130-2). Most steel rules are divided into sixteenths of an inch. For more accurate measuring, steel rules are sometimes used that are divided into thirty-seconds and sixty-fourths of an inch.

Circumference Rule. A circumference rule made of metal is 3 feet long (Fig. 130-3). It has a scale directly below the rule. This scale shows the circumference

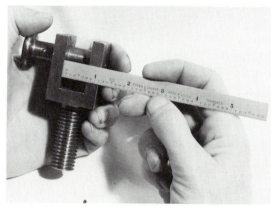

Fig. 130-1.   Accurate measurement is important in layout work.

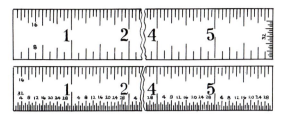

Fig. 130-2. Steel rules.

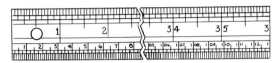

Fig. 130-3. A circumference rule for sheet-metal work.

Fig. 130-4. A steel tape with customary and metric graduations.

needed for any given diameter. This rule is used mainly for sheet-metal layout.

Steel Tape. This tape is used for making long measurements or for making inside measurements. It can also be bent to measure curved surfaces (Fig. 130-4).

Metric Rules

Common metric rules are available in 150-millimeter and 300-millimeter lengths. The rules are divided into full millimeters (1 mm) for most work and into half millimeters (0.5 mm) for more precise work (Fig. 130-5).

Layout Tools

Try Square. The try square is used for laying out a line across a piece of stock. It is also used for checking the squareness of a corner or for testing a 90° angle.

Combination Square. The combination square has a 12-inch (300-millimeter) metal rule and a square head. It is used for laying out, for testing 90° angles, and for many other purposes (Fig. 130-6).

A MILLIMETER IS ABOUT HALF WAY BETWEEN THE $\frac{1}{32}$" & THE $\frac{1}{16}$".

$\frac{1}{2}$ MILLIMETER IS ABOUT HALF WAY BETWEEN THE $\frac{1}{32}$" & THE $\frac{1}{64}$".

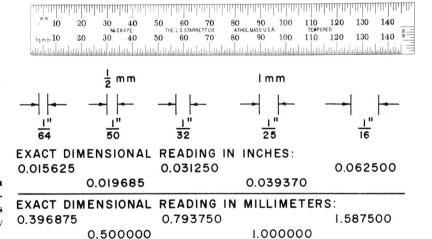

Fig. 130-5. A 150-mm rule showing a comparison of graduations with the customary rule.

EXACT DIMENSIONAL READING IN INCHES:
0.015625   0.031250   0.062500
0.019685   0.039370

EXACT DIMENSIONAL READING IN MILLIMETERS:
0.396875   0.793750   1.587500
0.500000   1.000000

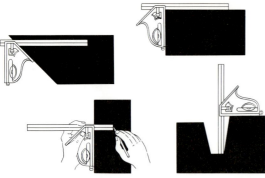

Fig. 130-6.   Using the square head of a combination square.

Fig. 130-7.   Dividers.

Fig. 130-8.   Scriber.

Steel Framing Square. The steel framing square, or carpenter's square, is used when large layouts must be made or large surfaces checked.

Dividers. Dividers are used to lay out arcs and circles (Fig. 130-7). Sometimes a simple pencil compass is enough. In drawing on stock, dividers are used in the same way as a compass.

Scriber. A scriber is a long, thin, pointed metal tool. It is used in much the same way as a pencil (Fig. 130-8).

Prick Punch. A prick punch, ground at an angle of 30°, is used for marking the places where all holes must be drilled (Fig. 130-9 a).

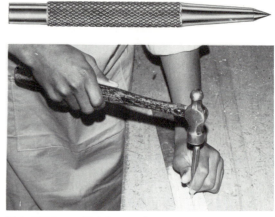

Fig. 130-9 a.   Prick punch.

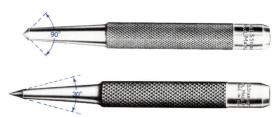

Fig. 130-9 b.   Note the difference between the point of the center punch and the prick punch.

Fig. 130-10.   The ball-peen hammer is used for pounding and for peening.

Center Punch. A center punch is ground at an angle of 90°. It is used to enlarge the dent made by a prick punch (Fig. 130-9 b).

Ball-peen Hammer. This hammer is used along with a prick punch for making dents (Fig. 130-10).

Measuring and Laying Out Metal

1. To check the thickness of a piece of metal use a sheet-metal gage (Fig. 130-11)

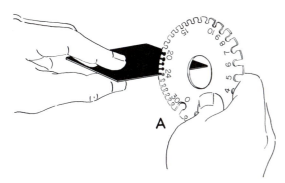

Fig. 130-11.   Using a sheet-metal gage.

Fig. 130-12.   Measuring the thickness of metal stock.

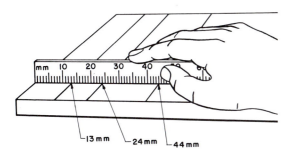

Fig. 130-13.   Measuring with a metric rule.

or hold the 1-inch or 10-millimeter mark of the rule over one side of the stock. Read the thickness directly over the other side (Fig. 130-12).

2. To measure the width of the metal, hold your thumb over one edge of the stock. Press the end of the rule against your thumb. Measure the width directly over the other edge.

3. To lay out short measurements, hold the end of the rule over the end of the stock. If the end of the rule is damaged in any way, hold the 1-inch or 10-millimeter mark over the end of the stock. For accurate measurement, hold the rule on edge. Use a scriber to mark the place for the length. Hold a try square or a combination square against the edge of the stock. Mark a line across the material to show the correct cutoff line (Fig. 130-13).

4. To lay out longer lengths, use a long rule or a steel tape. Hold the end of the rule over the end of the stock. Mark out the correct length. If the stock is quite wide, mark the place of the cut with a square (Fig. 130-14).

5. To lay out for cutting to width, mark the width at several points. Join these points with a straightedge.

6. To lay out arcs and circles, set the dividers or the compass. Place one leg over the 0, 10-millimeter, or 1-inch mark of a rule. Open the dividers until the distance between the legs is the same as the radius of the arc or the circle. Place one leg on the center of the arc or the circle. Tip the dividers slightly and twist clockwise (Fig. 130-15). Dividers are also used to lay out equal lengths along a straight or a curved line.

7. Follow the suggested steps when making a complete layout.

Fig. 130-14.   Marking the length of the stock.

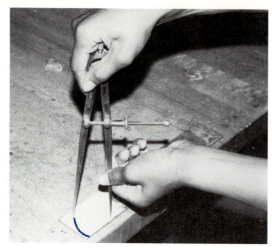

Fig. 130-15. Using dividers.

Making a Layout Directly on Metal

Since the layout will vary for each different project activity, only general suggestions can be made for making a layout directly on metal (Fig. 130-16). Before starting, be sure that you have the right size and kind of metal.

1. Cut or file one edge and end at right angles to each other. Use this edge and end as reference lines, and make all measurements from them.
2. Apply some kind of layout fluid.
3. Lay out all straight lines to show the widths and lengths of the various parts.
4. Lay out all angles and irregular lines.
5. Lay out all arcs.
6. Mark all holes to be drilled.
7. Lay out all internal lines.

Transferring a Design

There are several ways of transferring an irregular or complicated design:

1. Draw the design on thin paper and glue the paper to the metal surface with household cement or tape.
2. Apply a coat of showcard white and then transfer the design to the metal with carbon paper. First place a piece of carbon

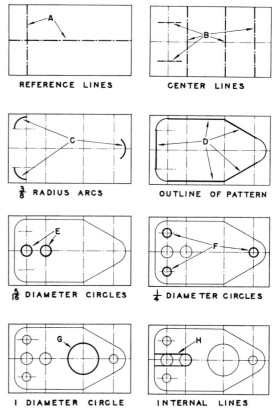

Fig. 130-16. Making a layout directly on metal.

paper and then the design over the metal, and clip the two in place. Next, outline the design with a pencil or a stylus, a sharp instrument used for marking.
3. If you are making many parts that have the same shape, you can cut a *template*, or pattern, of plywood or sheet metal. Hold the template on the metal, and trace the pattern with a scriber.

Sheet-metal Gages

Sheet-metal gages are round disks of metal with slots cut around the outside. (See Fig. 130-11.) They are used to measure the thickness of sheet metal and wire in the customary system. (No gages are used in the metric system. A metric micrometer is used to measure the thickness or diameter in millimeters.) Each slot of a gage is numbered. The numbers represent a certain thickness or diameter in decimals of an inch. The gage of the sheet metal or wire is the number of the slot in which it fits.

There are two common customary gage systems used for metal (see Fig. 129-7). The *United States Standard* (USS) gage is used for black and galvanized mild-steel sheets, steel plates, and steel wire. The *Brown & Sharpe*, or *American*, gage is used for measuring nonferrous metals such as copper, brass, and aluminum. To measure a piece of galvanized sheet, select a gage that is stamped United States Standard. Then lay the metal in the various slots until you find the one in which it just slips (Fig. 130-11). If the metal is 16 gage, it will be 0.0598 inch thick. To measure brass, use the Brown & Sharpe, or American, gage. If it is 16 gage, it will be 0.050 inch thick.

Unit 131

Cutting Heavy Metal

Heavy metal often must be cut or sheared to length, width, or thickness (Fig. 131-1). A hacksaw or cold chisel may be used.

Tools

Hacksaw. A hacksaw has a U-shaped frame. *Replaceable blades* (new ones that can be put in) of hard or flexible steel are fastened into this frame (Fig. 131-2). The blades most often used are 8, 10, or 12 inches (200, 250, or 300 millimeters) long with up to 32 teeth per inch. Figure 131-3 shows the number of teeth per inch needed for different metal shapes. For general cutting, use a blade with 18 teeth per inch. For thin sheet stock or tubing (Fig. 131-3), use a blade with 32 teeth per inch. The teeth are bent *alternately* (in an every-other pattern) to right and left so that the *kerf* (cut slot) will be wider than the blade itself.

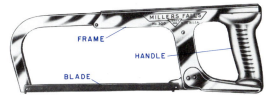

Fig. 131-2.   Hacksaw.

14 TOOTH: SOFT MATERIALS, LARGER CROSS SECTIONS

18 TOOTH: GENERAL USE, SAME BLADE SEVERAL JOBS

24 TOOTH: CROSS SECTIONS 1/16" TO 1/4"

Fig. 131-1.   The legs for this table are cut to length by one of the methods shown in this unit.

32 TOOTH: CROSS SECTIONS 1/16" OR LESS

Fig. 131-3.   Correct hacksaw blades for cutting.

Cold Chisel. A cold chisel (Fig. 131-4) is ground at an angle of 60 to 70° (Fig. 131-5). It may be used for cutting and shearing heavy stock.

Power Saws. Several kinds of power saw can be used to cut heavy metal including:

1. The horizontal band saw has a continuous blade that can cut metal of almost any size.

2. The power hacksaw uses a single blade and cuts like a hand hacksaw (Figs. 131-6 and 131-7).

3. The radial-arm saw can be equipped with an abrasive blade to cut metal (Fig. 131-8).

4. The reciprocating saw operates like a hacksaw without a frame (Fig. 131-9).

Fig. 131-4.   Cold chisel.

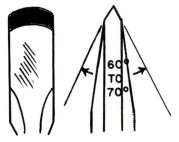

Fig. 131-5.   Correct angle for a cold chisel.

Fig. 131-6.   A portable power hacksaw.

Fig. 131-7.   The power hacksaw operates like a hand hacksaw.

Fig. 131-8.   Using a radial-arm saw for cutting metal.

Fig. 131-9.   This reciprocating saw can cut heavy metal that is held in a vise.

Cutting with the Hacksaw

1. Fasten the blade with the teeth pointing away from the handle. Then tighten the blade until it is *taut* (tight and stiff).

2. Fasten the stock in a vise. Be sure the layout line is close to the jaws of the vise. Put soft copper or aluminum over the vise jaws to protect a finished metal surface.

3. Start the saw by drawing back on the handle with your right hand. Guide the blade with the thumb of your left hand.

4. Operate the saw with both hands. Use a steady, even motion of about 30 strokes per minute. Apply pressure on the forward stroke. Release on the return stroke. Never drag the saw (Fig. 131-10).

5. When making the last few cuts, hold the stock to be cut off in your left hand as you operate the saw with your right hand. Stock must always be cut square (Fig. 131-11).

6. Cut thin stock by placing it between two pieces of plywood.

7. Make long cuts with the blade turned at 90° to the frame (Fig. 131-12).

Cutting with a Chisel over a Lead Plate

1. Place the metal over a soft iron or lead plate. Never place it over an anvil.

2. Hold the chisel as shown in Fig. 131-13. Strike it firmly with the flat head of a ball-peen hammer.

3. With light blows, go over the layout line once to outline the cut.

4. Strike heavy blows to cut through the metal.

Shearing Metal in a Vise

1. Place the metal in a vise with the layout line just above the jaws.

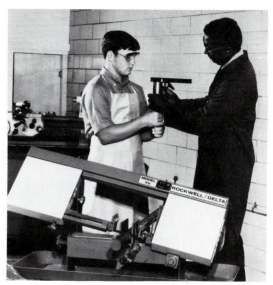

Fig. 131-11.   Check cut stock for squareness.

Fig. 131-10.   Cutting metal with a hacksaw.

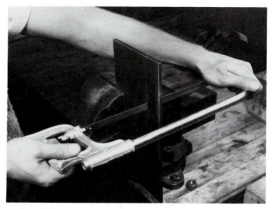

Fig. 131-12.   Turn the blade at a 90° angle when making long cuts.

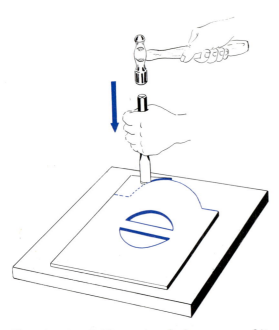

Fig. 131-13. Cutting out a design over a plate with a cold chisel.

Hold the chisel at an angle of about 30° (Fig. 131-14). Strike it firmly with a hammer to shear the metal. If the chisel is held too high, it will cut into the vise jaw. If the chisel is held too low, it will tear or *score* (mark) the metal.

Fig. 131-14. Shearing metal in a vise.

Unit 132

Drilling Holes and Countersinking

Holes are drilled for putting together projects with rivets, bolts, or screws. Holes are also cut to produce *internal* (inside) threads (Fig. 132-1). Drilling is also done on projects that must have small holes as a part of their construction.

Twist drills are made of tool, or high-speed, steel. The most common ones have straight shanks. Twist drills come in millimeter, inch, number and letter sizes (Fig. 132-2). The most common customary drill sizes are from $1/64$ to $1/2$ inch in diameter, increasing in size by sixty-fourths. From $1/2$ inch to 3 inches, the size increases by thirty-seconds. The drill size is stamped on the shank (Fig. 132-3). If the size has worn off, it can be checked with a gage or micrometer. A drill press (Fig. 132-4), a

Fig. 132-1. Drilling is one of the common metalworking processes.

CONVERSION TABLE FOR DRILLS

METRIC DRILLS (mm)	DEC. EQUIV.	DRILLS FRACTIONAL	NUMBER & LETTER
0.79	0.0312	1/32	
....	0.0320		67
....	0.0330		66
....	0.0350		65
....	0.0360		64
....	0.0370		63
....	0.0380		62
....	0.0390		61
1.0	0.0394		
....	0.0400		60
....	0.0410		59
....	0.0420		58
....	0.0430		57
....	0.0465		56
1.19	0.0469	3/64	
....	0.0520		55
....	0.0550		54
1.5	0.0591		
....	0.0595		53
1.59	0.0625	1/16	
....	0.0635		52
....	0.0670		51
....	0.0700		50
....	0.0730		49
....	0.0760		48
1.98	0.0781	5/64	
....	0.0785		47
2.0	0.0787		
....	0.0810		46
....	0.0820		45
....	0.0860		44
....	0.0890		43
....	0.0935		42
2.38	0.0937	3/32	
....	0.0960		41
....	0.0980		40
2.5	0.0984		
....	0.0995		39
....	0.1015		38
....	0.1040		37
....	0.1065		36
2.78	0.1094	7/64	
....	0.1100		35
....	0.1110		34
....	0.1130		33
....	0.1160		32
3.0	0.1181		
	0.1200		31
3.18	0.1250	1/8	
....	0.1285		30
....	0.1360		29
3.5	0.1378		
....	0.1405		28
3.57	0.1406	9/64	
....	0.1440		27
....	0.1470		26
....	0.1495		25
....	0.1520		24
....	0.1540		23
3.97	0.1562	5/32	
....	0.1570		22
4.0	0.1575		
....	0.1590		21
....	0.1610		20
....	0.1660		19
....	0.1695		18
4.37	0.1719	11/64	
....	0.1730		17
....	0.1770		16
....	0.1800		15
....	0.1820		14
			13
4.76	0.1875	3/16	
4.8	0.1890		12
....	0.1910		11
....	0.1935		10
....	0.1960		9
5.9	0.1968		
....	0.1990		8
....	0.2010		7
5.16	0.2031	13/64	
....	0.2040		6
....	0.2055		5
....	0.2090		4
....	0.2130		3
5.5	0.2165		
5.56	0.2187	7/32	
....	0.2210		2
....	0.2280		1
....	0.2340		A
5.95	0.2344	15/64	
6.0	0.2362		
....	0.2380		B
....	0.2420		C
6.25	0.2460		D
6.35	0.2500	1/4	E
....	0.2570		F
....	0.2610		G
6.75	0.2657	17/64	
....	0.2660		H
....	0.2720		I
7.0	0.2756		
....	0.2770		J
....	0.2811		K
7.14	0.2812	9/32	
....	0.2900		L
....	0.2950		M
7.5	0.2953		
7.54	0.2968	19/64	
....	0.3020		N
7.94	0.3125	5/16	
8.0	0.3150		
....	0.3160		O
....	0.3230		P
8.33	0.3281	21/64	
....	0.3320		Q
....	0.3390		R
8.73	0.3437	11/32	
....	0.3480		S
9.0	0.3543		
....	0.3580		T
9.13	0.3594	23/64	
....	0.3680		U
9.53	0.3750	3/8	
....	0.3770		V
....	0.3860		W
9.92	0.3906	25/64	
10.0	0.3937		
....	0.3970		X
....	0.4040		Y
10.32	0.4062	13/32	
....	0.4130		Z
10.72	0.4219	27/64	
11.0	0.4330		
11.11	0.4375	7/16	
11.51	0.4531	29/64	
11.91	0.4687	15/32	
12.0	0.4724		
12.30	0.4843	31/64	
12.7	0.5000	1/2	

Fig. 132-2.   Table of drill sizes.

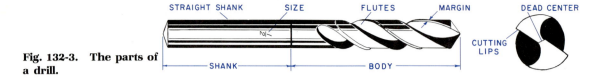

Fig. 132-3. **The parts of a drill.**

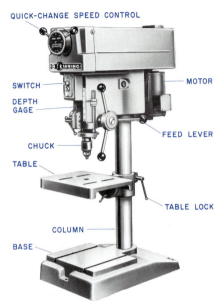

Fig. 132-4. **Parts of a drill press.**

Fig. 132-5. **Hand drill.**

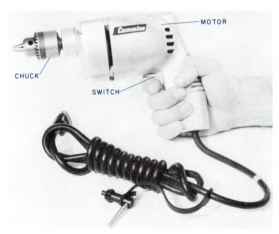

Fig. 132-6. **Major parts of an electric hand drill.**

hand drill (Fig. 132-5), and a portable electric drill (Fig. 132-6) are the usual drilling devices.

Using the Drill Press

1. Choose a drill of the correct size. Put it into the chuck. Make sure that it is straight. Lock the chuck tightly and remove the key (Fig. 132-7).

2. Adjust the speed of the drill press by changing the belt on the pulleys. Use a fast speed for small drills and soft materials. Use a slow speed for large drills and hard materials.

3. Mark the place you want the hole to be with a center punch. This will help start the drill.

4. Lock the work in a drill-press vise or use one of the methods shown in Fig. 132-8 a through e.

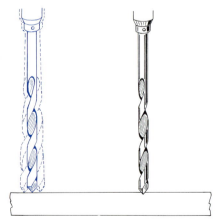

Fig. 132-7. **Make sure the drill is installed correctly in the chuck. The drill on the left has been installed incorrectly.**

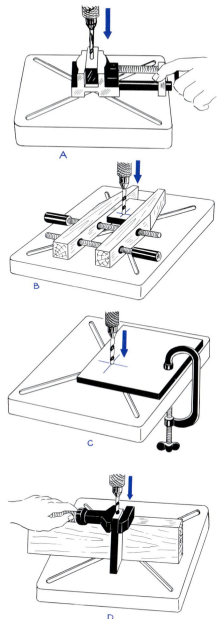

A

B

C

D

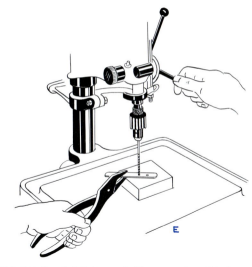

E

Fig. 132-8. (*a*) Using a drill-press vise. (*b*) Clamping metal between parallel clamps of a hand screw. (*c*) Clamping metal to the table with a C clamp. If the table is not perfectly centered, place a piece of scrap wood under the stock. (*d*) Using a monkey wrench to hold the stock. (*e*) Holding thin stock with a pair of pliers.

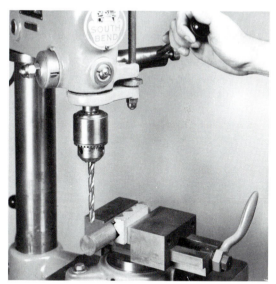

Fig. 132-9. Round stock clamped in a vise for drilling.

5. Turn on the power. Apply a slight pressure with the feed handle (Fig. 132-9). When drilling steel, place a little cutting oil at the drill point. Do not put too much pressure on the metal. This will make the drill burn. Too little pressure will make a scraping action. When steel is being drilled, a thin chip will curl out of the hole.
6. Ease up on the pressure slightly when the point of the drill pierces the lower surface of the metal. This will keep the drill from catching. Catching can damage the work or break the drill.

Using a Hand Drill

1. Hold the shell of the chuck in your left hand. Turn the handle backward until the chuck is open far enough to let the drill enter it. Then tighten the drill in the chuck.

2. Lock the work in the vise. Most drilling is done in a horizontal position (Fig. 132-10) with the metal held vertically. A piece of scrap wood should be placed behind the metal.

3. Hold the point of the drill in the center-punch hole. Then do the drilling. It is a good idea to have someone help you sight it. In this way, you can be sure that the hole will be drilled squarely with the metal surface.

Using a Portable Electric Drill

1. Lay out and mark the correct center of each hole with a center punch.

2. Open the chuck. Slip in the twist drill until it is at the bottom of the chuck. Then tighten it securely.

3. Clamp the work in a vise or to the bench.

4. Hold the control handle firmly. Point the drill toward the center-punch mark. Use your left hand to control the feed.

5. Start with the power off. Make sure that the tool is straight (Fig. 132-11).

Fig. 132-11.   Using a portable electric drill.

6. Turn on the switch. Guide the tool into the stock. Remember that any sidewise movement will break a small drill.

Countersinking

When you are putting in flat-head rivets, stove bolts, or machine screws, a metal countersink is needed (Fig. 132-12). However, a large drill can be used instead.

1. Install the countersink in the drill press, a breast drill, or a hand drill.

2. Use a slow speed. Cut the conical hole.

3. Check the depth of the countersink by holding the rivet, screw, or bolt upside down over the hole.

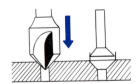

Fig. 132-10.   Clamp the workpiece in a vise and drill with a hand drill at right angles to the work surface. If the metal is thin, put a piece of wood behind it to give it extra support.

Fig. 132-12.   Countersinking a hole.

Unit 133

Filing

Filing smooths, shapes, and finishes metal (Fig. 133-1). *Single-cut* files have one row of teeth. *Double-cut* files have two rows that form a crisscross pattern (Fig. 133-2 a and b). The grades of coarseness most often used are the bastard (rough), second cut (medium), and smooth (fine). There are hundreds of different kinds, shapes, and sizes of files. But only a few will be discussed here (Fig. 133-3).

Types of Files

Flat and Hand Files. The flat and the hand files (Fig. 133-4 a and b) in 10- or 12-inch (250- or 300-millimeter) lengths are doublecut for rough work.

Mill File. The mill file (Fig. 133-5) is a rectangular single-cut file. It is used for fine, flat filing and for lathe filing.

Rattail File. The rattail, or round, file is single-cut or double-cut. It is used for internal filing of round holes or curved surfaces (Fig. 133-6).

Fig. 133-1. **Hand filing is a skill needed even in mass-production industries.** (Allis-Chalmers Company)

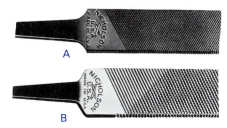

Fig. 133-2. (*a*) Single-cut file. (*b*) Double-cut file.

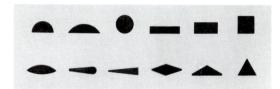

Fig. 133-3. Common shapes of files.

Fig. 133-4. (*a*) A flat file. (*b*) A hand file.

Fig. 133-5. Mill file.

Fig. 133-6. Rattail or round file.

Square File. The square file (Fig. 133-7) is double-cut and is used for slots or square openings.

Half-round File. The half-round file (Fig. 133-8) is a double-cut file used for concave or convex filing.

Triangular, or Three-square, File. The triangular file (Fig. 133-9) is either single-cut or double-cut. It is used for finishing out an internal rectangular opening.

Jewelers' File. The jewelers' file (Fig. 133-10) is also called a *needle file.* These files come in different shapes, usually as a set, for fine work in art metal and jewelry.

All files except jewelers' files should have good handles. Also, a file card is needed to keep the files clean.

Fig. 133-7.   Square file.

Fig. 133-8.   Half-round file.

Fig. 133-9.   Triangular, or three-square, file.

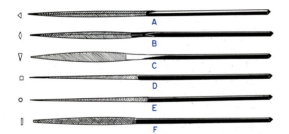

Fig. 133-10.   Typical jewelers' files: (a) three-square, (b) ratchet, (c) knife, (d) square, (e) round, (f) warding.

Crossfiling

1. Select the best file for the work. Clean the file by brushing a file card across the teeth.

2. Put on a handle (Fig. 133-11). Hold the file with the handle in your right hand. The point should be held lightly in your left hand. Follow the method shown in Fig. 133-12 for light filing and the method shown in Fig. 133-13 for heavy filing.

3. Lock the work in the vise with the surface close to the top of the jaws.

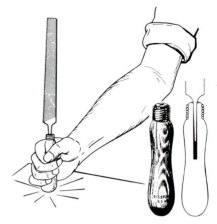

Fig. 133-11.   Always fit a handle on a file before using it. It is very dangerous to use a file without a handle.

Fig. 133-12.   Method of holding a file for light filing.

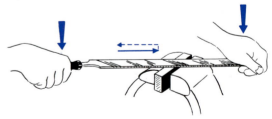

Fig. 133-13.   Method of holding a file for heavy filing.

4. Hold the file straight. Apply pressure on the forward stroke. Then release the pressure. Return to the starting position. Do not rock the file. Clean it often.

5. With jewelers' files, use only light pressure. When you use a round or oval-shaped file, twist the file slightly on the forward stroke (Fig. 133-14).

Drawfiling

1. Use a mill file. Hold it with the handle in your left hand and the body of the file in your right (Fig. 133-15).

2. Start at one end of the file. Draw the file toward you or push it away, making a smooth shearing cut. Then move to a new surface of the file. Repeat the action. When all of the surface of the file has been used, clean it.

3. For soft metals, such as copper or brass, cover the surface of the file with chalk. This will keep the metal from sticking in the teeth.

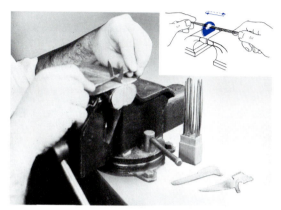

Fig. 133-14.   Using a jewelers' file.

Fig. 133-15.   Draw filing.

Unit 134

Smoothing a Surface with Abrasives

Abrasives are used to produce a fine finish on any product (Fig. 134-1). Abrasives are small grains or powders used for smoothing metal surfaces. Some abrasives, such as emery and corundum, are found in nature. However, the two abrasives used most often are not natural. These are silicon carbide and aluminum oxide. *Silicon carbide* is bluish-black in color. It is used for soft metals, such as copper or aluminum. *Aluminum oxide* is white or brown. It is used on hard metals, such as steel.

The grain or powder abrasive may be used as it is. But it also comes in the form of cloth, sheets, strips, and belts. It is graded by number: coarse is 60 to 80, medium is 80 to 100, and fine is 100 to 140. For handwork, 8 × 11 inch (200 × 275 millimeter) sheets or 1- to 2-inch (25- to 50-millimeter) strips are best.

Smoothing can be done by following these steps:

1. Tear a sheet into four or six equal parts or tear off a 2-inch (50-millimeter) strip of the grade needed for your work.
2. Wrap the abrasive around a file or a stick.

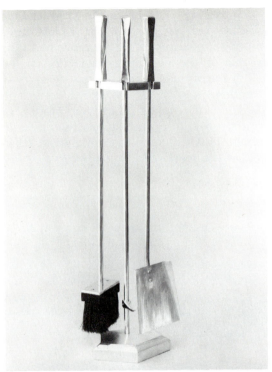

Fig. 134-1. Abrasives were used to produce this finish on the fireplace set.

3. Apply a little cutting oil to the metal surface. Work the abrasive back and forth as if you were sanding wood (Fig. 134-2).
4. For a really smooth surface, leave the oil on the metal. Turn over the abrasive and repeat the rubbing.

Fig. 134-2. When you smooth with a file, use a piece of abrasive cloth wrapped around the file. Apply a little oil to the surface and rub the file back and forth.

You should remember to start with a rough-grade abrasive. Then change to finer and finer abrasives until you get the smoothness you want. Figures 134-3 through 134-5 show abrasive dressing on power machines.

Fig. 134-3. An iron casting can be dressed and smoothed on the belt sander. (Rockwell Manufacturing Company)

Fig. 134-4. Edges of round sheet may be dressed and smoothed on the belt sander.

Fig. 134-5. Smoothing the edges of spoons on a belt abrasive machine. (Norton Company)

Grinding Metal and Sharpening Tools

A grinder is used for grinding metal and sharpening tools (Fig. 135-1). Grinding wheels are made from abrasive materials. Coarse wheels are use for general-purpose work. Fine wheels are for sharpening.

Grinding Metal

1. Make sure that the grinder has a safety shield. Wear goggles or a face protector. Adjust the tool rest to a position ⅛ inch (3 millimeters) from the wheel.

2. Check the edge of the wheel. It should be smooth and true. If it is not, dress the wheel with a disk or a diamond-point wheel dresser. Hold the wheel dresser firmly against the wheel. Move it back and forth until the wheel is true (Fig. 135-2).

3. Grind the metal by holding it firmly on the tool rest. Move it back and forth across the face of the wheel (Fig. 135-3 a, b, and c and 135-4). *Never use the side.* Do not hold the metal on the wheel too long, or it will burn. Keep the metal cool by dipping it in water.

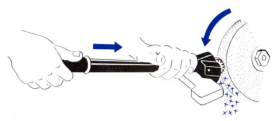

Fig. 135-2.   Straightening the outer edge of an abrasive wheel with a wheel dresser.

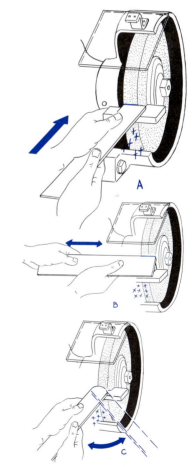

Fig. 135-3.   Common grinding operations: (a) grinding an end, (b) grinding a beveled edge, (c) rounding an end.

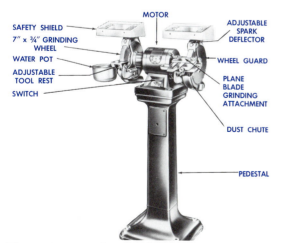

MOTOR

SAFETY SHIELD

7″ x ¾″ GRINDING WHEEL

WATER POT

ADJUSTABLE TOOL REST

SWITCH

ADJUSTABLE SPARK DEFLECTOR

WHEEL GUARD

PLANE BLADE GRINDING ATTACHMENT

DUST CHUTE

PEDESTAL

Fig. 135-1.   A pedestal grinder.

Fig. 135-4. Grinding the edge of a casting. Move it back and forth on the tool rest.

Sharpening Tools

Cold Chisel. To sharpen a cold chisel, adjust the grinder rest to an angle of about 30 to 35° to the wheel. Hold the chisel against the rest. Grind the edge to a slight arc. The chisel should have an included angle of 60 to 70° (Fig. 135-5 a, b, c, and d).

Center and Prick Punch. To sharpen a center punch, hold it near the point with the thumb and forefinger of your left hand and near the head with your right hand. Then turn the tool as you grind it to an included angle of 90°. Grind a prick punch at an angle of 30° (Fig. 135-6 a and b).

Screwdriver. Grind a screwdriver with a long taper on either side and square across the end.

Twist Drill. Follow the curve of the cutting edges to sharpen the drill (Fig. 135-7).

Fig. 135-5 a. This cold chisel is a hazard and is dangerous to use.

Fig. 135-5 b. Grinding the mushroom head.

Fig. 135-5 c. The edge of the cold chisel can be ground to shape and then dressed with a file.

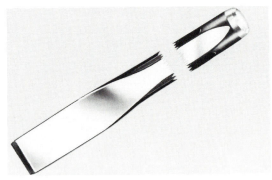

Fig. 135-5 d. This is a cold chisel that is in good condition and ready for use.

Fig. 135-6 a. The center punch should be ground to an angle of 90°, and the prick punch to an angle of 30°.

Fig. 135-6 b. **Grinding a center punch or a prick punch.**

Fig. 135-7. **Sharpening the cutting lips of a twist drill.** (Ridge Tool Company)

Unit 136

Bending Heavy Metal

There are many ways of making angular and circular bends (Fig. 136-1). The tools needed are a vise and a ball-peen hammer. You also need several pieces of pipe and some rod of different diameters.

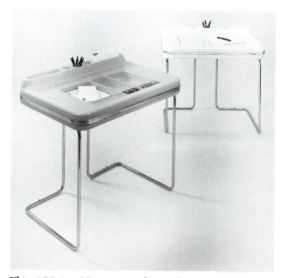

Fig. 136-1. **Many metal products have parts that are bent to shape.**

Making Angular Bends

1. Add an amount the same as one-half the thickness of the metal for each bend to be made. Mark the place of the bends.

2. Place the stock in the vise with the extra metal above the vise jaws. Check with a square to see that the metal is in a vertical position.

3. Force the metal over the vise jaws with your left hand. At the same time, strike it near the bend line with the flat of a hammer (Figs. 136-2 a and 136-3).

4. To square off the bend, place it in the vise, as shown in Fig. 136-2 b. Then strike several blows.

5. Make bends of less than 90° with a monkey wrench or an adjustable wrench (Fig. 136-4). You can also use a bending jig in a vise (Fig. 136-5).

Making Circular Bends

Method 1. Select a bending device (rod or pipe) with a diameter the same as the in-

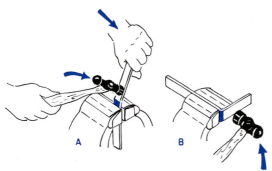

Fig. 136-2. (*a*) Starting a 90° bend, (*b*) finishing a 90° bend.

Fig. 136-5. Using a bending jig in a vise.

Fig. 136-3. Bending strap metal in a vise. (Reynolds Metals Company)

side of the circle to be bent. Place the metal and the bending device in a vise. Pull the metal toward you around the pipe (Fig. 136-6). Keep on feeding the metal around the pipe until the circle is complete.

Method 2. Select a rod or pipe. Fasten it in a vise. Rest the metal on the pipe. Strike it just beyond the point of contact (Fig. 136-7). Move the stock along slowly and keep on striking the metal until the circle wanted is complete.

Fig. 136-6. Bending a curve over a pipe.

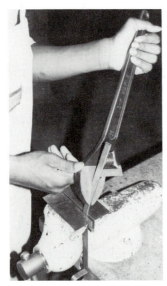

Fig. 136-4. Making a bend with an adjustable end wrench.

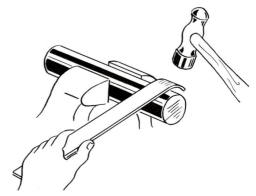

Fig. 136-7. Forming metal over a rod.

Bending a Scroll and Twisting Metal

Band-iron strips are often bent to form scrolls or twists in ornamental or wrought-iron project activities (Fig. 137-1).

Tools

The anvil, ball-peen hammer, simple bending jig, adjustable wrench, and metal vise are the tools you will need.

Bending a Scroll

1. Make a full-size pattern of the scroll to be bent.
2. Check the length of metal needed by laying a string or wire over the pattern.
3. *Flare* (widen) the end of stock over the edge of the anvil. Begin to bend the scroll (Fig. 137-2 a and b).
4. Fasten a simple bending jig (Fig. 137-3) in a vise.
5. Insert the flared end in the jig. Bend the scroll a little at a time (Fig. 137-4). Check the bending often by holding the scroll over the full-size pattern (Fig. 137-5).

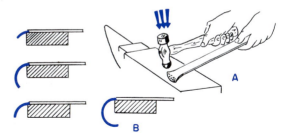

Fig. 137-2. (*a*) Flaring, (*b*) steps in bending.

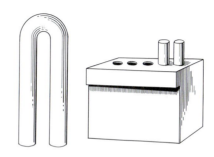

Fig. 137-3. Two types of bending jigs.

Fig. 137-4. Most of the forming is done with the left hand while the right hand guides the scroll.

Fig. 137-1. A scroll bent, band iron project.

Fig. 137-5. Checking the scroll over a full-size pattern.

6. Open the scroll slightly or bend it a little more at certain points if necessary. Do not try to bend too much of the curve at once.

7. When two scrolls are formed on the same piece, make sure there is a smooth curve from one scroll to the other. The scroll should lie flat after forming.

Twisting Metal

Twisting metal shortens its length. Because of that, you need extra material. Whenever possible, the piece is twisted first and then is cut off to the proper length.

1. Mark the beginning and the end of the twist.

2. For short twists, place the metal vertically in a vise. One layout line should be just above the vise jaws.

3. Place a monkey wrench or an adjustable wrench at the other end. Hold the wrench firmly near the jaws with your left hand. Then twist the metal the desired number of turns (Fig. 137-6).

4. For long twists, place the metal horizontally in a vise. Slip a piece of pipe over the area to be twisted. This will keep it from bending out of shape (Fig. 137-7 a and b).

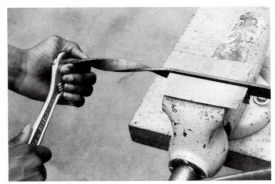

Fig. 137-7 a.   Twisting metal in a horizontal position. The twist may be right-hand or left-hand depending on the way you turn the wrench.

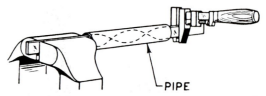

Fig. 137-7 b.   A piece of pipe over the area to be twisted will keep the metal straight.

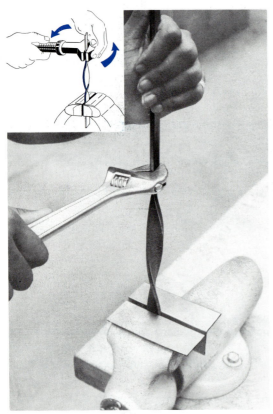

Fig. 137-6.   Twisting the metal held in a vertical position.

Unit 138

Peening and Planishing

Peening is done by striking even blows on metal with the peen end of a ball-peen hammer (Fig. 138-1). This process gives the metal a smooth-textured surface. If the metal is struck too hard, it will be stretched out of shape. You should work from the edges toward the center. Use small, overlapping, cone-shaped dents.

If both sides of the metal must be peened, place a piece of annealed copper under the finished surface. Peen the other side. A different look may be gotten with a cross-peen or a straight-peen hammer.

Planishing is done mostly on art project activities. It is like peening except that a special planishing hammer is used.

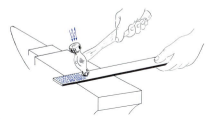

Fig. 138-1.   Peening the surface with a ball-peen hammer.

Unit 139

Buffing Metal

Buffing removes oxide, marks, and irregularities (scratches, for example) from a metal surface. Hard buffing wheels are made of rope, felt, canvas, or leather. These are put on a grinder. Buffing is done as follows:

1. Coat the edges of the buffing wheel with an adhesive, such as hot liquid glue. Pour out a row of abrasive powder or grains (100 for medium, 200 for fine, and 300 for extra fine) on a clean piece of paper. Roll the wheel in the powder to cover the edge. Let the wheel dry for 24 hours.

Another method is to coat the wheel after it is mounted. If you do this, hold a stick of greaseless compound (an abrasive mixed with glue in stick form) against the revolving wheel.

2. Mount the wheel on the *arbor* (shaft) of a buffing head or grinder.

3. *Put on a pair of goggles.*

4. Turn on the power. Hold the metal on the lower surface of the wheel. Buff the metal by working it back and forth (Fig. 139-1).

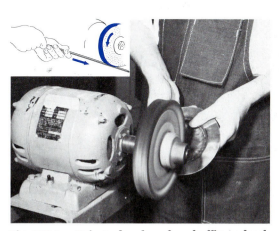

Fig. 139-1.   Using a hard-surface buffing wheel. Always wear goggles or use a safety shield when buffing.

Unit 140

Cutting Threads

Threads on the inside of a hole are cut with a hardened steel tool called a *tap*. The taper tap is used most often (Fig. 140-1). Threads on the outside of rods or pipes are cut with a *die* held in a *diestock* (Fig. 140-2). The parts of a thread are shown in Fig. 140-3.

Types of Customary Taps and Dies

There are taps for cutting *fine* threads (National Fine, or NF, series), *coarse* threads (National Coarse, or NC, series), and *pipe* threads. Coarse and fine threads are exactly the same shape. However, fine threads have more threads per inch. For example, a ½-inch National Coarse has 13 threads per inch, while a ½-inch National

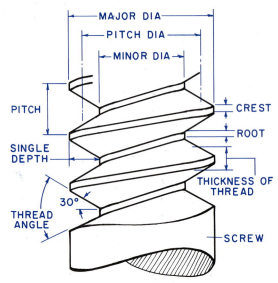

Fig. 140-3.   The parts of a thread.

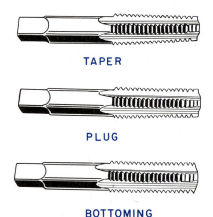

TAPER

PLUG

BOTTOMING

Fig. 140-1.   To tap through an open hole, use a taper tap. To tap almost to the bottom of a closed hole, use a plug tap. Finish with a bottoming tap.

Fig. 140-2.   Adjustable, round, split die.

Fine has 20 threads per inch. Below ¼ inch, taps are made in machine-screw sizes and are marked 6−32, 8−32, and so forth. An 8−32 thread is for a machine screw made from No. 8 wire and with 32 threads to the inch. The NF and NC series have now been absorbed into a unified system known as UNF (Unified National Fine) and UNC (Unified National Coarse) series.

Pipe taps are either sharply tapered or straight. The taper pipe tap is used for water and gas pipe. Pipe taps are much larger than the size shows. For example, a ⅛-inch pipe tap looks as large as a ⅜-inch regular tap. The ⅛-inch straight pipe tap is used most often since it is the size for most electric connections. This tap requires an ¹¹/₃₂-inch tap drill.

A set of taps and dies of the most common sizes is called a *screw plate*. The adjustable round split die can be made slightly larger or smaller by turning a small screw (Fig. 140-2).

Cutting Customary Internal Threads

1. Select the correct-size tap drill. Drill the hole straight. See Fig. 140-4 for tap-drill sizes.

2. Fasten the tap in the tap wrench. Place the work in the vise so that tapping will be done in a vertical position (Fig. 140-5).

NATIONAL FINE		NATIONAL COARSE	
SIZE AND THREAD	TAP DRILL	SIZE AND THREAD	TAP DRILL
4 — 48	$3/32$	4 — 40	$3/32$
5 — 44	$7/64$	5 — 40	$7/64$
6 — 40	$1/8$	6 — 32	$7/64$
8 — 36	$9/64$	8 — 32	$9/64$
10 — 32	$5/32$	10 — 24	$5/32$
$1/4$ — 28	$7/32$	$1/4$ — 20	$13/64$
$5/16$ — 24	$9/32$	$5/16$ — 18	$17/64$
$3/8$ — 24	$11/32$	$3/8$ — 16	$5/16$
$7/16$ — 20	$25/64$	$7/16$ — 14	$3/8$
$1/2$ — 20	$29/64$	$1/2$ — 13	$27/64$
$9/16$ — 18	$33/64$	$9/16$ — 12	$31/64$
$5/8$ — 18	$37/64$	$5/8$ — 11	$17/32$
$3/4$ — 16	$11/16$	$3/4$ — 10	$21/32$
$7/8$ — 14	$13/16$	$7/8$ — 9	$49/64$
1 — 14	$15/16$	1 — 8	$7/8$

Fig. 140-4.   Tap-drill sizes.

Fig. 140-5.   Make sure the top is straight before cutting the thread by checking this with a square.

3. Hold the tap wrench in the center with your hand to start the cutting. Apply pressure. Turn the wrench clockwise. Check to make sure that the tap is square with the work.

4. Apply a little cutting oil on the tap. Cut the threads by turning the tap about three-fourths of a turn forward and then about one-fourth of a turn back (Fig. 140-6). This will let the chips drop out through the *flutes* (the grooves cut lengthwise along the tap). Never force a tap. The smaller sizes break easily. It is almost impossible to remove a broken piece if it breaks below the top of the hole without using a special tool known as a tap extractor.

5. Back out the tap when the cutting is finished.

Cutting Customary External Threads

1. Grind or file a small *bevel* (slant) on the edge of the pipe or rod. Fasten this in a vise in either a vertical or a horizontal position.

2. Select a die of the correct size. Fasten it in the diestock. Place the tapered side of the die toward the guide.

3. Place the die over the pipe or rod. Apply a little cutting oil and start the cutting (Fig. 140-7). Follow the same general steps used for tapping.

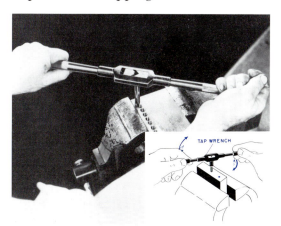

Fig. 140-6.   Using a tap.

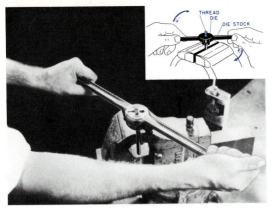

Fig. 140-7. Cutting external threads with a die.

4. Remove the die when the threads are cut to the length you want. Try the threaded rod or pipe in the threaded hole or nut. If it is too tight, adjust the die to a smaller size and recut.

Cutting Metric Threads

A set of metric taps and dies is needed to cut metric threads. The right tap drill for cutting internal threads is found by subtracting the *pitch* from the *diameter*. For example, the correct tap drill for M6 × 1 is 5 millimeters. See Unit 126.

Unit 141

Riveting Heavy Metal

Strap, or band, iron and heavy black iron sheet are often put together to last by riveting. Black iron countersunk or round-head rivets, ¹/₈ to ³/₁₆ inch (3 to 5 millimeters) in diameter, are the most commonly used (Fig. 141-1). For art metal, ¹/₈-inch (3-millimeter) countersunk rivets or round-head aluminum, brass, or copper rivets are best. A riveting plate, a rivet set, and a hammer are needed to rivet heavy metal.

Riveting

1. Choose rivets of the correct size.
2. Locate and drill the holes. Countersink one surface if the shank of the rivet is to be *flush* (even with the surface) or if a countersunk rivet is to be used. Countersink both surfaces for a flush countersunk rivet.
3. Put the rivet in the hole. Cut off the extra shank with a hacksaw or clippers. For a rounded shank, leave 1¹/₂ times the diameter of the shank. For a flush rivet, only a small amount should show.
4. Place the round head of the rivet over the conical hole in the rivet set or rivet plate (Fig. 141-2 a and b). Countersunk rivets can be placed on any flat surface.
5. Strike the shank of the rivet several times with the flat of a hammer. Do this until the rivet is seated on the metal.

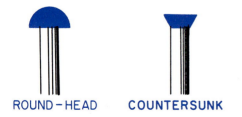

Fig. 141-1. **Round-head and countersunk rivets are common types of heavy rivets.**

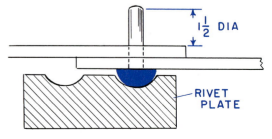

Fig. 141-2 a. **Rivet plate.**

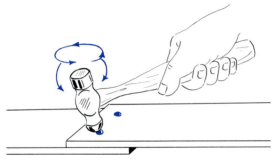

Fig. 141-2 b. **Rounding off the head of a rivet.**

Fig. 141-3. **Note the small metal rod in the vise that supports the rivet.**

6. For a round-head rivet, shape the shank to the contour you want by using the peen end of the hammer (Fig. 141-3). For a flush rivet, file off the extra stock.

Unit 142

Finishing Metal

A finish makes metal more beautiful (Fig. 142-1). It also protects it from rust and stains. You may finish metal by any of the methods described in this unit.

Wrought-iron projects that have been peened are often painted with black enamel. Sometimes a flat black paint is used. When dry, the surface is rubbed lightly with abrasive cloth to bring out highlights. Then a coat of clear lacquer is put on. Many project activities are finished by painting with colored enamels or lacquers. Sometimes bronze or silver powder is blown on the wet surface of clear or colored lacquer.

A wrinkle finish is made after putting on a wrinkle-finish paint and baking the project activity. This is done by following the directions given by the paint manufacturer.

To color copper from brown to black, use a solution of liver of sulfur. Dissolve a piece about the size of a small peanut in 1

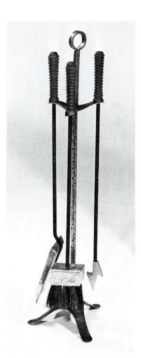

Fig. 142-1. **Clear lacquer will protect the finish on this fireplace set.**

gallon (4 liters) of water. Place the clean copper in the solution until it turns to the color you want. Allow it to dry. Then highlight certain areas by rubbing with a piece of steel wool.

For most art-metal project activities, a *transparent* (clear) finish is best. After a project activity has been polished, handle it with a cloth or a paper towel. Put on a coat of paste wax as a temporary finish. For a more permanent finish, brush or spray a clear metal lacquer on the project activity. Do not put on wax after lacquer. For the brush finish, warm the project activity slightly. Then apply a thin coat of lacquer with a brush. Using clear lacquer in a spray can is also a good way of putting on a finish.

Unit 143

Cutting Sheet Metal

Metal must be cut to size for the parts of sheet-metal projects. Tin snips and squaring shears are used for most metal cutting.

Tools

Tin Snips. There are many different types of tin snips. They are used for cutting sheet metal 18 gage (1 millimeter) or less in thickness. Straight lines and outside curves are cut with straight snips. Duckbill (circular) snips or aviation snips are needed for irregular cutting and for inside work (Fig. 143-1 a, b, and c).

Squaring Shears. Squaring shears are used mostly for cutting large sheets into smaller pieces. This machine has a sharp blade. It cuts off metal when the foot pedal is pressed down. Most shears are 30 inches (750 millimeters) wide. They have a front and a back gage for cutting many parts of the same size (Fig. 143-2).

CAUTION: Never let anyone stand near the squaring shears when you are using them.

Cutting with the Tin Snips

1. Hold the snips as in Fig. 143-3. The cutting edge should be at right angles to the work.

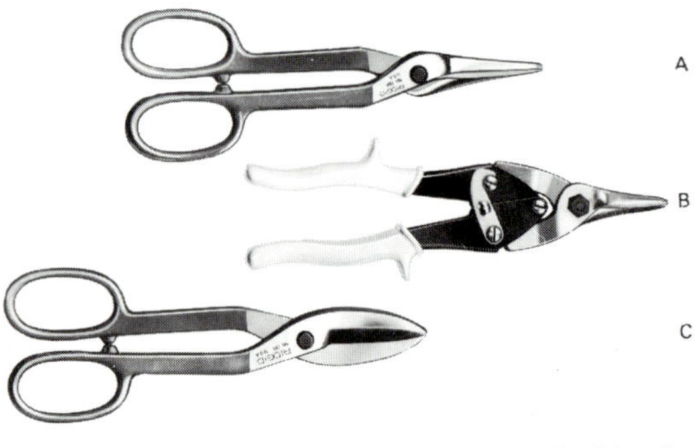

Fig. 143-1. Types of snips: (*a*) duckbill, (*b*) aviation, (*c*) straight. (Ridgid Tool Company)

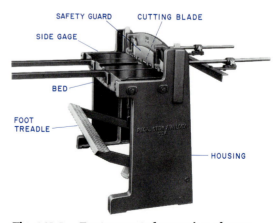

Fig. 143-2. Foot-operated squaring shears.

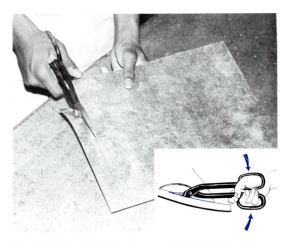

Fig. 143-3. Cutting sheet metal with tin snips.

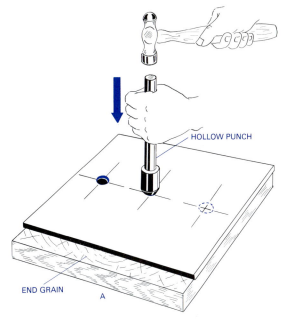

2. Open the snips as far as possible. Cut the metal to within $1/2$ inch (12 millimeters) of the end of the blade. Open and repeat the cutting. Cutting is done most often on the right edge of the sheet.

3. To cut into corners or to cut notches, use the point of the blade.

4. To cut an irregular shape or a curve, first rough-cut to within $1/8$ inch (3 millimeters) of the line. Then cut right to the line. This thin edge of metal will curl out of the way so that you can get a smooth edge (Fig. 143-4).

5. To cut an internal opening, punch or drill a hole in the waste stock. Complete the cut with hawksbill snips (Fig. 143-5 a and b).

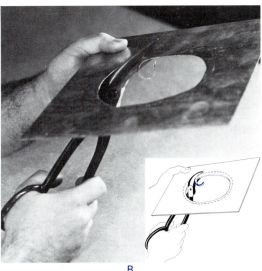

Fig. 143-5. (a) Punching a hole with a hollow punch, (b) cutting an internal opening with hawksbill tin snips.

6. Never cut nails, wires, rivets, or heavy bands with tin snips. Use a cold chisel, a hacksaw, or a bolt cutter.

CAUTION: Keep your fingers away from the blade at all times.

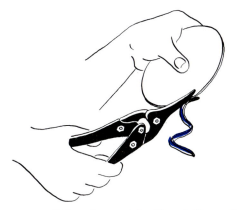

Fig. 143-4.   Trimming metal with aviation snips.

Cutting on the Squaring Shears

1. Slip the metal under the blade until the layout line is right under the cutting edge. Hold the left edge of the metal firmly against the left edge of the machine.

2. Apply even pressure to the hand pedal or the foot pedal to cut the stock (Fig. 143-6). Never cut wire, band iron, or metal heavier than 18 gage (1 millimeter) with squaring shears.

3. Adjust the front or the back gage to the needed length to cut several pieces the same size. Use the gage as a cutting guide.

Fig. 143-6. **Cutting sheet metal with a squaring shears.**

Unit 144

Bending Sheet Metal

Many sheet-metal parts must be bent and formed on stakes (Fig. 144-1 a and b). Others can be bent on a hand-worked bar folder. Examples of the types of bends made in sheet metal are shown in Fig. 144-2.

Tools

Sheet-metal Stakes. Some of the most commonly used sheet-metal stakes are shown in Fig. 144-3. Funnels and other cone shapes can be bent on a blowhorn

Fig. 144-1 a. **Simple bending produces this litter container.**

Fig. 144-1 b. **Some common sheet-metal bends.**

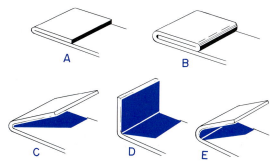

Fig. 144-2. Types of bends: (*a*) single hem, (*b*) double hem, (*c*) sharp-angle bend or fold, (*d*) 90° angle bend or fold, (*e*) rounded bend or fold.

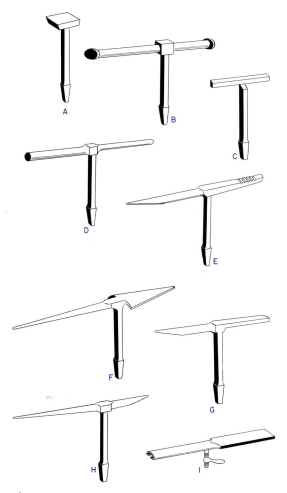

Fig. 144-3. Common sheet-metal stakes: (*a*) square, (*b*) double-seaming, (*c.*) hatchet, (*d*) conductor, (*e*) creasing, (*f*) blowhorn, (*g*) beakhorn, (*h*) candle-mold, (*i*) hollow-mandrel.

stake. A hatchet stake is used for sharp right-angle bends. Most sheet-metal bending, however, can be done over pieces of pipe or between wooden or angle-iron jaws. Always choose a rawhide or a wooden mallet to do sheet-metal bending. Metal hammers will mar the stake.

Bar Folder. The bar folder has many uses besides bending. It can be used to make a hem for stiffening an edge. Or it can be used to make an open fold for a folded or a grooved seam. The bar folder can also be used to make a rounded fold when you are getting ready for wiring an edge.

Most bar folders are 30 inches (750 millimeters) wide and have two adjustments (Fig. 144-4). The adjustment for making the bend sharp or round is made by raising or lowering the back wing. The bar folder is most often set to bend a sharp edge. The second adjustment is made by setting the gage to the correct amount of bend.

Box-and-pan Brake. This machine has an upper jaw that is made of removable fingers. These fingers will fit any width of bend (Figs. 144-5 and 144-6).

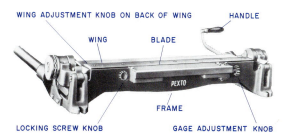

Fig. 144-4. A bar folder.

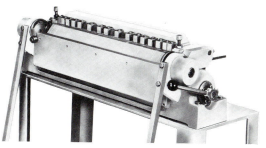

Fig. 144-5. A box-and-pan brake.

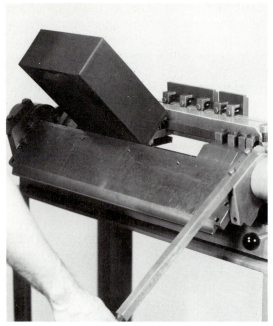

Fig. 144-6. Bending metal on a box-and-pan brake.

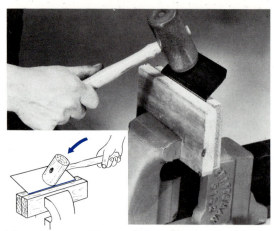

Fig. 144-7. Making a right-angle bend with the metal held between wooden jaws.

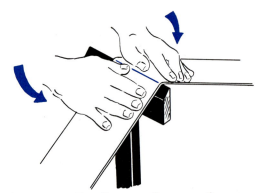

Fig. 144-8. Bending a sharp angle over a hatchet stake.

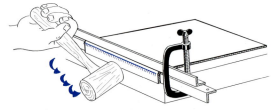

Fig. 144-9. Bending a hem over the edge of a table. The edge is bent at 90°.

Making Angular Bends

Method 1. Hold the metal with the layout line over the sharp edge of the bench. Force the metal over by hand. Square off the bend by striking it with a mallet.

Method 2. Fasten the metal between two long wood or angle-iron jaws. Clamp it in a vise. Force the metal over by hand. Strike it with a mallet to make the bend sharp (Fig. 144-7).

Method 3. Bend a sharp angle over a hatchet stake (Fig. 144-8).

A *hem* is a folded edge of metal for strengthening the edges of boxes and other containers. A single fold is called a *single hem*, and a double fold is called a *double hem* (Fig. 144-2). To bend a hem by hand, turn an edge down over the corner of a bench. Then reverse (turn over) the metal. Clamp it to the top of the bench. Close the hem by striking it with a wooden mallet (Fig. 144-9).

Making Circular Bends

Fasten a pipe, a rod, or a stake in a vise or a stake plate. Hold the metal over the bending device. Strike the metal with a mallet as you move it forward to form the curve (Fig. 144-10). To form cone-shaped objects, use the large end of a blowhorn stake. Force the metal around it with your hands. On heavy-gage metal, do the forming with a mallet (Fig. 144-11).

Making Bends on the Bar Folder

1. To bend a hem, adjust for the correct amount of bend. Then check the machine by making a practice bend on a scrap piece of metal. Insert the metal in the bar folder. Hold it in position with your left hand. Lift up on the handle with your right hand (Fig. 144-12). Release the handle. Remove the material. To close a hem, turn the metal with the folded edge up. Then bring down the handle with enough force to close the edge.

2. To bend a rounded edge, lower the back wing slightly so that a curve is bent. Adjust the gage for a fold equal to about $2^1/2$ times the diameter of the wire. Proceed as before.

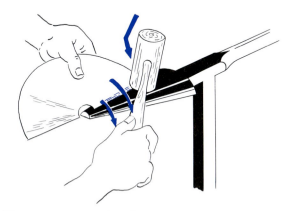

Fig. 144-11.   Forming the body of a funnel over a blowhorn stake.

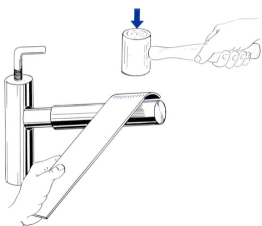

Fig. 144-10.   Making a circular bend.

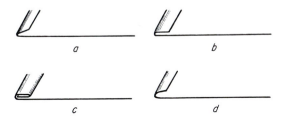

Fig. 144-12.   Bends made on the bar folder.

Unit 145

Rolling Sheet Metal

Many products have parts that are circular in form (Figs. 145-1 and 145-2).

The quickest way to form a sheet-metal *cylinder* (tube) is to use a slip-roll forming machine (Figs. 145-3 and 145-4). Follow these steps:

1. Adjust the lower roll to let the metal slip between it and the upper roll under slight pressure.

2. Adjust the idler (back) roll parallel to the front rolls at a distance wide enough to form the correct-size cylinder.

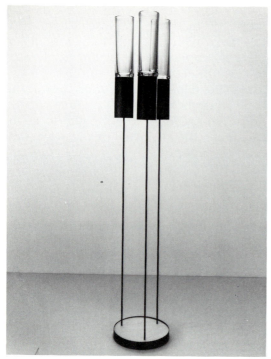

Fig. 145-1. The circular metal parts on this lamp were rolled into shape from pieces of flat sheet metal.

Fig. 145-2. Many sheet-metal products are circular in shape. (Niagara Machine and Tool Works)

Fig. 145-3. Using a slip-roll forming machine to form sheet metal. (Houdaille Industries, Inc.)

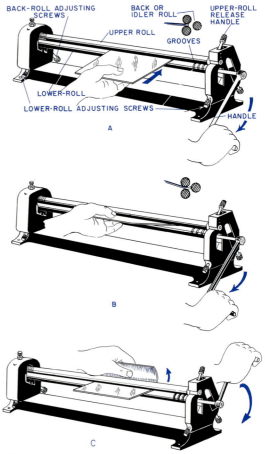

Fig. 145-4. (a) Starting the metal, (b) lifting up the metal, (c) completing the cylinder.

3. Put the metal in between the front rolls. Lift it slightly to begin forming the edge. Turn the handle to finish rolling the cylinder (Figs. 145-4 a, b, and c).

4. If needed, adjust the idler roll slightly to make the cylinder smaller. However, if it is too small, open it slightly by hand.

5. After the cylinder is formed, slip it off the right end of the upper roll, which can be swung free.

6. If the metal has a folded edge or end, insert this just inside the front rolls before forming.

7. If the metal has a wired edge, place this edge down in one of the grooves at the right end of the lower roll before doing the rolling.

Making Seams

Seams are one way of fastening sheet metal together. There are many kinds of metal seams. The simplest metal seams to make are *butt, lap, folded,* and *grooved* seams (Fig. 146-1). You will use these most often.

Tools

A hand groover and a hammer are needed to lock a grooved seam (Fig. 146-2). A butt seam most often is soldered or welded. A lap seam is most often riveted and then soldered if it is to be waterproof.

Making a Folded or a Grooved Seam

1. Allow extra material equal to three times the width of the seam. Usually, one-half of this extra material is added to either end.

2. Make a sharp open fold at either end of the material on the bar folder or by hand. On closed objects, make sure that the folds at the ends are in opposite directions (Fig. 146-3).

3. Hook the ends together. Place them over a solid surface, such as a stake or the top of a bench.

4. To make a folded seam, strike blows all along the seam with a wooden mallet.

5. To make a grooved seam, use a hand groover with a groove slightly wider than

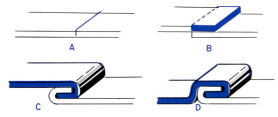

Fig. 146-1. Types of seams: (*a*) butt seam, (*b*) lap seam, (*c*) folded seam, (*d*) grooved seam.

Fig. 146-2. Hand groover.

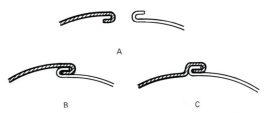

Fig. 146-3. The steps in forming a grooved seam: (*a*) The ends are folded in opposite directions. (*b*) The seam is closed. (*c*) The seam is locked.

the seam. Place the groover over the seam at one end. Lock by striking the tool firmly with a hammer. Then start at the other end to finish the seam by sliding the groover forward as you strike it with the hammer (Fig. 146-4).

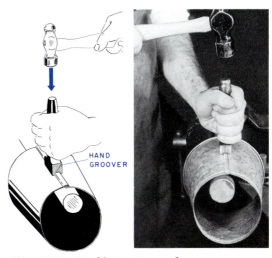

HAND GROOVER

Fig. 146-4. Locking a grooved seam.

Wiring an Edge by Hand

The edges of many containers are stiffened (made harder) by putting on a wire edge. On a rectangular object such as a small pan, the edge is bent first. Then the pan is formed. The wire is put in last. On most round containers, the wired edge is put on when the metal is flat. The container is then rolled into shape on the forming rolls. On a cone-shaped container such as a funnel, the edge is formed first. The wire is put in after the cone is shaped. The following steps are suggested:

1. Choose the right size of wire. Cut a piece to length. Usually, a 10- to 18-gage (1.2 to 3.5 millimeter) wire is used. Use 10-gage (1.2 millimeter) wire for the heaviest sheet metal and 18-gage (3.5 millimeter) wire for the lightest.

2. By hand or on the bar folder, make an open bend or a fold. Allow about 2½ times the diameter of the wire for the fold. This gives enough metal to cover the wire completely.

3. Form the wire to shape for rectangular and cone-shaped objects. Start at the end, corner, or seam to put the wire in the rounded fold.

4. Start to close the wired edge by striking it with a wooden mallet or squeezing it shut with pliers (Figs. 147-1 and 147-2). If pliers are used, tape the jaws to keep them from marking the metal.

5. Close the wire edge with a setting-down hammer (Fig. 147-3).

Fig. 147-2. Holding the wire in place with pliers.

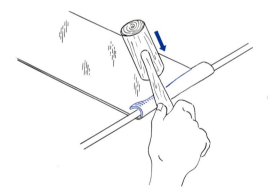

Fig. 147-1. Starting to close a wire edge. Use a wooden mallet first and then a setting-down hammer.

Fig. 147-3. A setting-down hammer.

Assembling Sheet Metal

Thin galvanized steel, black iron, tinplate, and aluminum may be put together in many ways. Riveting and spot welding are two ways of putting together sheet metal. Sheet-metal screws and epoxy cement can also be used to put together sheet metal.

Rivets

Rivets are measured by their weight per thousand (Fig. 148-1). For example, 1000 1-pound rivets weigh 1 pound (454 grams). Use a 1-pound rivet for 24- to 28-gage stock. Use a 2-pound rivet for 18- to 22-gage stock. A solid punch, a rivet set, and a riveting hammer are needed.

Riveting

1. Select a solid punch or drill that has the same diameter as the rivet shank. Choose a rivet set with a hole that will just slip over the shank.

2. Locate the holes a distance from the edge equal to 1½ times the diameter of the rivet.

3. Drill or punch the holes (Fig. 148-2). To punch, place the metal over a lead block or the end of a piece of hardwood. Hold the punch firmly and strike it a solid blow (Fig. 148-3).

4. Put the rivets in the holes. Place the work over a firm surface to hold the head.

5. Insert the hole of the rivet set over the shank. Strike the set several blows. This will draw the sheets together and flatten the metal around the hole (Fig. 148-4 a).

6. Strike the shank with the flat face of the hammer. This will fill up the hole and flatten the shank (Fig. 148-4 b).

7. Place the cone-shaped hole of the rivet set over the shank. Strike it several times to round off the shank (Fig. 148-4 c).

Fig. 148-2. Holes are drilled before installing aircraft rivets. These students are learning on the job. (McDonnell Douglas Corporation)

| 10 OZ | 12 OZ | 1 LB | 2 LB | 3 LB |

Fig. 148-1. Common sizes of sheet metal rivets based on weight. A 10-ounce rivet is approximately 2.5 mm in diameter, while a 1-pound rivet is about 3 mm in diameter.

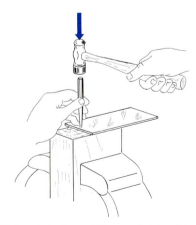

Fig. 148-3. Using a solid punch to cut holes in sheet metal.

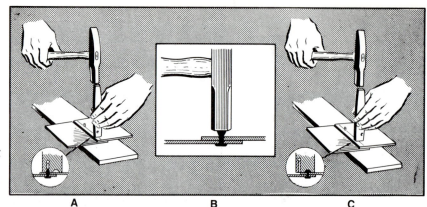

Fig. 148-4. (*a*) Drawing the sheets together, (*b*) upsetting, (*c*) rounding off the head of a rivet.

A   B   C

Riveting with a Pop Riveter

The easiest method of riveting is to use a tool called a *pop riveter* (Fig. 148-5 a and b). This tool eliminates the need for setting and flattening rivets by hand. And it requires working only one side of the piece (Fig. 148-6).

Unlike hand-installed rivets, pop rivets are hollow. On one end they have a *flange*, or head, which grips the work. On the other end is a pin, or *mandrel*, which is set in the riveter. The body of the rivet is placed in the hole. When you squeeze the handles of the riveter, the mandrel is pulled out of the rivet body until it breaks away from the ball-like head. By this time, the head is firmly embedded in the rivet body and the rivet is firmly fastened.

Always use pop rivets of the same metal as the metal in the product to be joined. Rivets are available in diameters of $3/32$ inch (2.4 millimeters), $1/8$ inch (3.2 millimeters), $5/32$ inch (4 millimeters), $3/16$ inch (4.8 millimeters), and $1/4$ inch (6.4 millimeters) in aluminum, steel, copper, and stainless steel.

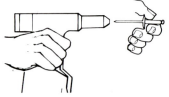

1. Insert rivet mandrel in rivet setting tool.

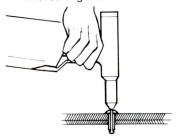

2. Using tool as a guide, insert rivet into prepared hole.

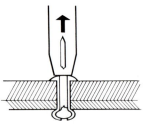

3. Squeeze trigger or handles to set rivet. Mandrel ejects after rivet is set.

Or insert rivet into prepared hole and then engage the mandrel with rivet-setting tool.

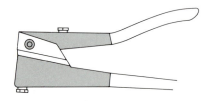

Fig. 148-5 a. Pop riveter.

Fig. 148-5 b. Steps in installing pop rivets.

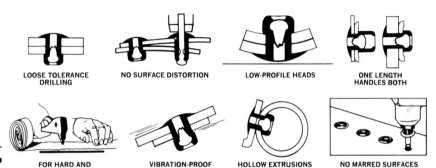

LOOSE TOLERANCE DRILLING   NO SURFACE DISTORTION   LOW-PROFILE HEADS   ONE LENGTH HANDLES BOTH

FOR HARD AND SOFT MATERIALS   VIBRATION-PROOF FASTENING   HOLLOW EXTRUSIONS AND TUBES   NO MARRED SURFACES

Fig. 148-6. Some advantages of using pop rivets.

Drill a hole in the two parts to be joined using the correct drill size.

Here is how to use pop rivets:

1. Insert the pointed end of the rivet mandrel into the rivet tool (Fig. 148-7).

2. Place the rivet in the prepared hole. As you operate the tool, the mandrel head is pulled into the rivet body. The rivet body expands, making it into a blind rivet holding the materials together (Fig. 148-8).

3. The mandrel comes apart from the rivet body and comes out of it.

4. The tool is now ready for the next rivet.

Self-tapping Sheet-metal Screws

Self-tapping sheet-metal screws cut their own thread in mild-steel sheet and soft aluminum alloys. These screws are especially useful when you want to be able to take the object apart. You will find them used in putting together many kinds of metal products. There are several kinds of head shapes with two types of points. Type A is pointed and is used to join metal up to 18 gage. Type B or Z has a flat point and is used for both light and heavy metals (Fig. 148-9).

Using Sheet-metal Screws

1. Choose and prick-punch the place for the hole.

2. Pick one of the most used sizes, such as No. 6 or No. 8.

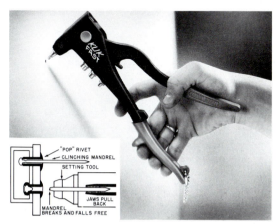

Fig. 148-7.   Fastening materials with pop rivets. First a hole is drilled where materials are to be joined. This shows the head of the rivet about to be inserted in one of the holes. (Reynolds Metals Company)

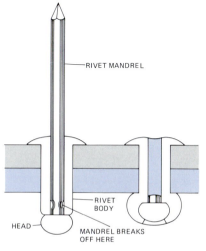

Fig. 148-8.   Here's how a pop rivet works.

■ **459**

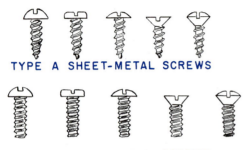

TYPE A SHEET-METAL SCREWS

TYPE Z SHEET-METAL SCREWS

Fig. 148-9.   Common sheet-metal screws.

3. Drill a hole equal to the root diameter of the screw. Use a $^7/_{64}$-inch (2.8-millimeter) drill for No. 6. Use a $^1/_8$-inch (3.2-millimeter) drill for No. 8. You can choose the drill size by holding the drill in back of the screw thread. The diameter of the drill should equal the inside diameter of the screw.
4. Drill the hole. Fasten the screw.

Spot Welding

Spot welding makes use of the heat given off by resistance to the flow of an electric current. This heat softens the metal pieces so that they will be welded together (Figs. 148-10 and 148-11). A spot welder will weld clean pieces of mild steel of up to $^1/_8$ inch (3 millimeters) combined thickness. The two pieces of metal are held together between the tips of the *tongs* (horns). Then a strong electric current makes the sheets weld by *fusion.* Most of the body parts of a car are put together by spot welding.

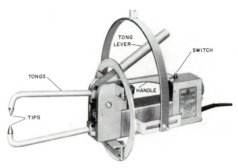

Fig. 148-10.   A portable spot welder.

Adhesives

Metal parts can be fastened to each other and to wood and plastic with an adhesive (Fig. 148-12). *Epoxy cements* are used in industry to fasten airplane parts together, for example. These cements are also used for such jobs as fastening metal electrical boxes to cement or wood buildings. To use an epoxy cement, follow these steps:

1. Clean and rough up the two surfaces to be joined.
2. Mix the two parts of the cement in correct proportion, following the manufacturer's directions.
3. Fasten the two pieces together. Clamp or weight them until they are dry.

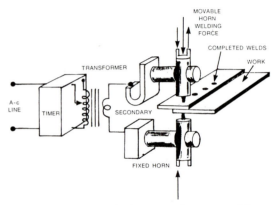

Fig. 148-11.   A spot weld is a resistance weld. No filler metal is needed. (Republic Steel Corporation)

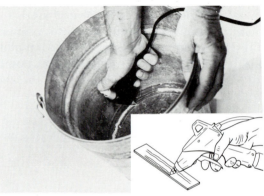

Fig. 148-12.   Using a hot melt glue gun to fasten metal together.

Soft Soldering

Soldering is a way of joining two pieces of metal. In soldering, you make use of an alloy that melts at a lower temperature than the metal being fastened. It is possible to make a waterproof container by careful soldering. Much soldering is done in sheet-metal work and in electrical construction. To do a good job, the metals must be very clean—free from dirt, grease, and oxide. The soldering copper must be large enough to give off enough heat. The soldering flux must be the correct kind (Fig. 149-1).

Materials

Solder is made of tin and lead. If they are in equal proportions, the solder is called *half-and-half.* Solder comes in the form of wire, a bar, or a hollow core. To help remove oxidation from the metal and to make the solder flow smoothly, a flux is used. An *acid* flux is used for black iron and galvanized iron. A *rosin* flux is used for tinplate, copper, and electric wire. If an acid flux is used, the article must be washed afterward with hot water to prevent *corrosion* (rust and decay). Wire solder with the flux in its core is used most of the time.

Tools

Ordinary soldering coppers are heated in a soldering furnace or with a blow-torch (Fig. 149-2). A 1-to 2-pound (0.5- to 1-kilogram) copper is usually used. A 150-watt electric *soldering copper* is very good for soldering heavy electrical parts (Fig. 149-3).

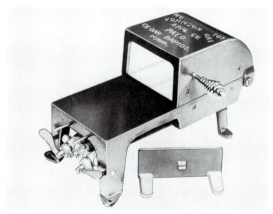

Fig. 149-2. A bench-model gas furnace for heating soldering coppers.

Fig. 149-1. Soldering electronic units together using a small soldering copper.

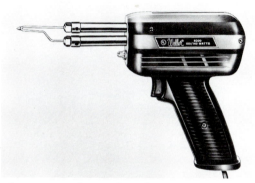

Fig. 149-3. A heavy duty electric soldering iron.

A 50- or 75-watt copper or gun is best for electronic work (Fig. 149-4).

Tinning a Soldering Copper

The tip of a soldering copper must be kept clean and *tinned* (covered evenly with solder). If it is overheated, the tinned point becomes corroded and must be tinned again, as follows:

1. File the point of the copper until it is clean and bright (Fig. 149-3).
2. Heat the copper until it is yellow to brown.
3. Rub the point on a bar of sal ammoniac. Add a little solder and rub until the point is tinned (Fig. 149-5).
4. Wipe the point with a cloth to take away the extra solder.

Fig. 149-4.   **Using a 50-W electric soldering copper to solder the connections on a transmitter-receiver.**

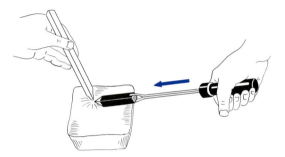

Fig. 149-5.   **Tinning the point of a soldering copper.**

Soldering

1. Place the two pieces to be soldered on a piece of brick or asbestos. The joint must be held tightly together. Clamp or wire it if necessary.
2. With a brush or a stick, put on a thin coat of the correct flux. No flux is needed with flux-core solder.
3. Heat the soldering copper until the solder melts freely when touched to the point. Never allow the copper to become red hot. If that happens, the tinned point will be burned off.
4. Place the face of the soldering copper at one end of the joint. Let the metal get hot. Then place a small amount of solder just ahead of the copper. Never melt the solder on the copper itself. Instead, make the metal hot enough so that it melts the solder.
5. *Tack* (spot-solder) the joint at several places to hold it well.
6. Begin at one end and move the copper slowly along in a forward direction. Add a little solder as needed (Fig. 149-6).
7. To sweat-solder two pieces together, cover one surface with a thin coat of solder. Then apply flux to the other surface. Clamp the two pieces together. Hold them with something other than your hands over a blue gas flame until the solder melts (Fig. 149-7).
8. To fasten feet or handles to a piece, apply a little solder at the point of contact. Hold the handle in place with pliers or tongs. Heat the area until the solder melts.
9. To solder a cylinder, first bind the cylinder together with oxide-covered soft

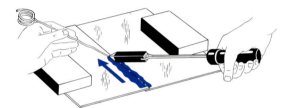

Fig. 149-6.   **Soldering a lap seam.**

iron wire. Hang it up several inches over a table. Apply the flux and solder to the inside of the joint. Hold the inner flame of a bunsen burner right under the joint. Move the flame back and forth (Fig. 149-8).

Fig. 149-7.   Sweat soldering.

Fig. 149-8.   Soldering a cylinder.

Unit 150

Metal Tapping

Good-looking designs for wall decorations, bookends, tie racks, and other projects (Fig. 150-1 a and b) can be made by tapping thin sheet metal (30-gage tinplate, copper, or brass). The tapping tools can be large common nails or punches with shaped points.

To make a wall decoration, follow these steps:

1. Cut a piece of metal to size.
2. Either transfer the design to the metal with carbon paper or glue the design to the metal.
3. Outline the design by tapping lightly with a nail or a punch. Do not tap so hard that you poke holes through the metal.
4. Tap in the background with a different tool. Always tap from the border inward and from the design outward (Fig. 150-2). This will keep the metal from stretching out of shape.
5. Clean, polish, and put a finish on the project. Metal lacquer will keep the metal from tarnishing.

Fig. 150-1 a.   Sheet-metal plaques made by tapping.

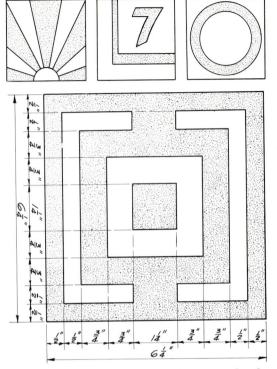

Fig. 150-1 b. Geometric designs can also be made by metal tapping.

6. Fasten the metal to a plywood back with aluminum, copper, or brass ¹/₄-inch (6-millimeter) *escutcheon* (round-head) pins or nails.

Fig. 150-2. Using a metal punch to do tapping.

Unit 151

Annealing and Pickling

When copper, brass, or aluminum is pounded, hammered, rolled, or formed, it becomes work-hardened and must be *annealed* (softened by heating). These metals also become dirty from oxides that form. Copper must be cleaned in an acid bath (pickling). Annealing and pickling are often done at the same time when metal is formed.

The pickling mixture for copper and its alloys must be kept in a covered stone or glass jar. The jar must be large enough to hold the project activity. The solution is made by adding one part hydrochloric acid to four parts water. Pour the correct amount of water into the jar first. Then *carefully* pour the acid into the water a little at a time.

CAUTION: Always pour acid into water, never water into acid.

To anneal and pickle copper at the same time, heat the copper piece in a soldering furnace or with a blowtorch to a low-red heat. Pick up the hot metal with copper tongs. Slip it into the pickling solution.

CAUTION: Be careful not to splash the liquid.

To anneal brass or nickel-silver, heat to a faint-red color. Avoid overheating these metals. Overheating will burn the zinc (a low-melting metal) in the alloy. Let the metal cool slowly in air.

To anneal aluminum, rub a little cutting oil on the metal surface. Heat it evenly until the oil begins to smoke. Cool the aluminum in air.

To clean copper or brass without annealing, place the cold metal in the pickling solution for 15 to 20 minutes. Then wash in clear water. Dry with paper towels or sawdust. Never handle clean metal with your bare hands. Fingerprints will cause spotting and mar a good finish.

Unit 152

Piercing with a Jewelers' Saw

A jewelers' saw is similar to a coping saw. It is used for fine cutting and *piercing* (internal cutting) (Fig. 152-1). The blades come in sizes from No. 8/0 (smaller than a thread) to No. 8 ($1/16$ inch in width). No. 0 is used for thin stock, and No. 2 or No. 3 is used for heavy stock. To cut with a jewelers' saw:

1. Fasten the blade into one end of the frame with the teeth pointing up and toward the handle. Apply pressure to the frame as shown in Fig. 152-2. Fasten the other end. If inside piercing is to be done, slip the blade through a hole drilled in the waste stock. Then fasten it.

2. Hold the work over a V block with your left hand. Operate the saw with your right hand (Fig. 152-3). The cutting is done on

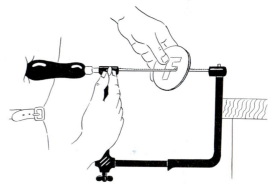

Fig. 152-2. Putting a jewelers' saw blade in a frame for internal cutting.

Fig. 152-1. The decorative cover for this bowl was pierced with a jewelers' saw.

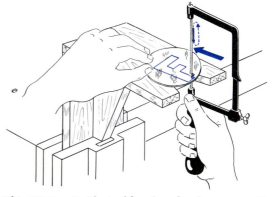

Fig. 152-3. Cutting with a jewelers' saw on a V block.

the down stroke. Beginning in the waste stock, cut up to the line. Do not twist the blade, as it breaks easily. To turn a corner, move the blade up and down with no for-

ward pressure as you turn the frame slowly.

3. Apply a little wax or soap if the blade tends to stick.

Unit 153

Forming Metal and Raising a Bowl

Shallow plates, trays, and dishes of copper, brass, or aluminum can be made by pounding metal into a wooden or metal form (Fig. 153-1 a, b, and c). This process is known as *raising*. The easiest way to raise a deep bowl is to use a raising block. This is a block of wood into which a small circular *depression* (hollow) has been cut (Figs. 153-2 and 153-3).

Industry uses this process (*cold forming*) in shaping many kinds of metal parts.

Many free-form designs may be shaped by raising. Products formed by raising are shown in Fig. 153-4.

Tools

Forming Hammer. A hammer of wood or metal can be used to do a very fine job of forming metal.

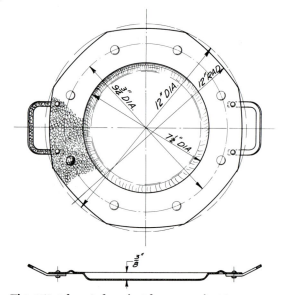

Fig. 153-1 b.   A drawing for a serving tray.

Fig. 153-1 a.   This serving tray was formed by pounding down part of the metal into a wood form.

Fig. 153-1 c.   This tray was formed by raising.

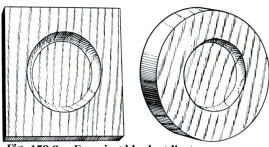
Fig. 153-2.   Forming blocks (dies).

Fig. 153-3.   A raising block.

Fig. 153-5.   Using a form hammer to form a tray in a wooden block.

Fig. 153-4.   These free form bowls and spoon were raised over a block.

Raising Hammer. Raising a bowl can be done very well with either a wooden or a metal hammer.

Forming Metal

1. Cut a piece of metal to the shape of the form and a little larger than the finished object. If the form is round, cut the metal in a disk shape.

2. Lay out a line to show where the *depression* (low part) is to start. On round objects, draw a circle with a compass.

3. Either hold the metal over the form or clamp it to the form. Start to stretch the metal by striking it just inside the layout line. Overlap the blows. Strike all the way around the metal. Form a little at a time (Fig. 153-5). The metal will become work-hardened very soon. Then it will need annealing.

4. Place the metal in the form again. Continue to stretch it until the bottom of the form is reached. Keep the edge of the metal flat by striking it often with a wooden mallet. Make sure that the bottom of the form is also flat.

5. Lay out and cut the outside edge to the shape you want. Then file it smooth and decorate it.

6. Clean the metal by pickling. Planish the edge if you want to. For an *antique* (old-looking) finish on copper or silver, color and highlight the metal with a liver of sulfur solution.

7. Clean the metal again. Apply a coat of wax or lacquer as protection against the metal tarnishing.

Raising a Bowl

1. Cut a disk of metal a little larger than the finished size. This disk may be made to any shape. If it is a round piece of metal, draw a number of equally spaced guide lines.

2. Fasten the forming block firmly in a vise.

3. Hold the edge of the disk at a low angle over the depression. Start to form the metal by striking it over the depression in the block (Fig. 153-6). Go all the way around the disk. Follow the guide lines. Be careful to keep the edge free of wrinkles.

4. Anneal and pickle the disk when it becomes hard.

5. As the bowl takes shape, lower the angle until it is formed to about the right shape. The bowl will look very crude at this time (Fig. 153-7).

6. Change to a round metal stake when you have the proper shape. Pound the shape smoother with a wooden mallet, as shown in Fig. 153-8.

7. Clean the bowl when it is completely formed. Then mark the bowl, trim the edge, and planish the outside surface (Fig. 153-9).

8. Clean again, polish, color, and lacquer the project activity.

Fig. 153-7.　Shapes of a bowl through the various stages in raising.

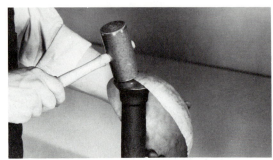

Fig. 153-8.　Removing the irregularities.

Fig. 153-6.　Starting to raise a bowl in a raising block.

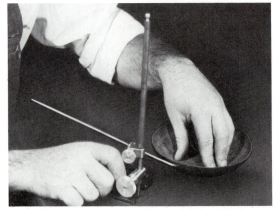

Fig. 153-9.　Marking a line around a bowl.

Etching

Etching is an industrial process used in producing many kinds of products, such as the "chips" used in computers. (See Unit 5.) *Etching* is also a way of decorating a metal surface (Figs. 154-1 and 154-2). A *resist* (acid-proof material) covers part of the metal surface in the form of a design. Then chemicals are put on to eat away the uncoated part of the metal (Fig. 154-3). To etch copper, brass, or aluminum, follow these steps:

1. Clean the metal in a pickling solution. Do not touch the metal after it is clean.
2. Transfer the design to the metal surface.
3. Cover the areas *not to be etched* with a resist. Asphaltum varnish is most often used (Fig. 154-3). Let the resist dry for 24 hours. Sometimes, for simple etching, the metal surface is covered with beeswax.

Fig. 154-2.   An etched tray.

Fig. 154-3.   Covering the area with a resist using a small brush.

Then the design is scratched or cut out. If the project activity is a dish, only the top needs to be coated with the resist. The acid can be poured into the dish. However, flat articles must have all edges and the back covered because they must be put right into the acid.

4. Place the metal in a solution of one part nitric acid and two parts water. For aluminum, the solution should be one part muriatic (hydrochloric) acid to one part

Fig. 154-1.   The surface of this bowl was decorated by etching away part of the metal surface to form the design.

CAUTION: Always pour acid into water, never water into acid. Be careful not to splash the liquid.

water. A nonacid etching solution can also be used for aluminum.

5. Keep the solution moving with a feather or a small bit of cotton on the end of a stick (Fig. 154-4). Allow the metal to stand in the solution about 1 hour for copper and 1/2 hour for aluminum. Remove the project activity often from the solution to check it.

6. When the etching is completed, wash the metal thoroughly to remove any traces of the etching solution. Then remove the resist with the proper remover as recommended by the manufacturer. Clean, pol-

ish, and finish the surface.

Fig. 154-4.   Pouring the acid into the object to be etched. Note that the acid is stirred with a wooden stick with cotton on one end.

Unit 155

Enameling on Metal

Enameling with ceramic materials adds color and decoration to a metal surface (Fig. 155-1). Stoves, refrigerators, and kitchen utensils are enameled. Thousands of items are treated in this way to add beauty and prevent rust and stains.

You can decorate small metal objects, such as jewelry items, bowls, and ashtrays. When *fused* (melted on) to copper, the ground glass creates colors resembling a glaze. You will need a kiln (Fig. 155-2). You will need the tools shown in Fig. 155-3.

Fig. 155-2.   A small, inexpensive kiln for metal enameling.

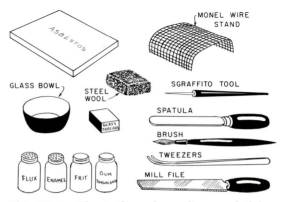

Fig. 155-3.   The tools and supplies needed for enameling on copper.

Fig. 155-1.   A decorative edge on this container was enameled.

Fig. 155-4. Clean the pieces of copper by dipping them in a commercial cleaning liquid. Rinse the pieces in water and then polish them with steel wool. Be sure to wear rubber gloves to protect your hands from the cleaning liquid.

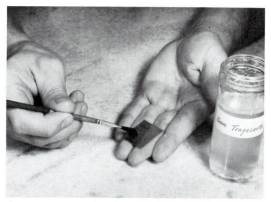

Fig. 155-5. Brush a light coat of gum tragacanth on the cleaned surface of the copper.

Enameling on Copper

1. Turn on the electric kiln at least ½ hour before you will use it. Heat it to about 1500°F (815°C).

2. Clean the copper pieces with liquid detergent and steel wool (Fig. 155-4). Wash off the pieces with water and dry them.

3. Polish the cleaned pieces with steel wool.

4. Apply a coat of gum tragacanth or liquid adhesive (Fig. 155-5).

5. Dust flux (first covering) over the top of the piece lightly (Fig. 155-6). A piece of nylon stocking fastened over the open container will produce an even coating of flux by sifting.

6. Put the copper piece in the kiln. Fire it until the flux melts. Handle it with a spatula (Fig. 155-7).

7. Remove the copper piece from the kiln. Let it cool. Then clean the back surface with steel wool.

8. Choose a background color. Brush another light coating of gum tragacanth on the melted-flux surface. Evenly dust the background color.

Before you fire this coating, you may add more decoration. Glass thread, *frit* (small broken glass lumps), or various col-

Fig. 155-6. Dust flux lightly and evenly over the coating of gum tragacanth. Then dust an even coating of enamel over the flux. The piece is ready for firing.

Fig. 155-7. Place the coated copper pieces carefully in the kiln. Use a spatula or knife. Place the pieces on a trivet when the bottoms are also coated. You will know that the enamel is properly fused when it becomes smooth and looks wet.

■ 471

Fig. 155-8. Definite designs can be created on the base coat by using glass threads. Coat the pieces again with gum tragacanth before adding the design pieces.

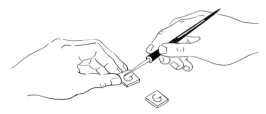

Fig. 155-10. To make sgraffito decoration, first dust on another coating of gum tragacanth. Then dust on a second, unfired coating of enamel of a contrasting color. Scratch a design through this second coating. Then fire the copper-enameled piece again, and cool it.

ors in glass powder may be used (Fig. 155-8). Let the decorated piece dry before firing it.

9. Place the piece in the hot kiln. Leave it there until the background color and the added decoration melt into each other.

10. Cool the piece by placing it on a wire rack or an asbestos sheet (Fig. 155-9).

11. Clean the back by putting the piece in a commercial cleaning fluid. Then use steel wool to polish it.

Enamel decoration may be added and refired as many times as you choose. Add gum tragacanth each time to hold the new decoration in place until fired.

If you want to do *sgraffito* decoration, dust a second coating of enamel on the fused flux. Then trace through it with a sharp-pointed tool. The undercoat will show through (Fig. 155-10). Fire the piece.

12. For finishing touches, clean the un-enameled underneath surface with a cleaning preparation. Polish it with steel wool and then file the edges to make them bright (Fig. 155-11). Brush the cleaned metal parts with a coating of metal lacquer to prevent tarnish on the project activity.

Fig. 155-9. Remove the enameled piece from the kiln. Place it on a wire stand to cool.

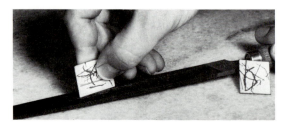

Fig. 155-11. The edges may be dressed on a flat file and then polished on emery cloth. Clean the backs as shown in Fig. 155-4.

Unit 156

Low-temperature Brazing and Hard Soldering

Low-temperature brazing and hard soldering are ways of joining two pieces of metal with an alloy that melts at tempera-

tures above red heat (Fig. 156-1 a and b). In art-metal work, a liquid or paste you can buy is used as a flux. A welding torch or a

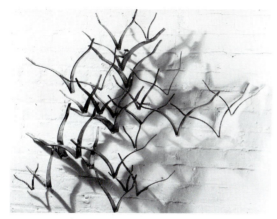

Fig. 156-1 a.   This wall sculpture was assembled by brazing the parts together. (The Sculpture Studio Inc.)

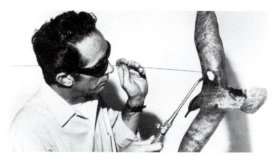

Fig. 156-1 b.   Forming a bird design by brazing.

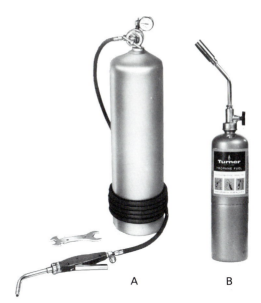

Fig. 156-2 (***a***) **An acetylene tank, hose, and torch needed for hard soldering.** (Ridgid Tool Company) (***b***) **A propane torch that can be used for hard soldering.**

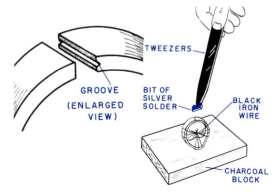

Fig. 156-3.   Placing bits of solder on the joint.

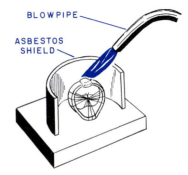

Fig. 156-4.   Applying heat to the joint.

propane torch with air pressure can be used to heat the metal (Fig. 156-2 a and b). Work in the following way:

1. Make a tight-fitting joint. If necessary, file a little groove along the joint for the solder to run into. Fasten the joint tightly. Tie the strips together with black iron wire.

2. Apply a small amount of flux to the joint with a brush.

3. For small objects, cut small bits of silver solder and place them with tweezers along the joint (Fig. 156-3). On larger project activities, feed the solder or brazing material into the joint as the heat is applied.

4. Preheat the area until the flux dries out. Then heat the joint until it is bright red (Fig. 156-4). If possible, heat it from underneath. Never apply heat right to the solder because the solder will form into balls and roll away. When the solder begins to melt, move the torch quickly along until the solder flows evenly.

5. Clean, file, and smooth the joint.

Unit 157

Polishing Metal

Most art-metal project activities must be given a high polish. This can be done by hand. But it is much easier to use a power polisher with soft cloth wheels. The three materials most often used come in stick form. *Tripoli,* a coarse compound, is yellowish-brown in color. *Rouge* is red and gives a fine polish. *Whiting* is a fine white polish. To do a good polishing job, follow these steps:

1. Fasten a soft, pliable polishing wheel of muslin, cotton, or wool to the grinder or buffing head.

CAUTION: Always wear goggles when polishing.

2. Turn on the power. Hold the polishing compound lightly against the revolving wheel to coat the edge.
3. Hold the project activity firmly against the lower edge of the wheel. Move it back and forth as the surface is polished (Fig. 157-1). Add more polishing compound as needed.

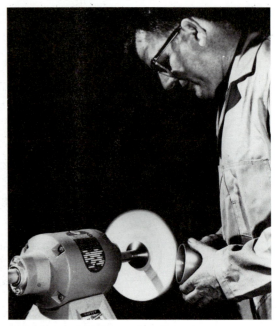

Fig. 157-1. Polishing a bowl.

4. Use a different wheel for each different kind of compound.
5. Use liquid metal polishes for hard polishing.

Unit 158

Forging Metal

Forging (also called hot forming) is another of the major ways in which industry shapes metal parts. Forgings are used in parts that must not fail, such as aircraft landing wheels or parts that hold the wings on airplanes.

When heavy stock is heated, it can be formed and bent in the same way as thin

Fig. 158-1 a. **Most hand tools are produced by the forging process.** (Ridgid Tool Company)

Fig. 158-1 b. **Forging an offset screwdriver or a punch is good experience in hot-forming metal.**

Fig. 158-1 c. **This candle snuffer would be a good project activity in forging.**

metal is worked when it is cold. It can also be stretched, upset, shortened, or changed in shape by forging (Fig. 158-1 a, b, and c).

Tools and Equipment

An anvil, a blacksmith's hammer, and a pair of tongs for holding the hot metal are needed (Fig. 158-2). The metal is heated in a gas, oil, or coal forge (Fig. 158-3).

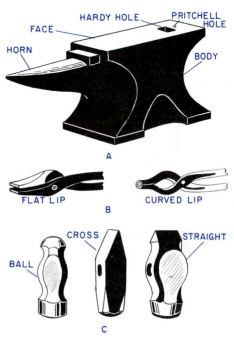

Fig. 158-2. (*a*) Anvil, (*b*) tongs, and (*c*) blacksmith's hammers.

Fig. 158-3. **Heating metal in a gas furnace for forging.**

Forging

1. Insert the metal in the forge. A gas or oil furnace should burn with a clear blue flame. This way, the metal does not oxidize.

2. Heat the metal until it is bright red. Hold the metal firmly with tongs to do the different forging operations.

3. To do *tapering*, as in making a cold chisel, hold the heated end at a slight angle on the face of the anvil. Strike it with a blacksmith's hammer to taper it. Always keep the metal at forging temperature. When the metal is shaped, reheat it to a dull-red heat and hammer out the rough parts.

4. *Drawing out* is done to make a piece of stock longer or to reduce its cross-sectional area. To do this, heat the metal to the proper temperature. Then strike it. For example, you can change its shape from

ROUND TO SQUARE   SQUARE TO OCTAGON   OCTAGON TO CONE

Fig. 158-4. Steps in forging the point of a prick punch or center punch.

Fig. 158-5. (*a*) Forming an angle over an anvil, (*b*) making a circular bend over the horn of an anvil.

square to round and then back to square as the drawing out is done. This will keep the metal from cracking or breaking (Fig. 158-4).

5. *Bending, twisting, and flaring* are done in the same way as for cold metal. To do these operations, heat the metal to bright red (Fig. 158-5 a and b).

Unit 159

Heat-treating

Small tools and parts are made of carbon tool steel. But they must be hardened and tempered before they can be used. This is called *heat-treating* (Fig. 159-1). Heat-treating is used for such tools as cold chisels, punches, and gages. To make a project activity such as a file from a piece of hard steel, the metal must first be annealed.

Annealing

This process softens metals. Bring the metal to a bright-red heat in a furnace. Then pack it in sand or some other non-conductor. Allow it to cool slowly.

Hardening

This process makes steel hard and brittle. Heat the metal to a bright-red heat in a fur-

Fig. 159-1. Using a heat-treating furnace. (Johnson Gas Appliance Company)

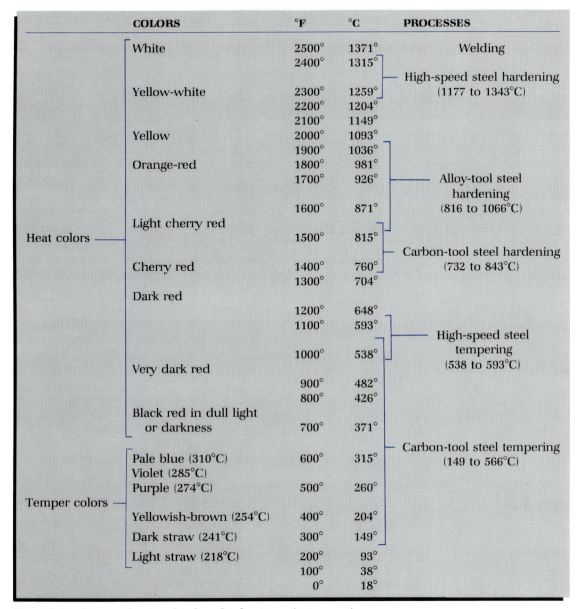

COLORS	°F	°C	PROCESSES
White	2500°	1371°	Welding
	2400°	1315°	High-speed steel hardening (1177 to 1343°C)
Yellow-white	2300°	1259°	
	2200°	1204°	
	2100°	1149°	
Yellow	2000°	1093°	
	1900°	1036°	Alloy-tool steel hardening (816 to 1066°C)
Orange-red	1800°	981°	
	1700°	926°	
	1600°	871°	
Light cherry red	1500°	815°	
Cherry red	1400°	760°	Carbon-tool steel hardening (732 to 843°C)
	1300°	704°	
Dark red			
	1200°	648°	
	1100°	593°	High-speed steel tempering (538 to 593°C)
	1000°	538°	
Very dark red	900°	482°	
	800°	426°	
Black red in dull light or darkness	700°	371°	
Pale blue (310°C)	600°	315°	Carbon-tool steel tempering (149 to 566°C)
Violet (285°C)			
Purple (274°C)	500°	260°	
Yellowish-brown (254°C)	400°	204°	
Dark straw (241°C)	300°	149°	
Light straw (218°C)	200°	93°	
	100°	38°	
	0°	18°	

Heat colors (White through Black red in dull light or darkness)
Temper colors (Pale blue through Light straw)

Fig. 159-2. Correct heat and colors for heat-treating operations.

nace or forge or with a welding torch. Then *plunge* (dip quickly) the metal into a pail of water. Keep the small end down so that all the parts will cool evenly. Move the metal back and forth to cool it quickly. After hardening, the metal will be brittle and will crack easily (Fig. 159-2).

Tempering

Tempering is the process of making metal less hard so that it can be used. Reheat the hardened metal to a lower temperature. Then dip it in water again.

To temper small tools on which a hard point and a soft handle are needed, follow this method:

1. Heat a large piece of scrap steel to red heat. Set it on a soldering or welding bench.
2. Use an abrasive cloth to clean off one side of the tool that has already been hardened.
3. Place the tool handle on the heated block. Watch the temper colors run from the handle to the point. When the proper color reaches the point (straw color for

°F	°C	COLOR	TOOLS
430°	220°	Yellow	Lathe tools
470°	243°	Straw	Punches and knives
500°	260°	Brown	Wood chisels
540°	282°	Purple	Cold chisels
570°	300°	Blue	Screwdrivers and springs

Fig. 159-3. Correct heat and colors for tempering tools.

center punches, for example), cool the tool in water again (Fig. 159-3).

Hardening and tempering can be done at the same time. First bring the tool to a bright-red heat. Then cool just the point in water. Then clean the area near the point with abrasive cloth. Again cool the whole tool when the proper temper color flows to the point.

Casehardening

This is a way of adding carbon to the outer surface of mild steel and then hardening this outer case. Casehardening is done on tools that are not to be ground. C clamps and hammer heads are two examples. Casehardening makes a hard exterior with a soft, tough interior.

To caseharden a surface, pack the tool in a metal box filled with a high-carbon material. Heat this box in a furnace for about 1 hour. Remove the tool from the box. Reheat it to a bright-red heat and plunge it in water.

Unit 160

Making a Casting

Castings are made in a foundry. Molten metal is poured into sand molds (Fig. 160-1). Simple objects, such as bookends, paperweights, and small trays, may be made in the same ways (Figs. 160-2 a and b and 160-3) used in factories for making larger industrial castings.

A pattern is needed for all casting. For simple castings, the original article may be used. A pattern may be made of white pine also. The pattern must be a little bit tapered so that it can be pulled out of the sand mold without breaking. This taper is called the *draft*.

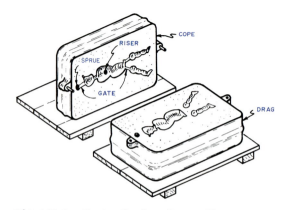

Fig. 160-1. Parts of a foundry mold.

Fig. 160-2 a. Castings are found everywhere. The base of this table is cast aluminum.

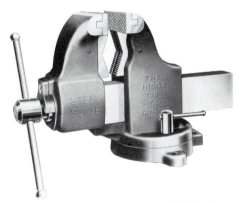

Fig. 160-2 b. The major parts of this vise were cast.

Fig. 160-3. Small foundry projects.

Tools and Equipment

The tools and equipment needed for making a casting include a molding board, a flask, a bellows, a shovel, a riddle, and molder's tools (Fig. 160-4). A melting furnace is also needed (Fig. 160-5).

Foundry Materials

The *foundry sand*, or *compound*, must be moist enough to show the shape of your fingers when a handful is pressed. Talcum powder or *parting sand* (fine beach sand)

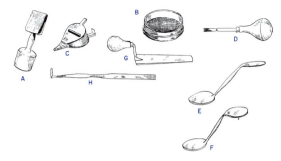

Fig. 160-4. Foundry tools: (*a*) bench rammer, (*b*) riddle, (*c*) bellows, (*d*) molder's bulb, (*e*) spoon-and-gate cutter, (*f*) slick and oval, (*g*) finishing trowel, (*h*) lifter.

■ **479**

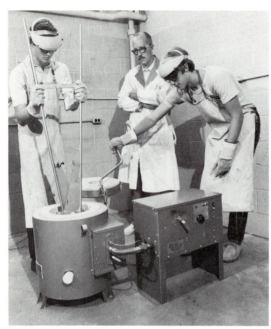

Fig. 160-5. A melting furnace. Note that the student is using tongs to lift the crucible out of the furnace. Both students are using safety protection.

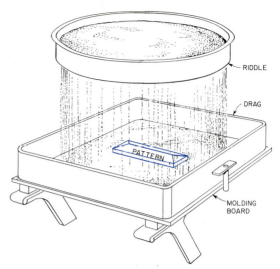

Fig. 160-6. Sand being riddled into the drag.

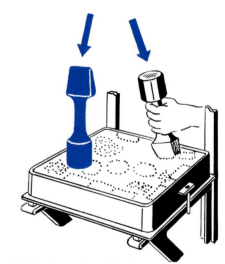

Fig. 160-7. Ramming the drag.

keeps the two parts of the mold from sticking together.

Making a Simple Mold

1. Place the pattern on a molding board. Keep the draft, or tapered side, up.

2. Place the *drag* (bottom part of the flask) over the board with the pins down.

3. Cover the pattern and the molding board lightly with parting sand.

4. Shovel some foundry sand into a *riddle* (screen). Shake it back and forth over the pattern until it is covered to a depth of about 1½ inches (40 millimeters) (Fig. 160-6).

5. Pack the sand around the pattern and into the corners with your fingers.

6. Shovel some sand just as it comes from the pile over the pattern. Pack the pattern firmly with a rammer. First use the peen end and then use the butt end for packing (Fig. 160-7).

7. Strike off or scrape the extra sand from the top of the drag with a straight stick or a rod (Fig. 160-8).

8. Place a second molding board on top of the drag. Turn the entire thing over (Fig. 160-9). Remove the top molding board. Blow off the loose sand with the bellows. Smooth the upper surface.

9. Place the *cope* (top half of the flask) in position. Insert a *riser pin* (a straight wood pin) about ½ to ¾ inch (13 to 19 millime-

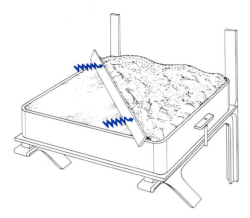

Fig. 160-8. Striking off the drag.

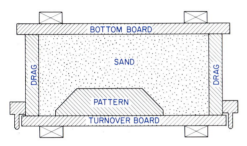

Fig. 160-9. The drag ready to be turned over.

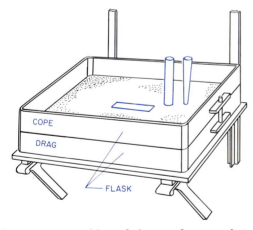

Fig. 160-10. Position of riser and sprue pins.

ters) from the pattern and about ½ inch (13 millimeters) into the sand. The riser will allow for any excess metal and leave enough for shrinkage (Fig. 160-10).

10. Now place the *sprue pin* (a tapered wood pin) about 2 to 3 inches (51 to 76 mil-

limeters) from the riser. This will form the hole. The hot metal will flow through it.

11. Sprinkle the upper surface with parting sand.

12. Repeat steps 3 to 7.

13. Punch small holes with a piece of wire to about ⅛ to ½ inch (3 to 13 millimeters) from the pattern. This is called *venting*.

CAUTION: Always wear gloves, goggles, and leggings when you pour the metal.

14. Remove the sprue and the riser pins. Wiggle and draw them out. Make a funnel-shaped hole at the top of each.

15. Lift off the cope. Set it to one side on its edge.

16. Remove the pattern. Insert a *metal pin* (draw spike) into the back of the pattern. Wet the edges around the pattern with a molder's bulb. Loosen the pattern by rapping or striking the pin on all sides with a metal bar. Then draw the pattern straight up (Fig. 160-11). If this is done the right way, there will be a clean impression in the sand. If needed, repair by adding bits of sand.

17. Bend a piece of sheet metal to a U shape. Cut a gate or groove from the sprue and riser pins to the impression. Blow out any loose sand.

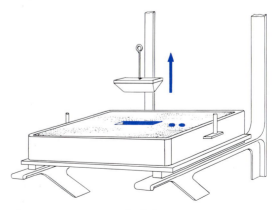

Fig. 160-11. Removal of the pattern.

18. Close the mold by replacing the cope and placing a weight on it.

19. Heat some lead, aluminum, or a low-melting alloy in a ladle in a furnace.

20. Slowly pour the hot metal into the sprue hole until it is full (Fig. 160-12 a and b).

21. Allow the metal to cool.

22. Break up the mold.

23. Cut off the metal that formed the sprue pin, riser pin, and gate. File the edges. Paint or decorate.

Fig. 160-12 a.   Removing the ladle full of hot molten metal from the furnace. Follow all safety rules as shown here.

Fig. 160-12 b.   Pouring the molten metal into the sprue hole. These students are practicing good foundry safety. Note that they are wearing face protection, gloves, sleeves, aprons, and leggings. (McEnglevan Heat Treating and Manufacturing Company)

Unit 161

Machine Shop

Machining is the process industry uses when the parts must be very accurate with a smooth finish (Fig. 161-1 a, b, and c). Metals are cut to shape on many different power tools in the machine shop. As you do more advanced work in metalworking, you will have a chance to use all these machines and learn more about the importance of the machine tools in everyday life. Some of these tools are:

1. The *drill press* is used for drilling holes, boring, reaming, and other cutting operations. You have used the drill press in making many of your bench-metal project activities (Fig. 161-2 a).

2. The *shaper* is used for finishing flat surfaces. It is also used for cutting grooves and keyways. It has a single-point tool on

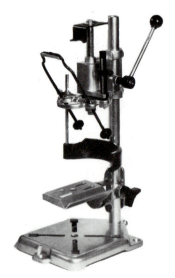

Fig. 161-1 a.   Many of the parts on this drill-press stand must be machined.

Fig. 161-1 b. All the parts of this brass balance scale are machined.

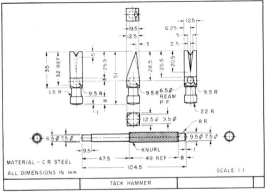

MATERIAL - C R STEEL
ALL DIMENSIONS IN mm

SCALE 1:1

TACK HAMMER

Fig. 161-1 c. This tack hammer is a typical beginning machine-shop project activity.

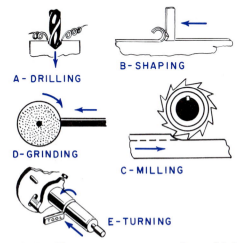

A - DRILLING

B - SHAPING

D - GRINDING

C - MILLING

E - TURNING

Fig. 161-2. Five common ways of machining metal: (a) drilling, (b) shaping, (c) milling, (d) grinding, (e) turning.

the end of a ram that moves back and forth to do the cutting. The *planer* operates in somewhat the same way, except that the work, rather than the tool, moves back and forth (Fig. 161-2 b).

3. The *milling machine* is used for cutting metal to shape with a revolving cutting tool (Fig. 161-2 c).

4. The *grinder* is used for smoothing and shaping work that is already been machined (Fig. 161-2 d). The two basic types are the surface grinder (for finishing flat work) and the cylindrical grinder (for grinding round work).

5. The *lathe* is used for turning metal to straight, tapered, or curved shapes (Fig. 161-2 e). A lathe holds the work securely and turns, or rotates, it (Fig. 161-3 a and b) while a cutting tool moves along to remove excess metal. The cutting tool can be fed lengthwise or crosswise.

Size and Parts of the Lathe

The size of the lathe is known by the swing and by the length of the bed. The *swing* is

Fig. 161-3 a. A student machinist operating a lathe. (Rockwell International)

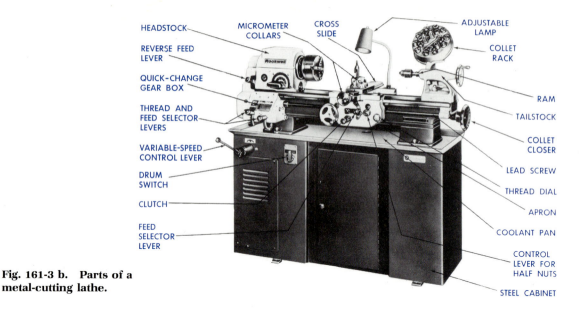

Fig. 161-3 b.   Parts of a metal-cutting lathe.

Labels on the figure:
HEADSTOCK
MICROMETER COLLARS
CROSS SLIDE
ADJUSTABLE LAMP
COLLET RACK
REVERSE FEED LEVER
QUICK-CHANGE GEAR BOX
THREAD AND FEED SELECTOR LEVERS
VARIABLE-SPEED CONTROL LEVER
DRUM SWITCH
CLUTCH
FEED SELECTOR LEVER
RAM
TAILSTOCK
COLLET CLOSER
LEAD SCREW
THREAD DIAL
APRON
COOLANT PAN
CONTROL LEVER FOR HALF NUTS
STEEL CABINET

the largest diameter that can be turned. It is important to learn the names of the parts of the lathe and what each part is used for. The parts of a lathe are shown in Fig. 161-3 b.

Accessories

1. Chucks are used to hold stock when facing the end and for drilling, boring, reaming, and other cutting operations (Figs. 161-4 and 161-5).

2. Lathe dogs are used for turning between centers. There are two types, the bent-tail and the clamped (Fig. 161-6).

3. Tool bits are small rectangular pieces of high-speed steel. These bits must be

Fig. 161-5.   A universal three-jaw chuck.

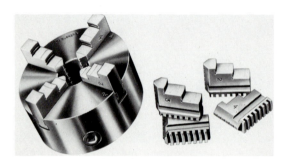

Fig. 161-4.   An independent four-jaw chuck.

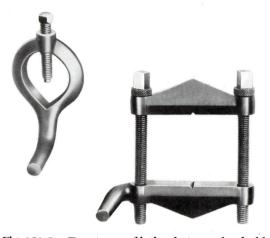

Fig. 161-6.   Two types of lathe dogs used to hold stock.

ground to different shapes for each different cutting operation (Figs. 161-7 and 161-8).

4. *Toolholders* hold the tool bits and are fastened in the tool post. There are the left-hand, the straight, and the right-hand. The straight and the right-hand toolholders are the most commonly used (Fig. 161-9 a and b).

5. There are many kinds of precision *measuring tools* that the machinist uses. Four tools needed by the beginner are a 6-

or a 12-inch ruler, inside and outside calipers, and a micrometer (Figs. 161-10 through 161-12). When you are working in metric measurements, a 150- or 300-millimeter rule and a metric micrometer are needed. (See Unit 126.)

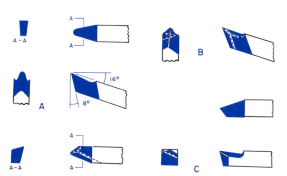

Fig. 161-7. The proper method of grinding some of the common cutting tools: (*a*) a round-nose tool, (*b*) a right-cut finishing tool, (*c*) a left-cut finishing tool.

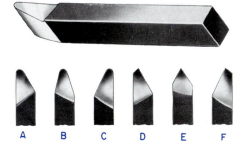

A   B   C   D   E   F

Fig. 161-8. Common cutting tools: (*a*) left-cut, (*b*) round-nose, (*c*) right-cut, (*d*) left-cut side facing, (*e*) threading, (*f*) right-cut side facing.

Fig. 161-9 a. Right-hand toolholder.

Fig. 161-9 b. A set of toolholders.

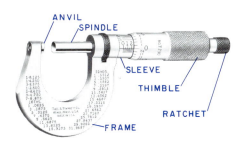

Fig. 161-10. Parts of a micrometer.

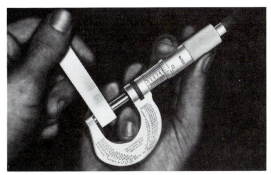

Fig. 161-11. The correct method of holding a micrometer.

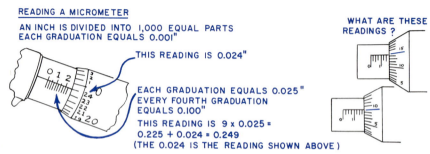

READING A MICROMETER

AN INCH IS DIVIDED INTO 1,000 EQUAL PARTS
EACH GRADUATION EQUALS 0.001"

THIS READING IS 0.024"

WHAT ARE THESE
READINGS ?

EACH GRADUATION EQUALS 0.025"
EVERY FOURTH GRADUATION
EQUALS 0.100"

THIS READING IS 9 x 0.025 =
0.225 + 0.024 = 0.249
(THE 0.024 IS THE READING SHOWN ABOVE)

Fig. 161-12. Reading a customary micrometer.

Lathe Operation

Turning between Centers. There are many operations done with the stock held between centers. The most common are rough and finish turning, cutting a shoulder, facing an end, taper turning, knurling, filing, polishing, and thread cutting. The first of these—rough turning—is basic to all other operations.

Choose metal about ⅛ inch (3 millimeters) larger in diameter and about 1 inch (25 millimeters) longer than the finished size.

Locating and Drilling Centers

1. Locate the center on either end of the stock by one of the methods shown in Fig. 161-13.

2. Select a combination drill and countersink of the correct size, usually No. 2, for small-diameter stock (Fig. 161-14).

3. Fasten this in a drill press. Hold the work on end to drill the hole (Fig. 161-15). The hole should be drilled until about three-fourths of the countersink has entered the metal. This can also be done on the lathe.

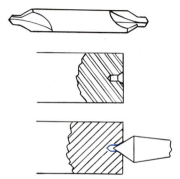

Fig. 161-14. A combination center drill and countersink forms the center holes.

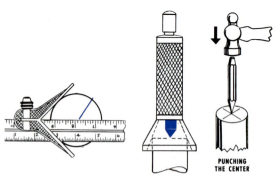

Fig. 161-13. Locate the center on the end of the stock with a center head. Then mark the center with a prick punch. The center may also be marked with a bell center.

PUNCHING
THE CENTER

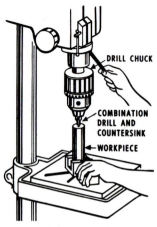

DRILL CHUCK

COMBINATION
DRILL AND
COUNTERSINK

WORKPIECE

Fig. 161-15. Drilling center holes on the end of the stock with a combination drill.

■ 486

Rough Turning

1. Fasten a lathe dog to one end of the stock.

2. Move the tailstock assembly until it provides an opening between centers slightly longer than the stock. Lock the tailstock.

3. Insert the tail of the lathe dog in the opening of the *faceplate* (drive plate). Place the work between centers. Lubricate the dead-center end with a small amount of white or red lead and oil.

4. Turn up the handle of the tailstock until the work is held snugly between the centers. Then lock it. It should not be so tight that the tailstock end heats up and burns. It should not be so loose that the lathe dog rattles.

5. Insert a tool bit in a toolholder. Fasten it in the tool-post holder. Adjust the cutting edge of the tool to a point directly on the center or slightly above it. The tool should be turned slightly away from the headstock.

6. Adjust to the correct speed for the kind of metal and the size of the stock.

7. Move the carriage back and forth by hand. The point of the tool should clear the right end of the work. When turning, the point should go over half the length of the stock without having the lathe dog strike the top of the carriage. Check to see that the carriage will move from tailstock to headstock when the power feed is on.

8. Set outside calipers $1/32$ inch (1 millimeter) larger than the finished diameter.

9. Bring the point of the cutting tool to the farthest right end of the work. Turn on the power (Fig. 161-16). Place your right hand on the handle of the cross feed. Your left hand should be on the handwheel. Then turn the cross slide in and the handwheel toward the headstock until the cutting edge begins to remove a small chip. When the chip is started, throw in the lever for the power longitudinal (lengthwise) feed. Allow the cut to move ahead

slightly more than half the length of the stock. At the end of the cut, turn the cross feed out. Release the power feed at the same time.

10. Return the carriage to the starting position. Turn off the power. Check the diameter with a caliper to tell how much more material to remove. Repeat the cutting operation until the first half is turned to rough size.

Finish Turning. Resharpen the cutting edge of the tool bit. Adjust the lathe to higher speed and a finer feed. Finish-turn the first half. Place a piece of soft copper under the lathe dog when finish-turning the second half. This will prevent marring the smooth surface.

Cutting a Shoulder and Facing the Ends

1. Mark off the various lengths for the different diameters.

2. Rough-turn and finish-turn to size. Leave rounded corners.

3. Select a right-hand or a left-hand cutting tool. Adjust it with the cutting edges at right angles to the work.

4. Cut out the sharp corner.

Fig. 161-16. Turning between centers. Note the way the work is held and how the cutter bit is set.

Use the same tool to face the end of the stock. You should place a half center in the tailstock for this operation (Fig. 161-17).

Filing and Polishing

1. Adjust the lathe to high speed.
2. Hold a fine mill file as shown in Fig. 161-18. Take long, even strokes across the revolving metal. Keep the file clean.

CAUTION: Keep your arm out of the path of the revolving lathe dog.

3. Smooth the surface with a fine abrasive cloth. This will give a very smooth polish.
4. Apply a small amount of oil to the cloth. Hold it around the revolving stock.
5. Move it slowly from one end to the other.

Knurling. The handles of many small tools, such as punches, hammers, screwdrivers, and clamps, are knurled to give the user a better grip. Figure 161-19 shows a knurling tool.

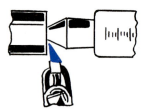

Fig. 161-17. Facing the end of the stock. Bring the cutting edge up to the work and then turn the crossfeed crank out, taking a light cut.

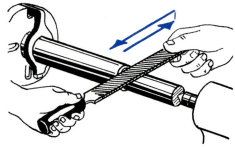

Fig. 161-18. The left-hand method of filing.

Taper Turning. There are two common ways of cutting a taper. One is by setting over the tailstock. The other is by using the compound rest.

1. Setting over the tailstock:
  a. To figure out the amount of setover, use the following formula:

$$\text{Setover} = \frac{TL}{L} \times \frac{LD - SD}{2}$$

where TL = total length
  L = length to be tapered
  LD = large diameter
  SD = small diameter

Total length should always be equal to the actual length of the stock being turned, not the finished article.
  b. Loosen the nut that holds the tailstock to the bed. Offset the tailstock by loosening or tightening the set screws found on both sides of the upper casting of the tailstock. Usually the tailstock is moved toward the operator. In that way, the smallest end of the taper will be toward the tailstock. Measure the amount of the offset. Hold a rule between the two witness marks on the tailstock or between the centers as shown in Fig. 161-20.

Fig. 161-19. A knurling tool.

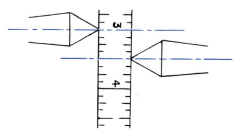

Fig. 161-20. Setting over the tailstock to do taper turning.

c. Install the work between centers. Lock it in position. Use a thin, round-nosed tool bit. Start the cutting about ½ inch (13 millimeters) from the end of the stock. Continue to make a light cut until the small end is the correct diameter.

Fig. 161-21. Facing the end of the stock held in the three-jaw chuck.

Fig. 161-22. Outside turning of stock held with a universal chuck.

2. The compound rest can be fitted to the correct angle of a short taper. The cutting is done by using the compound cross feed.

Chuck Work. Many operations, such as drilling, boring, and reaming, are done with the work held in a chuck. For round stock, a three-jaw universal chuck is used. For irregularly shaped work, a four-jaw independent chuck is used (Figs. 161-21 through 161-24).

Fig. 161-23. Drilling a hole on the lathe.

Fig. 161-24. Boring a hole on the lathe.

Unit 162

Welding

Welding is a way of joining metals by means of heat and sometimes pressure (Fig. 162-1). Industry uses many different methods of welding. In a car, for example, over 3500 individual welds are used to join the parts together. The three kinds of welding described in this unit are *oxy-acetylene, arc,* and *spot welding.*

Fig. 162-1. Welding heavy metal pipe for a boiler in a new building.

Fig. 162-3. A welding table.

Oxyacetylene Welding

Oxygen and acetylene are burned to make a heat of about 6300°F (3480°C) that easily melts metals. To do oxyacetylene welding, you need a complete welding outfit (Fig. 162-2). In addition, a welding table (Fig. 162-3), a spark lighter, and welding goggles

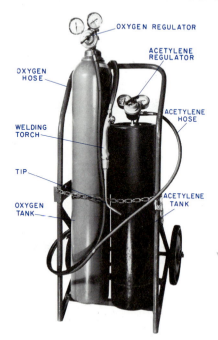

Fig. 162-2. A portable oxyacetylene welding outfit.

are needed. You will also need some mild-steel welding rod. For example, a $1/16$-inch (1.5-millimeter) welding rod should be used for $1/16$-inch (1.5-millimeter) sheet metal. The size of the welding tip should also change with the thickness of the metal. Use a small tip for thin metals and a medium tip for medium thicknesses.

Safety

1. Make sure that you know what you are doing before you start. If necessary, check with your instructor.
2. Use a spark lighter, never matches, to light the torch.
3. Always wear welding goggles when welding.
4. Always turn off the torch when you are done with the weld.
5. Never look directly at a welding flame.
6. Wear a flame proof outer garment and gloves since sparks can ignite clothing.

Making a Practice Weld

1. Place two pieces of scrap stock on the welding table (Fig. 162-4).
2. Place the safety goggles over your eyes.
3. Make sure that the handles on the pressure gages nearest the hose are backed off. In that way, no gas can get

through them. Then open the *cylinder valve* on the top of the oxygen tank as far as it will go. Turn the *tank handle* on the acetylene tank about ½ to 1½ turns.

4. Turn on the regulator handles on both the oxygen and the acetylene pressure gages to the pressure recommended by your teacher for the equipment you are using and the weld you are making (Fig. 162-5).

5. Open the acetylene valve about ¼ inch (6 millimeters) and light the torch with a scratch lighter (Fig. 162-6 a).

6. Adjust until the acetylene flame just jumps away from the tip. Now turn on the oxygen until you get a neutral flame, as shown in Fig. 162-6 b.

7. Hold the torch with the inner cone of flame about ¹⁄₁₆ inch (1.5 millimeters) away from the metal. Zigzag back and forth to

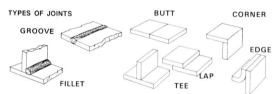

Fig. 162-4. **Common types of joints used in welding.**

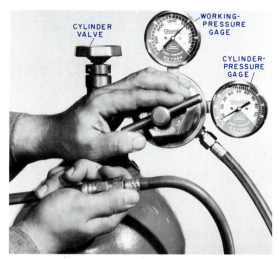

Fig. 162-5. **Adjusting the pressure gages on the tanks.**

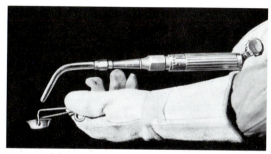

Fig. 162-6 a. **Lighting a torch.**

Fig. 162-6 b. **A neutral flame is obtained by burning an equal mixture (the same amount) of oxygen and acetylene. The pale blue core is called the *inner cone*. At the tip of this inner cone is the highest temperature. This is the right flame for most welding.**

tack the pieces at either end. Then start at one end. Work along the edge with a zigzag torch movement. Try to form a puddle of molten metal. Then move the torch along slowly.

8. Cut two pieces of ³⁄₁₆-inch (4.5-millimeters) mild steel. Grind the edges to form a V. Then place the two piecs of metal flat on the table. Tack the ends together. Start a puddle of molten metal on one end. Now put the welding rod into the middle of the puddle (Fig. 162-7). The rod should add the needed material for the bead.

9. To turn off the torch first close the acetylene valve and then close the oxygen valve.

Arc Welding

In arc welding, the heat needed for joining the two surfaces of metal comes from an electric arc. The two types of arc welders most often used are the *transformer* (Fig. 162-8) and the *motor-generator*. Both of

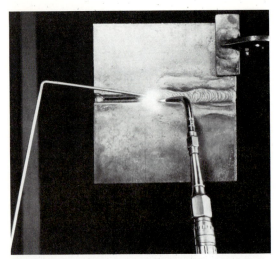

Fig. 162-7. Welding a butt joint with a rod. Note that the butt weld is tack-welded together before making the weld. The rod is placed just ahead of the oxyacetylene flame.

Fig. 162-8. A transformer arc welder. (Miller Electric Manufacturing Company)

these supply an electric current that jumps an air gap between the *electrode* (the metal wire that is melted into the weld) and the joint to be welded. The extremely hot arc melts the surface in a small spot. In that way, the metals to be joined are really fused together (Fig. 162-9).

The electrode is placed in the electrode holder. A ground wire is connected to the welding machine from either the metal itself or the metal table. The other wire runs to the electrode holder. An arc is started by scratching the tip of the electrode on the surface of the metal. Before you do this, however, you should put on a helmet that will shield your eyes from the arc. When the arc is started, it is moved along in a zigzag pattern. The electrode holder is lowered so that the electrode is fed into the weld.

Spot welding is a type of *resistance welding*. It is used to join sheet-metal parts together. (See Unit 148, "Assembling Sheet Metal.")

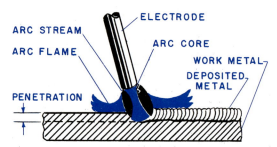

Fig. 162-9. The two metals are fused (melted) together by the very high heat of the arc.

Unit 163

Industrial Processes and Systems

Computers and robots (see Unit 5) are only two of the types of machines that are part of high technology in metalwork. However, industry continues to use the traditional (usual) techniques (cutting, forming, fastening, and finishing) that you are learning about in this book. Industry has also developed many nontraditional machining techniques that are needed to meet modern requirements for accuracy and finish. Today's metal manufacturers must also have equipment that can be adjusted to the need for frequent changes in parts and models (Fig. 163-1). To adjust to these changes, two systems—computer-aided design (CAD) and computer-aided manufacturing (CAM)—were developed. All these technological developments have been combined into the flexible manufacturing system (FMS).

Nontraditional Machining Processes

Some of the most widely used nontraditional machine processes include the following:

The EDM Process. Electrical discharge machining (EDM) is a form of metal removal in which pulsating direct current is applied to a shaped tool (electrode) and a workpiece, both of which are capable of conducting electricity (Fig. 163-2 a and b). The two are held close together with a dielectric (nonconducting) fluid that serves as an insulator between them. When the voltage is high enough to break down the insulator, a spark jumps the gap between the tool and the workpiece. This spark removes a small portion of material. This material is removed from the workpiece in

the form of melted or vaporized particles (Fig. 163-3).

The ECM Process. In electrochemical machining process (ECM), electrical energy is used to bring about a chemical reaction that causes metal to dissolve from a workpiece solution. Figure 163-4 shows how an ECM machine operates. Basically,

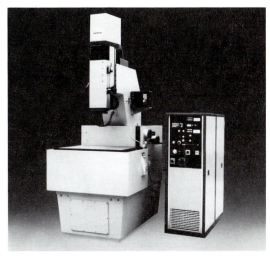

Fig. 163-2 a. A medium-size EDM machine. Note how it resembles a drill press or vertical milling machine. This process is used to make dies and molds. It is also used to cut odd-shaped openings in super-tough metal alloys. (Elox Corporation)

Fig. 163-1. Manufactured products, such as cars and other transportation vehicles, have frequent changes in design. (Honda Motor Co.)

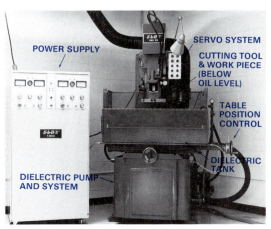

Fig. 163-2 b. A small EDM unit with the parts named. (Elox Corporation)

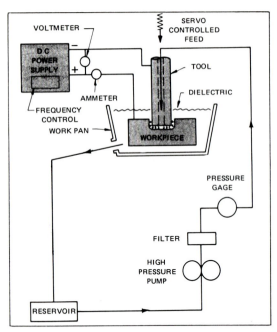

Fig. 163-3. A diagram showing the operation of an EDM machine.

the tool (cathode) is brought very close to the workpiece (anode). The distance between the tool and the workpiece may range from less than 0.001 to 0.010 inch. A low-voltage, electrically conductive electrolyte solution is then pumped between the tool and the workpiece. This solution is under high pressure, at a temperature of about 100°F. The current that passes through the electrolyte solution is usually quite high. As this current passes from the workpiece to the tool, metallic particles on the surface of the workpiece (ions) are forced into solution because of the electrochemical reaction. Then they are swept away by the rapidly flowing electrolyte.

The ECG Process. Electrochemical (electrolytic) grinding (ECG) is very much like electrochemical machining. The process is the same; only the application is different (Fig. 163-5). Metal is removed from the sur-

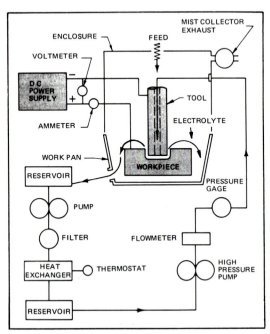

Fig. 163-4. A diagram showing the parts of an ECM machine. This tool removes metal by electrochemical reaction.

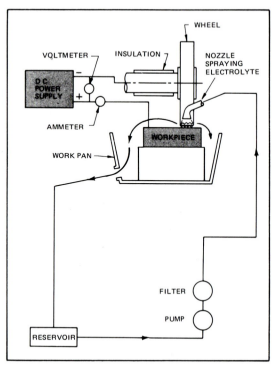

Fig. 163-5. A diagram showing the parts of an ECG machine.

face of the workpiece by a combination of electrochemical decomposition and the action of abrasive particles embedded in a metal-bonded wheel.

Computer-aided Design

Computer-aided design (CAD) is a fast and accurate method of creating and changing drafting design with the aid of a computer and the necessary software. A CAD system consists of three major elements: the *hardware* (computer system), the *software* (programs), and the *personnel* (engineer or technician) trained to use the total system. The hardware for CAD includes the typical units found in any computer system. In addition, the CAD unit must have a plotter for plotting the drawings. The plotter produces the drawing on paper after the trained personnel have developed the design on the computer. The software is of two types: the general operational software that tells the computer what to do and how to do it, and the specialized software that is used for such specific areas as mechanical, electronic, architectural, and civil (Fig. 163-6).

CAD has several advantages over the traditional method of drafting a new design. With CAD, once a design is developed, it never has to be completely redrawn again. Also, CAD eliminates the routine work, such as drawing lines and letters, using symbols, and doing other steps, that a conventional drafter must do over and over again. CAD is like instant photography, because the machine can create new drawings using elements that are in the storage or memory system of the computer.

Before you can use a CAD system, you must first be a good drafter. Then you must learn how to operate the computer. Once you have these skills, you can learn computer-aided drafting quite fast. Industry has discovered that CAD is a very powerful drafting and design tool.

Computer-aided Manufacturing

The term *computer-aided manufacturing* (CAM) is used to describe many kinds and levels of automated manufacturing. For many years, all machine tools in factories, such as the lathe and the milling machine, were operated by skilled workers (Fig. 163-7). To increase production, the numerical-controlled machine was developed.

Numerical control (N/C) is simply a means of directing some or all of the functions of a machine automatically by means of coded instructions. Numerical control is the automatic control of machines by means of electrical devices that receive operating instructions from a prepared tape (usually a punched tape) instead of a human operator (Fig. 163-8). This tape supplies, in coded form, dimensional data taken from a drawing of the part that is to be produced.

With N/C, additional steps have been inserted between the preparation of an engineering drawing and the machining of the part. These steps are programming and tape preparation. *Programming* is another name for writing down every move the machine is to make in producing a part. It requires a person who knows good machining and tooling practices and is fa-

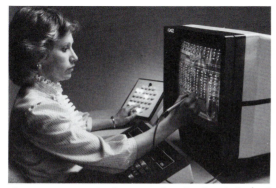

Fig. 163-6. Designing a printed circuit board using the CAD system. (Adage, Inc.)

Fig. 163-7. **This machinist is using a drilling machine. Most machining in small plants and repair shops is still done by all-around machinists.** (National Steel Company)

Fig. 163-8. **A manually programmed numerical-control (N/C) milling machine.** (Bridgeport Tool Company)

miliar with the cutting characteristics of the material from which the part will be made.

The worker must think through each step in machining the part, determine speeds and feeds, specify the path the tool will take, instruct the machine to turn on coolant or turn it off, and do all the other operations that a good machinist would do. The instructions, called a *format*, are written in a special way so that they can be readily converted into punched tape. This punched tape in turn is read by a control until that operates the machine according to the instructions it sees.

Numerical-control equipment not only is capable of doing complex and accurate work, it also offers many economic advantages. Tooling and fixture requirements are minimized. Machine flexibility is increased. Changing workpieces may require only changing to a new tape.

The next step was to add a computer to the machine tool so that it became a computer-assisted numerical-control machine (CNC) (Fig. 163-9). With this system, the computer programmer described the size and shape of the part, the steps in machining it, the tools required, and other details. The computer then controlled the ma-

Fig. 163-9. **A computer assisted numerical-control machine (CNC).** (DoAll Company)

chine directly without the use of tapes. These machines, however, were designed to perform one major kind of machining such as milling.

The next step was to develop a more versatile tool called a computer numerical-control machining center (CNCMC) (Fig. 163-10). These machines can do all the operations needed to finish a part and can automatically change the cutting tools used for each step of the operation. Many of these machines offer a high degree of flexibility, with automatic tool handling and storage systems and extreme ease of changeover. Until recently, most assembly lines were designed to produce only one part or product (Fig. 163-11). When that

part or product was changed, the whole assembly line had to be changed. This was expensive and time-consuming.

With the development of computer numerical-control machine centers, robots, and automatic handling devices, a better kind of assembly line has been developed, called a *flexible manufacturing system* (FMS). This consists of CNCMC machines grouped together to form manufacturing cells. The cells then become part of an automated production system that can manufacture many different kinds of workpieces at midvolume rates. The typical FMS consists of a number of machine tools tied together by a workpiece-handling system (including robots) and controlled by a central computer (Fig. 163-12).

However, not all manufacturing is done in this manner. Many factories find it just as efficient to use skilled workers who can do a variety of tasks in making the parts and assembling the product. Many small manufacturers who can't afford the large expense of a flexible manufacturing system can still produce a good product that is competitive with the products of a highly automatic manufacturing system.

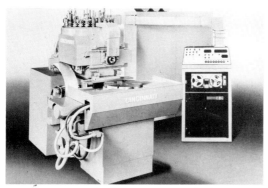

Fig. 163-10. A computer numerical-control machining center (CNCMC) with direct input from a computer. (Cincinnati Milacron)

Fig. 163-11. A typical automated system for machining and checking camshafts for an automobile engine. Note that this system has a manually controlled electronic system to control production. (The Cross Company)

Fig. 163-12. A small section of a flexible manufacturing system using robots to control the flow of parts. (Unimate, Inc.)

Discussion Topics on Metals Technology

1. Why is metalworking an important industry in the United States?

2. Describe a flexible manufacturing system.

3. About how many types of occupations are related to the iron and steel industry?

4. List 12 classifications of workers in the metals industries.

5. Choose and describe the qualifications and duties for one of the professional careers in the metals industries.

6. Describe the general qualifications, responsibilities, and titles of three careers in the craftsworker classification.

7. Tell how to dress and how to act to promote safety in the metal shop.

8. What is the difference between a metal and an alloy? Name three nonferrous metals. Is steel a metal or an alloy?

9. Name three common metals. In what different shapes are they used in shopwork?

10. In what ways can metal be cut by hand? What are the machine methods?

11. Describe the correct method for drilling a hole.

12. Name three other ways of shaping metal besides cutting.

13. What tool cuts threads on the inside of a hole? What tool cuts threads on the outside of a rod or pipe?

14. How can metal parts be put together?

15. Name as many methods as you can for bending cold metal.

16. Name four common seams used for sheet-metal projects.

17. Name two ways of strengthening the edge of sheet metal.

18. What are the different kinds of surface decoration and finishing?

19. Describe the soft soldering process.

20. Why must art metals be annealed and pickled during forming?

21. What is hard soldering?

22. How can such shapes as bowls, trays, and plates be formed or raised?

23. What is forging?

24. Why is heat-treating very necessary in making tools such as cold chisels?

25. How is a casting made?

26. List the five ways of machining metal.

27. Tell how to turn metal between centers on a lathe.

28. Name two methods of welding.

29. Name five important safety precautions you should always observe when working in the metal shop.

30. What are the four raw materials needed to produce iron?

31. What is the difference between a ferrous and a nonferrous metal or alloy?

32. Name several common forms of ferrous metals.

33. Define the abbreviation CAD.

34. Explain how the EDM process machines metal.

35. Name the three elements of a CAD system.

36. What is numerical control?

ENERGY, POWER, AND TRANSPORTATION TECHNOLOGIES

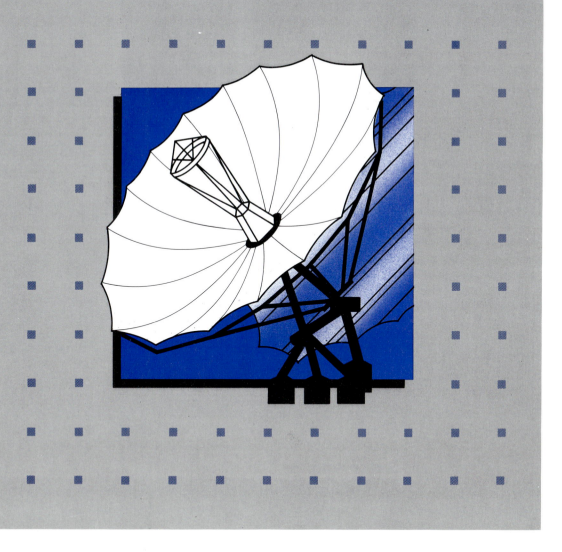

Section 9

Energy, Electricity, and Electronics Technologies

Unit 165

Energy and Power Systems

In the next two sections, you will learn a great deal about the energy and power systems that are so important to our industrial society and the development of technology. Energy and power systems consist of four major elements: (1) the sources of energy, (2) the conversion devices and machines that change energy from one form to another, (3) the transmission systems used to transport energy from one location to another, and (4) the systems used to store energy and control its use. Given the limited supply of our fossil fuels, conservation of energy is an extremely important part of the total system (Fig. 165-1). All devices that use energy must be made more efficient to reduce energy usage.

Energy, Work, and Power

Energy is defined as the ability to do work. When work is done, energy is given off to a body; when energy is released, work is done. Therefore, energy and work are the same thing. *Power*, on the other hand, is the rate of doing work or using energy. *Work* is measured by two factors: force and distance. The amount of work done is equal to the force times the distance. If either force or distance is zero, no work is done. When you lift a bundle and place it on your head, work is being done. However, if the bundle is merely held on your head, no work is being done. If you push a car any distance, work is being done. However, if the car is too heavy to push, no work is being done no matter how hard you push.

There are two kinds of energy: kinetic and potential. *Kinetic energy* is energy in motion. A body has kinetic energy because of its velocity or because it is changing its state. A bullet fired from a rifle, a strong wind, burning coal, and running water are examples of kinetic energy. *Potential* en-

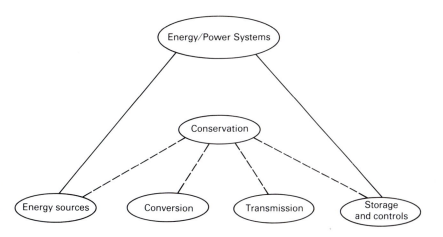

Fig. 165-1. A complete energy and power system includes all of these elements.

ergy is stored energy or energy in a fixed position. A rock resting on the edge of a mountain has potential energy. If it starts to roll, this becomes kinetic energy. A coil spring and gasoline have potential energy. If the coil spring is released, this becomes kinetic energy. When gasoline vapors and air are mixed in an engine and a spark explodes the vapor, this becomes kinetic energy and is a source of mechanical power. Remember that power is the rate at which work is done or energy is consumed.

Sources of Energy

All energy comes directly or indirectly from the sun. The sun is like a huge ball of fire or a great furnace without walls. The temperature on the sun is much higher than anyone can imagine. All plants, animals, and people need energy from the sun to live. Plants, including trees, store energy as food through a process called photosynthesis. Light energy from the sun causes the photosynthetic process in the leaves of green plants to take place. This is a chemical reaction that makes it possible for plants to utilize and store the nourishment essential to growth. When you eat vegetables and fruit, you are feeding on the sun's energy that has been converted to chemical energy. Also, meat, eggs, and milk that come from animals that have fed

on vegetation are sources of energy. Thus, all living things get energy from the sun.

The solar energy available from the sun is enormous. For example, for the entire world, 2 weeks' worth of solar energy is equal to all the fossil fuel (coal, oil, and gas) stored in the entire earth's known reservoirs. Of the huge amount of energy that is given off each day by the sun, only a very small part reaches the earth. Of that amount, only a very much smaller part is used directly or indirectly for many other energy needs.

For millions of years, people's major source of energy was their own muscle power and that of animals. The huge pyramids of Egypt, for example, were built by human labor. It took thousands of years before people discovered other sources of power. The water wheel, which used falling water, was one of the first major sources of mechanical energy. The early Chinese used the water wheel to water their crops and to power many of their machines. Today, and for many years to come, the major source of energy for our industrial society is the fossil fuels, including coal, oil, and natural gas. It is estimated that 87 percent of all of the energy used in the world in the next decade will come from fossil fuels, 8 percent from nuclear energy, and only 5 percent from all alternative energy sources (Fig. 165-2).

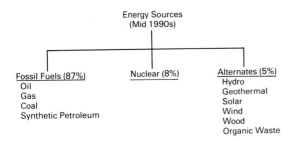

Energy Sources
(Mid 1990s)

Fossil Fuels (87%)
Oil
Gas
Coal
Synthetic Petroleum

Nuclear (8%)

Alternates (5%)
Hydro
Geothermal
Solar
Wind
Wood
Organic Waste

Fig. 165-2. Major sources of energy in the mid 1990s.

Fossil Fuels

Millions of years ago, during a turbulent era of the earth's development, plants, trees, and other tangled vegetation dropped to the ground, decayed, and were buried. This material was exposed to great amounts of heat and pressure and eventually became a concentrated kind of chemical energy in the form of coal, oil, and natural gas. Remember, however, that the energy stored in these materials is actually energy from the sun. These fossil fuels provide the major source of energy for our industrial society and make our technology function successfully.

Coal is the most abundant of the fossil fuels, but it is not the most commonly used source of fossil fuel. The United States has huge amounts of coal. Much of the ready coal supply is a rich source of synthetic gas and liquid fuels. These are needed to generate (make) electricity. These fuels, in the form of electricity, are used by industries, transportation, and homes. Plants for generating electricity will depend more and more on the use of coal.

Oil and gas are the easiest forms of energy to use and are, therefore, in great demand. Oil and natural gas are easy to store and transport. Both burn cleanly when good equipment is used. Other oil energy sources include off-shore (ocean) drilling and drilling deeper to find more oil on land (Fig. 165-3). Synthetic petroleum includes gas and petroleum liquids from coal, oil from tar sands, and oil from shale in addition to oil from organic wastes. There has been considerable research on synthetic fuels, but they will contribute less than 1 percent of the nation's energy needs in the next decade.

Once fossil fuels are gone, there will be no more. They are not renewable like wood from trees. That is why, some time during the twenty-first century, alternative energy sources, in particular nuclear and solar, will have to replace these dwindling supplies as sources of energy. The need for other energy sources will be essential for human survival and industrial development. Of the three fossil fuels, oil will continue to provide 40 percent of our needs, coal will provide 25 percent, and natural

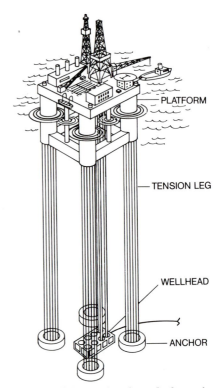

— PLATFORM

— TENSION LEG

WELLHEAD

— ANCHOR

Fig. 165-3. This tension-leg platform is an example of the oil industry's latest deep-water production technology. Mooring lines or tension legs keep the platform in place. (Phillips Petroleum Company)

gas will provide 22 percent over the next ten years. As you can see, over 85 percent of our energy in the immediate future will continue to come from fossil fuel sources.

Nuclear Energy

Nuclear energy is developed by changing matter into energy. In this process—nuclear fission—the nucleus, or center, of the atom of the fuel (uranium) is split (Fig. 165-4). It gives off radiant energy that produces heat. This heat is then used to change water into steam in order to drive the turbines that generate electricity (Fig. 165-5).

Until recently, nuclear power plants were considered to be the main source of future energy for the United States. However, nuclear power is now the most con-troversial of the basic energy sources used to produce electricity. The first nuclear power plants were built in the late 1950s, and it was estimated at that time that by the end of this century there would be hundreds of nuclear power plants in use. However, the many problems that developed caused people to question the safety and efficiency of this source of electric power. The costs of building the plants skyrocketed, and the cost of the electricity produced by nuclear energy soared. Also, nuclear reactors produce waste material, and there is a problem of disposing of this dangerous and radioactive by-product.

Many nuclear power plants have been abandoned before completion, and others in the planning stage are on hold. Opponents of nuclear power have pushed for a moratorium on nuclear power projects, warning of the possibility of disasters re-

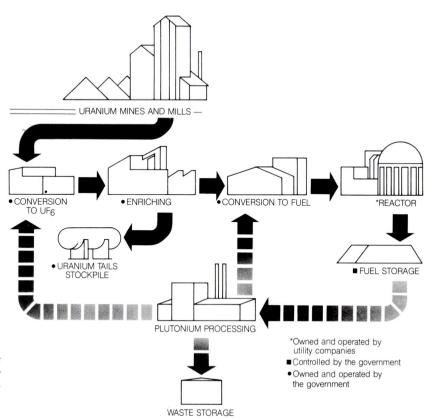

Fig. 165-4. Here is how the fuel for nuclear power plants is produced. (Phillips Petroleum Company)

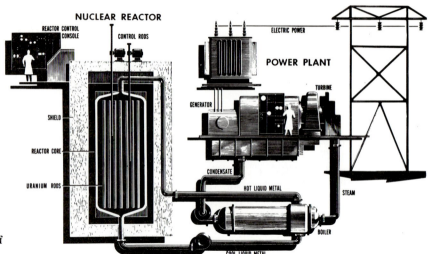

Fig. 165-5. Diagram of a nuclear reactor.

sulting from nuclear accidents or even sabotage. However, sometime in the future, the demand for electrical power will be so great and the supplies of fossil fuels so limited that there will be a renewed interest in nuclear energy.

Alternative Energy Sources

While the alternative energy sources represent only 5 percent of the energy that will be used in the immediate future, these are important sources that will eventually increase in usage. Hydroelectric power is produced when water falling from a higher elevation to a lower elevation is used to operate the turbine wheels (a so-phisticated water wheel) that generate electricity (Fig. 165-6). The force of gravity causes the water to fall, providing the source of energy. The kinetic energy (moving energy) of the falling water is turned into mechanical energy, which is converted into electrical energy by the generator. Hydroelectric plants are placed near huge dams or where the greatest possible drop of water is available. Most economical hydroelectric sites in the United States have already been used. As you will see later, hydroelectric power is one important way of generating electricity.

Geothermal (underground steam) processes employ heat from molten rock beneath the earth's surface to provide

Fig. 165-6. Hydroelectric power plants are located near large dams.

Fig. 165-7. Geothermal energy is obtained by drilling deeply enough to release heat from the earth.

energy. The energy source is limited to those areas of the world where steam and hot water are near the surface of the earth. Wells can be drilled, and steam and hot water are used to drive turbines that produce the electricity (Fig. 165-7). Most of the geothermal energy sources in the United States are in the western part of the country. Geothermal generator plants can be built near geysers and hot springs or over underground reserves of hot brine.

Solar energy from the sun is radiated in all directions into space as electromagnetic waves. These waves contain energy that can be used directly and indirectly. The sun's energy rays are used directly in passive or active solar energy heating systems (Fig. 165-8). This is the most common type of solar energy system because it is relatively simple and inexpensive.

With passive solar energy, the basic design includes south-facing windows or glass walls that invite in the winter sun and a large overhang that keeps out the summer sun. The house must be built with heavy absorbing floors or walls that will store the heat. When the sun goes down, curtains can be closed, and the house will then be heated from the heat stored in the walls.

In an active solar energy system, specialized equipment is needed, including collectors, pipes, pumps, fans, ducts, and storage walls (Fig. 165-8). These systems have not proved to be cost-effective because of the complicated nature of the equipment and the expensive maintenance required. Solar energy can also be converted directly into electricity with solar cells.

Generating electricity directly from the sun depends on the photovoltaic effect. This is the process that occurs when light hits certain sensitive materials and creates a flow of electrons (an electrical current). The devices that accomplish this change are solar cells. The first practical solar cells were manufactured in the mid 1950s and

Fig. 165-8. An active solar energy system requires large solar energy panels for heating and cooling. (Georgia-Pacific Company)

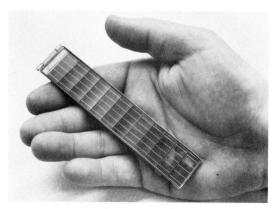

Fig. 165-9. This solar cell is just one of the many that are placed on the solar panel of a spacecraft to supply the necessary electricity. (Western Electric Company)

were used to supply small amounts of power for remote weather stations (Fig. 165-9). The most familiar application to date is in photography, where these cells are used in light meters. In space vehicles, solar cells provide most of the electricity for a variety of needs (Fig. 165-10). The typi-

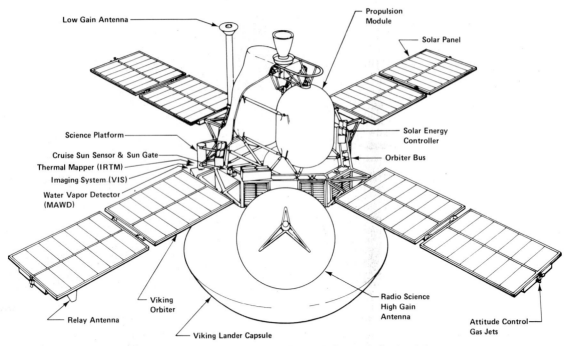

Fig. 165-10. A solar-powered spacecraft with four large solar panels. (NASA)

cal solar cell contains two very thin layers of materials with outside wires attached. Sunlight falling on the cells forces electrons to move along the wire from the phosphorus-silicon layer to the boron-silicon layer (Fig. 165-11). Solar cells can be connected electrically to form solar modules, which are the basic building blocks for solar electrical systems.

Wind power is one of the world's oldest forms of energy. Windmills were used for many years to pump water from the earth. They were also used for centuries to grind grain into feed and flour. Wind can produce usable amounts of electricity. A windmill works in a simple way. As the blades turn in the wind, power is created to operate a generator that changes mechanical energy into electrical energy (Fig. 165-12). The electricity that is generated can be distributed directly for use, or it can be stored in batteries. Some of the

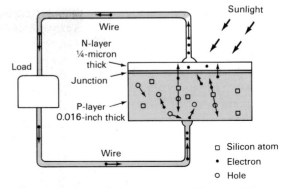

Fig. 165-11. A silicon solar cell has a thin n-layer (phosphorus-silicon) and a p-layer (boron-silicon). When sunlight delivers energy to the p-layer, electrons are knocked out of some of the silicon atoms, leaving "holes" in the electronic structure. These free and energetic electrons move across the junction to the n-layer and then through the wire to the load, where their energy is converted to useful work. The electrons then go to the p-layer and reenter its electronic structure at the "holes." (NASA)

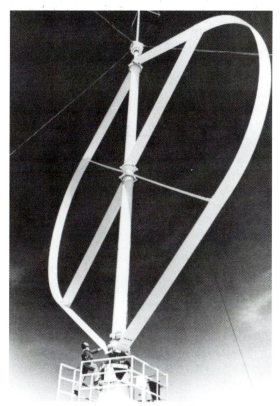

Fig. 165-12. This unusual wind machine is a tall rotating tower with two curved, extruded aluminum blades attached at the top and the bottom. The entire assembly spins from the aerodynamic lift provided by the wind moving through the rotating blades regardless of wind direction. (U.S. Department of Energy)

Fig. 165-13. Tidal waves are a good source of energy, but the process of capturing the power produced is expensive because of the complicated machinery needed.

Fig. 165-14. Organic waste can be used to supply energy and also to help keep the environment cleaner.

western states use windmills (wind farms) extensively to produce electricity.

The waves (tidal power) of the ocean are also a source of energy. Dammed tidal basins can direct the flow of ocean water to turn paddle wheels on driveshafts in order to transmit mechanical power that will turn machines (Fig. 165-13). Some countries have developed tidal power plants to produce electricity.

Wood is one of the few renewable energy sources. It can be burned as a clean fuel to produce steam to operate a turbine that turns the electrical generator.

Another source of energy is the United States' 2½ billion tons of organic waste produced annually (Fig. 165-14). This waste can be converted into oil or burned with the dual purpose of disposing of the waste and capturing the heat for other uses.

Electricity as a Source of Power

Just what is electricity? Where does it come from? Electricity does not flow from the ground or the air. It must be produced in power plants, which in turn must consume other forms of energy to operate (Fig. 165-15). Electricity is made in turbine generators that convert mechanical energy into electrical energy. This is the electricity that cooks dinner and lights our homes. Nearly all electricity is generated in this manner. The only difference between hy-

SOURCES OF ENERGY FOR ELECTRICAL POWER

SOURCE	PERCENT
Coal	45%
Gas and Oil	30%
Hydroelectricity (Water)	16%
Nuclear Power (Uranium)	9%
Geothermal (Underground Steam)	Less than 1%
Organic Waste (Rubbish)	Less than 1%

Fig. 165-15. The major sources of energy used to produce electricity.

droelectric, fossil fuel, and nuclear power plants is the source of the mechanical energy that turns the turbine.

Electricity is produced when tiny particles of atoms called electrons are put into motion. Electricity is produced in power plants by spinning a magnet inside coils of wire in a turbine generator. This starts the electrons moving, creating electrical flow. Different kinds of energy are needed to turn the turbine. In hydroelectric plants, the force of falling water turns the turbine. In other plants, fossil fuels including coal, gas, and oil, are burned to fire up boilers that make steam to turn the turbine. In nuclear power plants, heat to make the steam is created by controlled fission (splitting atoms) of nuclear fuel (uranium) in a reactor. Still other sources of energy for producing electricity are rubbish (waste), underground steam (geothermal), and solar. The use of falling water power at hydroelectric plants such as those at Niagara Falls and the Hoover Dam is one of the most efficient ways of making electricity. About 16 percent of all the electricity used in the United States is generated in this way.

A great majority of power plants rely on fossil fuels such as coal, oil, and natural gas as basic sources of energy. Coal-fired boilers generate about 45 percent of all the electrical power used. Oil and natural gas each account for 15 percent, and about 9 percent is produced by nuclear power. The United States has a huge supply of coal, but there are problems involved in developing these vast coal resources. Among these problems are the objections of environmentalists to strip mining, air pollution, and the huge financial investment necessary to use coal. Oil and natural gas account for 30 percent of the electricity generated in the United States.

Conversion Devices

According to the first law of thermodynamics, energy cannot be created or destroyed. Energy can only be converted from one form to another. For example, with fossil fuel, chemical energy in the form of oil, coal, and natural gas can be used to operate a gasoline engine that can drive a generator to produce electricity. The most common conversion device used in transportation is the internal-combustion engine. The transportation industry consumes about 25 percent of all of the oil used in this country. In the internal-combustion engine, fuel in the form of gasoline is used to operate the engine that provides mechanical power to operate all of the parts of the machine and drive it. Internal-combustion engines may be either two-cycle or four-cycle. Diesel engines are another kind of internal combustion engine. External-combustion engines such as steam engines use energy in the form of steam to operate machines. Other kinds of engines include jet, rocket, and nuclear energy. All of these are discussed in detail in Units 190 and 197.

Transmission

Energy can be transported, or moved, from one place to another by various means, in-

Fig. 165-16. Look at this illustration carefully. Then list the many ways energy is converted, stored, and controlled by this machine.

cluding chemical, electrical, and mechanical systems. Fossil fuels (chemical energy) such as oil, gas, and coal are transported by ships, railroads, trucks, and pipelines. Electricity is transmitted over long distances as alternating current, over short distances as direct current, or as electromagnetic waves. Solar energy can be transmitted directly from the sun to solar cells or to a passive or active solar energy system to heat and cool homes, offices, and factories.

There are many different mechanical devices that transmit energy. For example, in automobiles and other machines, devices such as gears, belts, chains, shafts, and pulleys are used (Fig. 165-16). Fluid power is used to transmit energy by pneumatic (air) and hydraulic (liquid) means. Fluid power transmission is used in automobiles (brakes), robots, heavy-construction equipment, jet airplanes, and many other places because it is more convenient than mechanical transmission. (See Unit 191.)

Storage and Controls

Energy can be stored for short or long periods of time as chemical, mechanical, thermal (heat), or electrical energy. Fossil fuel (chemical energy) is stored in the ground in huge tanks, in oil tankers, and in many other ways. Mechanical energy can be stored as compressed gasses, in springs, in flywheels, and by placing a heavy object in a position where it can be released to fall by gravity. Examples of the first three are found in the automobile, which uses shock absorbers (compressed gases), springs, and the flywheel of the engine to improve the performance of the machine. The drop forge press used in metalwork to shape materials is an example of the last method of storing energy. Thermal (heat) energy can be stored in walls, as in the passive solar energy systems used in homes, or as hot gases. Electrical energy can be stored for short periods of time with capacitors and inductors. A capacitor is a device that opposes any change in circuit voltage. The condensor used in a gas engine is a capacitor. (See Unit 195.) A good example of an inductor is the transformer used in electrical power systems and automobiles. (See Units 176 and 197.)

Controls for energy can be mechanical, electrical, or fluid. For example, in the automobile, the mechanical controls include the differential, clutch, and hand brake. In electrical systems, the controls include switches, relays, and solenoids. A relay is used to control a circuit from a remote source or to serve as an automatic control device. A solenoid is a variation of a relay. Fluids are controlled by valves and other regulators. Two high-technology control systems are the central processing unit (CPU) used in computers, calculators, and watches and the laser, which can be used for many purposes, including controlling electronic devices. (See Unit 5.) As you study the next two sections, refer back to this unit and try to identify which of the four major elements of energy and power systems are involved.

Introduction to Electricity and Electronics Technologies

What would life be like without electricity? Just about a century ago, electricity was like a toy. Or it was used for experiments only. But today there are thousands of uses for electricity. Homes, farms, businesses, and factories all need and use electricity. These needs are based on five important things that electricity does:

1. Electricity can produce light (in light bulbs, for example).
2. Electricity can produce heat (in stoves and arc welding) (Fig. 166-1).
3. Electricity can produce motion (in motors).
4. Electricity can produce electronic effects (in radio, television, and control devices such as an electric eye) (Fig. 166-2).
5. Electricity can produce electrochemical changes (in electroplating and battery-charging).

Learning how electricity works will help you in many ways. You use electric appliances and electric service. You should be able to make simple repairs on toasters, lamps, or other similar items in your home. You may someday even earn your living in electricity or electronics. You may work at one of many different kinds of jobs: as an engineer, a technician, a salesperson, an operator, a service person, a finance specialist, or a clerk.

Electrical engineering is the second largest engineering field. Four to five years of college is needed. The engineer designs and acts as a leader in building and caring for all kinds of electrical devices. These may be huge generators and transformers or tiny electric motors.

Construction electricians put wires in homes and buildings of all kinds. Those

Fig. 166-1. Arc welding depends on electricity as the source of energy. (United States Steel)

Fig. 166-2. Modern machinery uses electricity for the computer controls and the motors that do the machining. (Moore Tool Company)

who work for telephone and telegraph companies *install* (put in), *maintain* (care for), and *repair* (fix) electric appliances.

Radio, TV, and other electronic devices offer many chances for electronic technicians to find good jobs.

Metrics in Energy and Power Systems

DERIVED SI UNITS WITH SPECIAL NAMES

hertz (Hz)	The hertz is the number of times something repeats in 1 second.
newton (N)	The newton is the amount of force needed to *accelerate* (move faster and faster) a 1-kilogram mass 1 meter per second squared.
pascal (Pa)	The pascal is the pressure made by a force of 1 newton pushing evenly over an area of 1 square meter.
joule (J)	The joule is the work done when a force of 1 newton is moved a distance of 1 meter.
watt (W)	The watt is the amount of power made by 1 joule in 1 second.
coulomb (C)	The coulomb is the amount of electricity carried by a current of 1 ampere in 1 second.
volt (V)	The volt is the difference of *electrical potential* (the ability to do work) between two points (places) on a wire carrying a current of 1 ampere at a 1-watt rate of power.
ohm (Ω)	The ohm is the electrical *resistance* (opposition) between two points on a wire when 1 volt put across these two points makes a current of 1 ampere flow in the wire.

Fig. 167-1. Common technical units used in electricity and power.

Changing over to the metric system means that people will have to learn several measuring units often used in power and energy (Fig. 167-1). Since electricity was one of the first fields to use the metric system, there will be only small changes in the units used in this area. In fact, many electrical units will apply to other areas of power. For example, the watt and the kilowatt, which are units of power, will apply to engines and motors as well as to electricity. There will be few changes in the units of measure used in electricity. For example, house current in the United States has run on 120 volts and 60 cycles per second. In the SI system, this will be listed as 120 volts and 60 hertz (Hz) (Fig. 167-2).

Fig. 167-2. This meter shows the change in the number of hertz units on which the electricity is operating.

kilometers per hour	miles per hour
160	100
150	90
140	
130	80
120	
110	70
100	
90	60
80	50
70	
60	40
50	30
40	
30	20
20	
10	10
0	0

Fig. 167-3.   Comparison of kilometers per hour with miles per hour.

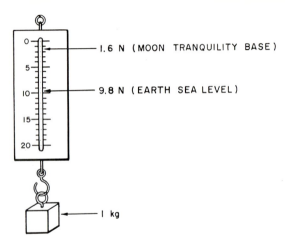

Fig. 167-4.   This shows the difference between mass in kilograms and force in newtons.

Length and Distance

Engine specifications will give all dimensions in millimeters. Travel distances will be shown in kilometers (1 kilometer = 1000 meters). Speed or velocity will be in kilometers per hour (km/h). To change miles per hour to kilometers per hour, multiply by 1.6. A speed of 50 miles per hour will equal approximately 80 kilometers per hour (Fig. 167-3).

Mass and Force

The base unit of mass is the kilogram (kg). Mass is the amount of material in an object. Mass does not change with the *location* (place) of the object. For example, the mass of an object is the same on earth, in space, or on the moon. For most people, the word *weight* tells something about the amount of such things as sugar or butter. Before the astronauts landed on the moon, it didn't make much difference whether the term *weight* was used for the force of gravity acting on a given body or mass. However, in the SI system, there is a big difference between *mass* and *force of gravity.*

Mass refers to the amount of material in a body measured in *kilograms.* The *force of gravity* acting on the mass is measured in *newtons.* For example, the force of gravity acting on a mass of 1 kilogram is about 9.8 newtons at sea level. It is only 1.6 newtons on the moon (Fig. 167-4). The word *weight* should *not* be used in technical work. When a person is talking about the amount of matter in a body, the kilogram, which is a unit of mass, should be used. The kilogram is a base unit. When you are talking about the *force* acting on that mass, the *derived* unit—the newton—should be used. For example, the force of gravity acting on a mass of 80 kilograms is approximately 784 newtons at sea level. It is only 128 newtons on the moon.

Area

Area is given in square meters (m²). However, for smaller areas, such as the opening of a cylinder, area will be given in square centimeters (cm²). For example, suppose a cylinder is 70 millimeters, or 7 centimeters, in diameter. To find the area in square centimeters, use this formula:

$$\text{Area} = \frac{d^2}{4} \times \pi$$

For this example, area would be figured as follows:

$$\frac{7 \times 7}{4} \times 3.1416 = \frac{49}{4} \times 3.1416 = 38.5 \text{ cm}^2$$

Volume

Volume is given in cubic meters (m³). This is a very large amount. Smaller volumes are shown in cubic centimeters (cm³). To find the piston *displacement* (volume) of a cylinder with a bore of 7 centimeters (70 millimeters) and a stroke of 10 centimeters (100 millimeters), first find the area of the piston or cylinder. Then multiply by the length of the stroke. *Displacement* equals *area* times *stroke* (Fig. 167-5). In the above case, 38.5 square centimeters times 10 centimeters equals 385 cubic centimeters (385 cm³). This would be the displacement

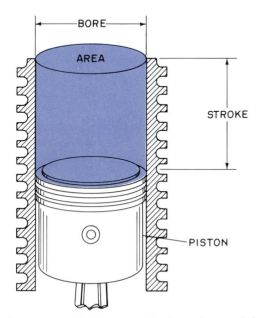

Fig. 167-5. Displacement is the volume of the cylinder when the piston is at its lowest point.

for the one cylinder. If there were four cylinders, the total displacement would be 385 × 4, or 1540 cm³. The liter (*l*) is the same as 1 cubic decimeter, or 1000 cubic centimeters. Engine displacement is therefore given either in cubic centimeters or in liters. The engine displacement above could be listed as 1540 cubic centimeters or 1.54 liters. Sometimes automobile manufacturers list the displacement as 1540 cc. This is not right in the SI system because cc really means "centi-centi," not cubic centimeters.

Liquid capacities, including measures for cooling fluids and oil, will be given in liters. Fuel use will be given in liters per 100 kilometers (l/100 km). For example, a car that operates at 40 miles per gallon will operate at about 100 kilometers to every 7 liters (7l/100 km).

Torque

Torque refers to rotational movement around a point. Torque is a product of *force* applied by the *length* of the lever arm. Force is measured in newtons and length in meters. Torque, then, is expressed (measured) in newton-meters (Nm). A torque wrench used to tighten engine bolts will have the scale *graduated* (marked off) in newton-meters. In the customary system, torque wrenches are graduated in foot-pounds. Roughly speaking, 1 foot-pound is about 1.4 newton-meters. A wrench with a scale of 5 to 100 footpounds is the same as a wrench with a scale of 7 to 140 newton-meters.

Energy (Work)

Energy is used up when work is done. Work happens when a force (measured in newtons) moves a body (mass, measured in kilograms) for a distance (measured in meters). The unit of energy or work is the joule (J). This is a small unit, and so work is most often expressed in kilojoules (kJ).

Power

Power is the rate of doing work. Power is equal to joules divided by seconds. The watt (W) is used as a unit of power. It is used not only for electrical work but also for mechanical work. The watt is a small unit, and so the unit most often used will be the kilowatt (kW). Horsepower will no longer be used. One horsepower (1 hp) is about 0.75 kilowatts (0.75 kW). Therefore, an automobile of 200 horsepower would be rated at 150 kilowatts (150 kW).

Pressure

Pressure is force times area. In SI, it is given in pascals (Pa). This is a small unit, and so most pressures will be given in kilopascals (kPa). In the customary system, pressure is shown in pounds per square inch (psi); 1 psi is about the same as 7 kilopascals (7 kPa). Therefore, all pressure for engine operations, tires, carburetors, and the like will be given in kilopascals. For example, a tire pressure of 20 pounds per square inch would be expressed as 140 kilopascals.

Temperature

Temperature will be given in degrees Celsius. Most engine thermostats will be rated at about 80 to 90° Celsius (80 to 90°C). That is slightly under the boiling point of water (100°C).

Unit 168

Related Careers in Energy, Electricity, and Electronics Technologies

A world without energy, electricity, and electronics? You might find it difficult to imagine what that would be like. Electricity and electronics are essential to power our lights, operate our machinery, and cook our food, as well as for thousands of other tasks.

Many thousands of workers are needed in the industries that produce, distribute, and manipulate electric power.

Electric Power

There are thousands of jobs in the electric-power-producing industry. Many people work in installations that *produce and generate* electricity. Others must see that the power is *transmitted and distributed* to customers. Some people are electrical *maintenance and repair* workers. There are also career opportunities in *administration* and in the *clerical and customer-service* areas.

The electric-power industry requires highly trained *scientists, engineers,* and other *technical personnel* for research and development. Many technical occupations require a college or university degree. Some careers require a person with a master's or doctoral degree. Work in other areas requires on-the-job training and apprenticeship for skilled workers. There are also openings for semiskilled and unskilled laborers in this industry.

Electronics Manufacturing

The United States depends on the science of electronics. Electronic devices guide missiles and control flights into outer space. Electronic devices also direct, control, and test production machines in the steel, petroleum, and chemical industries. Electronic data-processing equipment makes it easy for businesses to handle tons of paperwork (Fig. 168-1). The robots controlled by electronics are used to manufacture cars and other products. The science of electronics helps people in many ways.

Many jobs and occupations are available in the electronics industry, and they require a broad range of training and skills. Some of these are *production, maintenance, transportation, service, engineering, scientific experimentation, finance, administration,* and *supervisory, clerical, and sales work* (Fig. 168-2).

Electronics and electrical fields require many highly technical people. Some of these workers are *physicists, mathematicians, electrical engineers, technicians,* and *highly skilled laboratory workers.* Most of these occupations require college degrees. Technicians and drafters must prepare for their jobs through vocational or technical programs.

Electricians

The job of laying out, assembling, installing, and testing electrical fixtures and wiring in buildings belongs to the construction electrician (Fig. 168-3). Electrical systems provide light, heat, power, air conditioning, and refrigeration in homes, offices, and other buildings. There are many career classifications within the general category of electricians.

A person who wants a career in construction electricity must start as an apprentice. This is like the apprenticeship in

Fig. 168-1. Electronic equipment such as this transmitter for AM radio broadcasting gives only one example of the importance of this industry. (RCA)

Fig. 168-2. Many skilled workers are needed in electronic manufacturing. (Western Electronic Company)

Fig. 168-3. This construction electrician is erecting a microwave tower. (Western Electric Company)

the other building trades. There are also many career opportunities for skilled electricians who want to hold supervising jobs for an electrical contractor. The pay is high, and the working conditions and hours are like those in other building trades. Construction electricians also sometimes start their own businesses.

Television and Radio Service

There are about 300 000 television and radio service technicians in the United States (Fig. 168-4). At least one-third of these people work for themselves. Skilled television and radio service technicians must know about electrical and electronic parts and circuits. They put in and care for a growing number of electronic products.

To get into an electronics career, a person must complete electricity and electronics courses in a vocational high school or technical institute and community college. The armed services offer fine training, too. Many TV and radio service technicians work in manufacturing centers that make electronic equipment. Job prospects for these areas seem very good.

Television and Radio Broadcasting

The areas of radio and TV broadcasting make up four big categories: (1) those who work with *programming*, to prepare and produce programs, (2) *engineering* workers, who work and care for the equipment that changes sounds and pictures into electronic *impulses* picked up on home sets (receivers), (3) *salespeople*, who sell time to advertisers, and (4) the rest of the workers, who take care of business matters.

A person who wants to get into these areas should have a college or university background. People often study in the liberal arts or in technical areas. The job opportunities for both radio and TV person-

Fig. 168-4. **Radio and television service technicians must do house calls wherever service is needed.** (Western Electric Company)

nel are expected to increase in the next 10 years.

Broadcast Technology

Broadcast technicians set up, operate, and care for electronic equipment. The equipment is used to record or *transmit* (send) radio and TV programs. They must understand microphones, sound recorders, lighting equipment, sound-effects devices, TV cameras, magnetic videotape recorders, and motion-picture projection equipment. They must also use transmitting equipment. That is because they usually start their careers in small stations, where they must do many different tasks.

A person interested in becoming a broadcast technician should plan to get a radiotelephone first class operator's license. This is given by the Federal Communications Commission. But first a person must pass written examinations. To get ready for these tests, a person must carefully study mathematics, sciences, and electronics. This program can be taken in vocational and technical schools or in community colleges. The number of broadcast technicians should grow significantly in the next 10 years.

Fig. 168-5. This person is installing a telephone line for a new customer. (Western Electric Company)

Telephone Industry

The general communications industry, including telephone service, should grow a great deal in the next 10 years. The telephone industry offers people many job opportunities for steady work. A huge network of cables and radio-relay systems is used for communication. Thousands of broadcast and TV stations are hooked together around the world. It would take far too many pages to talk about all the new kinds of equipment and careers in this industry. But communications satellites and relay stations are among the newest and most interesting areas (Fig. 168-5).

Engineers, technicians, business administrators, clerical assistants, service persons, and *telephone operators* are areas that have hundreds of job titles within them. Among these are office-equipment installers, line-workers, and cable splicers.

Many of these jobs have union *affiliation* (membership). Others do not. Getting ready for each of these jobs requires special training. Very careful preparation or apprenticeship must be done. Career openings should grow because of new technology that offers many new telephone services to people everywhere.

Miscellaneous Electricity and Electronics Careers

Electronics technology has grown quickly during the past 20 years. Many new kinds of careers will come up in the next 10 years. New kinds of job preparation are needed. Career patterns often change. Careers include *programmers, systems analysts,* and *electronic-computer operators. Business-machine service people* are upgrading their jobs to cope with the new technological changes. *Appliance service people* repair such items as coffeepots, toasters, waffle irons, and the usual kitchen appliances. They must get special training to care for the microwave ranges (ovens) now being made for use in homes and in restaurants. Careers dealing with these devices require more training than they did before.

Unit 169

Electricity Safety

Electricity is a safe and willing servant. But it must be treated with respect and handled properly. Most electrical accidents are caused by carelessness. They could be avoided by following safety rules.

Safety Rules

1. Always remember that electricity can produce heat. Do not burn yourself when you solder or check heating elements.

2. Use ground fault outlets for power equipment (Fig. 169-1).

3. Never turn on a switch, connect wires to meters, or make any electrical changes until you are sure that everything is in order. Ask your teacher to check your project activity or wiring job before you test it.

4. Never mix water and electricity. Always be sure that your hands are dry. Be sure you are standing in a dry place before checking any electric wiring or equipment. Water is a conductor of electricity.

5. Never use worn or broken electric equipment. If a wire is frayed or a switch is broken, replace it.

6. If you are not sure of what you are doing—*do not do it*. Get help from someone who is trained to do the job.

Fig. 169-1. **This device will protect power tool users from electric shock. It automatically disconnects the electricity when any hazard develops, such as an overload on the electrical circuit.** (3M)

Unit 170

The Electron and the Atom

To understand electricity, you must know something about the atom. Every solid, liquid, and gaseous substance is made up of small particles called *atoms.* Each atom is made up of three particles: the *electron,* the *proton,* and the *neutron.* Atoms of the same kind are called an *element.* There are 92 natural elements. Among these elements are hydrogen, oxygen, copper, and iron. Other elements have been made artificially. Elements can be combined in many ways to form molecules of other substances. A *molecule* is a *combination* (joining) of two or more atoms. For example, hydrogen and oxygen combine to form water.

Atoms are very small. It would take billions of atoms to cover the head of a pin. If the simplest atom (hydrogen) could be seen it would look like Fig. 170-1 a.

The *nucleus* (center) of a hydrogen atom contains a proton, which has a positive electric charge. A single electron, which

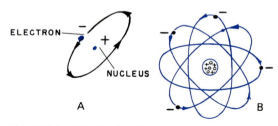

Fig. 170-1. **Parts of a simple atom.**

has a negative charge, *revolves* (moves on a path) around this nucleus. An atom is like a *miniature* (tiny) solar system, with electron "planets" whirling around a nucleus "sun." In electricity, opposite charges (negative and positive) attract each other. The proton is a positively charged particle. The electron is a negatively charged particle. And so the two attract each other.

Atoms of other elements have more protons, neutrons, and electrons (Fig. 170-1 b). In some substances, the outer electrons of the atoms are only loosely held. They can

■ **518**

be easily knocked off. These are known as *free electrons*. When they get loose, the atom does not have electrons. It becomes positively charged and then tries to become neutral again by attracting other electrons. Materials such as silver, copper, and lead have loosely held electrons. Since the electrons in these materials are free to move, they make good *conductors* of electricity (Fig. 170-2). Other materials, such as rubber, porcelain, and glass, have tightly held electrons. The electrons are not free to move. These materials do not conduct electricity well. They are called *insulators*.

Electricity, then, is the flow of free electrons through a material, or the effect of electrons moving from one point to another. A generator or battery produces more electrons on one *pole* (side) than on the other (Fig. 170-3). When a wire is hooked up between the uneven poles, electrons are forced along the wire. They go from the side with too many (negative charge) to the side with too few (positive charge). You can see why scientists say that electrons flow from the negative pole to the positive pole.

Some form of energy must be used to produce electricity. The six basic sources of energy are *friction*, *pressure*, *heat*, *magnetism*, *light*, and *chemical action*. The most common are magnetism (as in a generator) and chemical action (as in a battery).

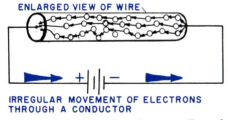

Fig. 170-2.   Notice how the electrons flow along a wire.

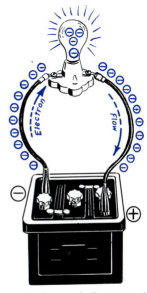

Fig. 170-3.   The flow of electrons is from the negative pole to the positive pole.

Unit 171

Magnetism

Magnetism is a force. It causes some materials to be *attracted* (pulled together). It causes others to be *repelled* (pushed away) by like materials. This force exists when the molecules of these materials are set up in a certain orderly way. Electricity and magnetism are related. Whenever an electric current is present, there is magnetism around it. For example, when electricity runs through a copper wire, a magnetic field is all around the wire. However, magnetism can exist without electricity (as in a bar magnet).

Magnets were known in ancient times. More than 2000 years ago, the Greeks and the Chinese found stones that had the

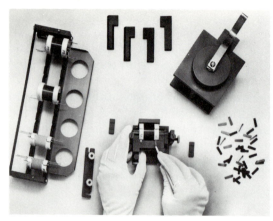

Fig. 171-1. Both permanent magnets and electromagnets are used in this transformer. (Western Electric Company)

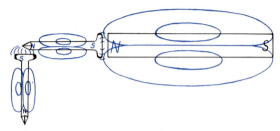

Fig. 171-2. A permanent bar magnet can magnetize nails.

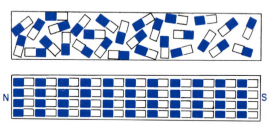

Fig. 171-3. Note that all the molecules in a magnetized bar magnet are arranged in one direction.

Fig. 171-4. If you cut a magnet into several parts, each smaller magnet will have two poles.

power to attract (pull) and hold other stones. These were called *lodestones*, or *natural magnets*. The Chinese may have been the first to use compasses made with natural magnets. Magnets made of steel alloys that hold their magnetism for a long time are called *permanent magnets* (Fig. 171-1). Metals such as soft steel, which lose their magnetism easily, are called *temporary magnets*. *Electromagnets* are pieces of metal temporarily magnetized by electricity.

Magnetic Effects

1. Magnets attract iron. Permanent magnets can lift iron particles or nails (Fig. 171-2).

2. Iron and steel may be magnetized. Soft steel is usually made into temporary magnets. Alloy steel is made into permanent magnets. A piece of iron or steel may be magnetized by rubbing it on another magnet. The second piece of metal becomes magnetized when most of the *individual* (single) molecules are lined up or turned in the same direction (Fig. 171-3).

3. All magnets have a north pole and a south pole. Even if a magnet were cut into many pieces, each piece would still have these two poles (Fig. 171-4).

4. *Like* poles repel. *Unlike* poles attract (Fig. 171-5). When a north pole is placed next to a south pole, there is an attraction between the two. But if two north poles or two south poles are placed close together, they repel each other. This basic law of magnetism is used in many devices.

5. A magnetic field surrounds a magnet.

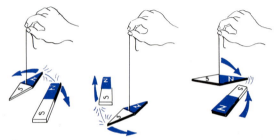

Fig. 171-5. Two north poles repel each other, two south poles repel each other, but a north pole and a south pole attract each other.

You cannot see the magnetic field. But there are lines of force running between the north and south poles (Fig. 171-6). This can be shown by placing a permanent magnet under a piece of paper. Shake iron filings on the paper. The filings will line up with the lines of force.

6. The earth itself is a huge magnet. It has a north magnetic pole close to the north (geographic) pole. A south magnetic pole is close to the south (geographic) pole. When we use a compass, therefore, the north pole of the compass points toward the north (magnetic) pole, which is located in the general vicinity north of Hudson Bay. It shifts about 6 miles every year. Its present location is about 930 miles south of the true North Pole.

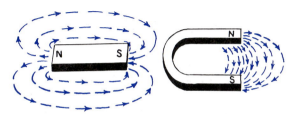

Fig. 171-6. Magnetic fields of permanent magnets.

7. Magnets can have their magnetism taken away (be demagnetized). When the molecules in iron or steel become mixed up, the piece is no longer a magnet. A magnet may be demagnetized in three ways: by heating it, by striking it with a hammer, or by placing it quickly into and out of a coil of wire carrying alternating current.

Unit 172

Forms and Effects of Electricity

Electricity exists in two forms—*static*, or electricity at rest, and *current*, or electricity in motion. Static electricity exists whenever too many electrons are together on a surface. When enough electrons get together, they often jump across an air space. This makes a spark. In this way, the atoms are neutralized.

Static Electricity

You have produced static electricity by *friction*. When you walk across a rug and touch a metal object with your finger, a spark appears. When this happens, static electricity is discharged. Static electricity may also be made by rubbing one object with another. For instance, you can make static electricity by combing your hair with a comb made of rubber. Lightning is the discharge of static electricity that is stored in clouds.

Current Electricity

Most electricity is current electricity. This is a flow of electrons. Before people knew very much about electricity, they thought that it flowed from *positive* to *negative*. However, now it is known that electrons flow from negative to positive, or from the point that has too many electrons to the point that has too few electrons. This principle is called the *electron theory*.

There are two types of electric current. Direct current (DC) flows in the same direction continuously. Alternating current (AC) changes direction, or reverses itself, all the time. It flows back and forth, with as much current going one way as the other.

Direct Current. Direct current comes from dry cells, storage batteries, and direct-current generators. It flows in a constant (even or regular) amount. It always flows in the same direction. It can be shown by a straight line. Direct current is used for automobile circuits, telephones, telegraphs, flashlights, most arc welders, and wherever batteries are the source of electricity. Many electronic devices, such as radio and television, use some direct current and some alternating current.

Alternating Current. Most electric current is alternating current. That is because alternating current can be produced at high voltages. It may be stepped up and down by means of transformers. It may be sent over long distances with smaller wire. Alternating current flows first in one direction for a short time. Then it reverses and flows in the opposite direction. It is shown by the wavy line in Fig. 172-1. It starts at 0 and flows to a maximum amount in one direction. Then it drops back to 0. It flows in the opposite direction in just the same amount. When electricity goes through

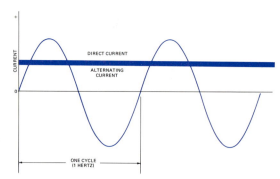

Fig. 172-1. The difference between direct current and alternating current.

this change, it is called a *cycle of current.* Most electricity used in houses is 60-hertz (60-cycle) alternating current. It changes in direction 120 times a second. In countries where alternating current operates at 30 hertz (30 cycles per second), a flicker in the light bulb can actually be seen with each change in the direction of current flow.

Unit 173

Conductors and Insulators

Perhaps you've noticed that electrical wiring is made of metal (usually copper) and wrapped with rubber or plastic. Have you ever wondered why wiring isn't made of rubber instead, with metal on the outside?

Any substance can conduct an electric current to some degree. Some carry it better than others. The substances that carry a current easily are called *conductors.* The best conductors are silver, copper, and aluminum. Other materials conduct electricity very poorly. These are called *insulators.* Because they are poor conductors, they are used for insulation in and around electric devices. Good insulators include porcelain, rubber, and glass. Figure 173-1 lists several common conductors and insulators. The materials used in making

transistors are called *semiconductors.* These are neither good conductors like copper nor poor conductors like rubber.

Electric Wires

Wire for electrical work is measured by a Brown & Sharpe, or American, gage. It ranges from the smallest wire, No. 40, to the largest, No. 4/0 (Fig. 173-2). There are hundreds of different kinds of wire. They are made for many different uses. Some types are solid, and others are stranded. Some of the most common wires used in the school shop include:

Magnet Wire. This is used in small-project work for building motors, tele-

GOOD CONDUCTORS	FAIR TO POOR CONDUCTORS	INSULATORS (VERY POOR CONDUCTORS)
1. Silver	1. Nichrome	1. Oil
2. Copper	2. Carbon	2. Porcelain
3. Aluminum	3. Salt water	3. Rubber
4. Zinc		4. Shellac
5. Brass		5. Glass

Fig. 173-1. Some examples of conductors and insulators.

graph sounders, and transformers. The common sizes are No. 18 to No. 28. Magnet wire with an insulating enamel may be bought. It is called *enameled magnet wire*. Cotton-covered magnet wire has one or more wraps of cotton yarn over the conductor. The letters scc mean that it is "single-cotton-covered." The letters dcc means that it is "double-cotton-covered."

Annunciator, or Bell, Wire. Size 18 wire is used often for bell, buzzer, or chime sys-

WIRE GAGE BROWN & SHARPE, OR AMERICAN	
GAGE NUMBER	SIZE
8	0.128
10	0.101
12	0.080
14	0.064
16	0.050
18	0.040
20	0.032
22	0.025
24	0.020
26	0.015
28	0.012

*Dimension of sizes in decimal parts of an inch. For example, 0.500 is 1/2 inch.

Fig. 173-2. Common sizes of electric wire.

tems. It is also used for other low-voltage wiring. It is like dcc wire except that it has a heavily waxed outer surface.

Building, or House, Wire. This wire is a single copper wire. It is covered with fabric and rubber or plastic insulation. It has a cover of various colors for identification in wiring. No. 14 is used for ordinary light circuits. No. 10 to No. 12 is used for large appliance circuits, and No. 8 is used for electric stoves. Often three or four strands of house wire are combined and wrapped together to form a *cable*.

Flexible Lamp Cord. This is used for extension cords and nonheating appliances, such as lamps. It is made of many fine copper wires so that it is *flexible* (can be bent). The wire may be covered with rayon dress braid, rubber, or plastic.

Heater-cord Wire. This type is made of many small, fine wires. It is covered with insulation, including a layer of asbestos and an outer layer of cloth. It is a flexible cord found on all heater appliances. All appliance cord should have the National Underwriters' approval band. This tells that it has been carefully inspected (checked) and is safe.

Making Splices

Wires are joined by making splices. These are usually soldered and taped if they are to be permanent connections. There are three basic types of splices: the *Western Union*; the *tap*, or *branch*; and the *pigtail*. These splices are used for both low-voltage and house wiring. The Western Union splice is made when two wires must be joined to make a single longer wire. It has little use today in house wiring (Fig. 173-3). The tap, or branch, splice is used to tap in or join one wire to a second at right angles. The pigtail splice is used mostly in house wiring. This splice is used for joining wires in outlet and switch boxes.

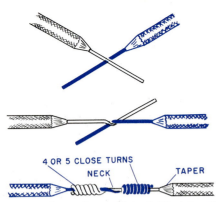

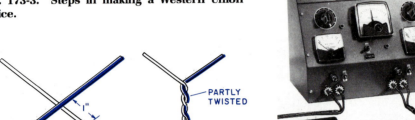

Fig. 173-3. **Steps in making a Western Union splice.**

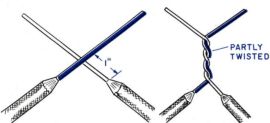

Fig. 173-4. **Steps in making a pigtail splice. The wire is twisted to within 6 mm (¹/₄ in) of the ends.**

Fig. 173-5. **Heat the splice with the soldering gun until the wire is hot enough to melt the solder. Soldered connections are used on most electronic equipment.**

Fig. 173-6. **A resistance-type soldering console for use in an electricity and electronics laboratory.** (American Heater Company)

Pigtail Splice

1. Carefully remove about 65 millimeters (2¹/₂ inches) of insulation from the ends of the two wires. Use a knife or a wire stripper. Make sure that the insulation is cut at a taper and that the wire itself is not nicked.

2. Scrape the wire with the back of a knife. Make sure it is clean and bright.

3. Cross the two wires about 25 millimeters (1 inch) from the insulation (Fig. 173-4) at an angle of about 60°. Twist and pull the wires. Wind them equally around each other to about 6 millimeters (¹/₄ inch) of the end.

4. Clip off the irregular ends.

5. Solder the connection (Figs. 173-5 through 173-7).

6. Cover the splices with insulation. Use electrician's tape. Tape the splice as you

Fig. 173-7. **Soldering connections on electronic equipment with a soldering center.** (American Heater Company)

would a sore finger. Lap the end over and then around until the exposed wires are completely wrapped. Always put back as much insulation as you have removed.

Fig. 173-8 a.   A solderless connector or a wire nut may be used to fasten two pieces of wire together.

Fig. 173-8 b.   These electrical spring connectors are used for splicing wire connections in all types of building construction. Some types do not require that the wires be twisted together. (3M)

7. To splice wires without solder or tape, fasten the twisted wires by using a solderless connector (or wire nut). Hold the ends of the wires together. Screw on the connector (Fig. 173-8 a and b).

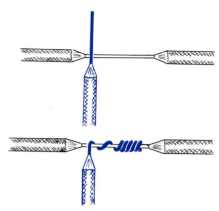

Fig. 173-9.   Steps in making a tap, or branch, splice.

Tap, or Branch, Splice

1. Remove about 38 millimeters (1¹/₂ inches) of insulation at the correct point along a main-line wire.

2. Remove about 75 millimeters (3 inches) of insulation from the end of the branch wire.

3. Hold the two wires at right angles, as shown in Fig. 173-9. Then wrap one wire around the other with one or two open turns and three or four closed turns (Fig. 173-6).

4. Solder the joint.

5. Tape the joint. Start the tape on the branch wire. Cover it and then cross over, first taping one side of the main-line wire and then taping back across the other side.

Unit 174

Electromagnetism

Electricity flows along a wire. It is surrounded by a magnetic field (Fig. 174-1). If the wire is made in the shape of a coil, it becomes a *solenoid* coil. If a soft steel core is placed near one end of the coil, the core will be pulled into the coil. This action is used to work switches and control devices.

The coil can be wound around a soft steel core that will easily magnetize and demagnetize. Then it becomes an *electromagnet* (Fig. 174-2).

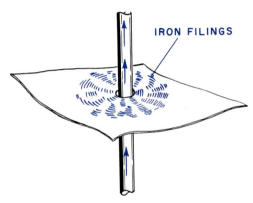

Fig. 174-1. A magnetic field can be shown by placing a piece of paper over the wire and sprinkling iron filings on it.

Fig. 174-3. This telegraph sounder has two electromagnets.

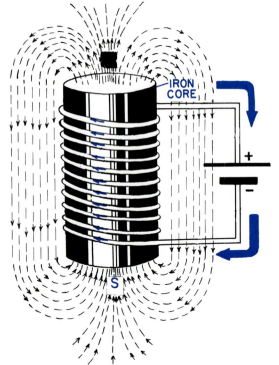

Fig. 174-2. A solenoid becomes a good electromagnet because it has a soft iron core.

Uses for Electromagnets

The electromagnet and the solenoid are found in bells, buzzers, telegraphs, telephones, relay devices, many motors, gen-erators, and almost every place where some movement is needed. The strength of an electromagnet depends on three things: the number of turns of wire, the kind of core, and how much current is flowing through the wire. To get the most from an electromagnet, the layers of wire must be wound evenly and smoothly.

Making an Experimental Electromagnet

Most small electromagnets are made by winding the wire on a large nail, a stove bolt, a machine screw, or a piece of band iron. This is called *winding an electromagnetic coil.* This electromagnet can be used for electrical project activities in the shop (Fig. 174-3).

To wind an electromagnetic coil on a nail, a stove bolt, or a machine screw:

1. Choose a metal fastener (bolt, machine screw, or nail) that will give you the size of electromagnet you need. Usually, the size is shown on your project activity drawing.
2. Lay out and cut two fiber washers that will hold the ends of the coil in place. Drill a hole in the center of the washers. Now they will slip over the metal fastener, one on either end.

3. Wrap the base of the metal fastener with rubber tape or friction tape or cover it with thin fiber. This covering protects the insulation on the wire. Cut insulation can short-circuit the electromagnet.

4. Select No. 18 to No. 28 scc or dcc wire, as shown on the project activity drawing.

5. Place one end of the wire under the lower fiber washer. This washer is the farthest away from the head of the metal fastener. Sometimes a small hole is drilled in this washer for the wire to pass through. Then fasten the point, or end, of the metal fastener in the chuck of a hand drill (Fig. 174-4).

6. Fasten the handle of the hand drill in a wood vise. Use a slow speed. Guide the wire. Carefully wind it around from one end to the other. In most cases, you should wind even layers. In that way, both ends of the wire are at the lower ends of the electromagnet. Fasten the outer end by covering the wire with tape or applying several coats of insulating varnish. If needed, slip the end through a small hole in the lower washer.

7. Attach a dry cell to the electromagnet. Check the polarity of each end with a com-

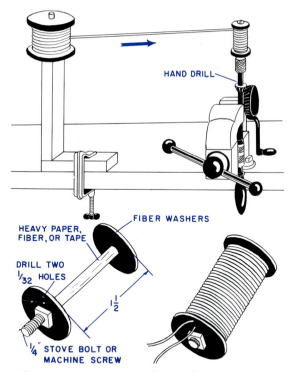

Fig. 174-4. **A way to make an electromagnet.**

pass. By reversing the wires on the battery, you also reverse the polarity of the electromagnet.

Unit 175

Batteries

Batteries are a source of direct current. A battery changes *chemical* energy into *electric* energy. It does not store electricity. There are two major kinds of batteries. In the *primary* cell, such as a flashlight battery, the materials are usually used up. The *secondary* cell, as in a car battery, can normally be recharged. In other words, the secondary cell can be used over and over again (Fig. 175-1 a and b).

Primary Cell

Most of you carry a primary cell around with you all the time. If you have a penny, a nickel, and a piece of paper, you have the makings of a simple cell. Wet the paper in your mouth and place the penny on one side. Put the nickel on the other. By hooking up wire from each coin to a sensitive meter, you can see a flow of electricity. An-

Fig. 175-1 a.   A mail truck powered by batteries.
(Lead Industries Association)

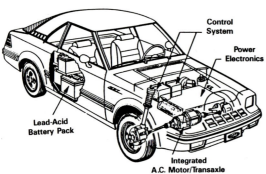

Fig. 175-1 b.   An electric and hybrid experimental car.

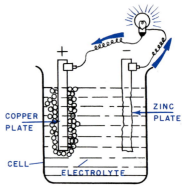

Fig. 175-2.   A simple cell is made by putting two different metals into an electrolyte that may be an acid, a base, or a salt.

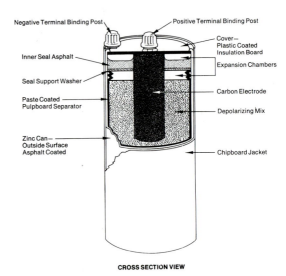

CROSS SECTION VIEW

Fig. 175-3 a.   A cutaway view of a dry cell.

other way to make a cell is to stick a copper nail in one side of a lemon. Place a *galvanized* (coated) nail in the other side. A simple cell is made up of two different metals and a material called an *electrolyte*. The electrolyte may be an acid, a base, or a salt (Fig. 175-2).

The most common primary cell is a *dry cell* (Fig. 175-3 a and b). All dry cells produce 1.2 to 1.5 volts. But a large dry cell delivers more current and lasts longer. Dry cells are the source of power for such items as flashlights, doorbells, toy motors, and transistor radios. Always keep a dry cell in a cool place. This prevents it from drying out. Never attach a wire right across

the two *terminals* (poles) of a dry cell. This can short-circuit it and wear it out quickly.

Dry Cells in Series and Parallel

Several cells connected together are called a *battery*. When the cells are connected in series, the positive pole of one cell is connected to the negative pole of the next cell. The full voltage is found by adding together the voltages of each cell. For example, if three dry cells, each producing 1.5

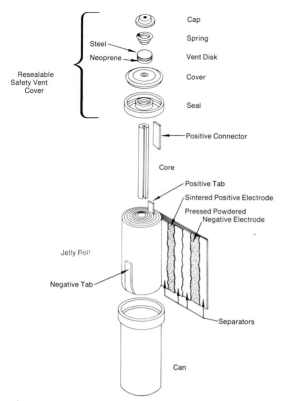

Fig. 175-3 b. **Exploded view of a sealed nickel-cadmium *rechargeable* cell.** (Union Carbide Corporation)

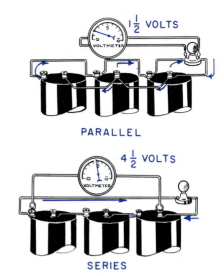

Fig. 175-4. **These dry cells are connected in parallel and in series.**

	SERIES	PARALLEL
Purpose	To build up voltage	To make battery last longer
Total voltage	Sum of cell voltages	That of single cell
Maximum current	That of single cell	Sum of cell amperages

Fig. 175-5. **Cells in series and parallel.**

volts, are connected in series, the total is 4.5 volts (Fig. 175-4). The *amperage* (rate of flow) is the same as in a single cell.

When cells are connected in parallel, the positive pole of one cell is connected to the positive pole of the next cell. Now the voltage or pressure is the same as in a single cell. Three dry cells hooked in parallel will produce 1.5 volts (Fig. 175-4). However, the amperes, or amount, will be the total of the individual cells. Figure 175-5 shows the differences between series and parallel.

Secondary Cell

An automobile storage battery is made up of three or more secondary cells. Each cell produces about 2 volts of electricity. Six

cells hooked in series produce 12 volts. Most modern cars have 12-volt batteries. When the battery is giving off energy, it is *discharging*. The process of putting energy into it is called *charging*.

There are two types of storage batteries. They are the conventional *lead-acid* battery and the *maintenance-free* battery (Figs. 175-6 through 175-8).

Conventional Lead-Acid Battery Operation and Construction

A lead-acid battery (Figs. 175-6 and 175-8) is a device for storing energy in a chemical

ONE-PIECE COVER
DELCO EYE
TERMINAL POST
FLAME ARRESTOR VENT PLUG
PLATE STRAP
SEPARATOR
PLATE
ELEMENT

Fig. 175-6. The parts of a lead-acid type of storage battery. (Delco-Remy, Division of General Motors)

form. The energy is stored in a chemical form so that it may be released as electricity when needed.

The necessary parts of the usual lead-acid battery are a number of negative (−) and positive (+) plates. These are sandwiched together. There is a thin, porous, nonconducting separator between them.

The positive plates contain lead peroxide. The negative plates contain sponge lead. In both plates, the material is attached to a grid structure. This is made of an *alloy*, or blending, of lead and antimony. The antimony is used to make the soft lead stronger. These grids direct current in the plates.

A *cell* is formed when the positive and negative plates (connected together) are put into a solution of sulfuric acid and water. This solution is called an *electrolyte*. The greater the number of plates in the cell, the greater the voltage during high discharge and low temperature rates.

The open voltage of a cell, no matter what the number of plates, is a little over 2 volts. When a number of cells are connected together in series, a *battery* is formed. The battery voltage is the sum of the voltages of each of the cells.

A charged cell is discharged when an outside circuit is completed (such as by switching on lights). The sulfuric acid in the electrolyte joins with the material on both the negative and the positive plates. This forms a new compound called *lead*

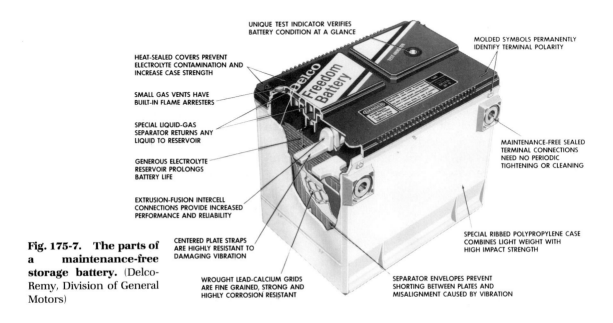

UNIQUE TEST INDICATOR VERIFIES BATTERY CONDITION AT A GLANCE

MOLDED SYMBOLS PERMANENTLY IDENTIFY TERMINAL POLARITY

HEAT-SEALED COVERS PREVENT ELECTROLYTE CONTAMINATION AND INCREASE CASE STRENGTH

SMALL GAS VENTS HAVE BUILT-IN FLAME ARRESTERS

SPECIAL LIQUID-GAS SEPARATOR RETURNS ANY LIQUID TO RESERVOIR

GENEROUS ELECTROLYTE RESERVOIR PROLONGS BATTERY LIFE

EXTRUSION-FUSION INTERCELL CONNECTIONS PROVIDE INCREASED PERFORMANCE AND RELIABILITY

MAINTENANCE-FREE SEALED TERMINAL CONNECTIONS NEED NO PERIODIC TIGHTENING OR CLEANING

SPECIAL RIBBED POLYPROPYLENE CASE COMBINES LIGHT WEIGHT WITH HIGH IMPACT STRENGTH

CENTERED PLATE STRAPS ARE HIGHLY RESISTANT TO DAMAGING VIBRATION

WROUGHT LEAD-CALCIUM GRIDS ARE FINE GRAINED, STRONG AND HIGHLY CORROSION RESISTANT

SEPARATOR ENVELOPES PREVENT SHORTING BETWEEN PLATES AND MISALIGNMENT CAUSED BY VIBRATION

Fig. 175-7. The parts of a maintenance-free storage battery. (Delco-Remy, Division of General Motors)

•IGNITION ARRESTOR®
INSPECTION HATCHES
act as recondensing chambers to return much of the water that would normally be lost, back to the system. Provide for easy, quick inspection and testing. Reduce hazards associated with battery gasses. Accessible in the event extreme conditions require the addition of water which dramatically extends life and consumer value.

•100% MORE ELECTROLYTE
Large reservoir allows for twice as much electrolyte. This battery is designed to "NEVER NEED WATER" under normal operating conditions.

•MANUFACTURED WITH MAINTENANCE-FREE MATERIALS — Exclusive TRI-LOY® grids are designed to deliver maximum power and reserve capacity and yet operate with substantially reduced water loss.

•EXCLUSIVE MINI-REST-UPS provide for adequate sediment space, yet allow the elements to be lower in the case providing for maximum electrolyte volume.

STANDARD VENT CAPS

•STANDARD ELECTROLYTE OVER PLATES

•STANDARD MATERIALS

•STANDARD SEDIMENT SPACE

Fig. 175-8.   Differences in standard lead-acid and maintenance-free batteries. (General Battery Corporation)

sulfate. More sulfuric acid is used to form lead sulfate on the plates as the discharge continues. When part of the sulfuric acid is used up, the cell can no longer give off useful voltage. It is then said to be *discharged.* You can measure the amount of charge left in a cell by finding out how much sulfuric acid is left in the electrolyte.

When a cell is being *charged,* current passes through it in the direction opposite to that of the discharge. The voltage of the charge must be more than the open-circuit voltage. This is done to overcome inside resistance. Lead sulfate from the plates is changed to sulfuric acid during charging. This slowly brings back the original strength of the electrolyte. The cell is now ready to give off electricity once again.

Extra current that is not used by the cell near the end of the charging period makes the plates break up water. This causes hydrogen and oxygen to be given off. Water must then be added to the cell from time to time to make up for the water loss.

The gases given off from the cell during charging are very explosive. Care must be taken to keep sparks, flames, and cigarettes away from the general area of the battery. Such care helps prevent a dangerous explosion.

Maintenance-free Batteries

Maintenance-free batteries (Figs. 175-7 and 175-8) are different from the usual older types. Maintenance-free batteries have a longer shelf-life. They do not have problems with *corrosion* (being eaten away by acid) at the terminals and inside plates. Under normal operating use, very little or no water needs to be added during their expected life.

In the older type of battery, water loss is a problem. This is caused by the use of antimony. Antimony hardens the lead grids used in the building of cell plates. When the amount of antimony is cut down, there is not so much loss of water.

Water loss can be reduced in a number of ways. Low-antimony (less than 3 percent), lead-calcium, and lead-strontium grids are now being made.

The cell plates sit much lower in low-antimony construction. This lets more electrolyte cover the plates. The plates sit lower in the cell, and there is less water loss. The battery can almost reach its expected life without water being added. There is less corrosion of the terminals because smaller amounts of gas are escaping. This battery does not overcharge at voltages higher than 14.4 volts. This is better than the performance of conventional batteries.

Lead-calcium grid batteries have plates that sit in the bottom of the case. There is no space left for sediment. The plates are in an envelope separator that is sealed on three sides. It traps any sediment that might short out the battery.

The calcium grid itself loses water at a slower rate than in those batteries we have already talked about. This small rate of water loss, along with the grids that sit in the bottom of the case covered with electrolyte, make the battery sealed but vented. The top of the battery case is smooth.

Lead-calcium batteries have a *minimum* (smallest amount) of terminal corrosion. They work better than conventional batteries at cold temperatures. If the battery is overcharged, it will not have water loss. The shelf life is much longer. However, lead-calcium batteries cannot stand deep cycle discharge such as that caused by leaving the headlights of an automobile on. A lead-calcium battery costs more than other batteries. Low-antimony batteries are maintenance-free and cost about the same as conventional batteries. Each has its good points. Try to choose the best battery for your purposes.

General Maintenance for Conventional Lead-Acid Batteries

1. The quality of a battery depends on the number and the size of the plates and the kind of separators used. Most car batteries have 13, 15, or 17 plates per cell. The best

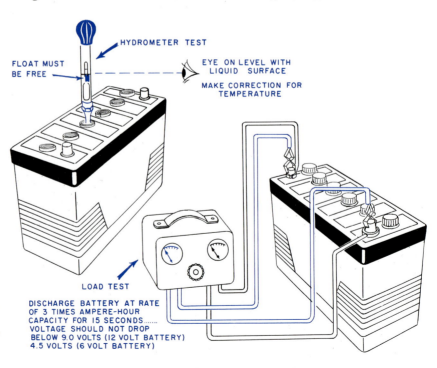

HYDROMETER TEST

FLOAT MUST BE FREE

EYE ON LEVEL WITH LIQUID SURFACE

MAKE CORRECTION FOR TEMPERATURE

LOAD TEST

DISCHARGE BATTERY AT RATE OF 3 TIMES AMPERE-HOUR CAPACITY FOR 15 SECONDS...... VOLTAGE SHOULD NOT DROP BELOW 9.0 VOLTS (12 VOLT BATTERY) 4.5 VOLTS (6 VOLT BATTERY)

Fig. 175-9. Two ways of checking the condition of a lead-acid storage battery.

separators are usually made of fiber glass or rubber.

2. A battery is rated by the number of amperes it will produce for a given length of time. For example, a 100-ampere-hour (A·h) battery will produce 100 amperes for 1 hour, or 25 amperes for 4 hours, before becoming discharged.

3. The condition of the battery can be learned by the use of a heavy *discharge* (load) test or a *hydrometer* (Fig. 175-9). This load test shows whether the battery can keep up the proper voltage while producing a heavy current. That is, it measures the *resistance* of the cell.

The hydrometer measures the condition of a battery by testing the density of the liquid. A fully charged battery will read 1.270. A discharged battery will read 1.150 (Fig. 175-10). When a battery is in a low state of charge, it can be recharged. The quick-charge way takes about 45 minutes. A long charge takes about 24 hours. Quick charging a fully discharged battery can harm it.

4. Batteries should be filled with distilled water at regular intervals.

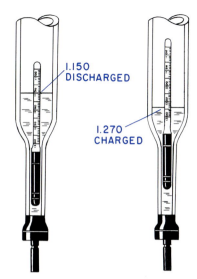

Fig. 175-10. A hydrometer test for a storage battery.

5. The terminals of an automobile battery can be *corroded*. If that happens, pour a mixture of vinegar or baking soda and water over the terminals. Allow this to stay a few minutes. Then wipe the terminals clean. Cover them with petroleum jelly.

Unit 176

Generators and Transformers

You have already learned that electricity and magnetism are related. Electricity can be changed into magnetism. Magnetism can be changed into electricity. If a wire that is connected to a sensitive meter is moved through a magnetic field, a current will flow (Fig. 176-1). This is called *electromagnetic induction*. It is used in large power plants that *generate* (make and send out) electricity. These plants have huge generators with many electromagnets and revolving wires that produce a high voltage.

For electromagnetic induction to take place, a magnetic field, a closed circuit, and movement are needed. Movement is a must. The magnetic field around a magnet produces electric current in a wire *only* when the wire is moving across the field. To produce electricity, you may move either the wire or the magnet. To get the continuous motion needed to keep producing a current, it was discovered, the wire must move in a circle inside the magnetic field. This is the method used to produce alternating-current electricity with a

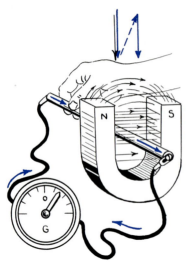

Fig. 176-1. **As the wire is moved up and down over the magnetic field, an electric current flows.**

generator (Fig. 176-2). For this reason, it is necessary to have huge generators to produce electricity for an entire city or area of the country.

The amount of electricity produced can be increased. This is done by moving the wire through the magnet faster, by increasing the strength of the magnet, or by making the wire longer.

Electricity is generated as alternating current at about 13 800 volts. There is less loss in transmitting the electricity if it is stepped up to a much higher voltage, such as 69 000 volts. Later, it may be stepped down to 13 800 volts at a substation, to 2400 volts at a distribution substation, and finally to 120 or 240 volts for household use (Fig. 176-3). This change in voltage is done by devices called *transformers*. They work on the principle of electromagnetic induction (Fig. 176-4).

The transformer is made of several pieces of *laminated* (layered) metal. Two separate windings, the *primary* and the *secondary*, are on them. If it is a step-up transformer (for example, if you want to put in 120 volts and step that up to 240 volts), there must be twice as many turns on the secondary winding as on the pri-

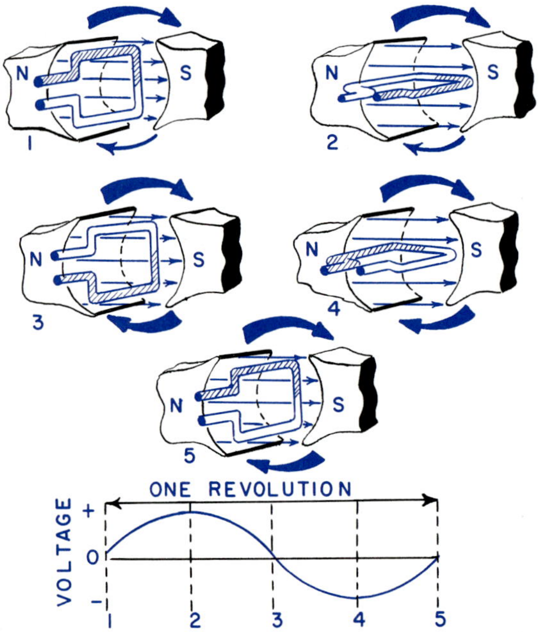

ONE REVOLUTION

VOLTAGE

Fig. 176-2. **How alternating current is made in a generator.**

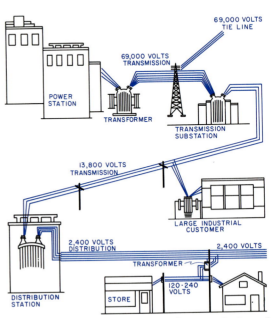

Fig. 176-3. **The generation and transmission of electricity.**

Fig. 176-4. Two of the largest 500-kV high-voltage transformers. (Westinghouse Corporation)

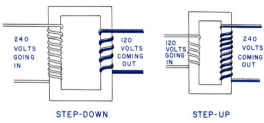

STEP-DOWN   STEP-UP

Fig. 176-5. A step-down transformer has more turns on the primary winding than on the secondary winding. A step-up transformer has more turns on the secondary winding than on the primary winding.

mary winding (Fig. 176-5). When the alternating current flows into the primary winding, the moving magnetic lines of force cut the wires in the secondary winding. This produces a flow of electricity from the secondary winding by electromagnetic induction (Fig. 176-5).

Small transformers are also found in houses. For example, when you want 6, 8, or 12 volts to operate a doorbell or a small electric train, you have a step-down transformer. This will reduce the voltage from the 110 volts at the outlet.

The minimum voltage in houses is 110 to 220; 110 volts for lighting circuits and 220 volts for large appliances, such as electric stoves or driers. Sometimes the voltage will be 115 and 230 or 120 and 240.

Unit 177

Measuring Electricity

Electricity is a flow of electrons through a wire. In many ways, electricity acts like the flow of water through a pipe. To have a flow of water, you need pressure from a tank of water. This pressure must overcome the resistance or friction from the sides of the pipe in order to force the flow through the pipe. Electricity acts in much the same way. The pressure is called *voltage* or *electromotive force*. The actual rate of *flow* (current) of electricity is measured in *amperes*. To cause an electron flow, the electric pressure must overcome the resistance of the wire. This is measured in *ohms*.

Wire resistance changes with four factors: (1) It increases as the wire gets smaller. (2) It increases as the wire gets longer. (3) It changes with the kind of wire. (4) It increases as the temperature of the wire increases. Copper wire, for example, has a low resistance. Iron wire has a high resistance.

Ohm's Law

Voltage or pressure, amperes or current, and resistance or ohms are always related to one another. This relationship is known as *Ohm's law*, which is stated as follows:

Voltage *(E)* equals amperes *(I)* times resistance *(R)*, or

$$E = I \times R \text{ or } I = \frac{E}{R} \text{ or } R = \frac{E}{I}$$

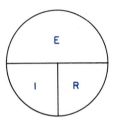

Fig. 177-1. Blocking out the unknown quantity shows the formula for Ohm's law.

This is the basic law of electricity. It is used in solving both simple and difficult electrical problems (Fig. 177-1).

If you know two of these three quantities (voltage, amperes, and resistance), the third can easily be found with Ohm's law. Suppose you connect a small electric bulb with a resistance of 2 ohms to four dry cells hooked in series, each producing 1.5 volts. These dry cells will have a total voltage of 6. To figure the number of amperes that will flow, you can use the formula

$$I = \frac{E}{R} \text{ or } I = \frac{6}{2} \quad \text{or} \quad 3 \text{ amperes}$$

Ohm's Law Problems

1. What will be the current flow through a small coil with a resistance of 10 ohms if the voltage is 100?
2. What is the resistance of a light bulb through which 2 amperes flows when it is operating on 110-volt circuit?
3. An electric toaster has 20 ohms of resistance. What current will flow when the toaster is connected to a 110-volt circuit?
4. If a coil of wire with a resistance of 5 ohms is connected to the terminals of a single dry cell, what current will flow?

Electric Meters

An electric *meter* is an instrument used to measure some quantity of electricity (Fig. 177-2). An *ammeter* is used to measure the number of amperes flowing through the

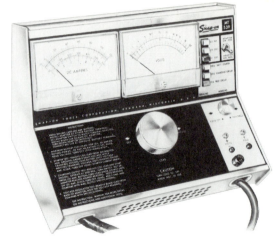

Fig. 177-2. This electric meter measures volts and amperes.

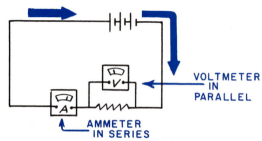

VOLTMETER
IN
PARALLEL

AMMETER
IN SERIES

Fig. 177-3. Notice that a voltmeter is placed in parallel with the circuit. An ammeter is placed in series with a circuit.

line. It is always connected *in series* with the circuit. A *voltmeter* is used to measure the difference in electrical potential (the ability to do work) between two points in a circuit. It is always connected *in parallel* with the circuit (Fig. 177-3). The voltmeter and the ammeter may be used to find resistance through the use of Ohm's law. An *ohmmeter* shows directly the resistance of a unit.

Combination meters are available.

Watts and Kilowatts

The units of electric power are *watts* and *kilowatts*. See Fig. 177-4. One kilowatt equals 1000 watts. Watts (*W*) equal volts (*E*)

TERM	DEFINITION	EXPLANATION
Ampere	Unit of electric current intensity	Commonly used to express an amount of electric current
Volt	Unit of electric pressure	Rate of pressure exerted on current flowing through a wire (similar to water pressure)
Watt	Unit of electric power	The product of amperes and volts—converted into working electric energy
Watt-hour	One watt used for one hour	The measurement of electric energy used
Kilowatt-hour	1000 watt-hours	The abbreviation kw·h is the term found on electric bills to show the amount of electric energy used

Fig. 177-4. Some definitions of common electrical terms.

times amperes *(I)*, or $W = E \times I$. You pay for electricity by the number of watts used. Since a watt is a rather small unit, electricity is really paid for per thousand watts per hour, or kilowatt-hours (kw·h). The electric meter of a house records the amount used. The meter is read about once a month or every 2 months by a meter reader from the electric company. You are charged for the number of kilowatt-hours used.

Reading a Kilowatt-hour Meter

An electric meter may have four or five dials (Fig. 177-5). One complete turn of the pointer of the right-hand dial represents

Fig. 177-5. A kilowatt-hour meter. (General Electric Company)

10 kilowatt-hours. Each division on that dial is 1 kilowatt-hour. One complete turn of this dial moves the second dial one division. Each division on this second dial is 10 kilowatt-hours. When you read from right to left, the dials show units, tens, hundreds, thousands, and ten thousands. The thousands and tens dials turn counterclockwise, and the ten-thousands, hundreds, and unit dials turn clockwise.

To read a meter, start with the first dial on the left. Read the number that the pointer has *just passed*. For example, in Fig. 177-6, a meter with four dials, the reading is 7. Then read the hundreds, tens, and units. In this case the complete reading is 7562 kilowatt-hours. Suppose that this is the reading at the first of the month and that the reading at the end of the month is 7873. The total number of kilowatt-hours used is 311. If the cost of electricity is $0.06 per kilowatt-hour, the bill will be $18.66.

Bills for the cost of electric power are the result of many additional cost factors. There are monthly service charges, federal and state taxes, and rate adjustments allowed by the local utility boards.

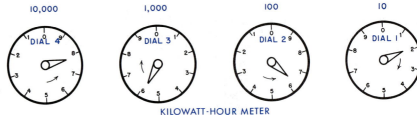

Fig. 177-6. Detail of kilowatt-hour meter dials.

KILOWATT-HOUR METER

Unit 178

Low-voltage Circuits

Most of your work in low-voltage wiring (less than 18 volts) will be completing bell, buzzer, and chime circuits. You will work with small light circuits to learn about series and parallel wiring. As you do these jobs, you will learn many basics that can be used later to do all types of electrical work.

Tools and Materials

Most low-voltage wiring done in school is put on a wiring board. The electric devices may be fastened either temporarily or permanently. Common tools needed are a 150-millimeter (6-inch) screwdriver, side-cutting pliers, a knife, and a small hammer. For bell and buzzer circuits, you will need bells, buzzers, push buttons, single-pole single-throw (spst) and single-pole double-throw (spdt) switches, dry cells or a low-voltage transformer, and No. 18 annunciator or dcc magnet wire. If the circuits are to be fastened to the board, No. 3 or No. 5 insulated staples may be used. For low-voltage lighting circuits, tiny light sockets and flashlight bulbs are also needed (Fig. 178-1).

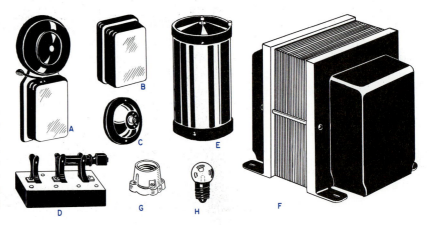

Fig. 178-1. Equipment used in low-voltage wiring: (a) bell, (b) buzzer, (c) push button, (d) single-pole double-throw switch, (e) dry cell, (f) transformer, (g) light socket, (h) flashlight bulb.

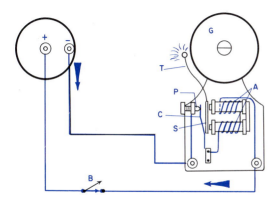

Fig. 178-2. The action of a simple circuit.

Action of a Simple Bell Circuit

The simplest low-voltage wiring circuit has four parts: a bell, a push button, a dry cell as a source of electricity, and a connecting wire (Fig. 178-2). The action of this circuit is as follows:

1. The dry cell changes chemical energy into electric energy. The excess of electrons gathers on the negative pole. This is the zinc plate. There is a lack of electrons on the positive pole, or carbon.

2. When the switch (B) is closed, the circuit is completed. Electrons flow from the negative pole through the wire, the breaker points (P), the contact spring (C), and the electromagnets (A). They then flow back through the wire and switch to the positive pole.

3. When this happens, the electrons energize or magnetize the coils (A). These in turn attract the armature (S), pulling it forward so that the hammer (T) strikes the bell (G).

4. This breaks the circuit at the breaker points (P), cutting off the flow of electrons. The electromagnets lose their magnetism right away, and the spring (C) carries the armature back to the first position. The circuit is again completed. The flow of electrons is started again.

All this happens so fast that the bell rings constantly. Buzzer circuits operate the same way.

Series and Parallel Wiring

There are two basic ways of wiring electric devices. They are done in *series* and in *parallel.* A good example of series wiring is seen in television sets using tubes. When one tube burns out, the whole set does not work. When several devices of any kind are hooked in series, none will work if one device breaks down. As you can see in Fig. 178-3, the devices are connected one right after another. The electrons must be able to flow through all of them. If one does not work, there is a break in the circuit, and all go dead.

In parallel wiring, the devices are wired side by side so that only some of the electrons flow through each device (Fig. 178-4). If one breaks down, there is still a complete circuit for the electrons to flow through. All homes and buildings are wired this way. If one lamp in your house burns out, the others still burn.

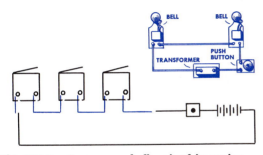

Fig. 178-3. Buzzers or bells wired in series.

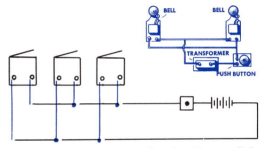

Fig. 178-4. Buzzers or bells wired in parallel.

When wiring bell and buzzer circuits in parallel, do the following:

1. Always run a wire from one side of the source of power to each of the electric devices.
2. Then run a wire from the other terminal of each electric device to the switch.
3. Finally, connect a wire from the switch to the source of power.

Wiring

There are many useful wiring jobs you can do to gain experience in planning, reading, and wiring electric circuits. Once you know how to do these low-voltage jobs, it is quite easy to do other wiring jobs. Follow these simple rules:

1. On a piece of 216 × 280 millimeter (8½ × 11 inch) paper, prepare a plan for a wiring job. Make it like the one shown in Fig. 178-5. Fill it out.
   a. Describe the kind of job. One example might be two bells connected in parallel controlled by a push button with dry cells as the source of power. Also tell what use can be made of this job. For example, this circuit might be used when someone wants a double door bell, one in the basement and another upstairs.
   b. Make a *schematic diagram* (drawing) of the wiring circuit. Use the correct symbols for the different electric devices. (See Unit 21, "Dimensions, Conventions, and Symbols.") Make a neat sketch or mechanical drawing.
   c. Make a sketch of the place where the wiring is to be used. This sketch can be made on a large piece of wrapping paper. Then it should be tacked over the wiring board. You will then see the electric devices, switches, and source of power in their proper positions before you wire the circuit.
2. Study the schematic drawing. Make sure you can trace the circuit with your finger. Follow the circuit from one side of the source of power, along the wires, and through the device and switch. Then trace it back to the other side of the source of power. Electricity never flows unless the circuit is complete.
3. Choose the electric materials you need (bells, buzzers, lamp sockets and bulbs, switches or push buttons, wire, and the source of power). Place the devices, switches, and source of power in the right places. Remove the insulation from the wire and scrape. Hook up the wire binding posts. Always put the wire around the posts in the same direction in which the nuts fasten down. If this is not done, the wire may push itself off. Make neat splices. Bend sharp corners in the wire. In that way, you can easily follow the circuits.
4. Test the operation. If it does not work correctly, retrace the circuits for incorrect wiring. Check for faulty devices or switches. Make sure that the connections are good. See that the switches work properly. Check the source of power.

WIRING JOB No. ______

Name ______

Kind of job ______

Schematic drawing

Sketch showing its application

How job operates—electrical principles involved

Checked by ______

Fig. 178-5. A wiring-job plan.

Typical Wiring Problems

1. *Problem:* One buzzer controlled by one push button. Two dry cells are the source

of power. *Situation:* Your next-door neighbor wants you to connect a push button at the front door and a buzzer mounted on the door frame near the kitchen. The system is to be operated by dry cells placed on a beam in the basement.

2. *Problem:* Two buzzers connected in parallel and controlled by a push button. A small transformer is the source of power. *Situation:* The principal wants a push button on her desk to call an office worker from either the outer office or the stockroom, which is some distance away. One buzzer is to be in the outer office. The other is to be in the stockroom.

3. *Problem:* Three buzzers connected so that they are operated by three push buttons. *Situation:* A three-family apartment house needs a buzzer system. One buzzer is to be located in each apartment and operated by pushbuttons at the front door.

4. *Problem:* Wiring a return-call system. *Situation:* You have a workshop in your garage. Your mother wants you to hook up a buzzer system so that she can call you and you can answer (Fig. 178-6).

5. *Problem:* Wire a two-door bell or chime system. *Situation:* A home should have a

Fig. 178-6.   A return-call system.

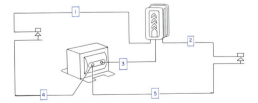

Fig. 178-7.   A bell or chime system for a house.

signal bell for the front door and a buzzer for the back door. Or it should have chimes with one sound for the front door and another sound for the back door. Connect wire 1 and wire 2 from the terminals of the combination bell and buzzer to the two push buttons. Then connect wire 3 to the transformer. Finally, connect two wires, 4 and 5, from the terminal on the transformer to each push button (Fig. 178-7). Push the buttons to see that the signal works.

Unit 179

House Wiring

A person not used to working with electricity should not try to put wiring in a house. However, the beginner should know about the different kinds of wiring devices available (Fig. 179-1).

Electricity in the Home

Electricity runs into a home through three wires. It runs first to a kilowatt-hour meter and then to a service switch box or distribution panel (Fig. 179-2). This box or panel

is hung in the basement or on the outside of the house.

The modern house should have at least a 100- to 150-ampere service. The box or panel provides a way of dividing the electricity among several circuits (Figs. 179-3 and 179-4). It also protects these circuits from overload. This is done by a safety device called a *fuse* or a *circuit breaker*. The box or panel also serves as a main switch.

In older distribution boxes, two types of fuses are used: *plug fuses* and the larger,

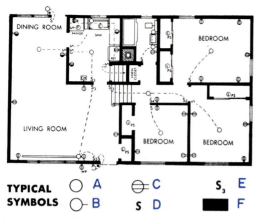

TYPICAL SYMBOLS: ○ A   ⊖ C S_3 E
           ○– B   s D   ▬ F

Fig. 179-1. A floor plan showing the wiring diagram with symbols: (*a*) ceiling light, (*b*) wall light, (*c*) convenience outlet, (*d*) single pole switch, (*e*) three-way switch, (*f*) main switch box.

appliances on that circuit. Do not use a fuse plug that is too large. It is your only protection against an overloaded circuit.

CAUTION: If a fuse burns out continuously, never replace it with a larger fuse. Find the trouble immediately.

heavier fuses called *cartridge fuses.* Cartridge fuses are used as main fuses or range fuses (Fig. 179-5). The newer boxes are equipped with *circuit breakers* (Fig. 179-6).

If something goes wrong with an appliance or if the line becomes overloaded, the fuse blows out or the circuit breaker trips open. This breaks the flow of electricity to that circuit. When the flow of electricity is broken, something is wrong. Check the appliance for trouble. Also check to see that there are not too many

If No. 14 wire is used for the lighting and general-purpose circuits in the house, a 15-ampere fuse or a circuit breaker should be used. For the kitchen or utility room, No. 10 or No. 12 wire is used with a 20- to 25-ampere fuse or a circuit breaker. Large appliances should have larger fuses or circuit breakers to protect them. Every home using fuses should have extra fuses on hand. Changing a burned-out or blown fuse is one of the most common repairs in the home.

There are many methods by which electricity is carried from the fuse box to the appliances in lighting outlets.

Nonmetallic sheathed cable is used for both open and closed wiring (Fig. 179-7 a and b). The two wires are protected by a flexible covering of insulating material.

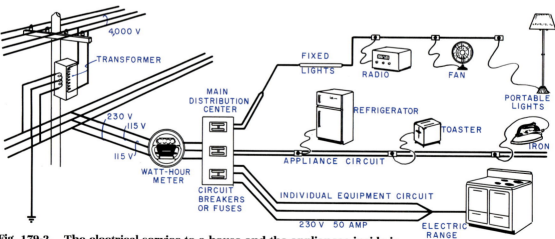

Fig. 179-2. The electrical service to a house and the appliances inside is shown. Note that the voltage to the house is 115 and 230 V. This is the average voltage. The minimum voltage in a house is 110 or 220 V.

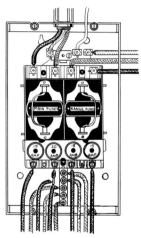

Fig. 179-3.   Inside a distribution panel or service switch box, the white wire is the ground wire. Grounding protects the circuits and eliminates much of the danger from shock.

Fig. 179-4.   Circuit breakers in a distribution panel.

Fig. 179-5.   These are common fuses. The small fuses are used in appliances and automobile circuits. The plug fuse is used for most lighting circuits. The cartridge is used for heavy electric appliances and distribution panels.

Fig. 179-6.   A circuit breaker. (General Electric Company)

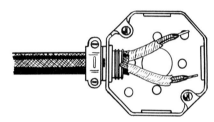

Fig. 179-7 a.   A nonmetallic sheathed cable.

Fig. 179-7 b.   Wiring a house using nonmetallic sheathed cable. The wire is stapled to the studs. The electrical outlet boxes are made of plastic, and a special plastic connector is used to join the connections. (3M)

This is made of plastic, fiber, rubber, or some other nonconductor. No supports other than staples are needed for the wiring. Only one hole has to be drilled for the cable. This cable runs to outlet boxes to which the switches, convenience outlets, and lights are attached. This method of wiring is used in many modern homes.

Flexible armored cable is installed in the same way as nonmetallic sheathed cable. The two wires are inside a *flexible* (bendable) metal cable (Fig. 179-8 a and b). This cable costs more money, but it is required by many local building codes. In some cases, armored cable is also prohibited by code due to rusting or corroding tendencies.

Electrical metal tubing is found in many industrial concerns, where open wiring is wanted. This tubing is sometimes called *thin conduit.* The two wires are run through the tubing. It can be bent to go around corners.

Rigid metal conduit is standard construction in most fireproof buildings (Fig. 179-9). It is like thin conduit except that the tubing cannot be bent. It needs connectors of various kinds. Rigid metal conduit can take very rough treatment without harming the wires.

Plastic conduit. Some codes require the use of plastic conduit PVC or ABS when wiring underground or outdoors.

Common House Wiring Repairs

Replacing a Fuse

1. Keep a candle or a flashlight at hand so that you can see the fuse box. When the fuse blows out, you may not have electricity near the fuse box. Also, be sure to have an extra fuse of the right size and kind.

2. Disconnect the defective appliance or any extra appliance on the circuit that may have blown out the fuse.

3. Open the door of the fuse box. Shine a light across each plug fuse. The thin covering of the burned-out fuse will be black.

4. Unscrew that plug fuse. Look at the size

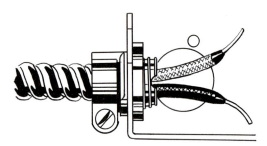

Fig. 179-8 a.   An armored cable.

Fig. 179-8 b.   A building wired with armored cable using electrical spring connectors to join the wires and make the connections in the electrical outlet boxes. These are metal boxes. (3M)

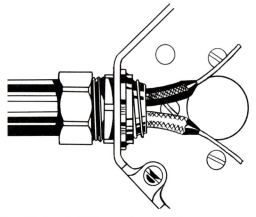

Fig. 179-9.   A rigid metal conduit.

stamped on the brass connector at the back of the plug.

5. Screw in a new fuse of the same size. Sometimes plug fuses come loose and cause a short circuit. Be sure that all are screwed tightly in place.

Resetting a Circuit Breaker. Modern distribution panels have a circuit breaker, which automatically trips over when the line is overloaded. Turn the toggle-switch handle to the extreme "off" position. Then return the handle to the "on" position.

Replacing a Switch or an Outlet. See Figs. 179-10 through 179-12. Two kinds of switches are most commonly used. *Single-pole* switches are for lights that are turned on and off in only one place. *Three-way* switches are for lights that can be turned on and off in two different places. To replace a switch or an outlet:

1. Obtain a switch of the type that has burned out.

2. Disconnect the electricity on that circuit. The easy way is to turn on a light on the same circuit. Then unscrew the fuses or trip the circuit breakers until the light goes out. You can also pull the main switch or circuit breaker to disconnect the electricity.

3. Remove the switch plate with a screwdriver (Fig. 179-13). Loosen the switch itself and pull it out of the box or wall (Fig. 179-14). Notice how the white and the black wires are attached to the switch. To replace a three-way switch that has three wires that must be connected, connect the red and white wires to the light-colored terminals. Connect the black wire to the dark-colored terminal (Fig. 179-15).

4. Remove one wire at a time. Attach it to

Fig. 179-10.　A recessed switch (below, left).

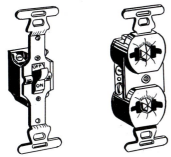

Fig. 179-11.　A convenience outlet (above, right).

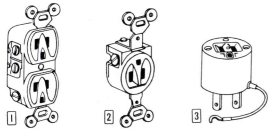

Fig. 179-12. UL-approved three-wire devices for equipment using 120 V—15 A or less, such as small air conditioners, gas driers, washers, power tools, and garden equipment. (1) Duplex receptacle with parallel blade and U-shaped ground. It fits a standard switch box. (2) Single receptacle with parallel blade and U-shaped ground. It fits a standard switch box. (3) An adapter for converting standard receptacles to take a three-prong plug (parallel blade and U-shaped ground).

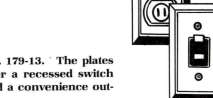

Fig. 179-13.　The plates over a recessed switch and a convenience outlet.

Fig. 179-14. Plate removed from the box, exposing the outlet.

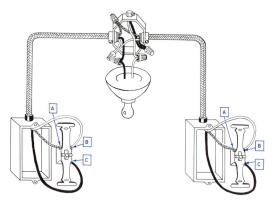

Fig. 179-15. Connections and wiring for three-way switches.

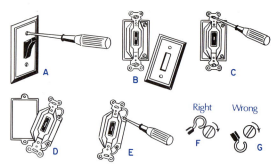

Fig. 179-16. Follow steps *a* to *f* in replacing a light switch. (*a*) Before removing screws, disconnect main switch. (*b*) Remove plate, exposing switch. (*c*) Remove screws holding switch to the box. (*d*) Pull switch out of box (wires are attached to switch terminals). (*e*) Unscrew terminals. (*f*) Right: turning screw closes loop. (*g*) Wrong: turning screws opens loop.

the new switch. The wire must go around the screw in the same direction in which it will be turned to be tightened. When the wires are replaced, fasten the switch and the plate in position (Fig. 179-16). Tighten the fuse or reset the circuit breaker. Test the switch to see that it works correctly.

5. Follow the same procedure when you replace a defective outlet.

Unit 180

Lighting with Electricity

Most people do not know what it is to be without electric lights. Yet less than a century ago, the only lighting known was some kind of open flame. Today all kinds of bulbs are used: flashbulbs for photography, infrared heat lamps for aching muscles, ultraviolet sunlamps that tan like the sun, and the "black light" used in crime detection, to name only a few (Fig. 180-1).

Incandescent Lighting

Thomas A. Edison invented the *incandescent* lamp in 1879. *Incandescent* means "glowing at white heat." He also devised the complete electric system for a city. This system was made up of generators powerful enough to light hundreds of bulbs. There also were meters to measure the amount of current used by each customer. In addition, a fuse to protect the system and a socket and a bulb base to make it easy to change bulbs were included.

When electrons flow along a wire, the wire becomes hot. In some appliances, such as toasters and stoves, the wires are red hot. A lamp bulb works like a heating element except that it becomes white hot and gives light. The parts of a bulb are shown in Fig. 180-2. The size of the bulb and the number of watts that it uses are stamped on the end. The *higher* the wattage, the *brighter* the light.

After experimenting for years with many materials to use for the lamp filament, Edison discovered that a carbon filament would work in a *vacuum* (a space without

Fig. 180-1. Two prize-winning lamp designs. One (left) is fluorescent, and the other (right) is incandescent. (Zinc Institute Inc.)

Fig. 180-2. Parts of an incandescent lamp bulb.

air). The incandescent lamp has been improved over the years (Fig. 180-3). Today lamp bulbs have a filament made of tungsten. The bulb is coated inside and is filled with gas. The light from a bulb burns with a reddish-white glow (Fig. 180-4).

Fluorescent Lighting

Another type of lighting is provided by the fluorescent tube. A fluorescent tube is less wasteful than an incandescent lamp. It also costs less to operate. A fluorescent lamp is a long glass tube with an electrode on either end. The tube is coated on the inside with fluorescent crystals. It is filled with mercury vapor gas (Fig. 180-5). When the switch is turned on, electrons shoot through the gas-filled space. They go from one electrode to the other to produce the light. There are many different kinds and shapes of fluorescent lights.

Wiring a Lamp

When you build a lamp, your big problem will be the wiring. Your experience with this project activity will help you rewire lamps at home, make new lamps, make an extension cord, or replace a plug on any appliance.

1. The materials needed include those shown in Fig. 180-6: a brass shell socket (available in four types), a key (a threeway-light key socket must be used for a three-way-light bulb), a push button, a keyless control or a pull chain, a flexible rubber-covered or fabric-covered extension cord of No. 14 wire, a rubber or plastic attachment plug, and a wire bushing. The cap of the socket has a 3.175-millimeter (1/8-inch) pipe thread cut on the inside. Therefore, you will need a short piece of 3.175-millimeter (1/8-inch) threaded pipe of black iron or brass. On some lamps, the pipe itself is the center of the lamp. Then only

Fig. 180-3. The development of the electric light bulb from Edison's original bulb (left) to the present-day bulb (right).

INCANDESCENT

TYPE	WATTS	USES
Standard 	40 to 300	For use in fixtures and portable lamps for task and general lighting
R (Reflector) 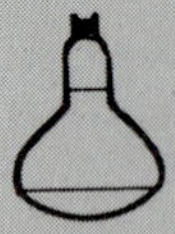	30 to 300	For use in recessed or surface-mounted round fixtures for task or general lighting
Par (Projector) (Weather Resistant) 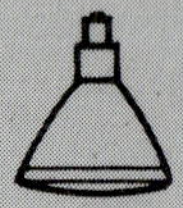	75 to 150	For use in outdoor flood fixtures for safety and security lighting; in garden lighting fixtures post lamps

Fig. 180-4.   Size of bulbs to use for lighting.

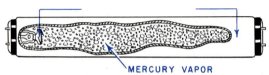

Fig. 180-5.   The wall of the tube is coated with fluorescent crystal.

one end has to be threaded. Whenever the wire goes through a metal, plastic, or wood base, a wire bushing should be used (Fig. 180-7).

2. Decide on the length of wire needed for the lamp or extension cord. The cord should be long enough to reach convenient outlets nearby. Too long a cord is a wasteful dust catcher.

3. Take the brass shell socket apart. Do this by pressing the upper part of the shell near the switch opening and pulling it apart. Never force the shell with a screwdriver.

4. Fasten the cap to the top of the lamp. Frequently, on wood or plastic lamps, a small metal plate is attached to the top of the lamp. This plate has a hole in the center with 3.175-millimeter (^{1}/8-inch) pipe

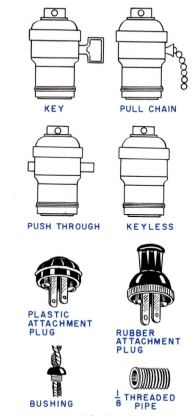

Fig. 180-6.   Materials for wiring a lamp.

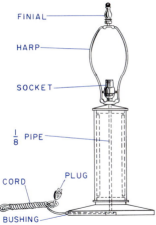

Fig. 180-7. The harp on this lamp must go underneath the socket. Another kind of harp screws onto the socket.

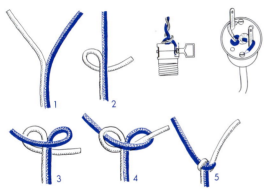

Fig. 180-8. Steps in tying an Underwriters' knot.

threads. The short pipe nipple fits into this hole before the cap is attached.

5. Thread the cord through the lamp base.

6. Remove the insulation about 19 millimeters (³/₄ inch) from the ends of the wire.

7. Tie an Underwriter's knot at the upper end (Fig. 180-8). This knot takes the pressure off the terminal screw of the socket. You may also use a strain knot. Make it out of a half width of electrician's tape wound around each of the wires and then around both. The Underwriters' knot is better on wires covered with fabric.

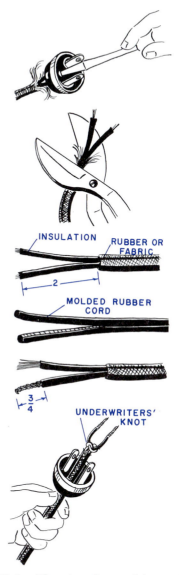

Fig. 180-9. The steps in repairing or installing an attachment plug.

8. Replace the outer shell.

9. Slip an attachment plug over the other end of the wire. Tie a knot. Fasten the ends of the wire to the terminal screws (Fig. 180-9).

Electricity as a Source of Heat

The resistance met by electrons flowing along a wire changes with the size, type, length, and temperature of the wire. Heat is produced from overcoming this resistance. Heat is produced in the iron shown in Fig. 181-1 by the use of resistance wires. Heat is one of the best-known services electricity performs in everyday living. Electric irons, stoves, toasters, roasters, heating pads, soldering coppers, waffle makers, hot-water heaters, and hundreds of other home and commercial appliances depend on this heat production. In large factories, heat is used for finishing and welding. Heat is used also in the manufacture of iron, steel, and aluminum. These are only a few of the many uses of electricity.

Several different kinds of wire are commonly used for heating elements. They are used because they have high resistance. Nichrome is used most often in small electrical project activities. You must figure the correct size and length of wire needed to build small project activities such as a portable stove, a soldering copper, a toaster, or a wiener roaster.

All heating devices are rated by the number of watts of electric power they use up. For example, a single burner of an electric stove may draw 500 to 750 watts. But an electric home clothes drier uses as much as 4500 watts. You will have to decide how much your heating element will use. This is often stated in the project activity plan. A small stove uses from 100 to 125 watts. A small soldering copper uses from 75 to 100 watts.

Suppose the device you are going to build is to be a 100-watt unit. Your first problem is to figure out the total resistance of the unit. Remember that $W = E \times I$, or watts equal volts times amperes. Since

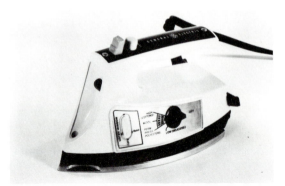

Fig. 181-1. Resistance wires are used to heat this iron.

the voltage (E) on which most home appliances operate is 110, you can easily find the amount of current (I) your device will require. In this case, it is $100 = 110 \times I$, or $I = 0.909$ amperes. To find the total resistance, use Ohm's law, which is $E = I \times R$. Since $E = 100$ and $I = 0.909$, you will find that the heating unit must have 121 ohms of resistance.

By checking Fig. 181-2, you can find the ohms of resistance per foot for all the common sizes of nichrome wire. For example, if you use No. 28 wire, which has a resistance of 4.10 ohms per foot, your heating

BROWN & SHARPE GAGE NUMBER	OHMS PER FOOT
18	0.406
20	0.635
22	1.017
24	1.610
26	2.570
28	4.100
30	6.500
32	10.170

Fig. 181-2. Resistance of common sizes of nichrome wire.

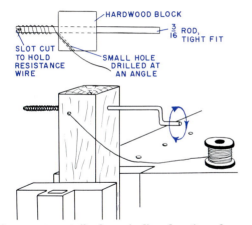

Fig. 181-3.   A jig for winding heating-element coils.

Fig. 181-4 a.   A wood-burning and heat transfer marking tool is an excellent product that involves a heating element.

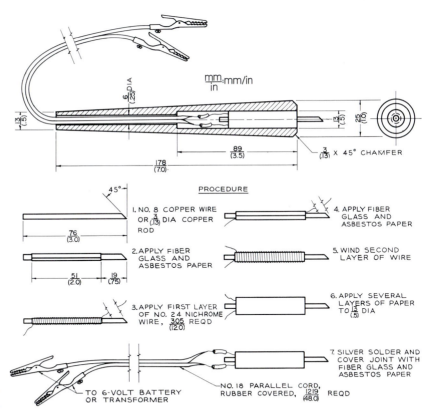

Fig. 181-4 b.   Plans for the wood-burning tool.

device will need 121 ÷ 4.10, or 29.5 feet, of wire. When using smaller wire, you need a shorter piece to get the same results.

Heating devices such as burning tools (Fig. 181-4 a and b), soldering coppers, and heating pads make use of wire in a straight form. Other devices, such as toasters and stoves, have coils. Figure 181-3 shows a small jig (device) for winding heating-element coils.

When nichrome wire is joined to heater-cord wire, the connection must be either silver-soldered or fastened with a mechan- ical connection. Brass rivets or screws are often used. Soft solder would melt.

Electric Motors

An electric motor changes electric energy into mechanical energy (Fig. 182-1). Motors are used for hundreds of purposes in homes, businesses, farms, and factories. There are two major types of motors: those which work on direct current and those which work on alternating current. It is important to study these types one at a time since each works on a different electrical principle. Most motors operate on alternating current.

Direct-current Motors

The parts of a direct-current motor are shown in Fig. 182-2 a. The diagram for this motor is shown in Fig. 182-2 b. These parts are made up of a field that is usually stationary, an armature that revolves in the field, brushes, and a commutator. The direct-current motor operates in this way: Electrons flow from either a battery or a direct-current generator through the brushes and then to the armature and field. Sometimes the armature and the field are wired in series. Other times they are wired in parallel.

The flow of electrons energizes these *electromagnets* (armature and field). This forms a north pole and a south pole on both. Because of the law of magnetism, the poles of the armature are attracted and repelled by the poles of the field. The commutator changes the direction of the electron flow in the armature at just the right moment. When a north pole of the armature just reaches the south pole of the field, the current is reversed. Then the

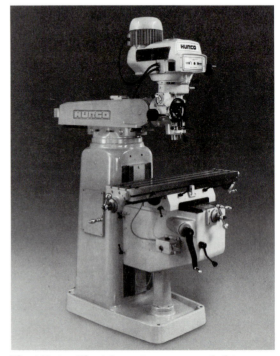

Fig. 182-1. Electric motors are needed to operate all kinds of machinery, such as this vertical milling machine. (Hurco Company)

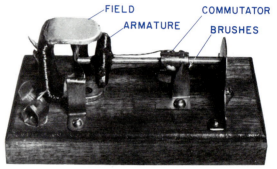

Fig. 182-2 a. A small direct-current motor.

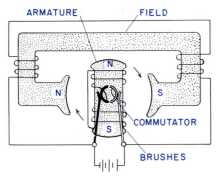

Fig. 182-2 b. Wiring diagram for a direct-current motor.

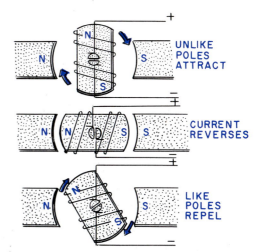

Fig. 182-3. The operation of a direct-current motor.

like poles repel each other. This cycle continues. It causes the armature to rotate quickly, producing mechanical energy (Fig. 182-3).

Alternating-current Motors

Most alternating-current motors are the induction type. These operate on a principle like that of the transformer. The field of an alternating-current motor is called a *stator* (primary). The armature, or rotating part, is called a *rotor* (secondary). There are no brushes, no commutator, and no electric connection between the rotor and the stator (Fig. 182-4).

When alternating current flows into the stator, it induces an electric current in the rotor, the same way as in a transformer. This current in the rotor produces magnetic poles that are attracted to the stationary magnetic poles of the stator. This causes the rotor to *revolve* (turn).

Building a Small Electric Motor

The parts of an electric motor are shown in Fig. 182-5 a and b. It is important to wind your electromagnet to form the field and armature as described in Unit 174, "Electromagnetism." The successful operation of your motor will depend on three factors: a properly wound armature and field, a well-made commutator, and

Fig. 182-4. An alternating-current motor.

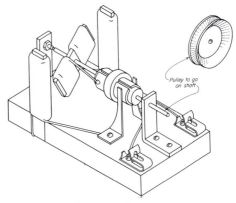

Fig. 182-5 a. This small direct-current motor is designed to have a pulley on the shaft to operate small toys.

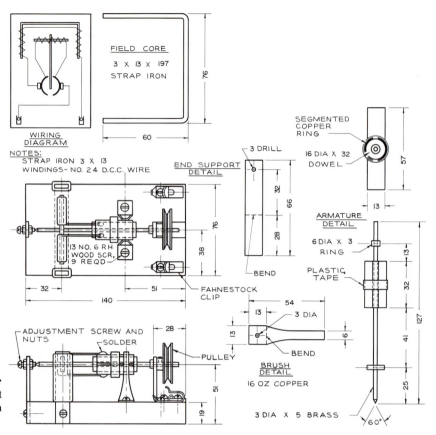

Fig. 182-5 b. Plans for the motor. Notice that all measurements are in metrics.

proper brushes. Figure 182-6 a and b shows two ways of making a split com-

mutator. Brushes may be made of thin spring brass.

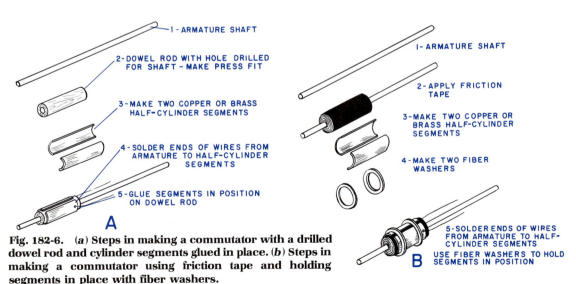

Fig. 182-6. (*a*) Steps in making a commutator with a drilled dowel rod and cylinder segments glued in place. (*b*) Steps in making a commutator using friction tape and holding segments in place with fiber washers.

Unit 183

Appliance Repair

It is important to know how to get the best service from the many appliances used in the modern house. Proper care and maintenance are the big factors. Some modern appliances are so complicated, however, that the beginner can make very few major repairs.

Appliances may be divided into two major groups: those which are mostly *motor-driven appliances* and those which are mostly *heating appliances.* Household motor-driven appliances include the refrigerator, vacuum cleaner, washing machine, and sewing machine. Some common heating appliances are the hand iron, toaster, waffle maker, frying pan, and electric heater.

The most common repair of heating appliances is the replacement of an appliance plug and/or cord. Frayed or worn cords and broken plugs should always be replaced (Fig. 183-1). If a new cord must be installed, it is often necessary to buy a new plug. This is because most plugs cannot be taken apart. Plugs are usually made of molded plastic or of two parts that are riveted together.

Always use heater-cord wire on heating appliances, never extension cord. Replace the cord following the instructions in Unit 180, "Lighting with Electricity." Sometimes a switch becomes defective and must be replaced. In most cases, however, only cord and plug repairs should be tried. If the heating element burns out, it should almost always be replaced by a repair service.

Repairs for motor-driven appliances are usually limited to installing new cords or

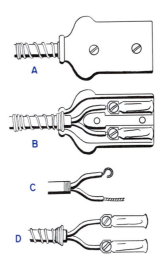

Fig. 183-1. **How to repair a heater-cord plug. First, be sure the cord is *not* plugged into a socket. (*a*) Take the plug apart by loosing the bolts. (*b*) Lift out the panels and spring guards. (*c*) Trim away any loose or worn insulation, or cut the cord off and start with new ends of the wire if necessary. Thread wrapped around the edge of the heater-cord cover will help keep the cover from raveling. (*d*) Attach the terminals of the new plug after putting the cord through the spring mount. Be sure that the plug end of the spring guard is toward the plug.**

plugs. The new cord for a motor-driven appliance should be a flexible, rubber-covered cord.

There is very little most people can do to repair an alternating-current motor. Most motors are permanently oiled. If not, the bearings should be oiled regularly according to the manufacturer's directions. The motor should be kept clean and free of oil or grease. If the pulley on the motor comes loose, it should be tightened or replaced if necessary.

Unit 184

Telegraph and Telephone

Modern communication devices are possible only because of electricity. The telegraph and the telephone are two of the most important (Fig. 184-1 a and b).

Telegraph

The modern telegraph was invented in 1844 by Samuel Morse. The simplest telegraph circuit includes a key, a sounder, a battery, and a wire. The sounder is a simple electromagnet. When the key is pressed down, the electromagnet is energized. The metal bar is attracted to it, causing a click (Fig. 184-2). By pressing the key in a certain way, dots and dashes (long and short sounds) can be produced. These have been developed into a code used for land and sea communication.

When electric energy is sent over a long distance, the signal becomes weak. It must be stepped up by a device called a *relay*. This is a very sensitive instrument that operates by electromagnetic action. The signal of the relay is made stronger by electric energy from batteries that operate the sounder. A telegraph key, a relay, and a sounder are shown in Fig. 184-3.

Teletypewriter

The teletypewriter (teletype) has replaced the hand telegraph key because it is much faster and easier to use. The machine op-

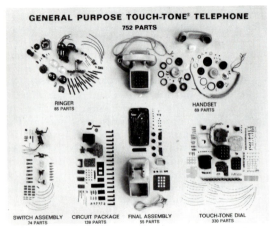

Fig 184-1 a. A general-purpose Touch-Tone telephone is made up of 752 individual parts. (Western Electric Company)

Fig. 184-1 b. A Touch-Tone telephone. You will need this kind of phone to use the services of many long-distance telephone companies (Western Electric Company)

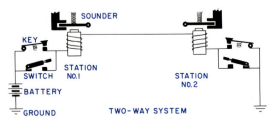

Fig. 184-2. A telegraph system.

Fig 184-3. A telegraph key, a relay, and a sounder.

erates like a typewriter for sending messages. At the receiving end, the message can be printed automatically on paper or paper tape.

Telephone

The telephone was invented by Alexander Graham Bell in 1874. It operates on the principle of changing sound waves to electric energy and then back to sound waves (Fig. 184-4). The simple telephone circuit shown in Fig. 184-5 operates like this:

The sound waves (1) of your voice strike the diaphragm of the transmitter (2). Just behind the diaphragm, there is a small box of carbon granules (3) that are very good conductors of electricity. As the diaphragm moves in and out, it changes the resistance of the circuit (4). This resistance allows varying amounts of electrons to flow. In the receiver end, the change in the flow of electrons causes a change in the strength of the electromagnet (5). This in turn causes the diaphragm of the receiver (6) to vibrate at various rates. This vibration produces the sound waves (7) that you hear.

Your telephone must be able to be connected with any other telephone. This connection is made in the central office with machine-switching equipment (Fig. 184-6). Although this equipment has many parts, the most important part is the electromagnet. Your telephone company may set up a trip through the central office. You will be able to find out more about how the telephone operates.

Fig 184-4. A cutaway of a telephone receiver and a transmitter. (American Telephone and Telegraph Company)

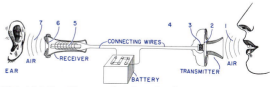

Fig. 184-5. How a simple telephone system operates.

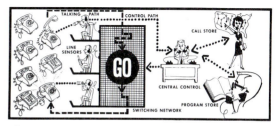

Fig 184-6. How it all works. Here's a cartoonist's view of the way an electronic switching system handles a telephone call. *Line sensors* scan all phones served by the electronic central office every tenth of a second, detect calls for service when a phone is lifted off the hook, and alert the "executive" section of *central control*. To set up a connection and provide other services, the "electronic executive" of central control uses the system's *call store*, or short-term memory, and *program store*, its permanent memory. With information from the memory sections, central control's "executive" can command the *switching network* to set up the *talking path* and order up other services.

Unit 185

Electronics: Radio and Television

Electronics is the study of electrons in motion. Although the electron is so tiny that no one has ever seen it, it is the basis of the mighty electronics industry.

Most people use electronic products every day. Examples are radio, television, and stereo. Radar guides planes through the skies so that they can land safely. Communications satellites are full of electronic equipment. Their success in getting into orbit depends on electronic guidance systems. Electronic products are used to automatically operate machinery in industry. Perhaps the most important thing that has happened in the twentieth century is the development of the electronics industry.

Some of the most important electronic devices are radar and sonar, which were developed during World War II.

Radar was used on ships and airplanes to detect other airplanes and ships. Today, radar is used for commercial purposes. By means of radio waves, it locates objects and determines their distances and speeds. Radar beams are sent out in all directions. When the beam strikes an object, it reflects back to the radar receiver and shows it on the radar screen.

Sonar was also used on ships and submarines to detect other ships or submarines. Since World War II, the greatest developments in electronics have been in television, electronic computers, and electronic equipment. The great progress in satellites and space exploration is due to the development of new power sources and electronic devices.

Radio

In radio broadcasting, someone must first speak into a microphone. This changes sound waves or energy into electric energy (Fig. 185-1). These waves are fed into a sender, or *transmitter*. Here they become a part of a carrier current, or wave. These carrier currents send out electromagnetic impulses from the transmitter at a certain frequency. A frequency is the number of cycles of alternating current completed in a certain period of time—usually 1 second. The transmitting antenna radiates the electric vibrations into space in the form of electric waves. These electric waves travel from the antenna in much the

Fig. 185-1. Sound waves are changed into electric waves in a microphone.

same way as the ripples of water do when you drop a stone into the center of a pool.

At the receiving end, these electromagnetic waves are picked up by the radio antenna. The receiver converts the radio waves into electric vibrations and sorts them out. Through transistors, the current is strengthened and sent to the speaker. The speaker changes the radio waves back into sound waves.

There are two ways of broadcasting. In *amplitude modulation* (AM), the *audio* (sound) wave changes the amplitude, or strength, of the carrier wave. The frequency remains constant. In *frequency modulation* (FM), the audio waves change the frequency of the carrier wave. The amplitude remains constant. Amplitude-modulation radio may be received over long distances. It gets static interference (Fig. 185-2). Frequency-modulation radio can be received over about 50 miles (80 kilometers). It is static-free. Also, it can reproduce sound almost exactly as it is broadcast.

Building an Experimental Radio Receiver

It is not necessary for you to understand everything about a radio in order to build a receiver. However, you should understand that a radio receiver must do the following jobs:

1. Pick up the radio-frequency signals from the transmitter

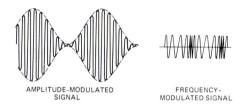

AMPLITUDE-MODULATED
SIGNAL

FREQUENCY-
MODULATED SIGNAL

Fig. 185-2. AM and FM audio waves. (J.A. Wilson and Milton Kaufman, *Basic Electronics: Theory and Practice*, McGraw-Hill, New York, 1977.)

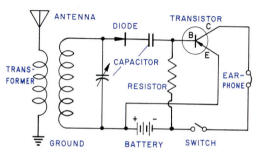

Fig. 185-3. A simple transistor-radio circuit.

2. Tune in one correct signal and *reject* (not pick up) the other signals that are in the air

3. *Amplify* (make louder) the radio signal you want

4. Separate the sound wave from the carrier wave

5. Amplify the audio signal and operate a speaker with it

If you want to build an experimental receiver set, buy the parts as a kit. You may want to build either a crystal set or a small transistor unit (Figs. 185-3 and 185-4). When you build a set, you must be able to identify the various parts. You must follow the schematic drawing that is supplied with the kit.

Many transistor radio sets today use an *etched circuit.* This is a piece of material that is a combination insulator and metal conductor. Part of the metal is *etched* (eaten by acid) away so that what is left serves as the wires between the parts. The parts need only be soldered to the etched metal that remains.

Television

Television operates by changing light and sound to an electron flow and sending it out over long distances. The three major parts of television are the *camera*, the *transmitter*, and the *receiver*. A television camera looks much like a movie camera, but it operates in a different way.

The heart of a television camera is the electron camera tube, which is called the

Fig. 185-4. Building a transistor radio.

image orthicon (Fig. 185-5). The picture image that enters the camera through the lens falls on a sensitive plate or surface (mosaic). This action produces thousands of tiny electric charges that vary in *intensity* (brightness) with the amount of light. An electron gun shoots electrons against this plate. It moves from left to right, in rows, to *scan* the picture on your television screen. In other words, the picture itself is broken down into tiny light or dark dots. These are changed into electric impulses and are taken off as the television signal. This happens so fast that when you see the picture, it looks almost as it would in life (Fig. 185-6).

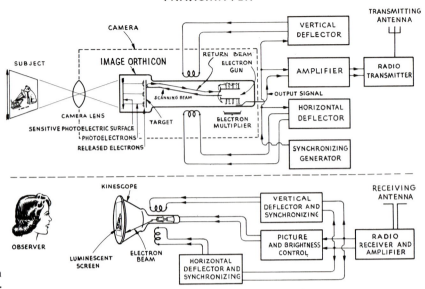

Fig. 185-5. The action of the camera transmitter and the receiver of a television set.

The changing electric current is then *amplified* (enlarged), attached to a carrier wave, and sent to a transmitter. There it is *telecast* (sent) by means of an antenna. The television *receiver* (set) receives these video waves and strengthens them. The picture tube itself is called a *cathode ray tube (CRT)*. It has an electron gun at one end and the face of the tube at the other. The face is covered with a fluorescent or luminescent screen.

The stream of electrons guided by electromagnets scans the screen the same way as the gun of the camera scans the picture. The fluorescent screen changes the electrons that strike it into light and dark spots

Fig. 185-7. Inspecting a color TV picture tube. (RCA)

Fig. 185-6. The action of a picture tube.

to produce the picture. The more electrons that strike the screen, the lighter the spot.

In television, the *audio* (sound) and *video* (picture) waves are separate. They are sent out on separate antennae. Audio waves are sent by frequency modulation.

Color television produces images in a way similar to black-and-white television. But three guns are used—red, blue, and green. Each color gun is aimed to hit a corresponding color phosphor dot on the inside of the screen. Dots combine to produce still other colors (Fig. 185-7).

Unit 186

Communication Systems

The word *communication* refers to sharing information. It is the exchange of thoughts, words, and ideas and the methods used for making these exchanges. All animal life has some form of communication. For example, a chimpanzee can learn dozens of distinct gestures, carrying on a limited communication with people and machines. People communicate in many ways by speaking, writing, and listening and by means of body movements. Machines also communicate. The instrument panel on the airplane shown in Fig. 186-1 shows the pilot the air speed, fuel level, horizon reference, compass direction, and information about the functioning of components. Ultimately, computers will help to speed and control all forms of communication between people, between people and machines, and between computers and other computers.

The first moon walk by our astronauts was seen live by over 125 million Americans; remember, the moon is over 238 000 miles away. Yet as late as 1815, a fierce battle was fought by mistake 16 days after the war had ended. It happened at the Battle of New Orleans when the two armies were still fighting because news of the treaty traveled too slowly to prevent the battle. Today, travelers from the United States who have an accident in Hong Kong can have their entire medical records

Fig. 186-1. The instrument panel of an airplane provides the pilot with many kinds of information. (Cessna Corporation)

available in seconds by satellite communication.

There are four major means or methods of communicating: graphic (visual), electronic, light, and acoustic (sound). Most types of communication, such as television, involve several of these methods (Fig. 186-2). You have already learned about many of these methods in earlier units.

Mass Communication Methods

Mass communication methods include newspapers, magazines, books, radio, television, motion pictures, and computers. Other means of mass communication include billboards, signs, and displays used by business and industry. Communication has made the world much smaller. Radio

COMMUNICATION SYSTEMS

GRAPHIC (Visual)	ELECTRONIC	LIGHT	ACOUSTIC (Sound)

Drafting (including designing and planning)	Telecommunication (microwave and satellite)
	One-way
	  Radio
	  Television
	  Radar
	Two-way
	  CB (citizens' band)
	  Radio
		Fiber optics and/or lasers
		  Telephones
		  Computers
		  Instruments
		Audio systems
		  EVR
		  Radar
			Means
			  Gas
			  Solids
			  Liquid
			Uses
			  Laser
			  Sonar

Graphic arts (including all systems of printing)	Television
	  Radar
	  Telephone
	Conductors
			Records
			  Tapes
			  EVR
			  Radio

Photography	One-way
	  Sound systems
	Two-way
	  Telephone
	  Telegraph
	  Teletype
	  TV systems
	Computers
	Videotex
			TV

Fig. 186-2. Communication systems.

can send one's voice around the world faster than people can shout from one end of a basketball court to the other. You can make a telephone call from Chicago or any other city in the United States to London and hear better than you would if the person were in the next room. Technical advances in communication systems have made the information explosion possible, changing the way people work, play, and live. Look how communication systems have changed our cars, homes, offices, and industries (Fig. 186-3).

Graphic (Visual) Communication

Do you remember the first great invention that changed the way people communicate? It was movable type. (See Unit 4.) In Sections 3, 4, and 5, you made an in-depth study of drafting (including design and planning), graphic arts (printing), and photography. These three methods are used to produce newspapers, magazines, books, visual aids, motion pictures, and even radio, television, and computer programs (Fig. 186-4).

Electronic, Light, and Acoustic (Sound) Communication

Many communication devices, such as the telephone, the television, and computers, use all of these methods. For example, transmitting information by telephone may be done by conductors (metal wires), microwaves, satellites, or light. Some of the recent technology includes the following.

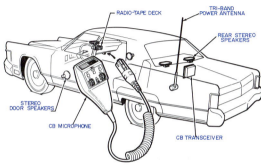

Fig. 186-3. Many cars are equipped with a radio, a tape deck, and a CB (citizens' band) radio so that the drivers can communicate with other people. The latest development is a portable telephone the driver can use to talk to anyone in the world.

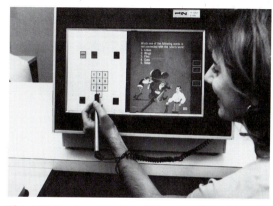

Fig. 186-4. A computer career-exploration system that will help you select an occupation. The image display units are linked by telephone to a computer. Typical questions about each occupation are displayed on the TV screen. You respond by touching the appropriate box with an electronic light pen. (IBM)

Microwaves. Microwaves are somewhat like radio waves except that they are of super-high frequency. These microwaves do not follow the earth's curvature but travel in fairly straight lines. They can be focused like a searchlight and aimed from point to point over a clear, line-of-sight path. Relay stations (microwave towers), ranging up to 400 feet in height and spaced an average of about 30 miles apart, catch the beams, amplify them, and send

Fig. 186-5. A microwave relay station. (Western Electric Company)

them on to the next station (Fig. 186-5). On top of each station are special antennas that concentrate the microwaves sharply into beams. This system can be used for telephone, radio, and television.

Satellite. In the 1950s and 1960s, a method was developed to send information over long distances by satellite. The first U.S. satellite, called *Echo*, simply reflected radio and television waves off the metal surface of the satellite from one ground station to the other. This was called a passive satellite. The signals sent to this satellite had to be strong since the waves weakened as they traveled to the satellite and back to earth. The first active satellite, the *Early Bird*, was powered by a solar battery; it was launched in 1965. This satellite weighed 88 pounds and had the capacity of 240 voice circuits and one television channel. In 1971, *Intelsat IV*, weighing nearly a ton, was launched, with a capacity of 6000 voice circuits and 12 television channels. Now this system has seven satellites in orbit and is used through 115 terminals on earth.

Since that time, many other satellites have been put into space. Most of the early satellite systems depended on giant rockets for launching. Now there is a series of moving satellites sending and receiving signals around the earth to complete the satellite communication system. A series of satellites are needed because a single satellite can transmit only to about one-third of the earth's surface at a time. Most systems use geosynchronized satellites; this means that they stay in one position over the earth.

With the development of the space shuttle, satellites are now being placed in orbit directly, with the astronauts serving as technicians. These satellites can also be retrieved, repaired, and returned into orbit without bringing them back to earth. Transmitting stations on the ground are an important part of the satellite system. They beam signals to the satellite by means of a special antenna. The satellite sends signals back to earth, where they are picked up by giant disks (Fig. 186-6).

Satellites have revolutionized communication, weather, navigation, and agriculture. Communication satellites provide live coverage of what is going on in the world to every nation. The most significant advance in weather reporting has been achieved through the use of meteorological satellites. These satellites provide observation for the entire surface of the earth so that weather can be predicted with greater accuracy (Fig. 186-7). Navigation of ships and aircraft is also controlled by satellites. Orbiting satellites can sight ships and aircraft positions much more accurately than any other system. With satellite assistance, ground controllers can pinpoint a jet liner's position within approximately 1 mile.

Aerial photography has also given the world more information about crops and forests. With the aid of satellites and re-

Fig. 186-6. Many earth stations (receiving dishes) receive communications from satellites that are in orbit around the earth. (Ford Motor Company)

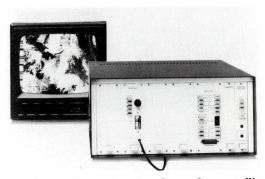

Fig. 186-7. This educational weather satellite receiving system provides students with instructions on receiving information from a weather satellite. (Feedback, Inc.)

mote sensors, we can better manage crops and timber resources. Satellites also help in pinpointing the earth's resources and minerals, such as iron, copper, gold, and nonmetallic deposits. Satellite communication has also revolutionized home entertainment, as you will learn in this unit.

Lightwaves. The latest development in communication technology is lightwave transmission. This system will eventually provide ultramodern telephone communication between all parts of the country.

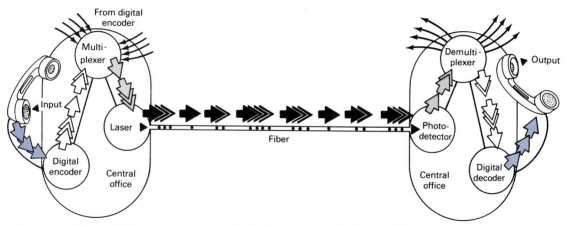

Fig. 186-8. How a lightwave system works. When you speak into a telephone a small microphone changes the sound wave from your voice into a continous electrical wave (shown here as blue arrows).

At a central office, an electronic circuit (*a digital encoder*) transforms the electrical wave into separate pulses (white arrows) coded to accurately represent the continuous electrical wave. A multiplexer combines pulses from many calls (light gray arrows).

The combined pulses turn a semiconductor laser on and off at a rapid rate so that the laser emits flashes of light in exact response to the electrical pulses. A glass fiber—a "lightguide"—carries the light pulses (solid black arrows in the center of the figure).

At another central office, perhaps miles away, a photodetector changes the light pulses back into electrical pulses (light gray arrows), and a demultiplexer separates the electrical pulses into individual calls (white arrows). A digital decoder circuit changes the electrical pulses back into the original continuous wave (blue arrows).

At the end of the line, the telephone converts the continuous electrical wave into a sound wave and the person listening hears your voice just as if it had been carried over copper wire—but perhaps more clearly.

Now you know how a lightwave system works. (Western Electric Company)

Until recently, metals, wires, and microwaves were the principal means of connecting telephones. With the development of fiber optics—a thin strand of glass covered with a special coating—all this is changing. With this system, up to 80 000 simultaneous calls can be carried on at the same time (Fig. 186-8). The light pulses carrying these calls will be transmitted at a rate of 90 million bits per second. Eventually, technology will double and then triple these speeds and capacities. Because this material is so light in weight and so efficient, fiber optics will eventually be used in most communications systems, including airplanes, satellites, and television cables.

Lasers. Next to nuclear power and integrated circuits, the laser is the most significant technological breakthrough of this century. Its capacity for information carrying is enormous. Lasers and fiber optics make it possible to increase enormously the effectiveness of telephone communication. Lasers can also be used to receive information on a memory disk. One million sheets of letter-size paper weighing 8500 pounds can be stored in an area smaller than a hatbox. The laser is used to

Fig. 186-9. This laser beam can carry the equivalent of the information in 200 Sunday newspapers.

Fig. 186-10. This student is using a telephone to call a computer for help with a reading lesson. (Bell System News)

record and store images on small phonographlike disks. Many of these can be retrieved almost instantly and viewed on a video screen. Laser beams will eventually provide communication between computers. A single laser beam can carry 1 billion bits of information (Fig. 186-9).

Computers. The computer has become the engine for the information revolution (Fig. 186-10). In the years ahead, everything that we do or say may be recorded and made available by means of the computer. There may be a computerized worldwide library network storing and offering for easy retrieval all of the accumulation of information throughout history. The computer can search through this storehouse of information and provide a readout for any kind of information an individual may desire. Computers may also eliminate almost all exchange of money and other currency. Practically all financial matters can be handled through the computer, including depositing, investing, and spending.

Personal Communications Systems

The greatest advances in communication have been in the area of personal communications systems that bring the world directly into the home. Magnetic tape recorders have been the common format for recording and playing music for many years. Four of the other important home communication systems include cable TV, satellite, ERV, and videotex.

Cable TV makes it possible to see full-length movies, sports events, and other forms of entertainment directly in the home. With this system, the signals are transmitted to the local television cable company and are picked up by huge receiving disks (antennas). Then wires are run to homes so that people can select from among many different channels. A computer in the cable TV center controls the lines into each home so that each of the services can be turned on or off automatically.

Communication satellites also make it possible to broadcast TV signals directly to homes anyplace on earth. This is made possible by having a disk-type antenna on the roof of the house or other structure. Today, these rooftop disk antennas are quite large and expensive. Eventually they will be much smaller and less expensive. The system is called direct or home satellite service, and it eliminates the need for cable TV. The system makes it possible for people living outside the cable system network to receive movies, sports, and other entertainment directly from the satellite.

Fig. 186-11 a. **Using a videotex system to obtain information from libraries, stores, and many other sources.** (Videodata Corp.)

EVR stands for electronic video recording. It is a means by which images and sound can be recorded on magnetic tape and then played back through a home television set. The use of video cassette recording (VCR) systems in the home is expanding rapidly. While the use of computers is growing at a rapid rate, the use of VCRs is growing even faster. With this system, in which the TV program is recorded in sound and color, the image can be stopped, replayed, moved forward at a fast rate, and erased. There are two different systems for video cassette recording: VHS and Beta. The tapes used in these systems are not interchangeable, and so the individual must choose one or the other of the two systems.

First came the telephone, and with it the instant communication that shrank our world to a more comfortable size. Then came television, the magic box that broadened our horizons and overnight turned living rooms into theaters, schools, and political forums. Then came the computer, whose electronic wizardry opened both outer space and the inner spaces of our minds. All three of these electronic devices have been brought together in a single system called videotex (Fig. 186-11 a and b). This is an electronic service that makes it possible for a person to call from home for virtually any kind of information and have it displayed on the home television screen. The family can also communicate directly with the source of that information—the computer. In other words, videotex offers two-way transmission communication for every aspect of our lives. With this system, a person can shop, do banking, and obtain information and many other services.

A videotex terminal used in the home consists of two units: a key pad very much like the keyboard of a typewriter and a control unit about the size of a tabletop radio. The control unit is connected to the TV set, and a telephone line is connected to the control unit. The unit becomes a high-speed communication link between a TV and the data base—the computer.

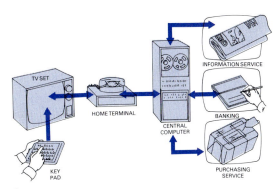

Fig. 186-11 b. **How the typical home videotex system operates.**

Discussion Topics on Energy, Electricity, and Electronics Technologies

1. Name the four major elements of an energy and power system.

2. What are the main sources of fossil fuel?

3. Name five sources of alternative energy.

4. Name five uses of electricity.

5. List six broad categories or areas of careers in the electricity and electronics industries.

6. State three reasons why career opportunities will expand in the electricity and electronics industries in the next 10 years.

7. What are the rules of safety in handling electricity and electronics projects?

8. What is magnetism? When does steel have magnetic properties?

9. What is an electromagnet? In what kinds of electrical equipment are electromagnets found?

10. Describe how an electromagnet is made.

11. Name the two forms in which electricity exists.

12. Name and describe the two kinds of electric current.

13. What is a generator? What does a transformer do?

14. Do all materials conduct electricity? What is a conductor? An insulator?

15. Name the three basic types of wire splices.

16. How is electric pressure measured? What is the flow of electricity called? What is resistance?

17. Explain Ohm's law.

18. What is the unit of electric power?

19. What kind of current is produced by a battery?

20. What is a primary cell? A secondary cell? What is the difference in voltage output of three dry cells hooked together in series and in parallel?

21. Describe the system of wiring in a home.

22. What is meant by low-voltage wiring? What kind of equipment uses this kind of wiring?

23. Name and describe some home-appliance repair jobs you will probably be able to do.

24. Explain the operation of the telegraph, telephone, radio, and television.

25. What is a transistor?

26. Name the six basic sources of energy used to produce electricity.

27. Explain the action of a battery during charging and discharging.

28. How is electromagnetic induction used to generate electricity?

29. What does an ammeter measure?

30. Name the four major means or methods of communication.

31. Why is fiber optics better than metal wire for a telephone system?

32. In the years ahead, which electronic device will store most of the known information?

33. What is electrical shock and how can it be avoided?

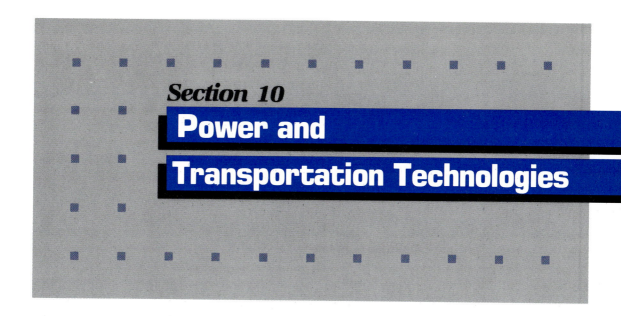

Section 10

Power and
Transportation Technologies

Unit 188

Introduction to Power and
Transportation Technologies

Power and transportation are an important part of the technology that has made it possible for people to put satellites into orbit and send space vehicles to the moon and other planets.

The study of power will give you a chance to understand some of the newer ideas of technological power (Fig. 188-1). You will learn about the following:

1. Maintenance and repair of small engines. These jobs include the work done by millions of people in the service industries. These people earn a living by doing maintenance (care) and repair work on mechancial products. There are, for example, over 30 million small engines used on power lawn mowers alone, and they often need service and repair. You will learn some of the skills by *disassembling* (taking apart), testing, repairing, replacing parts in, reassembling, and retesting engines.

Fig. 188-1.   Small internal-combustion engines are used in many kinds of vehicles.

2. Tools. There are tools needed for working with many mechanical devices.
3. Operation of internal-combustion engines. The knowledge of the basic operation of internal-combustion engines will be very important if you plan to study auto mechanics.

4. Various kinds of transportation vehicles.

5. The importance of our transportation system in a technological society.

Unit 189

Power and Transportation Technologies-related Careers

Careers in the power and transportation areas make up a new grouping of occupations in our technology. Space exploration and landings on the moon have opened up a very keen interest in occupations in all phases of aircraft, missile, and spacecraft production. Close to those are careers in the newly developed areas of atomic power. There are also new careers in motor-vehicle and equipment manufacture. And the automobile service occupations continue to be an important power industry.

Aircraft, Missile, and Spacecraft Manufacturing

The fast-growing field of aircraft, guided missiles, and spacecraft manufacturing makes up one of the largest and most complicated industrial areas in the nation. *Scientists*, *engineers*, and *technicians* account for a larger part of the total employment in these industries than they do in other manufacturing fields.

Types of aircraft range from small private planes that cost several thousand dollars to multimillion-dollar giant transports and airliners. Over half of the airplane production is for military use; the rest is for commercial use, private business, and pleasure (Fig. 189-1).

Missiles and spacecraft are produced mostly for the military. Some of them travel only a few miles. Others have intercontinental ranges of 7000 miles (11 265 kilometers) or more. One type of missile is designed for launching from land or underground sites. Others are put into orbit and become artificial satellites around the earth.

The manufacturing of most aircraft, missiles, and spacecraft is done under the leadership of a *prime* (overall) contractor. Many thousands of subcontractors make parts and subassemblies. Workers with different kinds of educational backgrounds and job skills are needed to design aircraft, missiles, spacecraft, and automobiles. Occupational needs and skills change from one subcontractor to the next. The work done by each requires different kinds of skills.

There are many professional and technical occupations in this field. A growing stress is placed on using people who can do research and development. About one-fourth of the employees are engineers, scientists, and technicians.

There are many jobs in administration and many clerical and plant occupations. Most plant operations are done by people with advanced technical skills. Those doing research and development usually have at least one college degree, and many have advanced (master's and doctoral) degrees. *Technicians, engineering aides, drafters, production planners*, and *tool designers* are semiprofessionals. They have a post-high school preparation of two or more years. Many of these trades are unionized, and entry is through apprenticeship. Other people gain their interests and skills by taking industrial arts and technology, vocational, and technical courses.

The occupational outlook in these many manufacturing industries will depend largely on the interest taken by governmental agencies. Job opportunities are projected to increase significantly.

Atomic Energy and Power

One of the most important uses of atomic energy is to produce electricity. Nuclear reactors are used as heat sources. Steam produced by this heat is used to generate (make) electricity. Nuclear reactors are also used to evaporate seawater to produce freshwater. This process is known as *desalinization*.

Nuclear reactors generate power for naval and commercial ships. One of the important benefits is the long period between refueling.

Most atomic-energy workers today are *scientists, engineers, technicians,* and *craftsworkers.* Many uses for atomic energy are still in the *experimental* (testing) stage. Many skilled workers are needed to build the special parts and equipment used in experimental work. Other craftsworkers keep in good repair the highly complicated machinery that produces atomic energy.

Fig. 189-1. Modern military aircraft provide jobs for many people in and out of the military service. (McDonnell Douglas Corporation)

Some of the other career opportunities in atomic energy are in *uranium exploration, mining, milling, refining,* and *enriching. Reactor manufacturing, operation, maintenance,* and *research and development,* will also provide career opportunities.

Careers in this field require people who have a high degree of education and ability. About 25 percent of the engineers and scientists have a Ph.D. (doctorate) degree. Employment will increase in the next decade.

Motor-vehicle and Equipment Manufacturing

Automobiles, trucks, buses, motorcycles, and other types of vehicles have become very important to all of us. Four out of five families own at least one automobile. One out of four own two or more. There are well over 150 million passenger cars, trucks, and buses on the highways and streets of the United States.

The *annual* (yearly) production of motor vehicles is about 10 million. Over 1 million people are employed in the motor-vehicle industry. Hundreds of thousands more work in the production of *components* (parts).

The three basic steps in the production of motor vehicles are (1) preliminary designing and engineering, (2) production of vehicle parts and subassemblies, and (3) the final assembly of units into completed vehicles.

Occupations in the motor-vehicle industry include the following categories: *administrative, professional,* and *technical; clerical* and *clerical-related; plant occupations, machining,* and *other metalwork; inspection, assembly, finishing,* and *materials handling;* and *custodial, plant protection,* and *maintenance work.* Employment opportunities in this industry will decrease during the next 10 years. More vehicles will be manufactured, but technology will automate many of the processes.

Automobile Service

There are hundreds of thousands of career and occupational opportunities in related services for automobiles and other vehicles. These include *automobile, truck,* and *bus mechanics; body-repair specialists, painters, parts clerks, trimmers,* and *installation specialists; and service advisers* and *service-station attendants.* This listing could also include the many thousands of people who work in petroleum production and distribution.

Over 170 000 automobile body-repair specialists maintain car bodies in safe working order. Their number will grow because of the greater number of traffic accidents. Auto mechanics are the second largest group of skilled workers, with over one million employed. Auto mechanics perform maintenance work, diagnose breakdowns, and make repairs. Employment in this field is expected to increase. Some mechanics belong to labor unions that often give workers training. Generally, however, the mechanic gets started in this skilled work in high school industrial arts automotive and power courses and vocational and technical programs.

Specialty areas such as body painters, trimmers, upholstery installers, and parts clerks learn their jobs almost as mechanics do. Service-station attendants are needed to service vehicles with petroleum products. They also do maintenance work. Some service centers have special mechanic facilities.

The special make-up of the modern automobile power plant and the need to control exhausts and fumes will require more training for workers. The number of service stations will grow, as will the kinds of services they offer.

Unit 190

Engines

From the beginning of time, people have searched for sources of energy apart from human muscle power. About 5000 years ago, they began to use the force of wind to sail boats. Later, people used windmills to pump water. Even later, the water wheel became a source of power. But it was only a few hundred years ago that the first real progress was made in the use of engines for power. James Watt, an instrument maker, developed the steam engine in 1765. This engine was first used to pump water out of coal mines. Soon after, Robert Fulton built a steamboat that worked very well.

Steam Engines

There are two types of steam engines: the *reciprocal steam engine* and the *steam turbine.* The engine developed by Watt was a reciprocal steam engine, which has a piston that moves back and forth inside a cylinder. The energy for operating the engine comes from steam supplied by a boiler.

A steam engine (Fig. 190-1) operates as follows. The steam from the boiler is allowed to enter the cylinder by means of valves. It enters first on one end and then on the other. The steam pushes the piston back and forth. The piston rod extends out from a hole in the end of the cylinder. The rod is connected to a crank. The crank itself is attached to a flywheel. The piston rod is also hooked to the connecting rod, which turns the crank. Another set of rods and cranks operates the valve inside the valve chest. This lets the steam in and out of the cylinder.

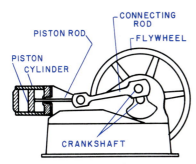

Fig. 190-1.　The steam engine.

Figure 190-2 shows the four steps in the operation of a steam engine: (1) A valve on the lower side of the piston opens and lets in steam from the boiler. This pushes the piston up. (2) At the same time, the "used-up" steam in the upper part of the piston is forced out in the opposite direction. (3) As the piston comes to the top of the cylinder, the lower valve closes. At the same time, the upper valve opens to let fresh steam into the upper end of the cylinder. (4) This pressure of steam pushes the cylinder back down. Old steam is forced out from the lower side of the piston.

Steam engines or locomotives were once the usual method of operating railroads. Today, however, most of these engines have been replaced by diesel-operated engines. Steam engines still operate many ships.

The second type of engine—the *steam turbine*—is a much newer invention. The steam turbine operates by forcing a stream of steam through the blades. The passage of steam forces the turbine to rotate (turn). A steam turbine engine has many rows of blades all fastened to the same shaft. As the steam passes through the first set of movable blades, the *stationary* (nonmoving) blades attached to the housing redirect the steam through the next set of movable blades to give the best possible force (Fig. 190-3).

Steam turbines provide one of the best ways of operating a ship. Sometimes the turbine drives the propeller. Other times, the turbine operates the electric generator. The generator in turn produces the electric current that is used by an electric motor to operate the propeller shaft.

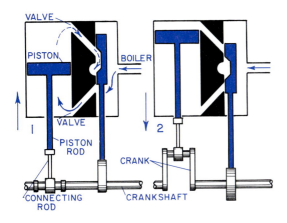

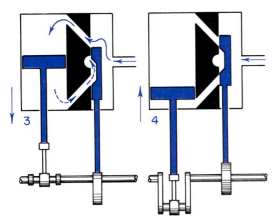

Fig. 190-2.　The operation of a steam engine.

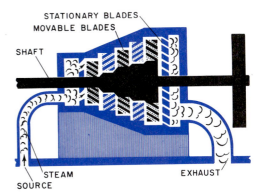

Fig. 190-3.　A steam turbine engine.

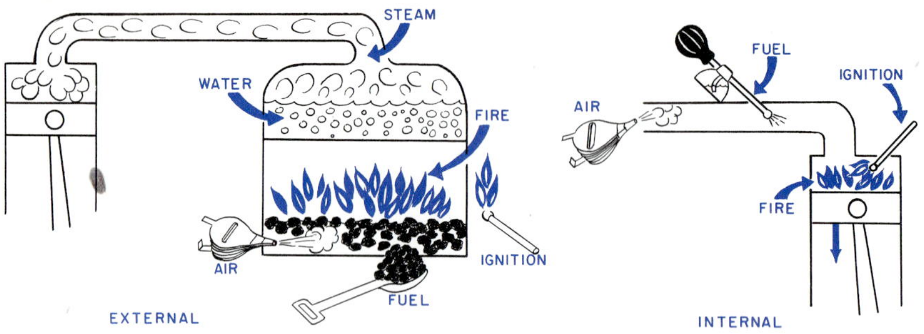

Fig. 190-4. The difference between an external-combustion engine and an internal-combustion engine.

EXTERNAL

INTERNAL

Internal-combustion Engines

In steam engines, the fuel can be burned outside the engine. The water is turned into steam, and the steam is fed into the cylinder. A steam engine is really an external-combustion engine (Fig. 190-4).

An internal-combustion engine is one in which air and fuel are mixed to cause the burning. In this type of engine, the fuel burns inside the machine. The fuel for most internal-combustion engines comes from *petroleum*, or crude oil. Petroleum is a heavy, natural oil that comes from underground. It is refined into many different materials, such as gasoline, kerosene, diesel oil, fuel oil, and lubricating oils.

The internal-combustion engine was first made in Germany. It operates as follows: A *piston* in a metal container called a *block* moves down and away from the closed end of the cylinder. At the same time, an *intake valve* opens and allows a mixture of fuel and air to enter the cylin-

der. When the piston is as far down as it can go, two valves close. Then the piston moves up and compresses the mixture. When the piston is at the top of the cylinder, an electric spark ignites the mixture and causes it to burn and expand. This action pushes the piston down and gives the power stroke. When the piston has gone down as far as possible, another valve, called an *exhaust valve*, opens. The piston again moves toward the top and forces out the mixture of air and burned gas. The piston is connected to a crankshaft by a *connecting rod*. Notice that there are four separate strokes in the operation of this engine (Fig. 190-5):

1. *Intake*—drawing in the gas mixture.
2. *Compression*—squeezing the fuel.
3. *Power*—exploding the gas mixture that forces the piston down, turning the crankshaft.
4. *Exhaust*—removing the burned gases.

Notice that the crankshaft has gone all the

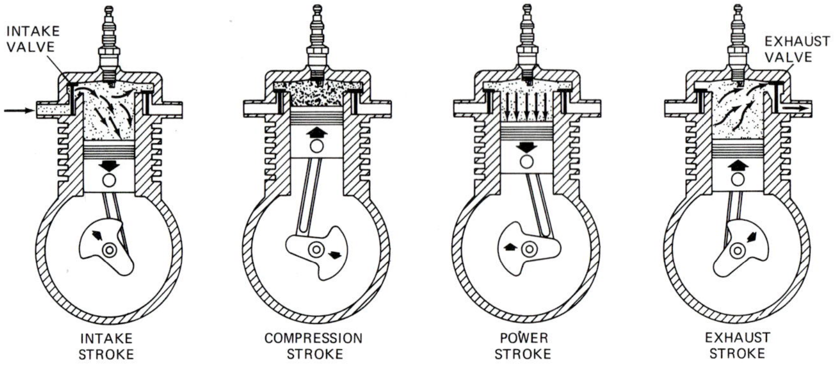

INTAKE VALVE

EXHAUST VALVE

Fig. 190-5. The operation of a four-cycle engine.

INTAKE STROKE

COMPRESSION STROKE

POWER STROKE

EXHAUST STROKE

way around twice while the engine has made four strokes, or cycles. This is why it is called a *four-stroke cycle engine*. This name is usually shortened to *four-cycle engine*. A *camshaft* alongside the *crankshaft* is connected to the crankshaft by a set of gears. This camshaft turns to open and close the valves at the correct time.

Two-cycle Engines

It takes four cycles to get one power stroke in a four-cycle engine. Many engineers thought that this was wasteful and set out to design a simpler engine. The result was a two-cycle engine that does three jobs at one time: (1) It allows the burned gas to escape. (2) It lets in a fresh mixture of fuel and air. (3) It compresses the mixture.

This engine needs only two strokes for the complete cycle: up for the compression stroke and down for the power stroke. The intake of the new fuel-air mixture and the removal of the exhaust of the burned gases happen just after the power stroke and during the compression stroke. This engine operates as follows (Fig. 190-6):

1. The piston is on the compression stroke. The fuel-air mixture is being compressed in the combustion chamber. At the same time, a fresh supply of fuel and air is drawn into the crankcase from the carburetor through the reed plate, which is the valve at the bottom of the crankcase. While the piston is on the compression stroke of the cycle, the reed valve is open.

The mixture used in two-cycle engines contains oil to lubricate the moving parts of the engine. The extra oil that enters the combustion chamber is burned with the gas.

2. Just before top dead center, the spark from the spark plug ignites the compressed charge.

3. The expanding gases force the piston down, causing the crankshaft to rotate. This downward push of the piston is called

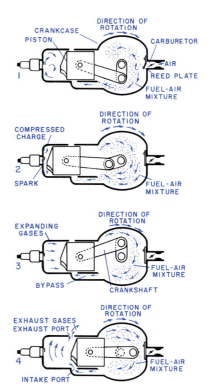

Fig. 190-6. The operation of a two-stroke cycle engine.

the *power stroke*. On this stroke, the piston compresses the fuel-air mixture in the crankcase and forces it up into the *bypass* (transfer). The reed valve must be closed during the power stroke so that the mixture cannot go back into the carburetor.

4. Near the end of the power stroke, the piston opens the intake port on one side of the cylinder and the exhaust port on the other side. This action allows the new fuel-air mixture to rush into the combustion chamber. At the same time, the exhaust gases left from the cycle before are forced out through the exhaust ports. Notice that in the process some of the fresh gas mixture may leak out, and some of the burned gases may stay in. Therefore, a two-cycle engine is less efficient than a four-cycle engine (Fig. 190-7).

Since the two-cycle engine does not use the camshaft and valves to control the in-

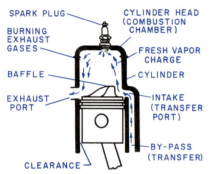

SPARK PLUG
CYLINDER HEAD (COMBUSTION CHAMBER)
BURNING EXHAUST GASES
FRESH VAPOR CHARGE
BAFFLE
CYLINDER
EXHAUST PORT
INTAKE (TRANSFER PORT)
BY-PASS (TRANSFER)
CLEARANCE

Fig. 190-7.   The new fuel-air mixture enters the combustion chamber as the exhaust gases are forced out.

INTAKE VALVE
INJECTOR
EXHAUST VALVE

Fig. 190-9 a.   In the diesel engine, air and fuel are forced into the cylinder and then are ignited.

take and exhaust, it is a somewhat simpler engine. Weight for weight, the two-cycle engine (Fig. 190-8) delivers more power than the four-cycle engine. For this reason, it is very popular for outboard motors, motor scooters, and lawn mowers.

Diesel Engine

The diesel engine was invented in 1890 by Rudolph Diesel. This engine is very similar to the four-cycle gasoline engine, but with a few important differences (Fig. 190-9 a). Basically, it is an engine in which air and fuel are forced into a cylinder and then ignited. In other words, the three elements are the same as in an automobile engine—air, fuel, and ignition. However, a different kind of fuel is used in the diesel engine, and there is no spark plug.

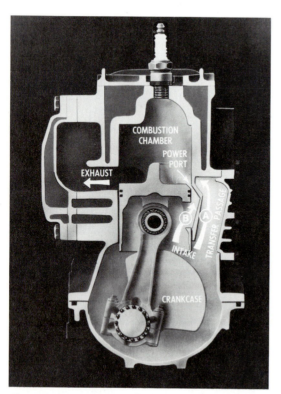

COMBUSTION CHAMBER
POWER PORT
EXHAUST
B A
INTAKE
TRANSFER PASSAGE
CRANKCASE

Fig. 190-8.   Cutaway of a two-stroke cycle engine.

Fig. 190-9 b.   Heavy duty trucks use diesel engines. (Caterpillar Tractor Company)

If you have ever pumped a tire, you know that as air is compressed, it becomes hot. This basic principle is applied in the diesel engine. A four-cycle diesel engine operates about as follows: An intake valve opens and draws a cylinder full of air into it as the piston goes down. Then the valve closes, and the piston moves up, squeezing the air into an extremely small space at the top of the cylinder. As this is done, the air becomes heated to a temperature of 1000°F (538°C) or more. At the top of this stroke, an injector squirts a fine stream of oil in to the cylinder. The oil burns as it mixes with the hot air. This burning mixture expands, forcing the piston down for the power stroke. On the next stroke, the exhaust valve opens, and the burned gases are forced out.

The main features of the diesel engine are as follows: (1) It mixes air and fuel after compression. (2) It does not have spark plugs. (3) It uses an injector rather than a carburetor.

There are also two-cycle diesel engines that operate in the same manner as a two-cycle gasoline engine (Fig. 190-10). Diesel engines generally must be built heavier than gasoline engines because of the high internal pressures. Diesel engines are often used for trains, ships, and trucks (Fig. 190-9 b). The use of diesel engines for cars is limited.

Jet Engine

Jet and rocket engines are forms of jet propulsion. To understand how these engines operate, you must know three laws of physics or motion described by Sir Isaac Newton:

1. A body remains at rest or in a state of motion in a straight line unless acted upon by an external force.

2. A force acting upon a body causes it to accelerate (move faster and faster) in the

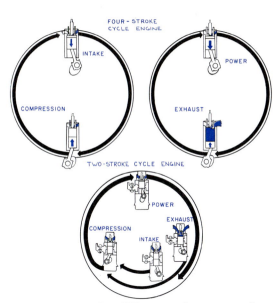

Fig. 190-10. **The difference between a four-stroke cycle engine and a two-stroke cycle engine is as follows. In the four-stroke cycle engine, two complete revolutions are needed for one power stroke. In the two-stroke cycle engine, one power stroke is needed for each revolution.**

direction of the force. The acceleration is proportional (equal) to the force and inversely proportional (opposite) to the mass of the body.

3. For every *action* there is an equal and opposite *reaction*.

The third law is really the basis on which jet and rocket engines operate. Here are a few examples of this law. If you turn on a water hose, the nozzle is forced in the opposite direction (Fig. 190-11). Suppose a person stands in the end of a boat and jumps toward the dock. The person will

Fig. 190-11. **This is an example of Newton's third law of motion: For every action there is an equal and opposite reaction.**

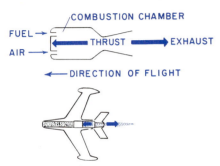

Fig. 190-12 a. **The forward thrust of a jet engine.**

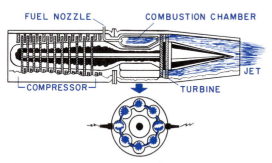

Fig. 190-12 b. **A cutaway of a jet engine.**

move forward, but the boat will move in the opposite direction. If the person jumps at a faster speed or if a larger person jumps at the same speed, the boat moves farther away.

This is how a jet engine operates (Fig. 190-12 a). Air enters the front end of the engine and goes to a compressor, where it is squeezed. The compressed air is then forced into the combustion chamber (Fig. 190-12 b). Nozzles in the front end of the combustion chamber carry fuel that is injected into the air as a spray. The fuel is ignited and keeps on burning like a blowtorch. This combustion causes the air and gas to expand rapidly.

The air and gas push toward the opening at the rear, but before they can get out, they must pass a turbine. This turbine is attached to the compressor so that both turn at the same high speed. The gases move out through the tail pipe. Since the end is smaller, the gases move at still faster

Fig. 190-13. **Jet engines are used on almost all major commercial airplanes.** (McDonnell Douglas Corporation)

speeds as they shoot out into the atmosphere. The jet really operates because of the reaction to the forces of this backward jet against the engine itself. This is why jet engines are called *reaction* engines.

A jet engine is different from an automobile engine in two major ways. Unlike the automobile engine, it operates with a continuous flow of power. That is, intake, combustion, compression, and exhaust go on continuously. All the major parts of the jet engine rotate. They do not move up and down, as in automobile engines except the Wankle engine.

Jet engines have the advantages of speed and altitude over other engines. They are able to carry a tremendous amount of power in a small space and with relatively little weight. They also make a plane go faster and higher than a propeller-driven plane (Fig. 190-13).

Jet engines are also used for another plane engine called the *turbo-prop* (Fig. 190-14 a and b). In this type, the use of propellers is combined with a jet engine. The turbo-prop is very economical to operate. It also has the advantages of the propeller for takeoff and landing on short airstrips.

Rocket Engines

The rocket engine differs from the jet engine in that it needs no outside air for combustion. Therefore, it can operate in outer space, beyond the earth's atmos-

Fig. 190-14 a. A turbo-prop engine.

Fig. 190-14 b. Turbo-prop engines are used on many smaller airplanes that service regional airports. (Cessna Aircraft Corp.)

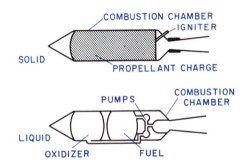

Fig. 190-15 a. Note the difference between a solid-propellant rocket and a liquid-propellant rocket.

Fig. 190-15 b. The space shuttle depends on rocket engines for power. (Rockwell International)

phere. The rocket engine carries its own fuel and oxidizer. The oxidizer takes the place of the oxygen in the air. These materials are burned in a combustion chamber that produces hot gases that are exhausted through a nozzle at temperatures of several thousand degrees. When this mass of hot gas is expelled as a jet, the *recoil* (kickback) is in the opposite direction. It works in the same way as a gun that recoils when a bullet is fired from it. The difference, however, is that in a gun the recoil is a sudden, single impulse caused by the ejection of the bullet. In a rocket engine, the recoil continues to accelerate the rocket as long as the fuel is burning.

The rocket fuels used can be *liquid* or *solid* (Fig. 190-15 a and b). With liquid fuel, the separate fuel is mixed with an oxidizer to make it burn. With solid fuel, the oxidizer-fuel mixture is a powdery or rubbery substance.

Nuclear Engines

Today nuclear engines are used in submarines and ships. Scientists are doing research on possible uses of nuclear en-

ergy for aircraft. Nuclear propulsion offers these advantages over other engines: (1) Lower-weight fuel may be used. (2) Fueling stops will be less frequent.

Nuclear reactors have been commercially used to generate electricity (Fig. 190-16). The heat produced by the atomic-fission process in the reactor is *absorbed* (soaked up) by a liquid metal passing through the reactor core. The liquid metal is then piped to a water boiler where steam is produced. The steam drives a turbine generator producing electricity.

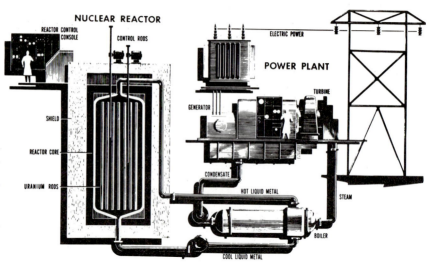

Fig. 190-16. An atomic-powered electric plant—an atomic energy reactor capable of producing electricity. About 8000 kW of electricity, enough to supply 2000 homes, can be produced.

Unit 191

Mechanics of Machines

Everyone knows that an engine is a machine. But do you know that *each part* of the engine is a machine? Do you know that all the tools you use for repair are also machines? Although the parts of the engine and the tools to repair it are simpler items, they are really machines just as much as the most complicated engines (Fig. 191-1).

What Is a Machine?

A machine is a device that is used to make work easier. Some of the properties of a machine include (1) the ability to change energy from one form to another, (2) the ability to transfer energy from one place to another, (3) the ability to change the direction of a force, (4) the ability to change the strength of a force, and (5) the ability to change the speed with which a force is applied.

A machine is really a device used to put force to good use. Machines are sometimes falsely called work-saving or labor-saving devices. To prove that this cannot

be true, think about the word *work*. Work is done when a force moves through a distance to make something move or stop

Fig. 191-1. The metal lathe is a good example of a complicated machine that is made up of many simple machines. The lathe has levers, inclined planes, wedges, screws, wheels and axles, and pulleys. Look at the machines in your laboratory and see if you can find the simple machines described in this unit. (Cincannati-Milcron)

moving. Work equals force times distance. You are doing work when you ride a bicycle. You are also doing work when you take an engine apart.

Force is the push or pull that does the work. A simple machine helps you work by multiplying the forces. This gain of force comes from a change in distance. For example, a small child cannot lift an adult without help. The child can use all the force he or she has, but will be unable to lift the adult. This job would be easy, however, with the help of a simple machine like a seesaw.

If the adult moves closer to the center of the board (nearer the *pivot*), the child can easily lift the adult. Why? The child moves through a greater distance, and the adult moves through a shorter distance (Fig. 191-2 a).

Some force is always lost through the *friction* of the machine. The more complicated the machine, the less efficient it is likely to be. More work or energy has to be put into the machine in comparison to how much comes from it. A good example of energy output is the small engine on which you will work. The energy of the gasoline provides *movement*. The engine also gives off a good deal of heat. In other words, the chemical energy in the gasoline produces mechanical energy and heat (wasted) energy.

Friction is always present whenever there is rubbing together of parts. While it is impossible to get rid of all friction, every effort is made to limit what friction does. In engines, the friction is lowered by the use of bearings and lubricants.

Types of Machines

There are two types of machines: *simple* and *complex*. Simple machines include the lever, the inclined plane, the wedge, the screw, the wheel and axle, and the pulley (Fig. 191-2 b).

Lever. The lever is a long rigid bar held at one point with a support called the *fulcrum*. There are three different classes of levers (Fig. 191-3).

The *first-class* lever has the fulcrum between the applied force and the output force. Pliers are a good example of a first-class lever. The fulcrum point is the place where the two parts of the pliers are

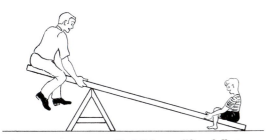
Fig. 191-2 a. A small child can lift a full-grown person by the use of a simple machine.

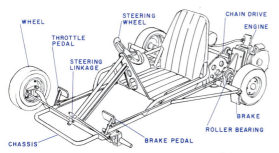
Fig. 191-2 b. How many simple machines can you identify in this go-kart?

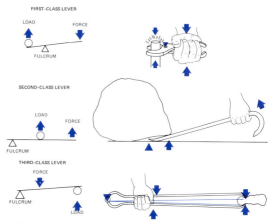
Fig. 191-3. Three classes of levers.

joined together. The handles of the pliers are longer than the nose ends. Therefore, a smaller applied force moves a longer distance to produce a greater force at the nose as it moves a shorter distance.

The *second-class* lever has the applied force at one end and the fulcrum at the other. The load is between the two. A good example of a second-class lever might be a crowbar that is used to move an object while one end of the crowbar rests on the ground.

The *third-class* lever is a machine in which the applied force is between the load and the fulcrum. Tweezers are a good example. Tweezers do not hold the thing you pick up as tightly as you are squeezing the handle. A gear is a spinning third-class lever.

Inclined Plane. An inclined plane is a tapered surface along which something moves (Fig. 191-4). An inclined plane makes work easy since a small effort can suffice to lift a heavy weight. However, the effort must be exerted farther along the incline than when a weight is lifted directly. A good example is the difference between climbing a flight of stairs and walking up a ramp. A ramp reduces the force but increases the distance that is needed to accomplish the work.

Wedge. A wedge is a form of inclined plane that has one or two sloping surfaces. A knife, a cold chisel, and the point of a nail are good examples of the wedge (Fig. 191-5 a and b).

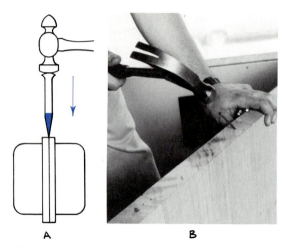

Fig. 191-5.   (*a*) A wedge. (*b*) The point of a nail is a wedge that helps the nail enter the wood.

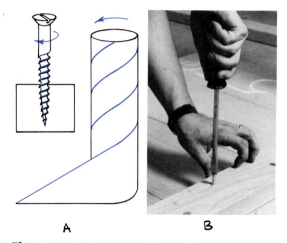

Fig. 191-6.   (*a*) A screw. (*b*) Installing a screw.

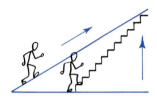

Fig. 191-4.   An inclined plane.

Screw. The screw is a spiral-shaped inclined plane that is wrapped around a rod. However, a screw does not work like an inclined plane. When a bolt or a nut is turned, the spiral inclined plane moves sideways like a wedge. This causes the object in contact with the bolt or nut to raise or lower vertically without moving horizontally (Fig. 191-6 a and b).

Wheel and Axle. The wheel and axle are connected. When one turns, the other also turns. The steering wheel of a car is a good example of the wheel and axle. Many tools, such as the screwdriver and the tap wrench, are wheels and axles (Fig. 191-7).

Pulley. The pulley is a wheel that turns around an axle. There are many arrangements of pulleys. Many belts and pulleys are used with engines (Fig. 191-8).

Ways of Putting Power to Work

There are many ways to put power to work so that it can be well used. The following mechanical devices are in common use.

Belts and Pulleys. Belts and pulleys are used to transmit power and to provide a way of changing speeds. The simplest belt-driven device is two pulleys belted together. One always drives while the other is driven. A pulley mounted on the shaft of an engine is the driver pulley. Driven pulleys are belted to the driver pulley to operate the transmission of a motor scooter, for example.

You should understand the relationship of size and speed between the driver and the driven pulleys. If both are the same size or diameter, both turn at the same speed. However, if the driver pulley is twice the size of the driven pulley, the *driven* (smaller) *pulley* turns at twice the revolutions per minute (rpm) of the *driver* (larger) *pulley.*

The general rule is that the smaller the pulley, the faster it turns. The larger the pulley, the slower it turns. In other words, the speed of pulleys is always inverse, or opposite, to the size. Pulley size also affects the power that is transmitted between driver and driven pulleys. If, for example, a small driver pulley is used with a larger driven pulley, it is possible to get more power in the driven pulley.

There are three kinds of pulleys and belts in common use (Fig. 191-9): flat belts and pulleys, V pulleys and V belts, and the

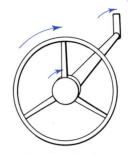

Fig. 191-7.   A wheel and axle.

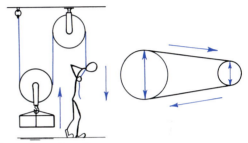

Fig. 191-8.   A pulley.

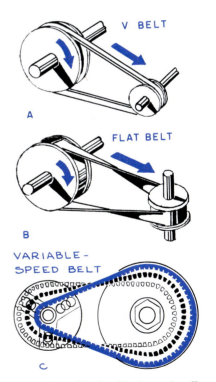

Fig. 191-9.   Three kinds of belts and pulleys: (a) V belt and pulley, (b) flat belt, (c) variable-speed belt.

variable-speed pulleys that have a V belt. The most common types for small gas engines are V pulleys and V belts and variable-speed pulleys. With a variable-speed pulley, it is possible to change the speed between the driver and the driven pulleys without stopping the engine. In fact, the speed should be changed only when the engine *is* running.

The driver pulley of a variable-speed belt is made of two parts having V-shaped sides. With the use of an adjusting screw attached to a crank wheel, one side of the pulley can be opened, or spread apart, from the other. As it spreads, the belt moves inward toward the smaller diameter. This action produces a slower speed on the driven pulley. When the sides of the pulley are brought together, the belt is forced outward toward the larger diameter. This increase adds to the speed of the driven pulley.

Friction and Bearings. Friction has both advantages and disadvantages. Many mechanical devices, such as a clutch or a brake, depend fully on friction. However, in transmitting power, the effects of friction should be avoided whenever possible.

When a shaft runs in a bearing, it produces friction that means wasted power. If there is too much friction, the parts get very hot, enlarge, and stick. To avoid this, various kinds of bearings are used. A *bearing* is a support for a revolving shaft or the moving part of an engine or machine. *Sleeve* (plain) *bearings* are commonly used for slow-running engines (Fig. 191-10).

The sleeve bearing is usually made of bronze. *Ball and roller bearings* are used in high-speed engine parts for workloads that are not too heavy (Fig. 191-11). *Roller* bearings may be either straight or tapered (Fig. 191-12). The small sizes used in engines are called *needle bearings* (Fig. 191-13).

Gears. A gear is a wheel with teeth around the outside (Fig 191-14 a and b).

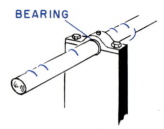

Fig. 191-10.   A sleeve bearing.

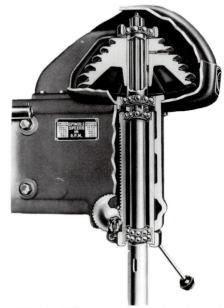

Fig. 191-11.   The use of ball bearings in the head of a drill press.

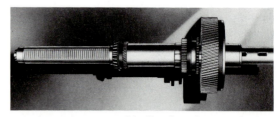

Fig. 191-12.   Tapered roller bearing used on a milling machine spindle.

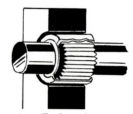

Fig. 191-13.   Needle bearings.

Fig. 191-14 a. Gears.

Fig. 191-16 a. Chain drive.

Fig. 191-14 b. A gear-cutting machine. (Allis Chalmers)

Fig. 191-16 b. Bicycles use chain drives and gears to transmit power to the rear wheel.

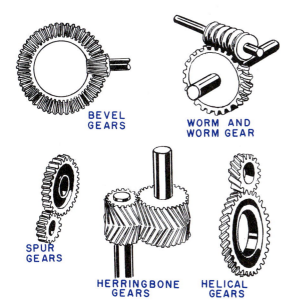

BEVEL GEARS

WORM AND WORM GEAR

SPUR GEARS

HERRINGBONE GEARS

HELICAL GEARS

Fig. 191-15. Some different kinds of gears.

These teeth give positive drive between gears. Belt drive is not positive because there is some slippage. Many kinds of gears are used in engines and machines. Some of the more common ones are spur gears, bevel gears, helical, or spiral, gears, herringbone gears, and worm gears (Fig. 191-15). Chain drives use gears and chains to operate machines, such as the wheel drive on the bicycle (Fig. 191-16 a and b).

Clutches. A clutch device is used to disconnect or connect a power line used in automobiles or small engines (Fig. 191-17). Most clutches are either *friction* clutches or *positive-acting* clutches. The friction clutch is good for being able to take up the pressure slowly while the parts are operating. For example, the friction clutch in a car disconnects the power between the engine and the transmission (Fig. 191-18). Positive-acting clutches have a big advantage in that there is no slippage. However,

Fig. 191-17.   A simple friction clutch.

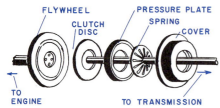

Fig. 191-18.   The main parts of a friction clutch in an automobile.

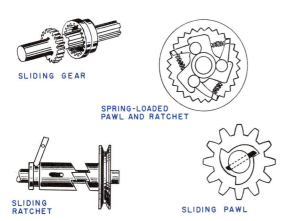

Fig. 191-19.   Several kinds of positive-acting clutches.

Fig. 191-20.   This carpenter is using pneumatic (air-operated) nailers to assemble the framework of a house. An air compressor is needed to provide the power in the form of compressed air.

there is also no protection. The parts must be engaged when everything is stopped (Fig. 191-19).

Fluid Power

Fluid power is a way of transmitting power through such materials as air, gases and fluids (oil). Oil is used in the hydraulic systems and air in the pneumatic systems. Air power tools are widely used in carpentry, auto mechanics, and many other construction and service industries. The carpenter, for example, uses air-powered nailers and staplers to assemble many parts of a house (Fig. 191-20).

Hydraulic systems are used in every area of technology from the braking system on a car to the operating systems of a robot. Hydraulic power is used on fork lift trucks, the landing gear of airplanes, bulldozers, and hundreds of other mechanical devices. The basic parts of a hydraulic system are a reservoir, a pump or compressor, tubing and piping, cylinders, and control valves.

This is how a hydraulic system works. Pressure on a confined fluid (oil) acts equally in all directions. The fact that the oil can't be compressed permits it to transfer power in an efficient manner. The transfer of force is done by converting the mechanical force on one piston to exert force on the liquid that provides force on another piston. In Fig. 191-21, note the closed system consisting of two pistons and the hydraulic fluid in tubes connecting them. A 10-pound weight acting downward on a 1-square-inch piston creates a pressure in the hydraulic fluid of 10 psi (pounds per square inch). This pressure is transmitted through the fluid and acts on the 10 square-inch piston developing an upward force of 100 pounds. As a result the 10 square-inch piston is able to support a 100-pound weight (Fig. 191-22).

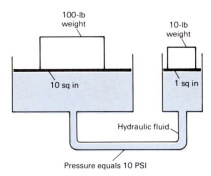

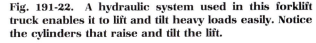

Fig. 191-21. A basic hydraulic unit.

Lubricants

Lubrication refers to the oiling or greasing of parts that move or rotate. It is very important to use the proper lubricants for all machinery. The purposes of lubricants are to (1) help overcome friction, (2) prevent excessive wear between parts, (3) act as a cooling agent to carry away heat, (4) protect surfaces from rust and corrosion, and (5) fill the space between metal parts, acting as a cushion.

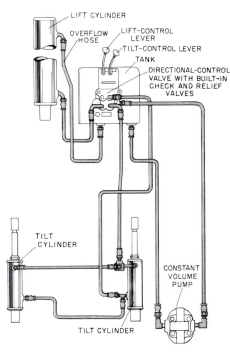

Fig. 191-22. A hydraulic system used in this forklift truck enables it to lift and tilt heavy loads easily. Notice the cylinders that raise and tilt the lift.

Unit 192

Tools, Wrenches, Gages, and Fasteners

An engine is made up of many separate parts. These parts are held firmly together by various kinds of metal fasteners, such as bolts, nuts, screws, washers, and pins. Although there are many types and sizes of fasteners, only the more common ones will be talked about in this unit. Threaded metal fasteners are also used on engines to make adjustments (Fig. 192-1 a and b).

Fig. 192-1 a. Mechanics must know about many different fasteners and know how to assemble engines using the correct kind. (Gulf Oil Corporation)

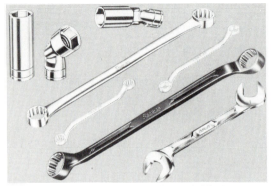

Fig. 192-1 b. Wrenches are available in many sizes and kinds.

Threaded Customary-size Fasteners

Machine Bolts. Machine bolts are made in diameters from $1/4$ inch to $1^1/2$ inches and in lengths from $1/2$ inch to 30 inches. They are made with either square or *hexagonal* (six-sided) heads. The threads are either National Fine or National Coarse. Both types are generally used with hexagonal or square nuts (Fig. 192-2).

When you take a machine apart, always be careful to keep matching bolts and nuts together. If you lose a nut, you can get a replacement by measuring the diameter of the bolt and checking the number of threads per inch with a rule, as shown in Fig. 192-3. Then by checking the table in Fig. 140-4, you can tell whether you need a National Fine or a National Coarse thread nut. However, the National Fine and National Coarse systems have been absorbed into the Unified National system. Threads in the unified system are designated as UNC or UNF and are compatible with the NF and NC systems.

Cap Screws. Cap screws are much like machine bolts, but they have a higher standard of accuracy (Fig. 192-4). They are made with hexagonal, round, flat, or fillister heads. The heads may be slotted for a screwdriver, or they may be the socket type. Socket heads may be either hexagonal or fluted. Cap screws are made in sizes from $1/4$ inch to $1^1/2$ inches in diameter and from $1/2$ inch to 6 inches in length.

Machine Screws. Machine screws are made with round, flat, oval, fillister, binding, pan, truss, or hexagonal heads (Fig. 192-5). Each type of head may be either slotted or recessed. Machine screws are bought with either National Fine or National Coarse threads. Below diameters of $1/4$ inch, the sizes are given by gage num-

Fig. 192-2.   A bolt and nut.

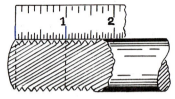

Fig. 192-3.   Measuring threads per inch with a rule.

Fig. 192-4.   Cap screw.

ber. For example, a 6−32 has a No. 6 diameter and 32 threads per inch. These screws are commonly made of brass or steel. Lengths are from ⅛ inch to 3 inches.

Machine screws are often used to hold parts together when one part has been tapped and the other part has a clearance or body-size hole. Either square or hexagonal machine nuts may be used with these screws. The heads of machine screws may be slotted for either a plain screwdriver or a Phillips screwdriver.

Set Screws. Set screws are made with or without heads. The point may be cup- or cone-shaped or may be one of the other shapes shown in Fig. 192-6. Set screws are made to hold two parts together when one part must keep the other from turning. For example, a set screw may be used on a pulley to hold it on a shaft. Heads of set screws are usually square. Headless set screws have either a slot for a screwdriver or a socket that needs a hexagonal or fluted key (Allen hex key).

Studs. Studs are short shafts with a threaded portion on one end or a continuous thread that runs along the entire length (Fig. 192-7) Usually, one end of the stud is screwed into a tapped hole. For example, the stud would be screwed into the cylinder block of an engine. The head is then slipped onto the block over the stud or studs. A nut is put in at the top of the stud to clamp the parts together.

Nuts. There are various types of nuts for use with bolts, machine screws, and studs.

Fig. 192-5.   Machine screws.

Fig. 192-6.   Set screws.

The most commonly used ones are the hexagonal and the square (Fig. 192-8). Jam nuts are hexagonal in shape but are much thinner than regular nuts. Jam nuts either are tightened against the regular nuts to lock them in position or are used in narrow spaces. Common washers used with jam nuts are flat, split-lock, or shakeproof (Fig. 192-9).

Cotter Pins.   A cotter pin is a split pin with a loop, or eyelet, at one end (Fig. 192-10). It is used to hold parts on a shaft or to keep a slotted nut from working loose. A cotter

Fig. 192-7.   A stud with a nut.

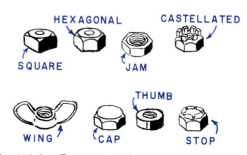

Fig. 192-8.   Common nuts.

Fig. 192-9.   Common washers.

Fig. 192-10.   A cotter pin.

pin is placed through a hole in the shaft. Then the split ends are bent outward, one end in one direction and the other end in the other direction, to keep the pin in place.

Metric Threaded Fasteners

Most imported cars, motorcycles, and bicycles use either the SI or ISO metric fasteners. Some engines made in the United States also have metric fasteners. The use of metric fasteners will grow quickly in the next 10 years, as the United States changes to the metric system. It is important to recognize the difference between metric and customary fasteners. Many seem the same, but they are not interchangeable. Check the diameter of the metric fastener with a gage or metric micrometer. The thread pitch for metric fasteners is identified by measuring the distance between each thread crest in millimeters—not by counting the number of threads per inch, as with customary fasteners (Fig. 192-11). Most metric fasteners will have hexagonal heads, the same as customary fasteners. However, some will have a 12-spline head and will be easy to recognize. (See Unit 126.)

Tools for Assembling with Bolts, Nuts, and Screws

Screwdriver. Screwdrivers are made in many sizes, lengths, and styles. The point is made for either a slotted head or a recessed Phillips head. Many machine parts have recessed Phillips-head screws. When you select a screwdriver, make sure that the blade fits the slot correctly. The *offset screwdriver* is used to drive screws in difficult places (Fig. 192-12).

Pliers. Pliers are used to hold small parts when you are making adjustments (Fig. 192-13). They cannot take the place of a wrench. They should never be used to

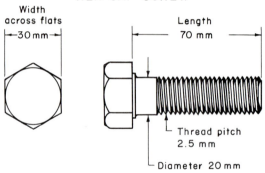

Fig. 192-11. A typical metric fastener. The capital M indicates that it is an ISO thread. The diameter is 20 mm, or slightly more than $3/4$ in. The pitch is 2.5 mm, or about 10 threads per inch.

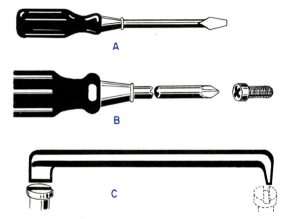

Fig. 192-12. Common kinds of screwdrivers: (*a*) plain blade, (*b*) recessed or Phillips, (*c*) offset.

Fig. 192-13. Slip-joint pliers.

hold bolts or nuts. The most frequently used types are slip-joint (combination) pliers, side-cutting pliers, needlenose (long-nose) pliers, and diagonal pliers.

Adjustable-end Wrench. The adjustable-end wrench (Fig. 192-14) is useful since two or three sizes can take care of a wide range of bolts and nuts (Fig. 192-15). However, a good mechanic never uses an adjustable wrench when there is another type available. When you use an adjustable wrench, always pull it so that the force will be against the solid jaw.

Open-end Wrench. Open-end wrenches are made either straight or S-shaped, with openings at each end to provide for two different sizes of bolts and nuts (Fig. 192-16). The head and the opening are made at an angle of 15° or 22½° to the body so that the nut or the bolt can be tightened when the space is small. By turning the wrench over, you can get a new grip on a nut or a bolt to turn it. The wrench openings range in size from ¼ inch to 1½ inches.

Box Wrench. A box wrench is made with an opening in the shape of a circle. Inside the circle are 6, 8, or 12 points or notches that fit over the bolt head or nut (Fig. 192-17). This makes it possible to move the nut a short distance before changing the grip. Box wrenches are made either straight or with an offset.

Socket Wrench. Socket wrenches come in sets made up of a number of individual sockets used with different handles (Fig.

192-18). The hinged offset handle is very useful. Other types of handles are the ratchet, the sliding offset handle, the key handle, and the speed handle. The universal joint may be used between the handle

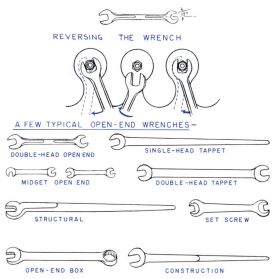

Fig. 192-16. **Open-end wrenches and how to use them.**

Fig. 192-17. **Box wrench.**

Fig. 192-14. **Adjustable-end wrench.**

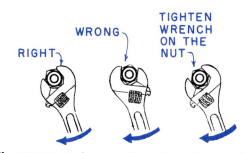

Fig. 192-15. **Using an adjustable-end wrench.**

Fig. 192-18. **A set of socket wrenches.**

and the socket for getting into difficult places. Torque-measuring handles can be used in engine work (Fig. 192-19 a and b). These handles can be set so that a certain amount of twisting force can be applied to the nut. This means that the nut will be neither too loose nor too tight.

All of the same types of fixed wrenches are made in *metric* sizes. Metric wrenches are stamped with the opening in millimeters, such as 12 mm. The fastener shown in Fig. 192-11 would need a metric wrench stamped 30 mm for the width across the flats—not 20 mm, the diameter of the threaded part.

Hints for Using Wrenches

1. Almost all nuts, bolts, and screws have right-hand threads. This means that you

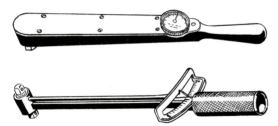

Fig. 192-19 a. Torque handles for socket wrenches.

Fig. 192-19 b. It is very important to use torque wrenches when assembling certain parts of an engine. The engine manual will tell you exactly how to set the wrench for a particular part assembly.

must turn the wrench or tool to the right to tighten it and to the left to loosen it. You sometimes may run into a left-hand thread. This thread is used, for example, to hold a blade on a lawn mower. In this case, you turn the handle to the right to loosen it and to the left to tighten it.

2. Select a wrench that fits the nut or the bolt correctly. Never use a loose-fitting wrench because it will jam the nut or round off its corners. A loose-fitting wrench may also be very dangerous.

3. Pull on a wrench. Never push it. By pulling, you can control its action.

4. Never strike a wrench with a metal hammer. If it is necessary to loosen a tight bolt or nut, tap the wrench with a soft-face mallet.

5. If there are a number of bolts or nuts to be removed, use a socket wrench set. For example, you might first loosen the nut slightly with a socket on a hinged offset handle. The handle should be bent at about a right angle for the necessary leverage. After the nuts are loosened, the hinged handle may be held in a vertical position and twisted between your fingers to remove the nuts. If you are removing several nuts from an engine and there is plenty of room to work, turn the nuts loose with an offset-handle socket. Then slip the socket onto a speed handle and spin the handle until the nuts come off.

6. If the space is very small, you should probably use a box wrench. In places where a threaded nut holds a line (such as a gas line to a carburetor), an open wrench must be used.

Gages

Gages are used to measure distances between parts. The operation of an engine depends on the correct adjustment of such parts as the spark-plug gap or points. The most common sets of gages for gas engines are *thickness*, or *feeler*, gages (Fig. 192-20). These gages are made up of a series of blades with specific thicknesses stamped on the surface. Each thickness is given in thousandths of an inch or in hundredths of a millimeter.

Feeler, or thickness, gages are made with up to 23 short or long blades. These blades should be used very carefully. When you are using a feeler gage, try to slip it into the opening so that there is only a slight drag on it. The blade should never be forced. A *wire* gage is best for setting a spark plug gap.

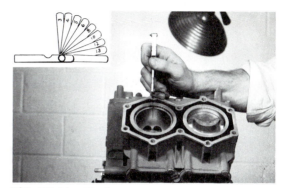

Fig. 192-20.   Feeler gage (upper left) and use.

Unit 193

Parts of a Small Gas Engine

Three basic systems make up an internal-combustion engine: the *power head system*, the *ignition system*, and the *fuel system*. This unit will review the major parts of the power head, or the mechanical parts of the engine (Fig. 193-1 a and b).

Mechanical Parts of the Four-cycle Engine

The cylinder block is a heavy casting with a round opening in it through which the piston moves up and down in an engine with a horizontal shaft (Fig. 193-2). Figure 193-3 shows an engine with a vertical shaft. Many one-cylinder engines are air-cooled, especially those used on small power units. On an air-cooled engine, the

cylinder block is cast with fins sticking out around it. These fins help take away the heat from the burning gases. The *head* is fastened securely to the block. On some two-cycle engines, the head and the cylinder block are one casting. The head also has fins to help in cooling. The *spark plug* is fastened to a hole in the head. *Studs* hold the head securely in place. *Gaskets* made of rubber, cork, or impregnated paper are placed between metal parts.

Fig. 193-1 b.   Using a chain saw to cut timbers. Besides a power lawn mower what other small power tools use a one-cylinder engine? (McCulloch Corporation)

Fig. 193-1 a.   The chain saw uses a one-cylinder engine as its source of power. (Skil Tool Company)

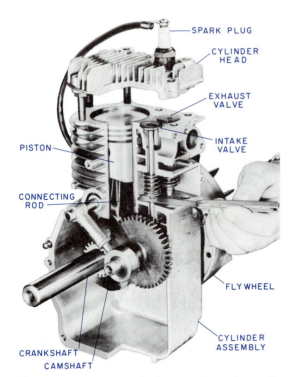

Fig. 193-2. Cutaway of a four-cycle engine with a horizontal shaft.

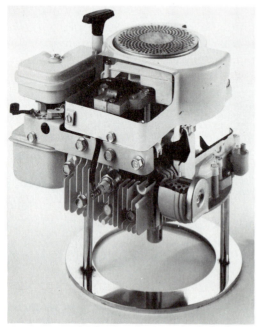

Fig. 193-3. A 3¹/₂-hp (2.6-kW) vertical-shaft engine. (Briggs & Stratton Corporation)

The *piston* is the tin-can−shaped part that moves up and down in the cylinder (Fig. 193-4). There are several grooves cut around the piston into which *rings* fit (Fig. 193-5). These metal rings prevent leakage of gas between the piston and the inside of the cylinder. A metal rod called a *connecting rod* is attached to the piston with a *piston pin.*

The other end of the connecting rod is attached to the *crankshaft.* The crankshaft has an offset so that the connecting rod will force it to turn as the piston moves up and down (Fig. 193-6). The crankshaft is enclosed in a *crankcase* attached to the cylinder of the engine. The crankcase holds the lubricating oil in which the moving parts work.

The *camshaft* operates through two timing gears from the crankshaft. On this shaft are two egg-shaped projections (*cams*) that open and close the valves. The *valves* are two metal rods with caps on them that fit into two holes in the cylinder block (Fig. 193-7). The *intake valve* opens the hole through which the air-fuel mixture enters the cylinder head. The *exhaust valve* is an outlet for the burned gases.

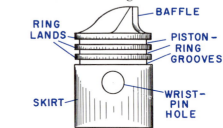

Fig. 193-4. A piston.

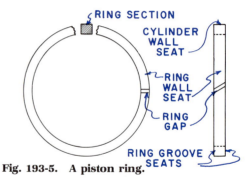

Fig. 193-5. A piston ring.

The Two-cycle Engine

The two-cycle engine differs from the four-cycle engine in that it does not have a camshaft and valves (Fig. 193-8). Instead it has a spring, called a *reed plate*, that covers a hole at the bottom or side of the crankcase. The reed plate springs open to let gases enter the crankcase and move up through a *port* opening toward the compression chamber (Fig. 193-9 a). As the piston nears the top of its stroke, it slows down, and suction decreases. This allows

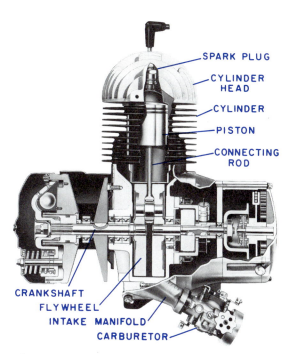

Fig. 193-8. Cutaway of a two-stroke cycle engine.

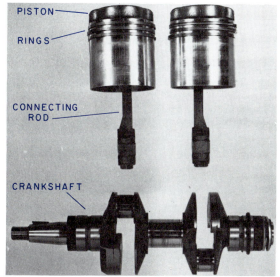

Fig. 193-6. The correct assembly of the piston, the connecting rod, and the crankshaft for a two-cylinder engine.

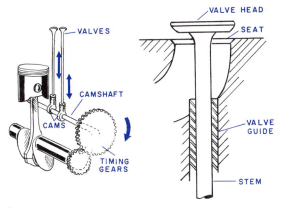

Fig. 193-7. The arrangement of valves in a four-stroke cycle engine.

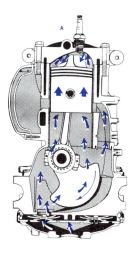

Fig. 193-9 a. The air/fuel mixture enters the crankcase from the carburetor through the reed plate and moves up through the intake part toward the combustion chamber. As the piston reaches top dead center, the spark ignites the compressed charge, forcing the piston down.

the reed to seat itself and prevent the flow of the air-fuel mixture from the crankcase to the carburetor as pressure is built up in the crankcase. Another port opening on the other side of the cylinder block allows the exhaust gases to escape (Fig. 193-9 b and c).

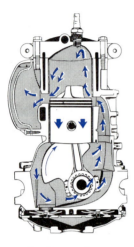

Fig. 193-9 b. As the burned gases expand and are exhausted through the exhaust port on the left side of the cylinder, a fresh charge of air/fuel mixture enters the combustion chamber through the intake port on the right.

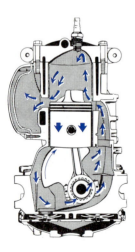

Fig. 193-9 c. The cycle is completed and begins again. The air/fuel mixture enters the crankcase as the charge is compressed in the combustion chamber. Each revolution of the crankshaft completes a full combustion sequence in the two-cycle engine.

Cooling System

Many small one-cylinder engines are air-cooled. Others, especially those in automobiles and outboard motors, are liquid-cooled (Fig. 193-10). The operating temperature of a liquid-cooled engine remains constant. This is in contrast to the operating temperature of an air-cooled engine,

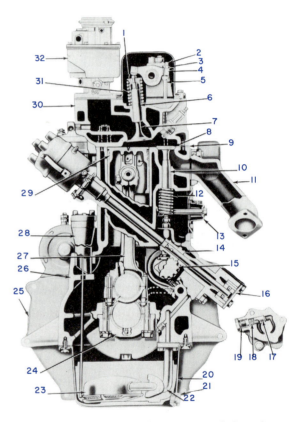

Fig. 193-10. Cutaway of a water-cooled engine: (1) inlet valve spring retainer, (2) adjusting screw, (3) nut, (4) rocker arm, (5) push rod, (6) inlet-valve guide, (7) inlet valve, (8) exhaust valve, (9) cylinder-head gasket, (10) exhaust-valve guide, (11) exhaust manifold, (12) exhaust-valve spring, (13) crankcase ventilator, (14) oil-pump gear, (15) camshaft, (16) oil pump, (17) relief plunger, (18) relief-plunger spring, (19) relief-spring retainer, (20) oil pan, (21) drain plug, (22) oil-float support, (23) oil float, (24) crankshaft, (25) engine rear plate, (26) cylinder block, (27) connecting rod, (28) oil-filter tube, (29) piston, (30) cylinder head, (31) inlet-valve spring, (32) carburetor.

which varies greatly with changes in air temperature, load, and speed. The big advantage of an air-cooled engine is that it does not need a complicated cooling system. Therefore, it is lighter, needs less space, and is easier to repair than a liquid-cooled engine.

Compression Ratio

The *compression ratio* of an engine is the relationship of the space in the cylinder when the piston is at the top of the stroke and the space when it is at the bottom of the stroke. For example, a compression ratio of 6 to 1 means that there is only one-sixth as much space in the cylinder at the top of the piston stroke as there is when it is at the bottom (Fig. 193-11).

High compression ratios usually mean greater efficiency than low ratios. How-

ever, high compression ratios require heavy engine parts to take care of loads and stresses.

Displacement is the volume of the cylinder opening when the piston is at the bottom of the stroke. It is given in cubic inches or cubic centimeters.

Valves

Good valves are important to good compression. If the valves leak or do not seat properly, there will not be enough compression at the top of the piston stroke. When an engine is operating at 3000 rpm, each valve opens and closes in about $1/50$ second. Valves need to be well sealed to stand pressures up to 500 pounds per square inch (34.5 megapascals) and temperatures exceeding 1200°F (650°C).

Figure 193-12 shows some common valve failures. One of the repair jobs most often needed on an old engine is replacing the valves. Regrinding valves so they seat properly is also a common repair operation.

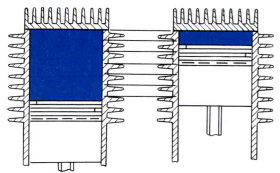

Fig. 193-11. A compression ratio of 6 to 1.

BURNED   DISHED   NECKED

Fig. 193-12. Common defects in valves of a four-stroke cycle engine.

Unit 194

Carburetion and the Fuel System

The fuel system of a small one-cylinder engine is made up of the fuel, or gas, tank, the fuel line, the carburetor, and the air cleaner. In an automobile and in many outboard motors, there is another device called the *fuel pump* (Fig. 194-1).

The purpose of the fuel system is to provide the proper proportions of fuel and air to make the mixture burn efficiently. The raw gasoline may be compared to large chunks of wood. It will burn, but not with an efficient, hot flame. Mixing air with the

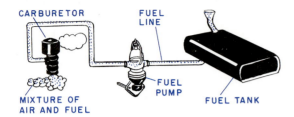

MIXTURE OF
AIR AND FUEL

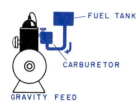

GRAVITY FEED

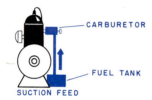

SUCTION FEED

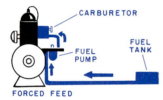

FORCED FEED

Fig. 194-2.   Three kinds of fuel feeding systems.

gas is like cutting the wood into fine shavings to make the mixture burn much hotter.

Parts of the Fuel System

The fuel system starts with a fuel tank in which a supply of gasoline is stored. The fuel line carries the gasoline from the tank of the carburetor by means of *gravity, suction feed*, or *force feed* (Fig. 194-2).

1. In the gravity system, the gas tank is placed above the carburetor. The fuel flows into the carburetor by gravity. A small air-vent hole in the tank lets the air flow in as the fuel flows out. Another vent hole in the carburetor lets the air flow out as the fuel flows in. Both of these vents must be kept open if the fuel is to get into the engine.

2. In the suction-feed system, the fuel tank is placed below the carburetor. It is connected right to the carburetor in such a way that the suction of the infeed stroke will draw fuel into the carburetor. A valve in the carburetor keeps the level of the fuel up when there is no intake stroke.

3. The force-feed system makes use of a mechanical fuel pump that sucks the gasoline from the tank and forces it into the carburetor.

Small gas engines have either a gravity-feed system or a suction-feed system. An outboard motor with a separate gas tank uses the force-feed system.

Functions of the Carburetor

The carburetor has several important functions:

1. It must *atomize* the fuel, or break up the gasoline into a spray of very small drops.

2. It measures the fuel so that the fuel will be mixed with the air in the correct proportion. There should be about 1 pound (454 grams) of gas to 15 pounds (6.8 kilograms) of air.

3. It directs this fuel into the combustion chamber through an opening called the *intake manifold.*

Operations of the Carburetor

When the gas enters the carburetor, it first flows into the *float chamber* (Fig. 194-3 a and b). This is a storage area for the gas. The float chamber has a *float* made of cork or hollow brass in it. The float opens and closes a valve when the gasoline has reached the correct level. As the gas enters the float chamber, it raises the float. The

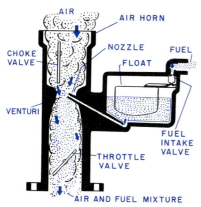

Fig. 194-3 a. Cutaway of a carburetor showing the essential parts.

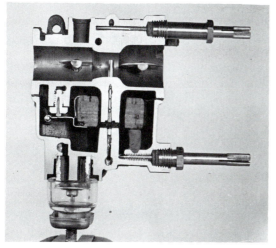

Fig. 194-3 b. Cutaway of a very simple carburetor found on a small two-cycle engine. (McCulloch Corporation)

float in turn raises the needle in the float valve, cutting off the fuel. As the fuel is used, the float drops down, opening the valve and allowing more fuel to come into the storage area. The float level is usually kept high enough to prevent flooding or leaking.

Next to the float chamber is a metal tube called an *air horn*. The air horn is open at the top so that air can come in. It is attached to the bottom of the engine. This hollow tube, which is the carburetor proper, sometimes has a smaller diameter

near its center. This area is known as the *venturi*. The venturi increases the velocity of air and decreases the air pressure at this point.

A venturi may be compared to a shallow, narrow place in a river. There the current is always faster than in the wider part of the river. The venturi is placed in a carburetor so that the intake air that is suddenly forced into a smaller space will speed up in order to maintain the same volume of air flow. At the same time, the air pressure at this particular point will decrease.

Carburetors for some small engines do not need the venturi since the air horn tends to do the same thing. Instead, there is a tube that runs from the float chamber into the carburetor opening at about the location of the venturi. As the piston in the cylinder moves down, it creates a partial vacuum in the cylinder. The air pressure on the outside forces the air through the carburetor. Then two things happen: (1) The fuel flows out of the small tube, and (2) the air rushing by this fuel atomizes it into small droplets. The result is a combustible mixture ready for firing.

If an engine were to run at only one speed, this process would be all that would be necessary. However, the speed of an engine must be controlled. This is done by adding a *throttle valve* (butterfly) between the venturi and the lower part of the carburetor. When this valve is opened, it does not affect the flow of air and fuel. However, when the throttle valve is closed, it begins to decrease the amount of air flowing into the cylinder. The throttle valve also decreases the power and speed of the engine. As the engine speed slows down to an idle, the throttle valve is nearly closed. Very little air then gets through the air horn. As a result, no fuel is picked up, and the combustible mixture becomes too lean to burn.

In order to supply fuel for idling or low speeds, another fuel passage from the float chamber is needed. This tube runs into the carburetor below the throttle valve (Fig. 194-4). In an automobile carburetor, there are other tubes, called *circuits*, for special operational conditions. For example, the *pump circuit* in a carburetor gives an extra supply of gas for quick acceleration.

Starting an engine when it is cold presents another problem. A cold engine will not start on the regular mixture used for running. It must have an extra-rich mixture, or more fuel in the air. To take care of this, a *choke valve* similar to the throttle valve is placed in the upper part of the air horn. With this choke valve, much of the air can be closed off so that a high vacuum is formed beneath it. This vacuum will draw in more fuel and less air, producing a very rich mixture for easy starting.

After the engine is started, the choke valve is opened to allow more air to enter. The engine will run more smoothly with the choke valve open. The choke valve on a small engine should not be kept closed too long, as this will cause raw gas to enter the combustion chamber. Raw gas can roughen the cylinder walls.

An *intake manifold* is located between the carburetor and the openings to the various cylinders. On a four-cycle engine, the fuel and air enter the carburetor by way of the manifold. The mixture then goes through the valve into the combustion chamber. On a two-cycle engine, the carburetor is attached to the crankcase. This arrangement allows the fuel to enter the crankcase through a reed valve. The fuel then flows into the combustion chamber by way of ports. For this reason, *plain gasoline is used on a four-cycle engine, and a mixture of gasoline and oil is used on a two-cycle engine.*

Because the crankcase of the two-cycle engine acts as a transfer pump for the air-fuel mixture, lubrication of the engine depends on the addition of oil to the gasoline. When the mixture passes through the carburetor, the gasoline becomes highly vaporized by the stream of air. The oil is broken into droplets that lubricate all the surfaces with which they come in contact. Some oil remains in the crankcase, but most passes through the combustion chamber, where it is burned and then forced out the exhaust port. The mixture is usually about 1 part oil to about 25 parts of gasoline.

Adjusting the Carburetor

If possible, follow the directions in the service manual for the basic carburetor settings. If you do not have a manual, use the following procedure:

1. Turn in both the high- and the low-speed needles carefully until they seat.
2. Open the high-speed (main) needle 1½ turns and the low-speed (idle) needle about 1 turn.
3. Choke the engine and then crank it once or twice. Open the choke halfway and crank it again to start.
4. After the engine is warmed up and is running at high speed, turn in the high-speed needle until the engine starts to slow down (Figs. 194-5 and 194-6). Then turn out the high-speed needle until the engine starts to falter.

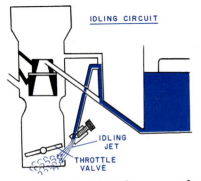

Fig. 194-4. The idling jet takes over when the throttle valve is closed to provide enough fuel for the engine to operate at low speed.

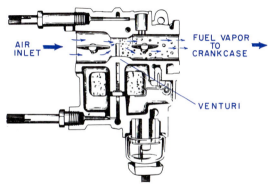

Fig. 194-5. This sectional view of the mixing chamber shows the choke valve and the throttle valve wide open. Note that the high-speed jet provides the fuel for the mixture, with a minimum of fuel coming through the slow-speed jet.

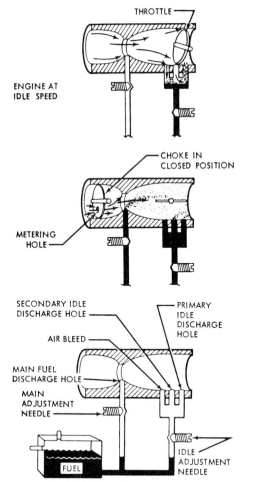

Fig. 194-6. A diagram of the sectional view of the mixing chamber.

5. Count the number of turns from the low points to the high points. Turn the needle back in about halfway and allow the engine to run for about 1 minute.

6. Close the throttle so that the engine is running at idling speed. Then adjust the low-speed needle the same way. Use step 4 (Fig. 194-7) to do it.

7. Repeat this process twice, allowing about 30 seconds to 1 minute between adjustments.

Air Cleaners

Air cleaners on a carburetor keep out the dust and dirt that is in the air. If no cleaner were on the carburetor, these *impurities* would be pulled into the engine through the carburetor and would cause extra wear. There are several types of air cleaners available, grouped as *dry* and *oil-bath* (Figs. 194-8 and 194-9).

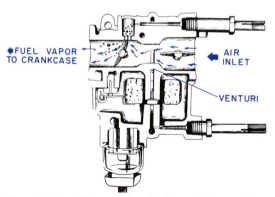

Fig. 194-7. This sectional view of the mixing chamber of the carburetor shows the throttle valve nearly closed. Note that the fuel is now entering through the slow-speed jet.

Fig. 194-8. The parts of a dry air cleaner.

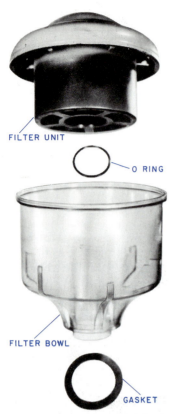

FILTER UNIT

O RING

FILTER BOWL

GASKET

Fig. 194-9. The parts of an oil-bath cleaner.

The dry air cleaner may be made of aluminum foil, a fiber element, or a metal cartridge. It acts as a filter to prevent dirt particles from entering the carburetor. This kind of air filter must be either cleaned or replaced.

The oil-bath oil cleaner draws the air through an oil bath to clean out impurities. The filter bowl in this kind must frequently be removed and washed thoroughly. Then new oil is added to the level shown. Follow the owner's manual for exact directions for cleaning and replacing air filters.

Engine Governors

On most small engines, an engine governor is used to control engine speed. The governor assembly varies with the kind of motor. It is most often operated by spring-loaded weights.

The governor assembly shown in fig. 194-10 operates as follows. As the crankshaft increases in speed, the weighted linkage is thrown outward by centrifugal force. This action causes the top ring to move toward the bottom ring. This shuts off the throttle valve. As the engine slows down, the centrifugal force of the weighted linkage decreases. This permits the governor spring to move the top ring up, thus opening the throttle.

There are many other kinds of governors. Another common type is operated by air vanes. Through a linkage, the air vane opens and closes the throttle valve of the carburetor.

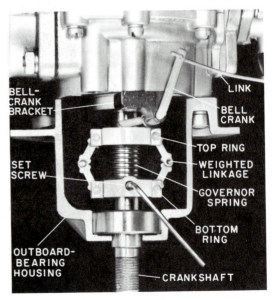

BELL-CRANK BRACKET

LINK

BELL CRANK

TOP RING

WEIGHTED LINKAGE

SET SCREW

GOVERNOR SPRING

BOTTOM RING

OUTBOARD-BEARING HOUSING

CRANKSHAFT

Fig. 194-10. A governor.

Ignition System

An internal-combustion engine has either a magneto or a battery electric system. The magneto system is used on most small two- and four-cycle gas engines. The battery system is used on all automobile engines and some larger outboard motors. Many similar devices are used in both systems.

Review of Basic Electrical Principles

1. A current of electricity is a flow of electrons through a wire.

2. There must be a complete circuit to have a flow of electricity.

3. When electricity flows through a wire, a magnetic field of force is created around it. The magnetic field travels in a circular pattern. This forms a magnetic cylinder the full length of the wire.

4. If several loops of wire are shaped in the form of the coil, the magnetic effect is greatly increased.

5. Voltage can be induced in windings by magnetism. If the magnetic lines of force of one coil cut a second coil, voltage is induced whenever the lines of force build up or collapse. In other words, with two coils, the electric energy can be transferred from one circuit to the other through a magnetic coupling. This transfer of energy is called *mutual induction.* Mutual induction is used in the ignition coil of both the magneto system and the battery system.

Magneto Ignition System

The magneto ignition system of a small gas engine is made of the following elements:

1. The *rotor* is a strong permanent magnet attached to the crankshaft that revolves inside the armature (Fig. 195-1).

2. The *armature* is made up of a lamination of metals with the coil wound around part of the armature (Fig. 195-1).

3. The *condenser* is a safety valve in a primary circuit (Figs. 195-2 and 195-3).

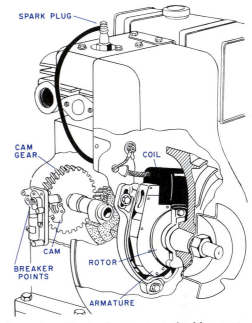

Fig. 195-1.   Parts of a magneto ignition system.

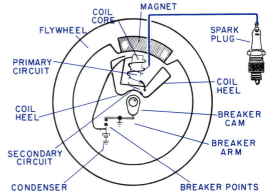

Fig. 195-2.   Diagram of a magneto ignition system.

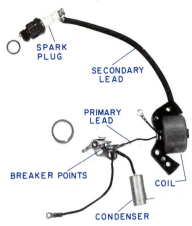

Fig. 195-3.   The magneto ignition system.

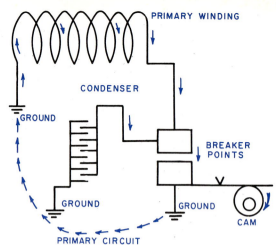

Fig. 195-4.   A schematic diagram of a primary circuit.

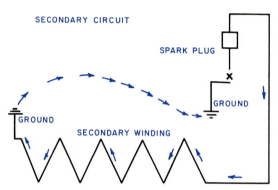

Fig. 195-5.   A schematic diagram of a secondary circuit.

4. The *coil* is made of a few heavy windings of wire over which many windings of finer wire are wound (Figs. 195-1 through 195-3).

5. The *breaker points* open and close to make and break the primary circuit (Figs. 195-1 through 195-3).

6. The *spark plug* produces the spark to *ignite* (begin to burn) the fuel. This is part of the secondary circuit (Figs. 195-1 through 195-3).

Notice that there are two circuits: the primary circuit and the secondary circuit. The parts of the *primary circuit* are the primary winding on the coil, the breaker points, and the condenser (Fig. 195-4). The secondary circuit is made up of the secondary, or fine, windings on the coil and the spark plug (Fig. 195-5).

Operation of the Magneto Ignition System

As the rotor turns inside the laminations of the armature, it causes a magnetic flux that cuts the coil of the armature. In this process, the moving magnetic flux builds up a voltage in the primary circuit.

At the beginning of the process, the breaker points are closed. The current flows freely through the primary circuit (Fig. 195-6). At just the right time, when the primary current is high, the breaker points are opened by a cam. The opening of the breaker points causes the magnetic field in the primary circuit to collapse. This action in turn causes a current flow in both the primary unit and the secondary circuit.

The amount of current flow depends on the proportion between the number of turns on the secondary circuit and the number of turns on the primary circuit. For example, suppose there are 100 volts in the primary circuit. Also suppose there are 10 times as many turns on the secondary circuit as there are on the primary circuit. Then the voltage in the secondary circuit will be about 10 000 volts. This high voltage flows through the secondary circuit, causing a spark to jump across the terminals of the spark plug (Fig. 195-7).

A condenser is placed in the primary circuit to act as a safety valve. Without a condenser, an *arc* (electric charge) would jump across the breaker points as they opened. The condenser temporarily stores the excess electric energy in the primary circuit just as the breaker points open.

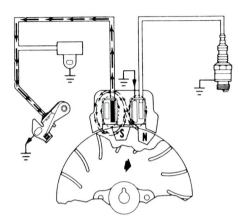

Fig. 195-6. **Current flow in the primary circuit; breaker points closed.** (McCulloch Corporation)

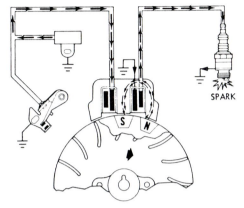

Fig. 195-7. **The breaker points are open, and the field collapses. The high voltage in the secondary circuit causes an arc across the points of the spark plug.** (McCulloch Corporation)

Unit 196

Servicing Small Engines*

A small gasoline engine must have three things in order to run. The correct air-fuel mixture must be able to enter the combustion chamber. The piston must be able to compress the fuel charge. The ignition system must be able to ignite the compressed fuel charge. If one of these elements is missing, the engine will not operate. Therefore, when an engine doesn't run or doesn't put out full power, the things to check are (1) compression, (2) carburetion, and (3) ignition or a combination of all three. Engine trouble can strike in two or three places at once.

Engine Maintenance

Follow these simple rules for proper engine maintenance:

1. Clean the air filter regularly. Cleanliness of fuel and oil is very important to the good operation of an engine.

2. Always use clean, fresh gasoline.

3. Use the correct fuel-oil mixture in the two-cycle engine. Never run a two-stroke cycle on gasoline alone. To do so would soon ruin it.

4. Never mix oil and gasoline for a four-cycle engine. Remember, in a four-cycle engine, you use gasoline in the carburetor and oil in the crankcase. Remember to check the oil in the crankcase after every 4 to 5 hours of operation.

5. Check the owner's manual for any special care and attention that you should give the engine.

---

*Adapted, courtesy of the McCulloch Corporation.

Compression

Compression can be checked by pulling the starter rope gently until the greatest engine resistance is felt, slackening the starter rope, and checking the amount of engine snapback. The starter rope also can be pulled and slacked to find out about "bounce" off the high end of the compression (Fig. 196-1).

Sometimes a two-cycle engine will lose compression because the exhaust port openings and reed plate are clogged with carbon (Fig. 196-2). Clear these openings when necessary. In a four-cycle engine, compression loss may be due to valves that are burned or defective. Of course, too much wear on the piston and rings will also cause loss of compression. It is usually necessary to overhaul the engine completely when this happens.

Ignition

Ignition can be checked by unscrewing the spark plug from the cylinder head and grounding it. This is done by holding the threaded part of the plug against bare metal on the engine. If there is no spark between the electrodes when the starter rope is pulled briskly, replace the broken spark plug with a good one. Then ground it on the engine and pull the starter rope again. If there is still no spark, the ignition system is at fault (Fig. 196-3).

Carburetion

Most carburetor troubles are caused by improper adjustment of the high (main) and low (idle) adjustment needles. The flow of the air-fuel mixture into the combustion chamber can be checked by holding your thumb over the spark-plug hole in the cylinder head and pulling the starter rope several times. If your thumb does not become moist with oil and gasoline, the trouble may be caused by an empty gasoline tank. It may also be caused by a defective reed valve or pump or by a clogged fuel filter (Fig. 196-4).

Carburetion problems can be caused by air-fuel mixtures that are either too rich or too lean. A lack of fuel and failure to run at full speed can also cause carburetion problems. Many of these problems can be traced to the fuel-adjustment needles.

When adjusted correctly, the high (main) needle maintains an air-fuel ratio that provides the greatest available power at full throttle. When the needle is turned in too far, the sound of the exhaust becomes sharp and barking. This sounds powerful and fools many people. However, it also results in a lean mixture that pre-

Fig. 196-1. **Checking the compression on a small engine.**

Fig. 196-2. **Make sure the reed plate is clean and operating easily. It may be necessary to replace this part if it is heavily corroded.**

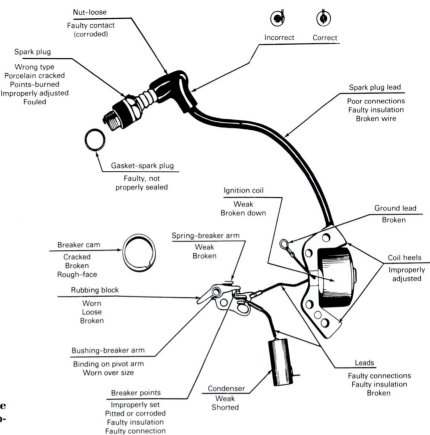

Fig. 196-3. Check these major ignition problems.

Labels in figure:

Nut-loose
Faulty contact (corroded)

Spark plug
Wrong type
Porcelain cracked
Points-burned
Improperly adjusted
Fouled

Incorrect   Correct

Spark plug lead
Poor connections
Faulty insulation
Broken wire

Gasket-spark plug
Faulty, not
properly sealed

Ignition coil
Weak
Broken down

Ground lead
Broken

Breaker cam
Cracked
Broken
Rough-face

Spring-breaker arm
Weak
Broken

Coil heels
Improperly
adjusted

Rubbing block
Worn
Loose
Broken

Bushing-breaker arm
Binding on pivot arm
Worn over size

Leads
Faulty connections
Faulty insulation
Broken

Breaker points
Improperly set
Pitted or corroded
Faulty insulation
Faulty connection

Condenser
Weak
Shorted

vents development of full power. When the main needle is turned out too far, a rich mixture results. That is, too much fuel is sprayed into the air as it passes through the venturi.

When the engine is running at idle speed, the fuel mixture is not broken down into as fine a spray as occurs when the engine runs faster. For this reason, a slightly richer mixture is provided at idle speed. The low (idle) adjustment needle controls this air-fuel mixture. When set correctly, the slightly richer mixture causes no trouble in the engine. If the idle needle is turned out too far, the mixture becomes so rich that too much carbon is formed. This fouls the spark plug, and the engine will run roughly or may die. If the idle needle is turned in too far, the mixture will be lean. Again, the engine may die

during idle, or it may miss and falter during acceleration.

Checking the Ignition System

There are three places in the ignition system where most troubles begin: the spark plug, the breaker points, and loose and shorted wiring. After these main trouble spots come the flywheel-to-lamination air gap, the condenser, the coil, the ignition switch, and the magnet.

Spark Plugs

Spark plugs can cause trouble by being dirty, having too large an electrode gap, having cracked or broken insulation, or being in the wrong heat range. Through normal use, carbon slowly builds up on

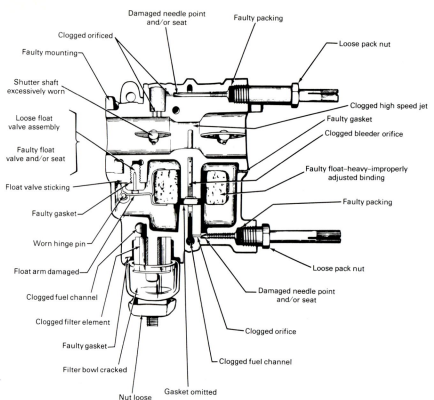

**Fig. 196-4. Major car-
buretor faults.**

the spark plug's insulator base and around
the electrodes. At the same time, the electrode tips become pitted and worn from
sparking. When this happens, the electrode gap slowly becomes wider. After a
while, the gap can be so large that not
enough current is produced by the ignition system for a spark to jump the gap.

The spark-plug electrodes can be
cleaned with emery cloth. The gap can be
adjusted by bending the side electrode.
The gap width should be checked with a
wire gage. Most spark plugs are gapped to
0.025 inch (0.63 millimeter) for best operation (Fig. 196-5). If the porcelain is cracked
or broken, throw the spark plug away.

Breaker Points

The breaker points can prevent operation
of the engine by being oily, dirty, pitted, or
worn. They can also cause trouble if they

Fig. 196-5. Setting the gap on a spark plug.

have a point gap that is too wide. Oily
points can be cleaned with a drop of
solvent followed by drawing a white card
(such as an index card) between them.
When the card comes away clean, the
points will be clean. Badly pitted and
worn-away points should be replaced
with a new breaker-point assembly. The
breaker points should have a gap of about
0.020 inch (0.51 millimeter) (Fig. 196-6).

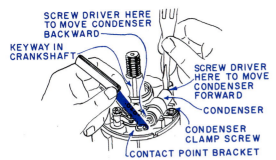

Fig. 196-6. Checking the gap on the breaker points.

Loose and Shorted Wiring

Check all electrical connections to make sure they are tight. Examine the wires running from the coil to the breaker points. Also examine wires from the coil to the ignition switch and from the coil to the spark plug. If any part of these three wires is bare or if the insulation is frayed or worn, short circuits may result, stopping the engine. Other checks of the ignition system are shown in Fig. 196-7. Most of these need special equipment, such as a coil and condenser tester.

CHECKING THE IGNITION SYSTEM
1 Spark plug
2 Breaker point
3 Loose and shorted wiring
4 Flywheel-to-lamination air gap
5 Condenser
6 Coil
7 Ignition switch
8 Magnet

Fig. 196-7. Checking the ignition system.

Unit 197

Your Car

Next to a home, an automobile is usually the most expensive item a person buys. The automobile is a highly complex machine. It is made up of over 15 000 individual parts (Fig. 197-1). The present-day car needs expensive tools and testing equipment to be kept in working order. However, if you understand the operation of your car, there are many adjustments you can do yourself.

Fuel System

The internal-combustion engine is a four-cycle engine such as the one you read about in Unit 193, "Parts of a Small Gas Engine." The engine runs by burning a mixture of gasoline and air. These are combined (mixed together) in just the right amounts for ignition (Fig. 197-2). The fuel is drawn from the tank by a *fuel pump*. It is filtered by a *fuel filter* and metered out by the *carburetor*. The air is drawn in through an air filter that cleans the air of dirt particles. The amounts of fuel and air are regulated by a *throttle* and, when the

Fig. 197-1. A car is a necessity for most families. (Copper Development Association Inc.)

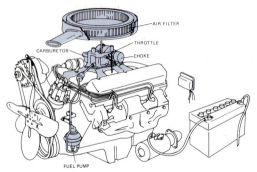

Fig. 197-2. The fuel system.

engine is cold, by a *choke.* An improperly adjusted carburetor or choke or a defective fuel or air filter can be the cause of problems.

Compression System

The fuel-air mixture is drawn into each of four or more engine *cylinders* (Fig. 197-3). In each cylinder, the mixture is compressed by a *piston* before being ignited. Each cylinder fires in a *prescribed sequence* (firing order). Leaks in this system cause poor operation. A malfunction here cannot be fixed by simple adjustment. Major repairs are needed.

Ignition System

In a properly tuned engine, the compressed fuel-air mixture in each cylinder is ignited at just the right moment by a spark from a *spark plug* (Fig. 197-4 a and b). Each cylinder has a spark plug. The *distributor* determines when an electrical charge goes to each spark plug. This is done by means of a *rotor* that spins around and makes electrical contact with the *spark-plug-wire terminals* inside the distributor. Before reaching the distributor, the electrical charge is boosted from 12 volts to as high as 40 000 volts by the *coil*. The distributor also has two *breaker points*, which control the flow of the electrical current by opening and closing. In recent models, the electronic equivalent of points is used. The distributor also has a *condenser*, which absorbs the overflow of electrical current.

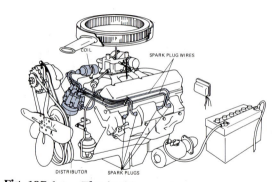

Fig. 197-4 a. The ignition system.

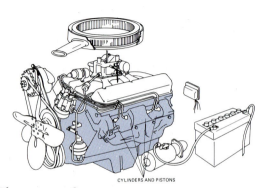

Fig. 197-3. The compression system.

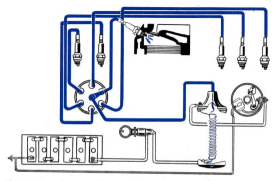

Fig. 197-4 b. A diagram of the ignition system showing the primary circuit in white and the secondary circuit in black.

Starting and Charging System

The electrical charge that sparks the spark plugs begins in the *battery*. The *alternator* (also called the *generator*) charges the battery by changing the mechanical energy of the engine to direct-current electricity. This is done by connecting the alternator to the crankshaft by means of the *fan belt*. The amount of charge is controlled by a *regulator* (Fig. 197-5).

The whole system starts when you turn the ignition key. This draws current from the battery to the *starter*, which cranks the engine. The battery also supplies the first charge to the spark plugs until the alternator can take over. A good tune-up makes sure that all parts of this system work properly.

Car Troubles

If you have trouble with your car, these are things to do before calling a mechanic.

I. Engine will not start, does not turn over.

   A. Headlights, domelight, or horn may not work.

      1. Dead battery. Use jumper cables and booster battery. Replace battery if necessary.

      2. Dirty, corroded battery connections. Disconnect, clean cable clamps and terminals; scrape with screwdriver or use sandpaper.

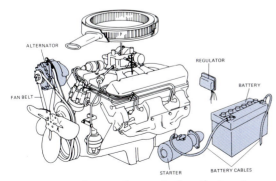

Fig. 197-5.   The starting and charging system.

B. Engine turns very slowly, and headlights or domelight dim out at same time.

      1. Run-down battery. If lights were left on, use jumper cables and booster battery.

      2. Loose or broken fan belt. Tighten or replace, and then use jumper cables with booster battery.

      3. Crankcase oil too thick because of cold weather. Use jumper cables and booster to start. Change to $10W-40$ multigrade oil after warmup.

C. Headlights or domelight do not dim at all when trying to start.

      1. Neutral safety switch not working properly. Put shift in *P* or *N* position and try to start. If no start, move shift in and out of *P* or *N* until engine starts. If still no start, see next step. (Some manual-shift cars require clutch pedal to be pushed to the floor.)

      2. Defective ignition or starter switch. Work switch back and forth several times.

II. Engine will not start, does turn over.

   A. Cylinders do not fire, strong gasoline odor present.

      1. Carburetor flooded, too much fuel. Hold accelerator pedal against floor and try to start. Do not pump accelerator pedal. If no start, remove top of air cleaner (Fig. 197-6). Push choke valve (Fig. 197-7). Opening should be resisted by spring force—strong if engine cold, weak if engine hot. Then try to start.

   B. Cylinders do not fire, no gasoline odor.

Carburetor too lean, not enough fuel. Depress accelerator fully two or three times, hold halfway to floor, and try to start. If no start, remove top of air cleaner and be certain choke closes. To

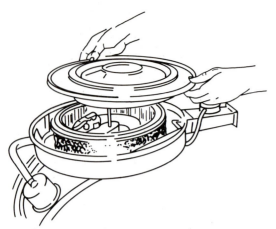

Fig. 197-6. **Removing the top of an air cleaner.**

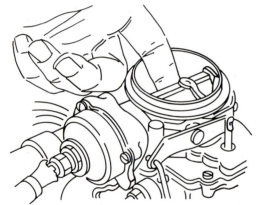

Fig. 197-7. **Checking the operation of the choke valve.**

check, depress accelerator, release, and watch carefully. If not closed, bounce choke several times with finger to free sticking linkage. Depress accelerator to set choke and try to start engine.

C. Cylinders do not fire.

No fuel to carburetor. With top of air cleaner off, hold choke open and look into carburetor throat for a stream of fuel while someone steps on accelerator.

If no fuel stream, check fuel gage. Add fuel if too low or empty. If no fuel stream and tank level is adequate, fuel

line may be plugged or fuel pump is defective. Plugged line must be blown out. A defective fuel pump must be serviced.

Check ignition system.

III. Engine will not start, ignition-system troubles.

A. Will not fire.

1. No spark at spark plug. Pull wire from any plug terminal. Slide protective boot back to expose metal clip, or insert paper clip, bobby pin, or key into boot to contact metal clip. Hold so end of wire (or metal extension) is about 6 millimeters ($^1/_4$ inch) away from the engine block.

Watch for strong spark between end of wire and engine when engine is turned over. If no spark, continue to test as follows.

2. Coil lead loose. Push coil lead down at center of distributor cap and at coil on firewall.

3. Distributor cap cracked. Wipe off the cap with clean rag to look for crack. If cracked, replace with new cap.

4. Inside distributor wet. Remove cap and wipe thoroughly with clean cloth or tissue. Replace cap and try to start engine.

5. Broken rotor. Remove rotor and inspect. If cracked or broken, replace with new rotor.

6. Dirty breaker points. Remove distributor cap and examine points. If dirty or oil-covered, remove and clean with a piece of stiff paper.

IV. Engine stops running while driving.

A. Engine stops, starts, then stops—perhaps a few times. Turns over when trying to start, fires, then dies.

Out of fuel. Look at fuel gage. Some gages may show slightly above *E* when tank is empty. To make sure, remove air-cleaner top, open and close throttle (accelerator pedal), and look for jet of fuel in carburetor throat. If none, add fuel to tank.

B. Engine stops (just dies).

Loss of ignition. Check if ignition switch was accidentally turned off. Check for other ignition-system troubles. Look for lead wires that may have become loose or disconnected at coil, distributor, or spark plugs. Reconnect or tighten loose leads. Look for leads that have burned or worn through insulation. Hold away from engine's metal parts, using string, cloth, or wood pieces to prevent grounding.

C. Engine slows down, labors, then stops.

    1. Overheated. Did "high temperature" or "hot" light come on? What does temperature gage show? If engine is very hot, raise the hood to increase circulation. Let it cool for 20 to 30 minutes. After it is cooled down, *very carefully* check radiator for coolant level. (Wear gloves and use a cloth. Keep head and face well away from cap.) When certain all pressure has been relieved, remove cap slowly. If steaming from filler neck, let cool longer before adding coolant. Some cars have plastic overflood tank for coolant, so *do not remove radiator cap.* Look for coolant in this tank. Thermostat in cooling system may be broken and closed. This must be replaced.

    2. Low oil or out of oil. Did oil-pressure warning light come on? Check dip stick for oil level; add oil if needed. If oil level is OK, oil pump may be defective.

D. Engine stops, will not turn over while trying to start; lights and horn do not work.

Dead battery. Did alternator/generator light come on (perhaps bulb is defective)? Check under hood for loose or broken fan belt. Check for loose wires at alternator/generator and voltage regulator. Push and pull connections to be sure they are tight. If the car is equipped with a multiwire connector on the firewall, jiggle the connector to test for tight connection.

E. Engine stops, will not turn over while trying to start; lights work but dim slightly while trying to start.

Engine seized (cannot turn over). Did oil-pressure or temperature-warning lights come on? Maybe one or both are defective. If engine is very hot, open hood and cool down 20 to 30 minutes. If engine does not turn over when starting, call service station for help (towing is probably needed). It will be helpful to make notes of just how the engine acted as it stopped. These may help the mechanic.

Unit 198

Transportation Systems

Transportation systems are ways of moving materials, goods, and people from one place to another using many different kinds of vehicles (Fig. 198-1). Materials must be moved to the places where they are needed or where they are used to produce goods. Goods must be moved because most are produced in one place and consumed or used in another. With our modern transportation system, everyone can have fresh fruits and vegetables throughout the year. Mail and packages

TRANSPORTATION SYSTEMS

MEANS OR METHOD

Land	Water	Air	Space	Commercial and Industrial
Above	*Above*	Airplane	Satellite	Elevator
Elevated rail	Hydrofoil	Helicopter	Spacecraft	Escalator
	Hovercraft	Blimp	Shuttle	Moving sidewalk
On		Balloon	Station	Pipeline
Bicycle	*On*		(Skylab)	Conveyor
Automobile	Passenger		Missiles	Off-the-road vehicles
Truck	Cargo			
Bus	Tanker			
Train	Barge			
Specialized (snowmobile,	Tugboat			
motorcycle, etc.)	Pleasure			
	Specialized			
Under	Military			
Subway				
Rapid transit	*Below*			
	Submarine			

TECHNICAL

Structure	Energy Sources and Conversion	Propulsion	Suspension	Controls	Guidance	Support
	Energy sources					
Vehicle	Fossil	Mechanical	Mechanical	Velocity	*Guideways*	Ticket
Body	Nuclear	Hydraulic	Fluid	Direction	Roads	offices
Fuselage	Electricity	Electrical	Magnetic	Altitude	Tracks	Parking
Cab	Solar	Reaction	Air	Performance	Cables	Baggage
Trailer	Magnetic				Monorail	handling
Coach					Belts	Harbors/
Car					Pipes	docks
						Rest stops
Guideways	*Conversion*				*On-board or*	Gas
Highways	Internal				*external*	stations
Runways	combustion				Dials	
Cables	engines —				Signals	
Tunnels	gas and				Computers	
Bridges	diesel				Radar	
Double rail	Steam				Flight	
Single rail	Turbine				controls	
Belts	Jet					
	Rocket					
Support	Hybrid					
Terminals						
Harbors						
Docks						
Storage						
Parking						
Maintenance						

Fig. 198-1. Transportation systems. Can you think of other ways or methods of transporting people, materials, or goods? What about a cable car?

can be delivered overnight to most places in this country. People, by choice or necessity, go from one place to another to satisfy business, social, cultural, and recreational needs. Our transportation system is one of the marvels of the American way of life. In terms of time, Moscow is as close to New York City today as Philadelphia was in 1833.

Elements of Transportation Systems

There are two main elements of any transportation system: the means or method used and the technical system utilized. Major means or methods of transportation include land, water, air, space, and the methods used in commercial and industrial complexes. Technical systems include the structure, energy source and conversion, propulsion, suspension, controls, guidance, and support. Let's look at the technical system of an automobile.

1. *Structure.* This consists of the frame and body in which people and baggage ride.

2. *Energy Sources and Conversion.* A fossil fuel (gas) is burned in an internal-combustion engine.

3. *Propulsion.* A mechanical and/or fluid drive delivers power to the wheels through the transmission, driveshaft, and axles.

4. *Suspension.* Springs and shock absorbers cushion the ride.

5. *Control.* Speed is increased or decreased with the gas pedal or by applying brakes. Direction is provided by turning the steering wheel.

6. *Guidance.* Roads and highways show drivers where to go. The operator must be able to steer the vehicle.

7. *Support.* This includes gas stations, rest areas, parking, and everything else the traveler needs.

Can you make the same kind of analysis for a bicycle or train?

Land Transportation

Vehicles can operate above, on, or below ground. A monorail built between lanes or along the side of a highway is a good example of aboveground transportation. Vehicles that operate on land include bicycles, automobiles, buses, trucks, and trains, plus specialized machines such as snowmobiles, motorcycles, riding mowers, and others designed to do specific jobs. The bicycle is the major means of transporting people in most third-world countries, while in the United States it is primarily a recreational vehicle. Cars transport people (usually only one or two) for work and pleasure. Buses transport large numbers of people from place to place. Trucks and trains are two major methods of transporting materials and goods. Subways in large cities are a good example of below-ground transportation, as are the specialized vehicles used in underground mining.

Automobiles. See Unit 197, "Your Car." Americans have had a love affair with cars for a long time. A car is not only a means of transportation but also a way of life and an important piece of personal property. For most people, the automobile is a necessity for both work and play. Because of the automobile, this country has built an efficient highway system that connects all major cities. With a car, a person can travel in 1 hour a distance that would have taken a day by wagon train. The automobile developed slowly from a gasoline carriage to a modern streamlined car. In recent years, cars have become smaller and more fuel-efficient. Some of the more recent technological advances include:

1. Emission control system to reduce air pollution

2. Smaller engines, primarily of two, three, four, and six cylinders

3. Front-wheel drive to increase fuel efficiency and make it possible for designers to retain interior dimensions while reducing length and width

4. Computers to control all functions of a car in order to improve gas mileage and provide the driver with information on performance

5. Lighter materials, including aluminum, plastics, and composites, to reduce weight and increase gas efficiency

Buses. Buses carry a large number of people on both short and long trips. Many cities depend on buses for their mass transit systems. Larger transcontinental buses have become increasingly important as the number of passenger trains has decreased. Most smaller cities and towns depend entirely on buses for long-distance travel.

Trucks. Trucks are heavy-duty vehicles designed to haul specific kinds of freight. Trucks used in and around cities deliver all kinds of materials, such as construction needs and consumer goods; they also have other uses, such as picking up refuse. Long-distance trucks haul all kinds of materials and goods. Most of these trucks consist of two parts: the cab or tractor section with the power plant and the trailer to haul the freight. Many different kinds of trailers are needed, such as tankers to haul liquids, refrigerated trailers to haul fresh produce, rack trailers to hold from 6 to 10 new cars, and enclosed trailers to haul general freight. Often the trailer section is moved from one part of the country to another by one of several different methods. For example, the trailer can first be shipped by rail to a port, where it is loaded on a container ship. When the ship docks at another port, the cab or tractor picks up the trailer to be delivered to its final destination.

Trains. For many years, trains were the major method of long-distance passenger travel. However, with the increase in air travel and the development of superhighway systems, only a few passenger trains

serve the major cities. Most trains are an efficient and low-cost method of moving freight. They consist of two types of units: the locomotives and the various freight cars.

Many different kinds of freight cars are available to handle the different kinds of freight. Boxcars are designed to hold cargo and protect it from the weather. Refrigerator cars hold perishable foods. Flat cars carry heavy machinery. Tanker cars hold liquids such as milk, chemicals, and fuel. Stock cars move animals to market. Auto rack cars transport as many as 15 to 30 new automobiles on one car. Hopper cars are large, open containers that carry coal and other bulk materials. The caboose is always added to the end of the freight train to serve as the conductor's office. The conductor is then in contact with the engineer in the locomotive by means of a two-way radio.

Other Rail Transportation. Too many cars, trucks, and other vehicles are jamming major city streets. This has created a need for people movers, or rapid transit systems. Rapid transit systems may move above, on, or below ground or use a combination of these methods. In congested areas, where land is at a premium, monorail systems are being built. This system has been used for many years at world's fairs and amusement parks and also is used to connect hotels and shopping centers. The cab of the monorail hangs from overhead guideways (tracks) or rides upright on one rail. Monocabs are self-operating since they are controlled by a complex computer system.

Water Transportation

For many years, ocean liners moved people by sea from one country to another. Today, most passenger ships are used for pleasure cruises. Water transportation is one of the most efficient methods of moving cargo at a very low cost. Cargo transportation has changed dramatically since the mid 1950s. Until that time, stevedores loaded cargo in the open holds of ships. Since then, container ships have been built; containers can be stacked one on another with the materials or goods kept safe inside. Once the ship reaches port, trucks move the containers to their final destination unopened. Container ships require special cranes and other loading and unloading facilities at the docks.

Other cargo ships are designed for specific purposes. Supertankers move oil from one part of the world to another. Specialized ships are designed to hold only one kind of cargo. Barges are used on rivers and canals and along coastal regions of the country to move bulk materials. The barges are either pulled or pushed by tugboats. The military has many different kinds of ships, including destroyers, aircraft carriers, cruisers, and supply ships. Pleasure boats, including power and sail, are among the most popular vehicles for recreation. Two specialized ships are the hydrofoil, which actually rides above the water on fins, and the hovercraft, which rides on a cushion of air.

Air Transportation

The Wright brothers described the birth of powered flight at Kitty Hawk, North Carolina, on December 17, 1903, as follows: "The first flight lasted only 12 seconds, a flight very modest when compared with that of birds but it was nevertheless the first in the history of the world in which a machine carrying a man had raised itself by its own power into the air in free flight, had sailed forward on a level course without reduction in speed, and finally landed without being wrecked." This was the beginning of our fantastic air transportation system. During World War I, the government encouraged the development of airplanes for military use. As more efficient power plants became available, larger airplanes proved to be the best and fastest method of transporting people.

There have been many famous planes throughout history. One of the first was the Douglas DC-3, which began service in 1936 with a top speed of 230 miles per hour and a range of about 1500 miles. For many years, this was one of the major commercial airplanes, and some are still in use.

During World War II, the jet engine was developed and this became the primary power plant for commercial travel. The first commercial jet flew in 1958. Since that time, commercial jets have become an increasingly important factor in transportation. Each new model of commercial jet is more fuel-efficient and can transport more people and cargo more rapidly from one location to another.

Air travel also depends on efficient airports and safe and efficient air traffic control and air navigation systems. Over 55 000 people work in some phase of air traffic control. They staff 400 airport tower controls, 25 air-route traffic control centers, and over 30 flight service stations.

Almost all airplane flights operate under instrument flight rules (IFR) regardless of weather conditions. This means that they are followed from takeoff to touchdown by air traffic control to make sure that each flies in its own reserved block of air space safely separated from other air traffic in the system. A typical transcontinental flight from Los Angeles to New York, for example, involves almost a dozen air traffic control facilities. From the tower at Los Angeles International Airport, the flight is transferred, or "handed off," first to the terminal radar control room and then to the air route traffic control at Palmdale, California. The Salt Lake City center takes control next; depending on the route, it may be followed by centers at Denver, Kansas City, Chicago, Cleveland, and New York. Approximately 30 miles from the Kennedy International Airport, the flight is handed off to the radar approach control facility serving all New York airports. Finally, the Kennedy tower issues final landing instructions.

When weather permits, many general aviation pilots (those who fly small planes on business or pleasure) follow visual flight rules (VFR), which means that they maintain separation from other aircraft on a see and avoid basis.

There are many types of air vehicles. The most common are the commercial airplanes serving approximately 500 airports in the United States. Most of these aircraft are powered by jet engines, while some of the smaller ones have turbo-prop jet engines. See Unit 190, "Engines." There are several different classifications of general aviation vehicles. Businesses use small jet aircraft to save time and money transporting their executives from one part of the country to another.

Users of personal aircraft include all the people who own or rent planes. Private pilots use the airplane much as other people use the family car for both business and pleasure. Special-purpose flying includes a variety of uses for airplanes and helicopters. Small airplanes are used in agriculture for crop dusting and spraying (Fig. 198-2). Helicopters are used as air ambulances and for police patrol, and they can move materials and goods in and out of areas that are inaccessible by land vehicles (Fig. 198-3).

The military uses a wide variety of aircraft. Passenger-type aircraft transport troops, while fighter planes are designed to be fast and maneuverable. Bombers are built to fly long distances and carry heavy bomb loads. The greatest technological developments in air transportation have been in military aircraft. Eventually, the United States will have an airplane made of composite materials that cannot be detected by radar (this is being called the stealth plane).

Space

Vehicles that travel in space include satellites, spacecraft (from *Gemini* to those which landed on the moon), shuttles, space stations, and missiles (Fig. 198-4). The space shuttle, the most recent of the vehicles operating in space, has a big advantage because it is reusable (Fig. 198-5). Space stations have been in orbit around the earth for limited lengths of time. Eventually, there will be permanent stations where people will live and work. Satellites travel in space to provide the earth with a

Fig. 198-2. Crop dusting with a small plane such as this is a hazardous occupation requiring great skill. (Cessna Corporation)

Fig. 198-3. The helicopter's bucket is being loaded with fertilizer to spread over the forests. (Western Wood Products Association)

great deal of information, including specific weather data.

Commercial and Industrial Transportation

Many people think of transportation only in terms of such vehicles as cars and planes. There are many other systems used for commercial and industrial purposes. Tall buildings, shopping centers, and airports have elevators and escalators to move people. Moving sidewalks (operating on belts) are found in airports and in many large cities. Industry uses pipelines to transport liquids, gases, and semi-liquids (coal slurry) over long distances. The Alaskan pipeline, for example, made it possible for all of us to utilize the great oil discoveries in the northern part of that state. Pipelines crisscross our country and even cross our borders to bring us essential energy. Conveyers operating on belts, rollers, or chains are used in manufacturing to move materials and products (Fig. 198-6). Off-the-road vehicles, such as fork-lift trucks and machines to clean floors and ice rinks, are examples of special-purpose vehicles.

Fig. 198-4. Have you seen pictures of a moon landing by our astronauts? Notice the special vehicle that was used to transport the people on the moon's surface. What was the source of power for this vehicle?

Future of Transportation

Future transportation will include such technological developments as automated highways, better rapid transit systems, and safety systems for air travel. Our transportation system of the twenty-first century will be dramatically different from what it is today.

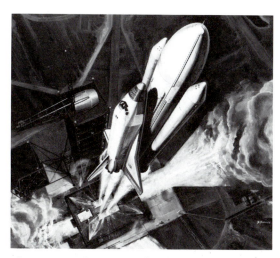

Fig. 198-5. The space shuttle blasts off on another trip around the earth. (NASA)

Fig. 198-6. This freshly cut lumber moves on a conveyer of rollers. (Western Woods Products Association)

Unit 199

Discussion Topics and Activities on Power and Transportation Technologies

Discussion Topics

1. Describe the operation of a steam engine.
2. How does a steam turbine work?
3. What is the difference between an internal-combustion engine and an external-combustion engine?
4. Where was the internal-combustion engine invented?
5. Describe the action of a four-stroke cycle engine.
6. How does a two-stroke cycle engine operate?
7. How does the fuel ignite in a diesel engine?
8. Why is a jet engine called a reaction engine?
9. What are the advantages of a turbo-prop engine?
10. What kinds of fuels do rockets use?
11. Name the six basic machines.
12. List places where the six basic machines are used.
13. What is the difference between a V belt and a flat belt?
14. What is friction?
15. Name several kinds of bearings.
16. What are some of the common gears?
17. List three kinds of threaded metal fasteners.
18. What is the difference between National Fine and National Coarse threads?
19. Describe several kinds of wrenches used in power mechanics.
20. What are the purposes of thickness, or feeler, gages?
21. Describe the parts of a four-stroke cycle engine.
22. How is the compression ratio of a small engine found?
23. Name three common fuel feed systems.
24. Describe the purpose of the venturi in a carburetor.
25. What kind of electric system is used on small engines?
26. How does the electric system of a small engine differ from that found on an automobile?
27. Name the three major forms of transportation.

Activities

1. Study the cutaways of a small four-stroke cycle and a two-stroke cycle engine. Describe the differences.
2. Write a report on the opportunities for a serviceperson of small engines.
3. Find out what the opportunities are in auto mechanics. Write a report.
4. Disassemble a small four-stroke cycle engine, and clean out the carbon from the head. Check the valves to see whether they need regrinding or replacing.
5. Build a model of some type of engine for school display. This may be an internal-combustion engine, a rocket engine, or any other type that you can find in a hobby store.
6. Recondition a power mower by sharpening the blades, cleaning the air cleaner, cleaning and readjusting the spark plugs, and adjusting the carburetor.
7. Take a carburetor apart and examine the parts. Outline the steps in assembling and disassembling it.
8. Write a report on how gasoline is processed from crude oil.
9. Build and fly a model powered airplane.

Appendix:
Suggested Project Activities

Suggested Projects in Woodworking: Folding Chair

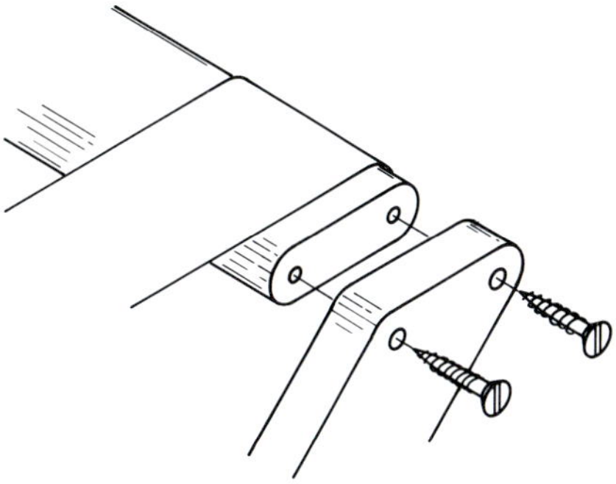

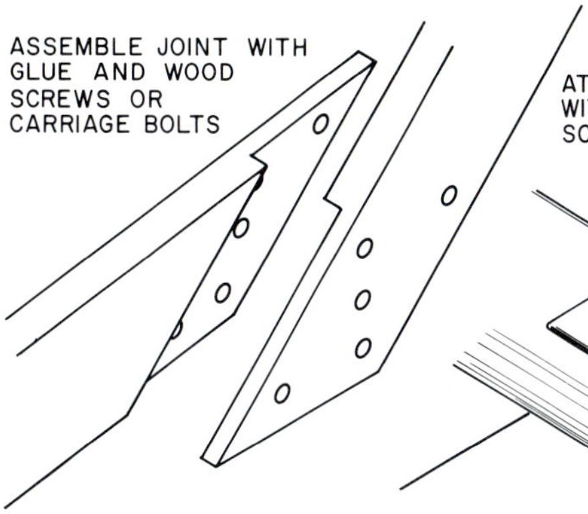

ASSEMBLE JOINT WITH GLUE AND WOOD SCREWS OR CARRIAGE BOLTS

ATTACH PLASTIC WEBBING WITH WASHERS AND WOOD SCREWS

TYPICAL JOINT
ASSEMBLY DETAIL

DETAIL "A"
HALF-LAP JOINT

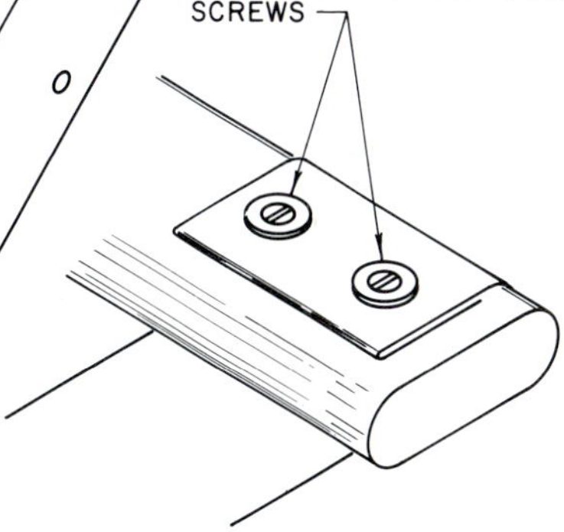

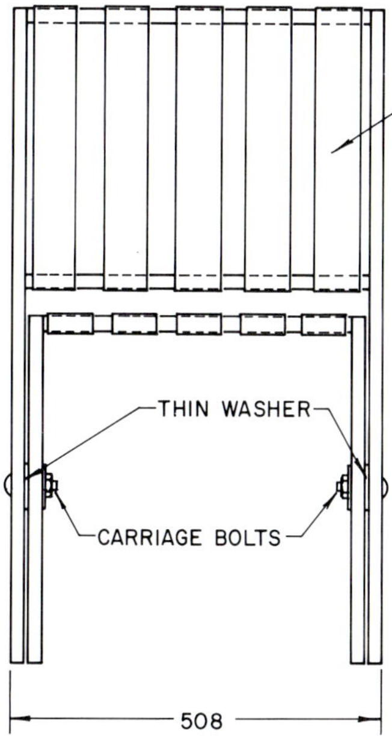

NOTE: USE 20 MM HARDWOOD FOR FRAME

PLASTIC WEBBING

THIN WASHER

CARRIAGE BOLTS

508

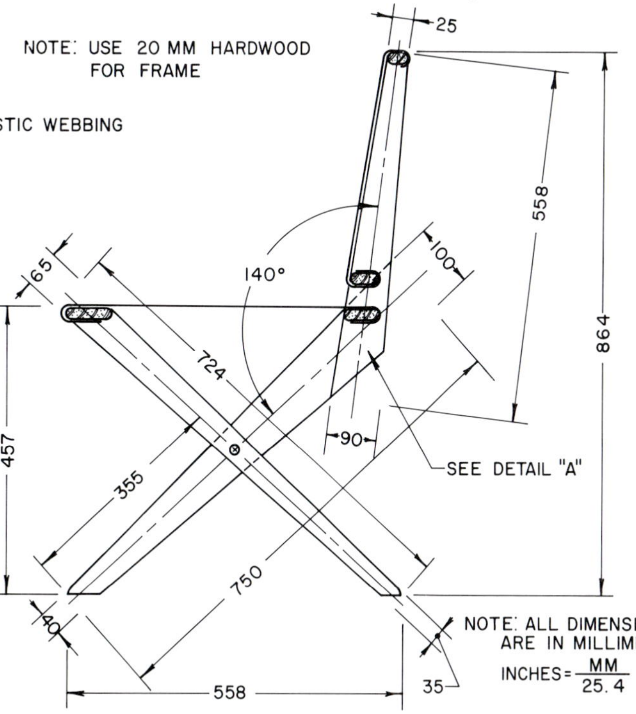

25

558

864

140°

65

100

724

457

355

90

SEE DETAIL "A"

40

750

558

35

NOTE: ALL DIMENSIONS ARE IN MILLIMETERS

$INCHES = \dfrac{MM}{25.4}$

Suggested Projects in Woodworking

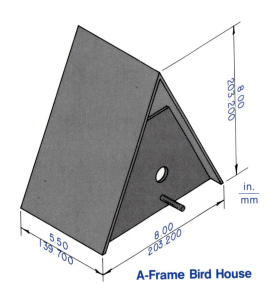

8.00
203.200

in.
mm

5.50
139.700

8.00
203.200

A-Frame Bird House

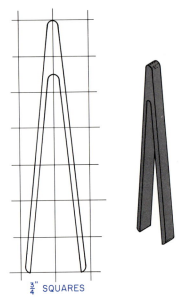

$\frac{3}{4}"$ SQUARES

Toast and Muffin Tongs

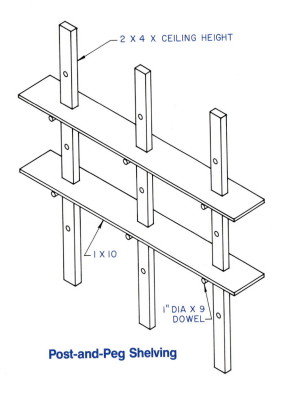

2 X 4 X CEILING HEIGHT

1 X 10

1" DIA X 9
DOWEL

Post-and-Peg Shelving

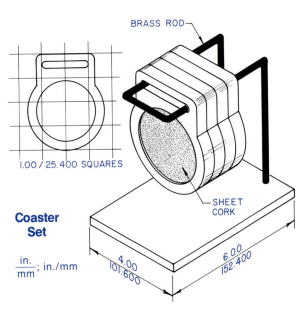

BRASS ROD

SHEET
CORK

1.00 / 25.400 SQUARES

**Coaster
Set**

$\frac{in.}{mm}$; in./mm

4.00
101.600

6.00
152.400

Suggested Projects in Metalworking: Racing Car

NOTES:

Fins may be made from wood or sheet metal.

Design and construction of cab optional.

All body parts to be made from sheet metal. Develop all body parts using standard surface-development techniques. Allow laps for joining parts.

Axles and wheels are available at hobby shops. Solder axles to housing.

Paint and decorate as desired.

This racing car may be propelled by using a CO_2 cartridge inserted in the opening in the tail end. A $3/4$-in. hole is necessary for this purpose.

TOP VIEW OF FRONT END

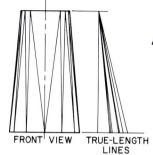

FRONT VIEW   TRUE-LENGTH LINES

STRETCHOUT FOR FRONT END

This development is for instructional purposes only. It is not shown at the size needed for the finished racer.

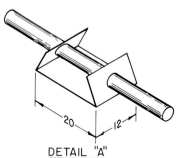

DETAIL "A"

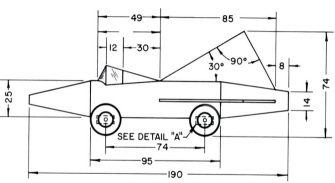

NOTE: ALL DIMENSIONS ARE IN MILLIMETERS

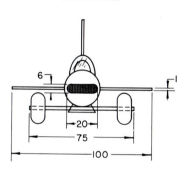

■ **623**

Suggested Projects in Metalworking

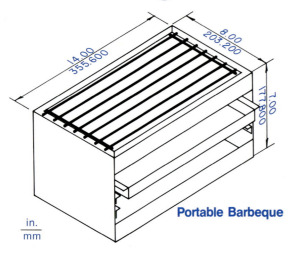

Portable Barbeque

in.
―――
mm

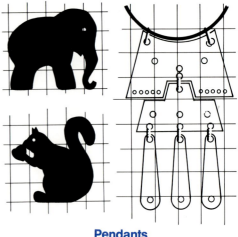

Pendants

These pendants can be made easily and quickly from thin-gauge aluminum or brass. Sizes are optional. Neck band is made from $1/8$-DIA brass or aluminum rod.

This project may be made from $1/8$-in. sheet aluminum, or it may be cast in aluminum. The extensions on the legs are used for supporting the dog in the ground. If it is to be wall mounted, eliminate the leg extensions. Numbers may be purchased, cast in place, or cut out and mounted.

Scotty-Dog House Number

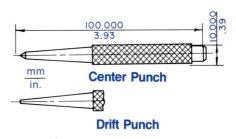

mm
―――
in.

Center Punch

Drift Punch

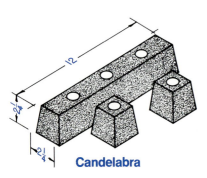

Notes:
Material: cast aluminum
Finish: wire brush or polish
Holes may be formed by casting or machining

Candelabra

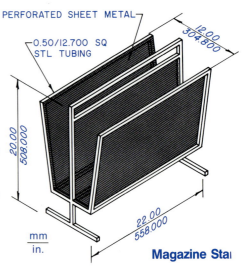

PERFORATED SHEET METAL

0.50/12.700 SQ STL TUBING

mm
―――
in.

Magazine Stand

Suggested Projects in Plastics: Photo Cube

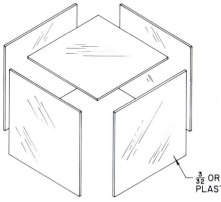

$\frac{3}{32}$ OR $\frac{1}{8}$ CLEAR PLASTIC

The photo cube is designed to display five photographs. Assemble the plastic box before cutting stock for the plywood box. Allow approximately $\frac{1}{32}$ in. clearance on all sides for pictures.

Attach the plastic base to the plywood box with flat-head wood screws.

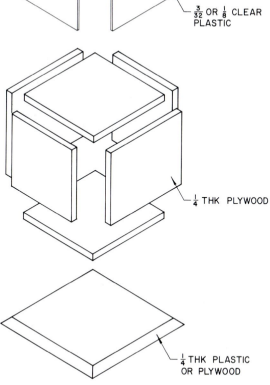

$\frac{1}{4}$ THK PLYWOOD

$\frac{1}{4}$ THK PLASTIC OR PLYWOOD

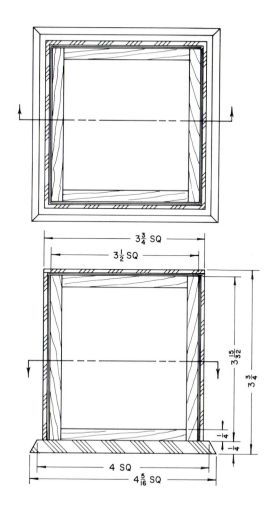

$3\frac{3}{4}$ SQ

$3\frac{1}{2}$ SQ

$3\frac{15}{32}$

$3\frac{3}{4}$

$\frac{1}{4}$

$\frac{1}{4}$

4 SQ

$4\frac{5}{16}$ SQ

Suggested Projects in Plastics

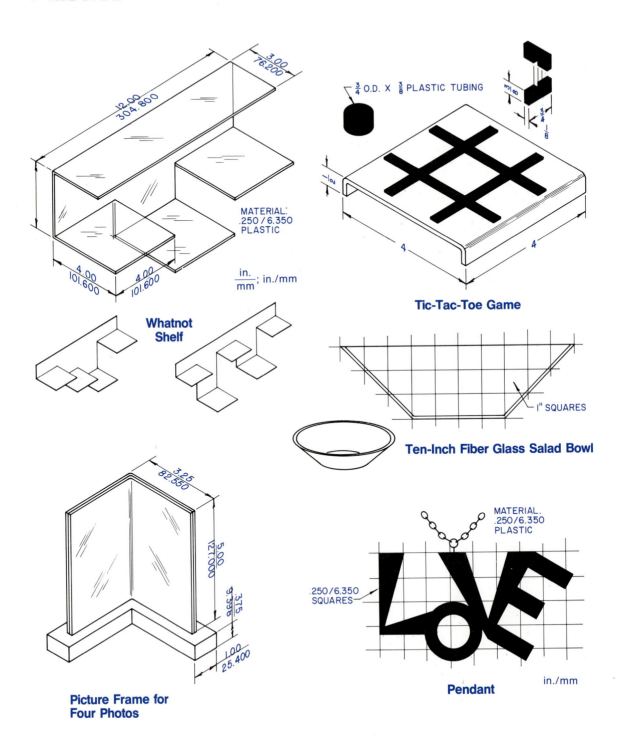

3.00 / 76.200

12.00 / 304.800

MATERIAL: .250 / 6.350 PLASTIC

4.00 / 101.600   4.00 / 101.600

$\dfrac{\text{in.}}{\text{mm}}$; in./mm

Whatnot Shelf

¾ O.D. X ⅜ PLASTIC TUBING

½

4   4

Tic-Tac-Toe Game

I" SQUARES

Ten-Inch Fiber Glass Salad Bowl

3.25 / 82.550

5.00 / 127.000

.375 / 9.396

1.00 / 25.400

Picture Frame for Four Photos

MATERIAL. .250/6.350 PLASTIC

.250/6.350 SQUARES

Pendant

in./mm

Suggested Projects in Electronics: Three-Transistor Radio

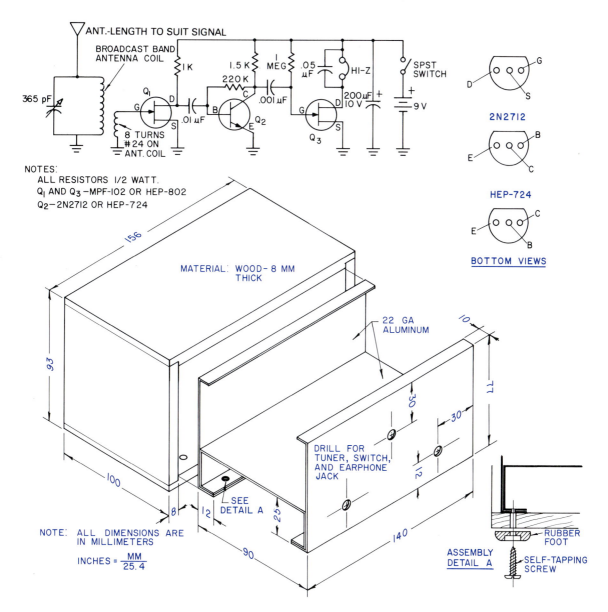

ANT.-LENGTH TO SUIT SIGNAL

BROADCAST BAND
ANTENNA COIL

365 pF

1K

Q₁

G D

S

.01 μF

8 TURNS
#24 ON
ANT. COIL

220 K

1.5 K

B

C

Q₂

E

.001 μF

1 MEG

.05 μF

HI-Z

G

D

Q₃

S

200 μF
10 V

SPST
SWITCH

9 V

D G
S
2N2712

E B
C
HEP-724

C
E B

BOTTOM VIEWS

NOTES:
ALL RESISTORS 1/2 WATT.
Q₁ AND Q₃ —MPF-102 OR HEP-802
Q₂—2N2712 OR HEP-724

156

MATERIAL: WOOD—8 MM
THICK

93

22 GA
ALUMINUM

10

77

30

30

12

DRILL FOR
TUNER, SWITCH,
AND EARPHONE
JACK

25

SEE
DETAIL A

140

100

8

12

90

NOTE: ALL DIMENSIONS ARE
IN MILLIMETERS

$$INCHES = \frac{MM}{25.4}$$

RUBBER
FOOT

ASSEMBLY
DETAIL A

SELF-TAPPING
SCREW

■ 627

Suggested Projects in Electronics

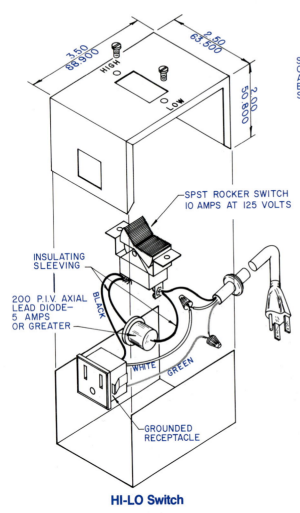

HI-LO Switch

- SPST ROCKER SWITCH 10 AMPS AT 125 VOLTS
- INSULATING SLEEVING
- 200 P.I.V. AXIAL LEAD DIODE— 5 AMPS OR GREATER
- BLACK
- WHITE
- GREEN
- GROUNDED RECEPTACLE
- 3.50 / 88.900
- 2.50 / 63.500
- 2.00 / 50.800
- HIGH
- LOW

USES

1. Lamp dimmer (cuts intensity to approximately half).
2. Motor speed control (drills, sabre saws, etc.)
3. Heat control (cuts heat to approximately half)

CAUTION

DO NOT allow contact between diode and metal case.
DO NOT use with split-phase motors. Examples: washing machine motors, stationary power tools such as table saws and drill presses.
DO NOT exceed current capacity of diode.

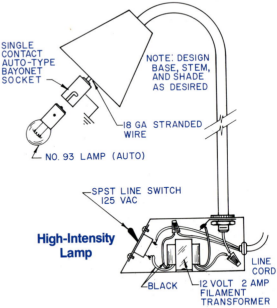

High-Intensity Lamp

- SINGLE CONTACT AUTO-TYPE BAYONET SOCKET
- NOTE: DESIGN BASE, STEM, AND SHADE AS DESIRED
- 18 GA STRANDED WIRE
- NO. 93 LAMP (AUTO)
- SPST LINE SWITCH 125 VAC
- LINE CORD
- BLACK
- 12 VOLT 2 AMP FILAMENT TRANSFORMER

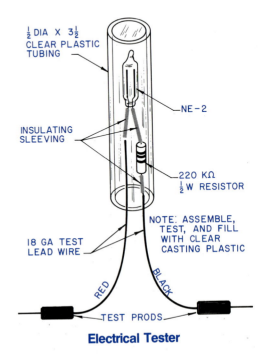

Electrical Tester

- $\frac{1}{2}$ DIA X $3\frac{1}{2}$ CLEAR PLASTIC TUBING
- NE-2
- INSULATING SLEEVING
- 220 KΩ $\frac{1}{2}$ W RESISTOR
- 18 GA TEST LEAD WIRE
- NOTE: ASSEMBLE, TEST, AND FILL WITH CLEAR CASTING PLASTIC
- RED
- BLACK
- TEST PRODS

Mass Production Suggestions

The projects shown on this page are designed to offer experience in combining materials and developing mass-production techniques. All projects are shown as basic designs only; they may be redesigned as desired.

Complete working drawings of the projects, as well as any necessary jigs and fixtures, should be made. A prototype should be made before mass-production procedures are developed. Notes should be taken during the construction of the prototype, and rough sketches of jigs and fixtures may be helpful at this stage.

A high degree of accuracy is necessary for the assembly of mass-produced parts. Take extra care in developing all phases of the production line so that all parts will fit together with a minimum of effort and time while providing a high-quality finished product.

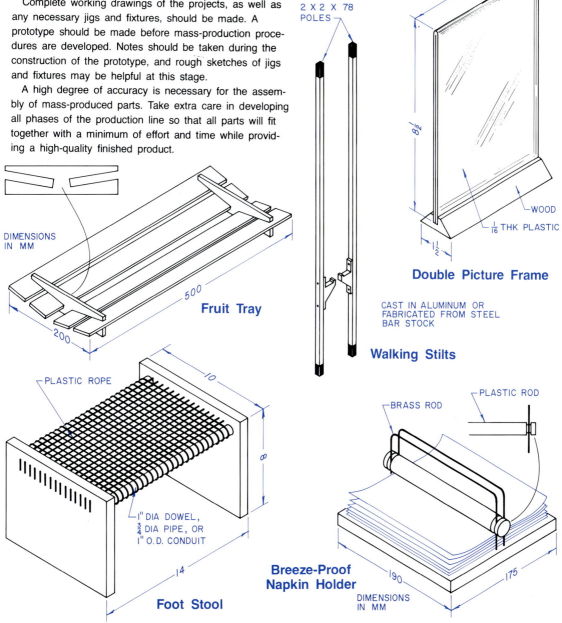

DIMENSIONS IN MM

500

200

Fruit Tray

2 X 2 X 78 POLES

CAST IN ALUMINUM OR FABRICATED FROM STEEL BAR STOCK

Walking Stilts

5

$8\frac{1}{2}$

WOOD

$\frac{1}{16}$ THK PLASTIC

$1\frac{1}{2}$

Double Picture Frame

PLASTIC ROPE

10

8

14

1" DIA DOWEL, $\frac{3}{4}$ DIA PIPE, OR 1" O.D. CONDUIT

Foot Stool

PLASTIC ROD

BRASS ROD

190

175

Breeze-Proof Napkin Holder

DIMENSIONS IN MM

INDEX